FORENSIC ENGINEERING

PUBLISHED AND FORTHCOMING TITLES IN THE *ADVANCED FORENSIC SCIENCE SERIES*

Published

Forensic Fingerprints
Firearm and Toolmark Examination and Identification
Forensic Biology
Forensic Chemistry
Professional Issues in Forensic Science
Materials Analysis in Forensic Science
Forensic Pathology
Forensic Anthropology

Forthcoming

Behavioral Analysis
Digital and Documents
Forensic Toxicology and Drugs

FORENSIC ENGINEERING

Advanced Forensic Science Series

MAX M. HOUCK, PhD, FRSC

Managing Director, Forensic & Intelligence Services, LLC, St. Petersburg, FL, USA

ELSEVIER

ACADEMIC PRESS

An imprint of Elsevier

Academic Press is an imprint of Elsevier
125 London Wall, London EC2Y 5AS, United Kingdom
525 B Street, Suite 1800, San Diego, CA 92101-4495, United States
50 Hampshire Street, 5th Floor, Cambridge, MA 02139, United States
The Boulevard, Langford Lane, Kidlington, Oxford OX5 1GB, United Kingdom

Library of Congress Cataloging-in-Publication Data
A catalog record for this book is available from the Library of Congress

British Library Cataloguing-in-Publication Data
A catalogue record for this book is available from the British Library

ISBN: 978-0-12-802718-9
ISSN : 2352-6238

For information on all Academic Press publications visit our website at
https://www.elsevier.com/books-and-journals

Publisher: Sara Tenney
Acquisition Editor: Elizabeth Brown
Editorial Project Manager: Joslyn Chaiprasert-Paguio
Production Project Manager: Lisa Jones
Designer: Matthew Limbert

Typeset by TNQ Books and Journals

CONTENTS

Senior Editor: Biography *ix*

List of Contributors *xi*

Foreword *xiii*

Preface *xv*

SECTION 1 INTRODUCTION **1**

Principles of Forensic Science 1
F Crispino, MM Houck

Forensic Classification of Evidence 7
MM Houck

Interpretation/The Comparative Method 13
MM Houck

Forensic Engineering/Accident Reconstruction/Biomechanics of Injury/Philosophy, Basic Theory, and Fundamentals 19
SC Batterman, SD Batterman

Key Terms 28

Review Questions 29

Discussion Questions 29

Additional Readings 29

SECTION 2 INVESTIGATIONS **31**

Collection and Chain of Evidence 31
F Poole

Accident Investigation—Determination of Cause 37
H Steffan

Human Factors Investigation and Analysis of Accidents and Incidents 47
SA Shappell, DA Wiegmann

Major Incident Scene Management 59
J Horswell

Crime Scene Analysis and Reconstruction 63
WJ Chisum

Key Terms 68

Review Questions 68

Discussion Questions 69

Additional Readings 69

SECTION 3 TRANSPORTATION INVESTIGATIONS 71

Railroad Accident Investigation and Reconstruction 71
RW Halstead

Aircraft Flight Data Recorders 79
VL Grose

Air-Bag-Related Injuries and Deaths 89
WS Smock

Electronic Data Recorders (EDRs, Black Boxes) 99
W Rosenbluth

Analog Tachograph Chart Analysis 109
RF Lambourn

Traffic Injuries and Deaths 121
M Graw, J Adamec

Airplane Crashes and Other Mass Disasters 129
O Peschel, W Eisenmenger

Key Terms 134

Review Questions 134

Discussion Questions 135

Additional Readings 135

SECTION 4 FIRE INVESTIGATIONS 137

Chemistry of Fire 137
E Stauffer, N NicDaéid

Physics/Thermodynamics 145
JHC Martin, RS Pepler

Thermal Degradation 151
K Grimwood

Types of Fires 157
JB Crippin

Evidence Collection at Fire Scenes 161
JJ Lentini

Fire Scene Inspection Methodology 167
JJ Lentini

Fire Patterns and Their Interpretation 173
JJ Lentini

Analysis of Fire Debris 185
N NicDaéid, E Stauffer

Interpretation of Fire Debris Analysis 191
E Stauffer, N NicDaéid

Suspicious Deaths 205
J Horswell

Key Terms 208

Review Questions 208

Discussion Questions 209

Additional Readings 209

SECTION 5 OTHER METHODS 211

Audio Enhancement and Authentication 211
C Grigoras, JM Smith

Investigation and Analysis of Electrical Accidents 225
TP Shefchick

Investigation and Analysis of Structural Collapses 243
RT Ratay, DB Peraza

Biomechanics of Human Gait—Slip and Fall Analysis 249
TE Lockhart

Forensic Chemical Engineering Investigation and Analysis 261
DD Perlmutter

Materials Analysis and Failure Analysis 269
AD Micheals

Key Terms 281

Review Questions 281

Discussion Questions 281

Additional Readings 282

SECTION 6 PROFESSIONAL ISSUES 283

Crime Scene to Court 283
K Ramsey, E Burton

Forensic Laboratory Reports 289
J Epstein

Health and Safety 293
N Scudder, B Saw

Key Terms 300

Review Questions 300

Discussion Questions 300

Additional Readings 301

Index 303

SENIOR EDITOR: BIOGRAPHY

Max M. Houck is an internationally recognized forensic expert with research interests in anthropology, trace evidence, education, and the fundamentals of forensic science, both as a science and as an enterprise. He has worked in the private sector, the public sector (at the regional and federal levels), and in academia. Dr. Houck has published in a wide variety of areas in the field, in books, book chapters, and peer-reviewed journals. His casework includes the Branch Davidian Investigation, the September 11 attacks on the Pentagon, the D.B. Cooper case, the US Embassy bombings in Africa, and the West Memphis Three case. He served for 6 years as the Chair of the Forensic Science Educational Program Accreditation Commission. Dr. Houck is a founding coeditor of the journal Forensic Science Policy and Management, with Jay Siegel; he has also coauthored a major textbook with Siegel, *Fundamentals of Forensic Science*. In 2012, Dr. Houck was in the top 1% of connected professionals on LinkedIn. Dr. Houck is currently the Managing Director of Forensic & Intelligence Services, LLC, St. Petersburg, FL.

LIST OF CONTRIBUTORS

J Adamec
Institute of Legal Medicine, Munich, Germany

SC Batterman
Batterman Engineering, LLC, Cherry Hill, NJ, USA

SD Batterman
Batterman Engineering, LLC, Cherry Hill, NJ, USA

E Burton
Greater Manchester Police Forensic Services Branch, Manchester, UK

WJ Chisum
(Retired) President of California Association of Criminalists and American Society of Crime Lab Directors, Elk Grove, CA, USA

JB Crippin
Western Forensic Law Enforcement Training Center, Pueblo, CO, USA

F Crispino
Université du Québec à Trois-Rivières, Trois-Rivières, QC, Canada

W Eisenmenger
University Hospital München, München, Bavaria, Germany

J Epstein
Widener University School of Law, Wilmington, DE, USA

M Graw
Institute of Legal Medicine, Munich, Germany

C Grigoras
University of Colorado Denver, Denver, CO, USA

K Grimwood
University of Technology Sydney, Sydney, NSW, Australia

VL Grose
US National Transportation Safety Board, Washington, DC, USA

RW Halstead
IronWood Technologies, Inc., Syracuse, NY, USA

J Horswell
Approved Forensics Sendirian Berhad, Selangor, Malaysia

MM Houck
Consolidated Forensic Laboratory, Washington, DC, USA

RF Lambourn
Transport Research Laboratory, Wokingham, UK

JJ Lentini
Scientific Fire Analysis, LLC, Big Pine Key, FL, USA

TE Lockhart
Virginia Tech, Blacksburg, VA, USA

JHC Martin
Institut de Police Scientifique et de Criminologie Université de Lausanne, Lausanne, Switzerland

AD Micheals
San Jose State University, San Francisco, CA, USA

N NicDaéid
University of Strathclyde, Glasgow, UK

RS Pepler
Institut de Police Scientifique et de Criminologie, Université de Lausanne, Lausanne, Switzerland

DB Peraza
Exponent, Inc., New York, NY, USA

DD Perlmutter
University of Pennsylvania, Philadelphia, PA, USA

O Peschel
Institut für Rechtsmedizin der Ludwig Maximilians Universität München, München, Germany

F Poole
Forensic Services Group, New South Wales Police Force, Parramatta, NSW, Australia

K Ramsey
Greater Manchester Police Forensic Services Branch, Manchester, UK

RT Ratay
Columbia University, New York, NY, USA

W Rosenbluth
Automotive Systems Analysis, Inc., Reston, VA, USA

B Saw
Australian Federal Police, Canberra, ACT, Australia

N Scudder
Australian Federal Police, Canberra, ACT, Australia

SA Shappell
Embry-Riddle Aeronautical University, Daytona Beach, FL, USA

TP Shefchick
Shefchick Engineering, Sunnyvale, CA, USA

JM Smith
University of Colorado Denver, Denver, CO, USA

WS Smock
University of Louisville, Louisville, KY, USA

E Stauffer
Commissariat d'Identification Judiciaire, Police Cantonale Fribourg, Fribourg, Switzerland

H Steffan
Graz University of Technology, Graz, Austria

DA Wiegmann
University of Wisconsin, Madison, WI, USA

FOREWORD

> "The best thing for being sad", replied Merlin, beginning to puff and blow, "is to learn something. That's the only thing that never fails. You may grow old and trembling in your anatomies, you may lie awake at night listening to the disorder of your veins, you may miss your only love, you may see the world about you devastated by evil lunatics, or know your honor trampled in the sewers of baser minds. There is only one thing for it then — to learn. Learn why the world wags and what wags it. That is the only thing which the mind can never exhaust, never alienate, never be tortured by, never fear or distrust, and never dream of regretting. Learning is the only thing for you. Look what a lot of things there are to learn". — T.H. White, *The Once and Future King*

Forensic science has much to learn. The breadth of the discipline alone should render any reasonably learned person dizzy with expectations; insects, explosives, liver functions, DNA, firearms, textiles, adhesives, skeletons, and so on the list goes on forever. That is because anything, truly *anything* can become evidence, from a single fiber to an entire ocean liner. Forensic science does not lack for specialized knowledge (some might stay too specialized), but what it is wanting is knowledge that is comprehensive, integrated, and foundational. Introductions to forensic science abound, and many highly specialized texts are also available, but a gap exists between the two: a bridge from novice to practitioner. As the 2009 NRC report noted:

> Forensic science examiners need to understand the principles, practices, and contexts of scientific methodology, as well as the distinctive features of their specialty. Ideally, training should move beyond apprentice-like transmittal of practices to education based on scientifically valid principles. NRC (2009, pp. 26–27).

The *Advanced Forensic Science Series* seeks to fill that gap. It is a unique source, combining entries from the world's leading specialists who contributed to the second edition of the award-winning *Encyclopedia of Forensic Sciences* and organizing them by topic into a series of volumes that are philosophically grounded yet professionally specialized. The series is composed of 12 volumes that cover the breadth of forensic science:

1. Professional Issues
2. Biology
3. Chemistry
4. Fingerprints
5. Firearms
6. Materials Analysis
7. Pathology

8. Anthropology
9. Engineering
10. Behavioral
11. Digital and Documents
12. Toxicology and Drugs

Each volume contains sections common to all forensic sciences, such as professionalism, ethics, health and safety, and court testimony, and sections relevant to the topics in that particular subdiscipline. Pedagogy is included, providing review questions, discussion questions, the latest references in additional readings, and key words. Thus, each volume is suitable as a technical reference, an advanced textbook, or a training adjunct.

The *Advanced Forensic Science Series* provides expert information, useful teaching tools, and a ready source for instruction, research, and practice. I hope, like learning, it is the only thing for you.

M.M. Houck, PhD, FRSC
Series Editor

Reference

National Research Council, 2009. Strengthening Forensic Science in the U.S.: A Path Forward. National Academies of Science, Washington, DC.

PREFACE

"Engineering is achieving function while avoiding failure." Henry Petroski

So much of what has been published in the forensic science literature and about the profession relate to the scientific methods that are used. The notion that forensic science is "only a programmatic set of various disciplines," as Crispino and Houck put it, meaning it is little more than a toolbox filled with useful things but no real coherence or underlying philosophy, sadly, persists. In that view, engineering is "only" an applied science, yet people drive over bridges, ride elevators, use electrical appliances, and fly in airplanes without a second's worry about engineering's status as science. Methods alone, however, will not build a plane or a building or a toaster. Some underlying philosophy or ethos has to exist for any science—any profession—to operate. Behind each method, regardless of the discipline, is the accumulated knowledge of successful discoveries, failed attempts, and the eventual establishment of reliable theories or even laws. Information is that which reduces uncertainty, and methods are designed to be reproducible ways to reduce uncertainty. Inherent in each method, therefore, is the supporting knowledge that makes the method reliable; "the concept is synonymous with the corresponding set of operations" (Bridgman,1928, p. 14).[1] An underlying guiding philosophy has yet to be properly articulated to frame forensic science as a peer to biology, chemistry, and the other sciences.

As a technical endeavor, forensic engineering gets a pass, so to speak: It is almost a pure application of existing engineering theory and methods with the most minimal smattering of "forensics." Perhaps it is because of the scope of engineering in modern life, or that it deals with failure of designed objects, that engineering can be used in a forensic investigation almost whole cloth from its native profession. Despite the fact that nearly all practitioners work in the private sector, forensic engineering plays a role in many criminal and civil cases, from small electrical fires to collapsed buildings, and is an integral part of the forensic enterprise

[1] Bridgman, P.W. 1928. *The Logic of Modern Physics*. Beaufort Books.

Section 1. Introduction

Engineering moves forward in time, innovating, designing, and creating things of all sorts. Add "forensic," however, and everything reverses: Forensic sciences are inherently historic in nature, revealing as they do past events. Time does not "flow," however, but rather is a series of events that form an irreversible unidirectional sequence. The arrow of time is said to point to the future but this is no more accurate than saying because a compass' needle points north if it is traveling north. The arrow of time, as it were, demonstrates that the world is asymmetrical *in* time and not that time *itself* moves in any one direction. For example, think of video recording an egg falling on the floor and breaking. Run backward, any viewer would recognize the sequence for what it is. Editing the video into individual frames and shuffling would not preclude a viewer from arranging the images into a proper timeline sequence. Even mixed, the stack of images retains the asymmetry of time, revealing it as a property of the stochastic world and not a property of time in and of itself. Eggs do not spontaneously reconstitute! Neither do devices or buildings or airplanes. Forensic engineering uses its knowledge of how things are designed and made in the present to help determine how they failed in the past.

Principles of Forensic Science

F Crispino, Université du Québec à Trois-Rivières, Trois-Rivières, QC, Canada
MM Houck, Consolidated Forensic Laboratory, Washington, DC, USA

Glossary

Abduction Syllogism in which one premise is certain whereas the other one is only probable, generally presented as the best explanation to the former. Hence, abduction is a type of reasoning in which we know the law and the effect, and we attempt to infer the cause.

Deduction Process of reasoning which moves from the general to the specific, and in which a conclusion follows necessarily from the stated premises. Hence, deduction is a type of reasoning in which, knowing the cause and the law, we infer the effect.

Forensic intelligence Understanding on how traces can be collected from the scene, processed, and interpreted within a holistic intelligence-led policing strategy.

Heuristic Process of reasoning by rules that are only loosely defined; generally by trial and error.

Holistic Emphasizing the importance of the whole and the interdependence of its parts.

Induction Process of deriving general principles from particular facts or instances, i.e., of reasoning that moves from the specific to the general. Hence, induction is a type of reasoning in which, knowing the cause and the effect (or a series of causes and effects), we attempt to infer the law by which the effects follow the cause.

Linkage blindness Organizational or investigative failure to recognize a common pattern shared on different cases.

Science The intellectual and practical activity encompassing the systematic study of the structure and behavior of the physical and natural world through observation and experiment. It is also defined as a systematically organized body of knowledge on a particular subject.

 http://dx.doi.org/10.1016/B978-0-12-802718-9.00001-6

Given that it identifies and collects objects at crime scenes and then treats them as evidence, forensic science could appear at first glance to be only a pragmatic set of various disciplines, with practitioners adapting and developing tools and technologies to help the triers of fact (juries or judges) interpret information gained from the people, places, and things involved in a crime. The view could be—and has been—held that forensic science has no philosophic or fundamental unity and is merely the application of knowledge generated by other sciences. Indeed, many working forensic scientists regard themselves mainly as chemists, biologists, scientists, or technicians, and rarely as practitioners of a homogeneous body of knowledge with common fundamental principles.

Even the 2009 National Academy of Sciences National Research Council Report failed to recognize such a concept, certainly blurred by a semantic gap in the terminology itself of field practitioners, who confuse words like "forensic science(s)," "criminalistic(s)," "criminology," "technical police," "scientific police," and so on, and generally restrict the scientific debate on analytical techniques and methods. An independent definition of forensic science, apart from its legal aspects, would support its scientific status and return the expert to his/her domain as scientist and interpreter of his/her analyses and results to assist the lay person.

What Is Forensic Science?

In its broadest sense, forensic science describes the utility of the sciences as they pertain to legal matters, to include many disciplines, such as chemistry, biology, pathology, anthropology, toxicology, and engineering, among others. ("Forensic" comes from the Latin root *forum*, the central place of the city where disputes and debates were made public to be solved, hence, defining the law of the city. Forensic generally means of or applied to the law.) The word "criminalistics" was adopted to describe the discipline directed toward the "recognition, identification, individualization, and evaluation of physical evidence by application of the natural sciences to law-science matters." ("Kriminalistik" was coined in the late nineteenth century by Hans Gross, a researcher in criminal law and procedure to define his methodology of classifying investigative, tactical, and evidential information to be learned by magistrates at law schools to solve crimes and help convict criminals.) In the scheme as it currently stands, criminalistics is part of forensic science; the word is a regionalism and is not universally applied as defined. Difficulties in differentiating the concepts certainly invited the definition of criminalistics as the "science of individualization," isolating this specific epistemologically problematic core from the other scientific disciplines. Individualization, the concept of determining the sole source of an item, enthroned a linear process—identification or classification on to individualization—losing sight of the holistic, variable contribution of all types of evidence. Assessing the circumstances surrounding a crime, where the challenge is to integrate and organize the data in order to reconstruct a case or propose alternative propositions for events under examination, requires multiple types of evidence, some of which may be quite nuanced in their interpretation. This is also true in the use of so-called forensic intelligence, which feeds investigative, police, or security needs, where one of the main reasons for failures is linkage blindness. Nevertheless, it seems that the essence of the forensic daily practice is hardly captured within the present definitions of both terms.

Forensic science reconstructs—in the broadest sense—past criminal events through the analysis of the physical remnants of those activities (evidence); the results of those analyses and their expert interpretation establish relationships between people, places, and objects relevant to those events. It produces these results and interpretations through logical inferences, induction, abduction, and deduction, all of which frame the hypothetico-deductive method; investigative heuristics also play a role. Translating scientific information into legal information is a particular domain of forensic science; other sciences must (or at least should) communicate their findings to the public, but forensic science is often required by law to communicate their findings to public courts. Indeed, as the Daubert Hearing stated, "[s]cientific conclusions are subject to perpetual revision as law must resolve disputes finally and quickly." This doubly difficult requirement of communicating to the public and to the law necessitates that forensic scientists should be better communicators of their work and their results. Scientific inferences are not necessarily legal proofs, and the forensic scientist must recognize that legal decisions based, in part, on their scientific work may not accord with their expert knowledge. Moreover, scientists must think in probabilities to explain evidence given possible causes, while jurists must deal in terms of belief beyond reasonable doubt. As Inman and Rudin state: "Because we [the scientists] provide results and information to parties who lack the expertise to independently understand their meaning and implications, it is up to us to furnish an accurate and complete interpretation of our results. If we do not do this, our conclusions are at best incomplete, at worst potentially misleading."

The Trace as the Basic Unit of Forensic Science

The basic unit of forensic science is the trace, the physical remnant of the past criminal activity. Traces are, by their very nature, semiotic: They represent something more than merely themselves; they are signifiers or signs for the items or events that are its source. A fiber is not the sweater it came from, a fingerprint is not the fingertip, soot in the trachea is not the victim choking from a fire, blood droplets are not the violence against the victim, but they all point to their origin (source and activity) to a greater or lesser degree of specificity. Thus, the trace is a type of proxy data, that is, an indicator of a related phenomenon but not the phenomenon itself. Traces come from the natural and manufactured items that surround us in our daily lives. Traces are, in essence, the raw material available at a crime scene which becomes forensic intelligence or knowledge. Everyday items and their traces become evidence through their involvement in criminal activities; the activities add meaning to their existing status as goods in the world; a fireplace poker is transformed into "the murder weapon" by its use as such. The meaning added should also take into account the context of the case, the circumstances under which the criminal activities occurred, boarding the trier of fact mandate.

Traces become evidence when they are recognized, accepted as relevant (if blurred) to the past event under investigation, and collected for forensic purposes. Confusing trace, sign, and evidence can obscure the very process of trace "discovery," which lies at the root of its interpretation. Evidence begins with detection by observation, which is possible because of the available knowledge of the investigator or scientist; unrecognized traces go undiscovered and do not become evidence. When the investigator's or scientist's senses are extended through instrumental sensitivity, either at the scene or in the laboratory, the amount of potential evidence considerably increased. Microscopes, alternate light sources, instrumental sensitivity, and detection limits create increases in the number of traces that can be recognized and collected. More evidence, and more evidence types, inevitably led to increases in the complexity not only of the search for traces but also to their interpretation. Feeding back into this system is the awareness of new (micro)traces that changed the search methods at scenes and in laboratories, with yet more evidence being potentially available.

Traces are ancillary to their originating process; they are a by-product of the source activity, an accidental vestige of their criminal creation. To be useful in the determination of associations, traces whose ultimate sources are unknown must be compared to samples from a known source. Comparison is the very heart of the forensic science process; the method is essentially a diagnostic one, beginning with Georges Cuvier, and is employed by many science practitioners, including medical professionals (including, interestingly, Arthur Conan Doyle, a medical doctor and author, whose Sherlock Holmes character references Cuvier's method in *The Five Orange Pips*). Questioned traces, or items, may have a provenance (a known location at the time of their discovery) but this is not their originating source; a few examples may help:

Trace (questioned)	*Source (known)*
Fiber on victim	Sweater
Gunshot residue	Ammunition discharge
Blood droplet	Body
Tool marks in door jamb	Pry bar used to open door
Shoeprint in soil	Shoe from suspect
Fingerprint on glass	Finger from suspect

The collection of properly representative known samples is crucial for accurate forensic analyses and comparisons. Known samples can be selected through a variety of legitimate schemes, including random, portion, and judgment, and must be selected with great care. Thus, traces are accidental and known samples are intentional.

Some of the consequences of what has been discussed so far induce the capacities and limitations of a forensic investigation based on trace analysis. A micro- to nano-level existence allows forensic scientists to plan physical and chemical characteristics in their identifications and comparisons with other similar data. This allows forensic science to be as methodologically flexible as its objects of study require. Because time is asymmetric and each criminal action is unique, the forensic investigation and analysis in any one case is wedded, to a certain degree, to that case with no ambition to issue general laws about that event ("In all instances of John Davis being physically assaulted with a baseball bat…"). Inferences must be drawn with explicit uncertainty statements; the inferences should be revised when new data affect the traces' relevancy. Therefore, the search for traces is a recursive heuristic process taking into account the environment of the case at hand, appealing to the imagination, expertise, and competency of the investigator or scientist to propose explicative hypotheses.

Two Native Principles

With this framework, two principles can be thought of as the main native principles that support and frame philosophically forensic science. In this context, principles are understood as universal theoretical statements settled at the beginning of a deduction, which cannot be deduced from any other statement in the considered system, and give coherence to the area of study. They provide the grounds from which other truths can be derived and define a paradigm, that is, a general epistemological viewpoint, a new concept to see the natural world, issued from an empiricist corroborated tradition, accepted by the community of practitioners in the field. Ultimately, this paradigm can even pilot the perception itself.

Although similar but nonequivalent versions are used in other disciplines, Locard's exchange principle exists as the central tenant of forensic science. The principle that bears his name was never uttered as such by Locard, but its universal statement of "every contact leaves a trace" stands as a universally accepted shorthand phrasing. Locard's principle embraces all forms of contact, from biological to chemical to physical and even digital traces, and extends the usual perception of forensic science beyond dealing only with physical vestiges.

One of its corollaries is that trace deposition is continual and not reversible. Increases in the number of contacts, the types of evidence involved, and cross-transfers (A–B and B–A) also increase the complexity of determining the relevance of traces in short duration and temporally close actions.

Even the potentially fallacious rubric of "absence of evidence is not evidence of absence" leads to extended discussions on the very nature of proof, or provable, that aims to be definitive, notwithstanding the explanations for the practical aspects of the concept (lack of sensitivity, obscuring of the relevant traces, human weakness, actual absence, etc.). Applying Locard's principle needs to address three levels. First, the physical level, which deals with ease of transfer, retention, persistence, and affinity of materials, which could better support the exchange of traces from one source to another. Second is the situational or contextual level, which is the knowledge of circumstances and environments surrounding criminal events and sets the matrix for detection, identification, and proximate significance of any evidence. Third, the intelligence level, which covers the knowledge about criminal behavior in single events or series, specific problems related to current trends in criminal behavior, and communication between relevant entities (police, scientists, attorneys, etc.); these components help the investigator in the field to focus on more meaningful traces that might otherwise go undetected.

The second, and more debated, principle is Kirk's individuality principle; again, Kirk did not state this as such beyond saying that criminalistics is the science of individualization. In its strongest form, it posits that each object in the universe can be placed demonstratively into a set with one and only one member: Itself. It therefore asserts the universal statement, "every object in our universe is unique." Philosophers like Wittgenstein have argued that without defined rules or limits, terms such as "the same" or "different" are essentially meaningless. There is little question that all things are unique—two identical things can still be numerically differentiated—but the core question is, can they be distinguished at the resolution of detection applied? Simply saying "all things are unique" is not useful forensically. For example, each fingerprint left by the same finger is unique, but to be useful, each print must also be able to be traced back to its source finger. Uniqueness is therefore necessary to claim individualization, but not sufficient. Thus, it is the degree of association that matters, how similar, how different these two things being compared are. Referring to Cole, "What distinguishes ... objects is not 'uniqueness'; it is their diagnosticity: our ability to assign traces of these objects to their correct source with a certain degree of specificity under certain parameters of detection and under certain rules governing such assignments," or as Osterburg stated, "to approach [individualization] as closely as the present state of science allows." Statistics, typically, is required to accurately communicate levels of comparison that are reproducible. In fact, Kirk noted that individualization was not absolute. ("On the witness stand, the criminalist must be willing to admit that *absolute identity is impossible to establish*. ... The inept or biased witness may readily testify to an identity, or to a type of identity, that does not actually exist. This can come about because of his confusion as to the nature of identity, his inability to evaluate the results of his observations, or because his general technical deficiencies preclude meaningful results" (Kirk, 1953; emphasis added).)

Nonnative Principles

Numerous guiding principles from other sciences apply centrally to forensic science, several of which come from geology, a cognate historical science to forensic science. That these principles come not from forensic science but from other sciences should not imply that they are somehow less important than Locard's or Kirk's notions. The first, and in many ways the most important, of the external principles is that of Uniformitarianism. The principle, proposed by James Hutton, popularized by Charles Lyell, and coined by William Whewell, states that natural phenomena do not change in scope, intensity, or effect with time. Paraphrased as "the present is the key to the past," the principle implies that a volcano that erupts today acts in the same way as volcanoes did 200 or 200 million years ago and, thus, allows geologists to interpret proxy data from past events through current effects. Likewise, in forensic science, bullets test fired in the laboratory today do not change in scope, intensity, or effect from bullets fired during the commission of a crime 2 days, 2 weeks, or 2 years previously. The same is true of any analysis in forensic science that requires a replication or reconstruction of processes in play during the crime's commission. Uniformitarianism offers a level of objectivity to historical sciences by posing hypotheses or relationships generally and then developing tests with respect to particular cases.

Three additional principles from geology hold as applicable to forensic science. They are as follows:

- *Superposition*: In a physical distribution, older materials are below younger materials unless a subsequent action alters this arrangement.
- *Lateral continuity*: Disassociated but similar layers can be assumed to be from the same depositional period.
- *Chronology*: It refers to the notion of absolute dates in a quantitative mode (such as "10:12 a.m." or "1670–1702") and relative dates in a relational mode (i.e., older or younger).

These three principles are attributed to Nicolaus Steno but were also formalized and applied by William Smith. A forensic example of applying the principle of superposition would be the packing of different soils in a tire tread, the most recent being the outermost. A good case of lateral continuity would be the cross-transfer of fibers in an assault, given that the chances of independent transfer and persistence prior to the time of the incident would be improbable. An example of absolute chronology in forensic science would be the simple example of a purchase receipt from a retail store with a time/date stamp on it. Examples of relative chronology abound but could range from the *terminus post quem* of a product no longer made to something hotter or colder than it should be.

See also: **Foundations:** Forensic Intelligence; History of Forensic Sciences; Overview and Meaning of Identification/Individualization; Semiotics, Heuristics, and Inferences Used by Forensic Scientists; Statistical Interpretation of Evidence: Bayesian Analysis; The Frequentist Approach to Forensic Evidence Interpretation; **Foundations/Fundamentals:** Measurement Uncertainty; **Pattern Evidence/Fingerprints (Dactyloscopy):** Friction Ridge Print Examination – Interpretation and the Comparative Method.

Further Reading

Cole, S.A., 2009. Forensics without uniqueness, conclusions without individualization: the new epistemology of forensic identification. Law, Probability and Risk 8, 233–255.

Crispino, F., 2006. Le principe de Locard est-il scientifique? Ou analyse de la scientificité des principes fondamentaux de la criminalistique. Editions Universitaires Européennes No. 523, Sarrebrücken, Germany. ISBN:978-613-1-50482-2(2010).

Crispino, F., 2008. Nature and place of crime scene management within forensic sciences. Science and Justice 48 (1), 24–28.

Dulong, R., 2004. La rationalité spécifique de la police technique. Revue internationale de criminologie et de police technique 3 (4), 259–270.

Egger, S.A., 1984. A working definition of serial murder and the reduction of linkage blindness. Journal of Police Science and Administration 12, 348–355.

Giamalas, D.M., 2000. Criminalistics. In: Siegel, J.A., Saukko, P.J., Knupfer, G.C. (Eds.), Encyclopedia of Forensic Sciences. Academic Press, London, pp. 471–477.

Good, G. (Ed.), 1998. Sciences of the Earth, vol. 1. Garland Publishing, New York.

Houck, M.M., 2010. An Investigation into the Foundational Principles of Forensic Science (Ph.D. thesis). Curtin University of Technology, Perth

Inman, N., Rudin, K., 2001. Principles and Practice of Criminalistics: The Profession of Forensic Science. CRC Press, Boca Raton, FL, pp. 269–270.

Kirk, P.L., 1953. Crime Investigation: Physical Evidence and the Police Laboratory. Interscience, New York, p. 10.

Kirk, P.L., 1963. The ontogeny of criminalistics. Journal of Criminal Law Criminology and Police Science 54, 235–238.

Kuhn, T., 1970. La structure des révolutions scientifiques. Flammarion, Paris.

Kwan, Q.Y., 1976. Inference of Identity of Source (Ph.D. thesis). Berkeley University, Berkeley.

Mann, M., 2002. The value of multiple proxies. Science 297, 1481–1482.

Masterman, M., 1970. The nature of a paradigm. In: Lakatos, I., Musgrave, A. (Eds.), Criticism and the Growth of Experimental Knowledge. Cambridge University Press, Cambridge, pp. 59–86.

Moriarty, J.C., Saks, M.J., 2006. Forensic Science: Grand Goals, Tragic Flaws, and Judicial Gatekeeping. Research Paper No. 06-19. University of Akron Legal Studies.

National Research Council Committee, 2009. Identifying the Needs of the Forensic Science Community, Strengthening Forensic Science in the United States: A Path Forward. National Academy of Sciences Report. National Academy Press, Washington, DC.

Osterburg, J.W., 1968. What problems must criminalistics solve. Journal of Criminal Law, Criminology and Police Science 59 (3), 431.

Schuliar, Y., 2009. La coordination scientifique dans les investigations criminelles. Proposition d'organisation, aspects éthiques ou de la nécessité d'un nouveau métier (Ph.D. thesis), Université Paris Descartes, Paris. Université de Lausanne, Lausanne.

Sober, E., 2009. Absence of evidence and evidence of absence: evidential transitivity in connection with fossils, fishing, fine-tuning, and firing squads. Philos. Stud 143, 63–90.

Stephens, C., 2011. A Bayesian approach to absent evidence reasoning. Informal Logic 31 (1), 56–65.

US Supreme Court No. 92-102, 1993. William Daubert et al., Petitioners v Merrell Dow Pharmaceuticals Inc. Certiorari to the US Court of Appeals for the Ninth Circuit. Argued 30 March 1993. Decided 28 June 1993.

Wittgenstein, L., 1922. Tractacus Logico-Philosophicus Gallimard Tel 311, Paris.

Relevant Websites

http://www.all-about-forensic-science.com—All-About-Forensic-Science.COM, Definition of Forensic Science

http://www.forensic-evidence.com—Forensic-Evidence.com.

http://library.thinkquest.org—Oracle ThinkQuest—What is Forensics?.

Forensic Classification of Evidence

MM Houck, Consolidated Forensic Laboratory, Washington, DC, USA

Glossary

Set Any group of real or imagined objects.

Taxon (pl. taxa) A group of one or more organisms grouped and ranked according to a set of qualitative and quantitative characteristics; a type of set.

Taxonomy The science of identifying and naming species with the intent of arranging them into a classification.

Introduction

Evidence is accidental: Items are transformed into evidence by their involvement in a crime regardless of their source or mode of production. By becoming evidence, their normal meaning is enhanced and expanded. Evidence is initially categorized much as the real world; that is, based on the taxonomy created by manufacturers. Forensic science adds to this classification to further enhance or clarify the meaning of evidence relevant to the goals and procedures of the discipline.

Methods of Classification

Set Theory

Any collection of objects, real or imagined, is a set; set theory is the branch of mathematics that studies these collections. Basic set theory involves categorization and organization of the objects, sometimes using diagrams, and involves elementary operations such as set union and set intersection. Advanced topics, including cardinality, are standard in undergraduate mathematics courses. All classification schemes are based on set theory, to a greater or lesser degree.

The notion of "set" is undefined; the objects described as constituting a set create the definition. The objects in a set are called the members or elements of that set. Objects belong to a set; sets consist of their members. The members of a set may be real or imagined; they do not need to be present to be a member of that set. Membership criteria for a set should be definite and accountable. The set, "All people in this room are over 5′5″ tall," is a well-defined, if currently unknown, set—the height of the people in the room would have to be measured accurately to populate the set. If the definition is vague then that collection may not be considered a set. For example, is "q" the same as "Q"? If the set is "The 26 letters of the English alphabet," then they are the same member; if the set is, "The 52 upper-case and lower-case letters of the English alphabet," then they are two separate members.

Sets may be finite or infinite; a set with only one member is called a single or a singleton set. Two sets are identical if and only if they have exactly the same members. The cardinality of a set is the number of members within it, written |A| for set A. A set *X* is a subset of set *Y* if and only if every member of *X* is also a member of *Y*; for example, the set of all Philips head screwdrivers is a subset of the set of all screwdrivers. Forensic scientists would term this a "subclass" but that is a terminological and not a conceptual difference. Two more concepts are required for the remainder of our discussion. The union of *X* and *Y* is a set whose members are only the members of *X*, *Y*, or both. Thus, if *X* were (1, 2, 3) and *Y* were (2, 3, 4) then the union of *X* and *Y*, written $X \cup Y$, would contain (1, 2, 3, 4). Finally, the intersection of two sets contains only the members of both *X* and *Y*. In the previous example, the intersection of *X* and *Y* would be (2, 3), written $X \cap Y$.

Taxonomy

Natural items, such as animals, plants, or minerals, often occur as evidence. These items are classified according to schemes used in other sciences such as biology, botany, or geology. It is incumbent on the forensic scientist to be knowledgeable about the classification of naturally occurring items.

In biology, taxonomy, the practice and science of classification, refers to a formalized system for ordering and grouping things, typically living things using the Linnaean method. The taxa (the units of a taxonomic system; singular "taxon") are sufficiently fixed so as to provide a structure for classifying living things. Taxa are arranged typically in a hierarchical structure to show their relatedness (a phylogeny). In such a hierarchical relationship, the subtype has by definition the

 http://dx.doi.org/10.1016/B978-0-12-802718-9.00002-8

same constraints as the supertype plus one or more additional constraints. For example, "macaque" is a subtype of "monkey," so any macaque is also a monkey, but not every monkey is a macaque, and an animal needs to satisfy more constraints to be a macaque than to be a monkey. In the Linnaean method of classification, the scientific name of each species is formed by the combination of two words, the genus name ("generic" name), which is always capitalized, and a second word identifying the species within that genus. Species names (genus species) are either italicized or underlined, for example, *Homo sapiens* (humans), *Sus scrofus* (pigs), *Canis familiaris* (domesticated dogs), and *Rattus rattus* (rats).

The term "systematics" is sometimes used synonymously with "taxonomy" and may be confused with "scientific classification." However, taxonomy is properly the describing, identifying, classifying, and naming of organisms, while "classification" is focused on placing organisms within groups that show their relationships to other organisms. Systematics alone deals specifically with relationships through time, requiring recognition of the fossil record when dealing with the systematics of organisms. Systematics uses taxonomy as a primary tool in understanding organisms, as nothing about the organism's relationships with other living things can be understood without it first being properly studied and described in sufficient detail to identify and classify it correctly.

In geology, rocks are generally classified based on their chemical and mineral composition, the process by which they were formed, and by the texture of their particles. Rocks are classified as igneous (formed by cooled molten magma), sedimentary (formed by deposition and compaction of materials), or metamorphic (formed through intense changes in pressure and temperature). These three classes of rocks are further subdivided into many other sets; often, the categories' definitions are not rigid and the qualities of a rock may grade it from one class to another. The terminology of rocks and minerals, rather than describing a state, describes identifiable points along a gradient.

Manufacturing

Manufactured evidence is initially categorized by the in-house or market-specific system created by one or more manufacturers. Manufacturers of economic goods create their classifications through product identity or analytical methods. Set methods of production ensure a quality product fit for purpose and sale; the classification is based on the markets involved, the orientation of the company production methods, and the supply chain. Explicit rules exist on categories recognized by manufacturers and consumers, as either models or brands. Materials flow downstream, from raw material sources through to a manufacturing level. Raw materials are transformed into intermediate products, also referred to as components or parts. These are assembled on the next level to form products. The products are shipped to distribution centers and from there on to retailers and customers.

Forensic Approaches to Classification

The supply network of raw materials, intermediate steps, production methods, intended consumer end use, and actual end use all contribute to the characteristics available for forensic taxonomic classification. While the forensic taxonomies are unique to that discipline, they are based on the production taxonomies used in manufacturing. These characteristics form the basis for statements of significance, that is, the relative abundance or rarity of any one particular item in a criminal context. Some objects are common but have a short-entrance horizon (e.g., iPods), but are essentially identical at the outset while others are common with long-entrance horizons (denim blue jeans), but have a high variance (regular, stone washed, acid washed, etc.). It is in the best interest of forensic scientists to understand the fundamental manufacturing processes of the items that routinely become evidence. This understanding can form the basis for statistical significance statements in courts and may provide the foundations for a more quantitative approach to testimony.

Forensic analytical methods create augmented taxonomies because the discipline uses different sets of methods and forensic scientists have different goals. Their taxonomies are based on manufactured traits, but also aftermarket qualities, and intended end use, but also "as used." The "as-used" traits are those imparted to the item after purchase through either normal or criminal use. Forensic science has developed a set of rules through which the taxonomies are explicated. For example, forensic scientists are interested in the size, shape, and distribution of delustrants, microscopic grains of rutile titanium dioxide incorporated into a fiber to reduce its luster. The manufacturer has included delustrant in the fiber at a certain rate and percentage with no concern for shape or distribution (but size may be relevant). The forensic science taxonomy is based on manufacturing taxonomy but is extended by incidental characteristics that help us distinguish otherwise similar objects.

Natural, manufacturing, and forensic classifications lead to evidentiary significance because they break the world down into intelligible classes of objects related to criminal acts. Forensic science has developed an enhanced appreciation for discernment between otherwise similar objects but has yet to explicate these hierarchies to their benefit.

Class-Level Information

Identification is the examination of the chemical and physical properties of an object and using them to categorize it as a member of a set. What the object is made of, its color, mass, size, and many other characteristics are used to identify an object and help refine that object's identity. Analyzing a white powder and concluding that it is cocaine is an example of identification; determining that a small translucent chip is bottle glass or yellow fibrous material and determining that they are dog hairs are also examples of identification. Most identifications are inherently hierarchical, such as classification systems themselves: In the last example, the fibrous nature of the objects restricts the following possible categories:

- Hairs
- Animal hairs
- Guard hairs
- Dog hairs
- German shepherd hairs

As the process of identification of evidence becomes more specific, it permits the analyst to classify the evidence into successively smaller classes of objects. It may not be necessary to classify the evidence beyond dog hairs if human hairs are being looked for. Multiple items can be classified differently, depending on what questions are asked. For example, the objects in **Figure 1** could be classified into "fruit" and "nonfruit," "sports related" and "nonsports related," or "organic" and "inorganic."

Sharing a class identity may indicate two objects that come from a common source. Because forensic science reveals and describes the relationships among people, places, and things involved in criminal activities, this commonality of relationship may be critical to a successful investigation. Commonality can show interactions, limitations in points of origin, and increased significance of relationships. What is meant by a "common source" depends on the material in question, the mode of production, and the specificity of the examinations used to classify the object. For example, the "common source" for an automotive paint chip could be the following:

- the manufacturer (to distinguish it from other similar paints),
- the factory (to determine where it was made),
- the batch or lot of production (to distinguish it from other batches at the same factory),
- all the vehicles painted with that color paint, or
- the vehicle painted with that color paint involved in the crime in question.

All of these options, and they are not exhaustive, could be the goal in an investigation of determining whether two objects had a "common source."

Uniqueness and Individualization

If an object can be classified into a set with only one member (itself), it can be said to be unique. An individualized object is associated with one, and only one, source: It is unique. Uniqueness is based on two assumptions. The first assumption is that all things are unique in space and, thus, their properties are nonoverlapping. The assumption of uniqueness of space is considered axiomatic and, therefore, an inherently nonprovable proposition for numerous reasons. The population size of "all things that might be evidence" is simply too large to account. In addition, conclusive evidence is not readily

Figure 1 A range of objects may be classified in a variety of ways, depending on the question being asked. For example, given the objects in this figure, the sets would differ if the question was, "what is edible?" rather than "what is sporting equipment?".

available in typical forensic investigations. Because of this, as Schum notes, statistics are required:

> Such evidence, if it existed, would make necessary a particular hypothesis or possible conclusion being entertained. In lieu of such perfection, we often make use of masses of inconclusive evidence having additional properties: The evidence is incomplete on matters relevant to our conclusions, and it comes to us from sources (including our own observations) that are, for various reasons, not completely credible. Thus, inferences from such evidence can only be probabilistic in nature (Schum, 1994, p. 2).

A statistical analysis is therefore warranted when uncertainty, of either accounting or veracity, exists. If an absolutely certain answer to a problem could be reached, statistical methods would not be required. Most evidence exists at the class level, and although each item involved in a crime is considered unique, it still belongs to a larger class. In reality, the majority of forensic science works at a class level of resolution. Indeed, even DNA, the argued "gold standard" of forensic science, operates with classes and statistics.

It has been argued that the concept of uniqueness is necessary but not sufficient to support claims of individualization. If it is accepted that uniqueness is axiomatic, then:

> what matters is whether we have analytical tools necessary to discern the characteristics that *distinguish* one object from all others or, in the forensic context, distinguish *traces* made by each object from traces made by every other object ... Every object is presumably unique at the scale of manufacture. The question is whether objects are distinguishable at the scale of detection. Since all objects in the universe are in some respects "the same" and in other respects "different" from all other objects in the universe, according to Wittgenstein, what really matters is not uniqueness but rather what rules we articulate by which we will make determinations of "sameness" and "difference" (Cole, 2009, pp. 242–243).

Although things may be numerically unique at the point of *production*, this does not help to distinguish between otherwise similar objects at the point of *detection* or *interpretation*. This is where forensic science adds value to the investigative and legal processes.

Relationships and Context

The relationships between the people, places, and things involved in crimes are central to deciding what items to examine and how to interpret the results. For example, if a sexual assault occurs and the perpetrator and victim are strangers, more evidence may be relevant than if they live together or are sexual partners. Strangers are not expected to have ever met previously and, therefore, would have not transferred evidence before the crime. People who live together would have some opportunities to transfer certain types of evidence (e.g., head hairs and carpet fibers from the living room) but not others (semen or vaginal secretions). Spouses or sexual partners, being the most intimate relationship of the three examples, would share a good deal of more information (**Figure 2**).

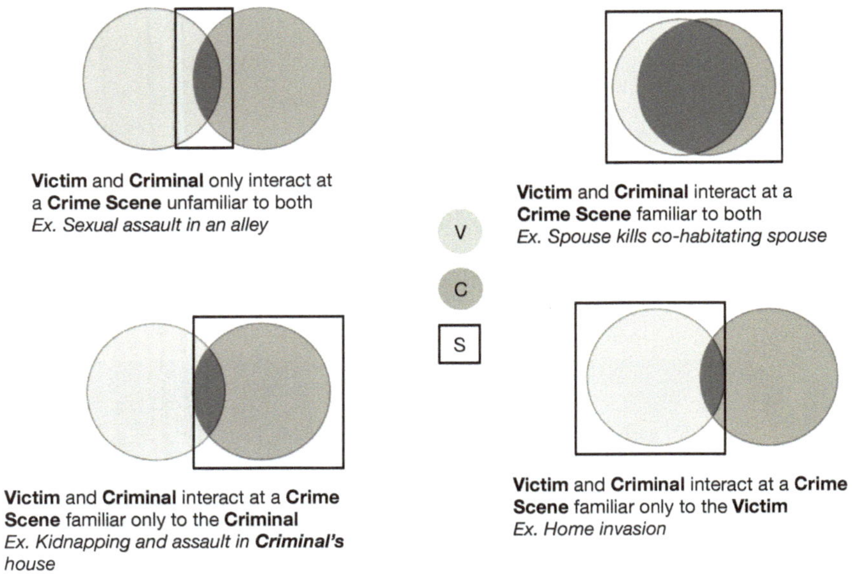

Figure 2 The relationships between suspect, victim, and scene influence what evidence is collected, and what its significance is.

Stranger-on-stranger crimes beg the question of coincidental associations, that is, two things that previously have never been in contact with each other have items on them, which are analytically indistinguishable at a certain class level. Attorneys in cross-examination may ask, "Yes, but could not [insert evidence type here] really have come from anywhere? Are not [generic class level evidence] very common?" It has been proven for a wide variety of evidence that coincidental matches are extremely rare. The enormous variety of mass-produced goods, consumer choices, economic factors, biological and natural diversity, and other traits create a nearly infinite combination of comparable characteristics for the items involved in any one situation.

See also: **Foundations:** Evidence/Classification; Statistical Interpretation of Evidence: Bayesian Analysis; The Frequentist Approach to Forensic Evidence Interpretation.

Further Reading

Cole, S., 2009. Forensics without uniqueness, conclusion without individualization: the new epistemology of forensic identification. Law, Probability and Risk 8 (3), 233–255.

Devlin, K., 1993. The Joy of Sets. Springer, Berlin.

Haq, T., Roche, G., Parker, B., 1978. Theoretical field concepts in forensic science. 1. Application to recognition and retrieval of physical evidence. Journal of Forensic Sciences 23 (1), 212–217.

Houck, M.M., 2006. Production Taxonomies as the Foundation of Forensic Significance. European Academy of Forensic Sciences, Helsinki, Finland.

Johnson, P., 1972. A History of Set Theory. Weber & Schmidt, New York.

Kwan, Q.Y., 1977. Inference of Identity of Source. University of California (Ph.D thesis).

Schum, D.A., 1994. Evidential Foundations of Probabilistic Reasoning. John Wiley & Sons, New York.

Thornton, J., 1986. Ensembles of class characteristics in physical evidence examination. Journal of Forensic Sciences 31 (2), 501–503.

Underhill, P., 2000. Why We Buy: The Science of Shopping. Simon & Schuster, New York.

Interpretation/The Comparative Method

MM Houck, Consolidated Forensic Laboratory, Washington, DC, USA

Glossary

Alignable differences Differences that are connected to the hierarchical system of relatedness of two or more things.

Analogous trait A characteristic that is similar between two things that is not present in the last common ancestor or precedent of the group under comparison.

Analogy A cognitive process that transfers information or meaning from one subject (the analog or source) to another subject (the target).

Diagnosticity The degree to which traits classify an object.

Homologous trait A characteristic shared by a common ancestor or precedent.

Nonalignable differences Differences with no correspondence at all between the source and the target.

Introduction

Analogy, and its more specific relative comparison, is a central component of human cognition. Analogy is the process behind identification of places, objects, and people and plays a significant role in many human mental operations, like problem solving, decisions, perception, memory, and communication. Some researchers, including Hofstadter, have even argued that cognition is analogy. Likewise, the cognitive process of analogy and the method of comparison lie at the heart of the forensic sciences. The ability to compare is predicated on some sort of classification (more properly, a taxonomy) that results in classes, groups, or sets.

Aristotle is considered the first to approach comparison as a way to arrange the world. His attempt to codify the process raised, however, an intractable problem that would only be addressed later: the classification of living things. Comparison, by itself, is a minimal technique, at best. A classification system—a taxonomy—is a prerequisite to a fuller comparative methodology. Comparative anatomy, one of the earliest formal applications of the method, goes beyond mere representation (mere comparison, that is) to explain the nature and properties of each animal.

The French naturalist Pierre Belon (1517–1564) compared the skeletal structures of birds to humans in his book *L'Histoire de la Nature des Oiseaux* (*History of the Nature of Birds*, 1555; **Figure 1**), and, along with the Flemish naturalist Andreas Vesalius (1514–1564), was one of the first naturalists to explicitly apply the comparative method in biology. Georges Cuvier (1769–1832) was the first to use comparative anatomy and taxonomy as a tool, not an end in itself, in his studies of animals and fossils. Cuvier was frustrated that biological phenomena could not be reconfigured into experimental conditions that would allow controlled testing, a difficulty common to many sciences (e.g., see Diamond). The intimate integration of a living organism's physiology with its anatomy created obstacles in teasing out and relating function to structure: Once an organism was dead and prepared for dissection, its function had ceased, thus confounding the relationship of form to function. Cuvier considered that careful examinations and the interrelating of structures between specimens might also prove to be useful in revealing principles of observation and comparison. Perhaps the original scientist-as-detective, Cuvier, used scattered, fractured bits of information to reconstruct the prehistory of the Earth and its animals. In a 1798 paper, Cuvier wrote on his realization of the form and function of bones as it relates to the overall identifiable anatomy of an animal, leading to the recognition of the creature from which the bone originated:

> This assertion will not seem at all astonishing if one recalls that in the living state all the bones are assembled in a kind of framework; that the place occupied by each is easy to recognize; and that by the number and position of their articulating facets one can judge the number and direction of the bones that were attached to them. This is because the number, direction, and shape of the bones that compose each part of an animal's body are always in a necessary relation to all the other parts, in such a way that—up to a point—one can infer the whole from any one of them, and vice versa (Rudwick, 1998, p. 36).

This has been called "Cuvier's Principle of Correlation of Parts" and is a central tenet in biology and paleontology. It is

 http://dx.doi.org/10.1016/B978-0-12-802718-9.00003-X

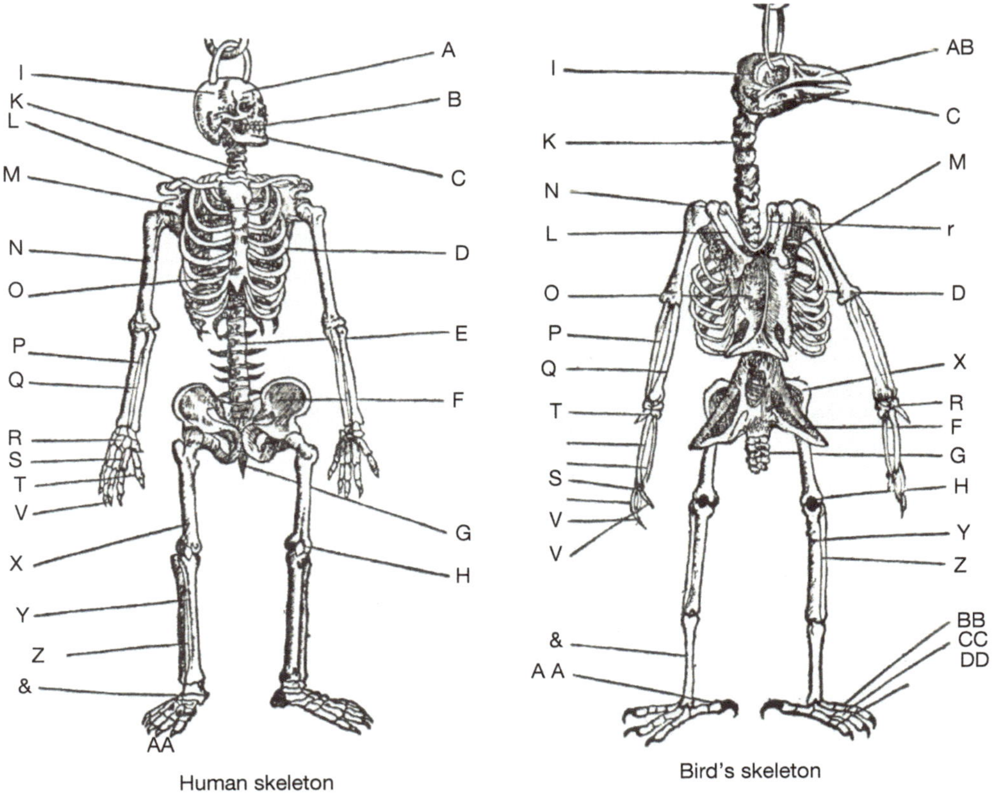

Figure 1 A drawing from Pierre Belon's 1555 book, *History of the Nature of Birds*, comparing the skeletal anatomy of birds to humans which is one of the first books using the science of comparative anatomy. Source: Wikimedia commons, open source.

important to note that Cuvier claimed to be able to *identify* an animal taxonomically from a single bone, but not completely *reconstruct* it, as the above quote might imply. The reconstruction would only be possible with a sufficient number of bones representing the animal in question. The comparative method has been a successful cornerstone of science ever since, with new or emerging sciences, such as ecology, moving from the purely observational or descriptive approach to that of comparison through experimental or analytical methods.

A short discussion of terms in biology will help clarify concepts used in biological comparisons. The concept of homology, the same structure under every variety of form found in different animals, is the organizing foundation for comparative anatomy. Animals share homologous traits because they also share a common ancestor with the same or related trait. By contrast, analogous traits are similarities found in organisms that were not present in the last common ancestor of the group under comparison; that is, the traits evolved separately. The canonical example of the difference between homologous and analogous traits is the wings of birds and bats: they are homologous as forearms but analogous as wings; the latter structures evolved their functions separately. A homologous trait is termed a homolog. In biology, evolution and natural selection formed the system within which these relationships developed and were maintained, homogenized, or differentiated.

In manufacturing, other external and internal constraints form the basis for homologous and analogous traits through design, function, form, and costs. Design follows from the product's intended end use, aesthetic concerns, and cost limitations. The function and form of an object tend to correlate and variances in design cluster around necessary and sufficient criteria. In **Figure 2**, for example, although the hammer heads, opposite sides, handles, materials, weight, shape, and components all vary, they are nonetheless identifiable as hammers. If **Figure 2** were finches, as Darwin studied in the Galapagos in his historic voyage with the *Beagle*, the base process of taxonomy would be the same but the criteria and foundations—the history and causes—would obviously vary because of the vastly different processes that produce hammers and finches.

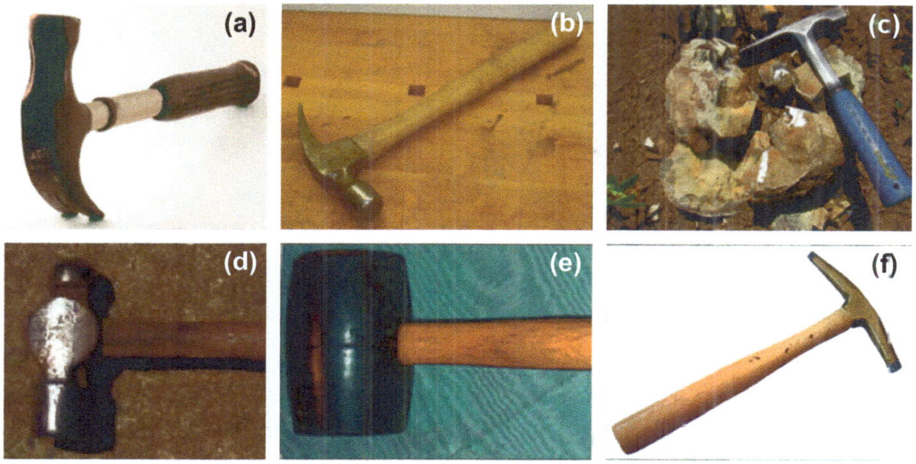

Figure 2 Hammers. All of the objects (a–f) are recognizable as hammers even though their components vary. (a) Claw hammer; (b) framing hammer; (c) geological hammer; (d) ball-peen hammer; (e) rubber mallet; and (f) upholstery hammer. Source: Wikimedia commons, open source.

Broadly speaking, the supply chains and distribution networks of material goods are like the phylogenetic trees based on evolutionary descent. Regardless of whether the items are biological or manufactured, the independence of traits should not be assumed. Comparative studies that do not control for historical relationships through phylogeny or supply chains may imply spurious relationships (coincidences). Forensic science is unique in its use of the comparative method to reconstruct past criminal events and sourcing of evidence, either biological or manufactured (in essence, reverse engineering to a level of distribution or manufacturing resolution).

Analogy and Comparison within a Forensic Process

Analogy is a cognitive process that transfers information or meaning from one subject (the analog or *source*) to another subject (the *target*); it thus implies at least two things: situations or events. The source is considered to be the more complete and more complex of the two and the target is thus less informative and incomplete in some way. The incompleteness may be due to any of several factors, alone or combined, such as damage, fracture, deterioration, or size. The elements or traits—including their relationships, such as evolutionary or supply chains—between the source and the target are mapped or aligned in a comparison. The mapping is done from what is usually the more familiar area of experience and more complete repository of information, the source, to the typically more problematic target.

Salience of the elements or traits is of prime importance: there are an innumerable number of arbitrary differences in either elements or relations that could be considered but are not useful given the question at hand ("Are both items smaller than the Empire State Building? Are they redder than a fire truck?"). Ultimately, analogy is a process to communicate that the two comparators (the source and the target) have *some* relationship in common despite any arbitrary differences. Some notion of possible or hypothetical connection must exist for the comparison to be made. As a forensic example, consider trace debris removed from the clothing of a suspect and the body of a victim: Although there may be no physical evidence (hairs, fibers, glass, soil, etc.) in common, the suspect's clothing and the victim's body have, at least prima facie, a common *relationship* (the victim is the victim and the suspect is a person of interest in the crime) until proven otherwise. Thus, common relations, not common objects, are essential to analogy and comparison.

The comparison process as a method makes several assumptions. First, the space in which the comparators are mapped is assumed to be Euclidean. Second, the method embeds the comparators in a "space of minimum dimensionality" (Tversky) based on all observed salient similarities. Each object, a, is detailed and described by a set of elements or traits, A. Any observed similarities between a and another object b, denoted as $s(a, b)$, are expressed as a function of the salient traits they are determined to have in common. The comparison and any observed familiarity can be expressed as a function of three arguments (Figure 3):

- $A \cap B$, the features shared by a and b
- $A - B$, the features of a that are not shared by b
- $B - A$, the features of b that are not shared by a

Psychological studies show that people tend to pay more attention to the target (the comparator with less information)

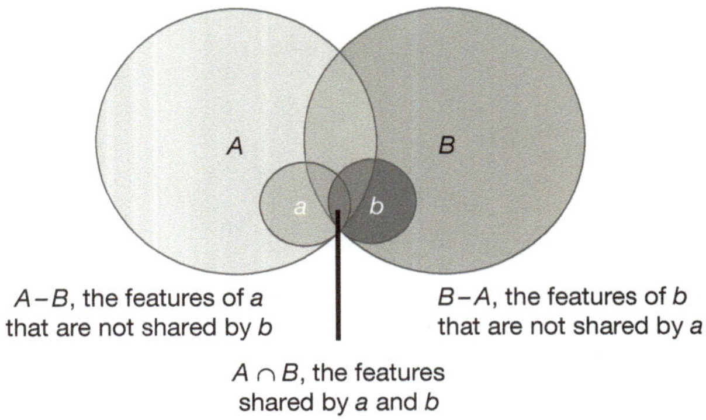

Figure 3 A comparison of observed familiarities can be expressed as a function of three arguments visualized here.

than to the source. In forensic science, this means that analysts would pay more attention to the samples from the crime scene or actors than to the known samples collected. This is true even though the known has more salience, because arguably it has more information and a documented provenance than the questioned sample. For example, a toy ship is quite similar to a real ship because most of the main features of the real ship are expressed in the toy (otherwise it might not be recognized as a simulacrum of its referent). A real ship, however, is not as similar to the toy ship because many of the features of a real ship are not expressed in the toy (due to function, scale, or safety, among other factors). The reason for paying more attention to the target is, first and foremost, to determine if there is sufficiency of salient information in the target for the comparative process to occur (see Vanderkolk for a discussion on this).

The main determinant of feature salience for comparative purposes is the degree to which they classify an object, that is, their diagnosticity. A feature that serves as the basis to reassign an object from one class to another class with fewer members is more salient than one that does not. Salience is hierarchical and is based on how many members of a class share that feature; the goal is thus to place an object, by successive comparative features, into classes with increasingly fewer members. Salience of a feature, therefore, should increase inversely with the number of members of a class into which it places an object; $A \cap B$ increases and may be thought of as an expression of diagnosticity. A comparative process that does not maximize diagnosticity or exploit features that do so will have low forensic utility.

The Comparative Method within Forensic Science

The comparative method involves the aligning of the relational structures between one or more targets (items of questioned source; Qs) and one or more sources (items of known provenance or source; Ks). This alignment, to work as a method, has three constraints or requirements:

- The alignment has to be *structurally consistent*, that is, it has to observe a one-to-one correspondence between the comparators in an argumentative structure that is the same between the comparisons (*parallel connectivity*). One point of comparison can be aligned with at most one other point of comparison in the target or source. Similarly, matching relationships must have matching arguments to support them (the reason for the proposed relationship cannot be based on an unrelated argument).
- The comparison has to involve *common relations* but does not have to involve common object descriptions. All the evidence that came from the crime scene, for example, need not have originated from only one source.
- Finally, comparisons are not made merely between the objects at hand but also include all of the higher order "constraining relations" that they may share (*systematicity*). In biology, this would relate to the evolutionary and genetic connections; for manufactured materials, this would be the design factors and the supply chain of raw materials and intermediate processes that lead to a finished consumer good. The deeper the relational history, the more higher order classes that two objects share, the stronger the relationship they share, and, therefore, the greater is the chance of a shared origin. This obviates the significance of coincidental matches between otherwise similar but unrelated objects: A series of coincidences between two objects are not a salient relationship, no matter how many of them exist. Type I and type II errors stem from these coincidences.

A comparison results in a type of cross-mapping of analogous traits or phenomena that have differential relational roles in two situations (e.g., victim's clothing and crime scene). A systematic mapping between source and target is a natural method for differentiating potentially ambiguous relationships. This relates to the classification of the target and source, the identification of traits or features each has that place them in one or more sets (classes) of items. The cross-mapping is of these traits within a class. Once a source has been aligned to a target, *candidate inferences*, based on the source, can be projected onto the target, such as a shared source or history. A handgun with blood on it, for example, can be compared to a bullet removed from a victim (through test firings of similar ammunition) and determined to have been the source (to some degree of certainty) of the bullet while the blood can be tested through DNA typing with the victim's known sample and be shown to have the victim as its source (again, to some degree of certainty); the fact that the victim's blood is on the handgun indicates a shared history of occurrence (lateral contemporaneity).

Comparison is selective. The requirement of systematicity is predicated on the idea that classes or sets are flexible and hierarchical. Higher order connections predict lower order relations, and commonalities that are not a part of the aligned system of relationships are considered inconsequential: A blue shoe and a blue car have little in common other than the stated color category; likewise, the fact that the source shoe and the target print might have the same kind of outsole design recedes in importance to the fact than none of the individual traits on the sole appears in the print. Differences that are connected to the hierarchical system of relatedness are called *alignable differences;* those differences with no correspondence at all between the source and the target are called *nonalignable differences.* Alignable differences are more meaningful and salient than nonalignable ones because they exist within the same relationship system making them more relevant to each other. The strange conclusion this observation leads to is that there should be more meaningful differences for comparators that are very similar (*toy train–real train*) than for ones that are less similar (*toy train–toy ship*) because the more similar comparators will have or be derived within more common systems of relationships and will have more alignable differences. As an example, consider all the possible differences for the pair *automobile–truck* and for the pair *duck–baseball.* More alignable differences could be found for the first pair than the second: After a few differences ("You don't play sports with a duck. You don't hunt baseballs "), the list seems pointless because the two are not aligned. The details that could be elicited by comparing *automobile* with *truck,* however, could go on for some time, depending on the level of detail desired. Most sets of comparators in the world are dissimilar (which is why forensic comparisons tend to be stronger in exclusion than inclusion) and this "nonconsideration" heuristic makes sense given humans' cognitive load: "Intuitively, it is when a pair of items is similar that their differences are likely to be important" (Genter and Markman). Psychological experiments support this statement and it seems to be an integral part of human cognition. Related to this idea is Wittgenstein's proposal 5.5303 in his work *Tractatus logico-philosophicus*: "Roughly speaking, to say of two things that they are identical is nonsense, and to say of one thing that it is identical with itself is to say nothing at all." This points to the need for a statistical evaluation of the *strength* of a comparison, either inclusive or exclusive.

See also: **Foundations:** Forensic Intelligence; Overview and Meaning of Identification/Individualization; Semiotics, Heuristics, and Inferences Used by Forensic Scientists.

Further Reading

Diamond, J., Robinson, J.A. (Eds.), 2010. Natural Experiments of History. Cambridge University Press, Cambridge, MA.

Gentner, D., Markman, A.E., 1997. Structure mapping in analogy and similarity. American Psychologist 52 (1), 45–56.

Hofstadter, D., 2001. Analogy as the core of cognition. In: Gentner, D., Holyoak, K., Kokinov, B. (Eds.), The Analogical Mind: Perspectives from Cognitive Science. MIT Press/Bradford Book, Cambridge, MA, pp. 499–538.

Markman, A.B., Genter, D., 2000. Structure mapping in the comparison process. American Journal of Psychology 113 (4), 501–538.

Pellegrin, P., 1986. Aristotle's Classification of Living Things. University of California Press, Berkeley, CA.

Rudwick, M., 1997. Georges Cuvier, Fossil Bones, and Geological Catastrophes. University of Chicago Press, Chicago.

Tversky, A., 1977. Features of similarity. Psychological Review 84, 327–352.

Vanderkolk, J., 2009. Forensic Comparative Science. Academic Press, New York.

Wittgenstein, L., 1922. Tractatus Logico-Philosophicus. Routledge, London (C.K. Ogden (1922), prepared with assistance from G.E. Moore, F.P. Ramsey, and Wittgenstein, trans).

Forensic Engineering/Accident Reconstruction/Biomechanics of Injury/Philosophy, Basic Theory, and Fundamentals

SC Batterman and SD Batterman, Batterman Engineering, LLC, Cherry Hill, NJ, USA

Introduction

Forensic engineering may be defined as the application of engineering principles toward the purposes of the law. It is important to remember that although the analysis and results are to be applied to the understanding and resolution of legal matters, they must follow from a rigorous application of the governing natural laws and application of scientific principles, consistent with the available information and data. As in all areas of scientific inquiry and endeavor, the quality and reliability of an analysis and the resulting solution are directly related to the quality and detail of the input data. At times, there may be considerable uncertainty associated with the results or insufficient information and data available with which to reach objective conclusions to a reasonable degree of engineering and scientific certainty. In such cases, this in and of itself may be an important conclusion for the finders of fact (i.e., the judge and jury).

It is important to note at the outset that although individuals may refer to themselves as forensic engineers, such individuals are typically formally educated and rigorously trained engineers applying their scientific and engineering backgrounds, education, and knowledge in one (or more) of the traditional engineering disciplines to the solution of engineering problems that arise in forensic settings. Although colleges, universities, and institutions of higher learning may offer select courses in forensic engineering, in the United States, currently, there are no degree granting programs in forensic engineering, per se. Qualified practitioners of forensic engineering are rigorously trained and formally educated in the traditional engineering disciplines and/or closely related sciences.

In performing a scientific accident reconstruction, the available physical evidence is utilized in conjunction with the governing laws of nature and physical principles to work backward from the aftermath of an event in order to determine the conditions at the commencement of, or immediately before, the subject incident sequence. With the final state, or condition known, the objective of any forensic investigation in general, or accident reconstruction specifically, is to determine the initial conditions or state of the system at the commencement of the event sequence. It is this knowledge of the final state of the system, with the objective being to solve for the initial conditions, that distinguishes investigations and analyses in forensic science from more typical, or mainstream scientific, investigations where the state of the system at some initial time (t_0) is known and the objective is to solve for the final state of the system at some later time (t_f).

In cases involving accidental injury and/or death allegedly caused by defectively designed and/or manufactured products (i.e., products liability), forensic engineers are often called upon to either help support or defend against allegations of a product defect being causally related to the subject accident and/or injuries. Suffice it to say that in order to be legally significant in matters of personal injury litigation, it is typically not sufficient to simply demonstrate, or prove, that a particular product is defective. It must also be demonstrated through the appropriate analyses that the defective condition was causally related to the resulting injuries and/or death(s). The goal of an automotive accident reconstruction is often the determination of impact velocities (speeds, i.e., the magnitudes of the velocities) and/or changes in vehicle velocities (delta-v's), which are important considerations in determining responsibility for an accident and also in assessing the severity of an accident in terms of its injury-producing potential. However, oftentimes there are also allegations of defective vehicle design and/or manufacturing being causally related to the happening of an accident and/or the failure of safety features to function properly and/or to protect an occupant as intended or designed. For example, in automobile crashes involving occupant injury and/or death, the issue often arises as to whether or not the air bag(s) should have deployed by design, and/or whether or not the proper use of seat belts and/or an air bag deployment would have mitigated and/or prevented the resulting injuries. Before these questions can be answered to a reasonable degree of engineering and scientific certainty, an engineering accident reconstruction must be performed along with the appropriate biomechanical analyses.

Basic Principles

Classical mechanics may be defined as the science that treats the response of material bodies to the forces, motions, and displacements imposed upon them. Basic concepts are *space*, *time*, *mass*, and *force*.

 http://dx.doi.org/10.1016/B978-0-12-802718-9.00004-1

Biomechanics, a specialty within the more general field of classical mechanics, is then defined to be the science that treats the response of living systems to the forces, motions, and displacements imposed upon them. Typically, as is the case in most accident reconstruction analyses, biomechanicians (or biomechanists as some refer to them) are concerned with the response of the human body to the forces, motions, and displacements experienced during a crash. However, as the biomechanical response of an occupant is highly dependent upon the crash response of the vehicle (or system) that he/she occupies, a rigorous engineering accident reconstruction must be performed before a biomechanical analysis of the occupant response to a crash can be performed, and the injury-producing potential of the accident evaluated. This chapter begins with a review of the relevant physical principles.

In 1687, Sir Isaac Newton published his, now famous, three natural laws of classical mechanics. These laws govern the motion of all bodies in the universe omitting relativistic effects, that is, they are the operative laws for all problems when speeds do not approach the speed of light (299 million ms^{-1} or approximately 186 000 miles s^{-1}). These laws, originally stated for a particle of mass, m, are summarized below.

First law: Every material body continues in its state of rest, or of constant velocity motion in a straight line, unless acted upon by external forces, which cause it to change its state of rest or motion.

Second law: The time rate of change of linear momentum of a particle (product of mass and velocity) is proportional to the external force acting on it and occurs in the direction of the force. An alternate form of this law, and the one that is most familiar, is that the resultant force $(\overline{F})$ acting on a particle is equal to the product of its mass (m) and acceleration $(\bar{a})$ or $\overline{F} = m\bar{a}$(where vector quantities are indicated with a bar above the symbol).

Third law: To every action there is an equal and opposite reaction; or, equivalently, the mutual forces of two bodies acting upon each other are equal in magnitude and opposite in direction.

It is important to realize that there is a considerable amount of sophisticated philosophy embodied in Newton's laws, which must be understood by the forensic engineer/accident reconstructionist and is critical to mention here. The following are noted.

First, the laws are not scalar statements (which consider magnitude only) but are vector statements. This means that both the magnitudes and directions of forces, velocities, and accelerations must be considered. For example, speed and velocity are not interchangeable and constant speed is not the same as constant velocity. A vehicle traveling at constant speed in a circular path has a different velocity at each instant of time because the travel direction of the vehicle is constantly changing. In fact, the vehicle is indeed accelerating toward the center of curvature. However, a vehicle traveling in a straight line at a constant speed is also traveling at a constant velocity and is not accelerating.

Second, "motion" and "rest" are relative terms when compared to a frame of reference. For accident reconstruction applications, the earth is typically a suitable, "fixed" frame of reference. However, for other applications such as rocket launchings, space flights, and general astronomical considerations, the earth would not be a suitable fixed reference frame and distant stars would then be used.

Third, Newton's laws mention forces without ever defining them. Force is an abstract concept and nobody has ever seen a force, that is, only the effects of forces are felt. Forces may be defined as the actions of one body on another and are classified as either contact forces (surface forces) or forces at a distance, such as those due to gravity (weight) or electrical and/or magnetic effects. Again, it is noted that force is a vector quantity, which has both an associated magnitude and direction. It is further worth noting that in automobile accidents, or other vehicular crash scenarios, forces on occupants only arise when the occupant contacts, or is restrained by, another surface such as a seat, seat belt, air bag, door, another person, or other vehicle structure. When this occurs, the occupant is accelerated (decelerated) and Newton's second law must be used to determine the forces acting on the occupant.

Fourth, Newton's laws are stated only for a particle, which is a mathematical idealization of a point with constant mass. It requires further rigorous treatment to extend Newton's laws to the motion of rigid or deformable bodies of finite dimension, the motion of the mass center of a body, or the motion of variable mass systems such as rockets, for example.

Fifth, the concept of units is embedded in Newton's second law. In the International System (SI), mass is given in kilograms (kg), acceleration is given in meters per second squared (m s^{-2}), and force is the derived unit given in Newtons (N). One Newton is the force required to impart an acceleration of $1\ \text{m s}^{-2}$ to a mass of 1 kg and has units of kg m s^{-2}. In the US–British system of units, force is given in pounds (lb), acceleration in feet per second squared (ft s^{-2}), and mass is the derived unit given in slugs. One slug is equal to $1\ \text{lb s}^2\ \text{ft}^{-1}$.

It is also worth noting that the second law introduces the concept of acceleration. Mathematically, the acceleration of a particle, $\bar{a}$, is the time rate of change of its velocity, $\bar{v}$, or in the language of calculus, the first derivative of velocity with respect to time. For the purposes herein, it is noted that acceleration can be considered the change in velocity $(\Delta\overline{V})$ divided by the short time interval (Δt) of a crash over which the change in velocity occurs. Defined in this manner, the ratio $\Delta\overline{V}/\Delta t$ rigorously defines the average acceleration over the time increment considered. As the actual crash is typically a very rapid event occurring over milliseconds, the change in velocity $(\Delta\overline{V})$ of the center of mass of a vehicle, which is often more easily calculated than the acceleration, provides useful information concerning the severity of a crash and is discussed further in the following sections.

This section is ended by noting that Newton's laws can be utilized to further derive three principles, which are often convenient for solving dynamic problems. These principles are *the principle of work and energy, the principle of linear impulse and momentum,* and *the principle of angular impulse and angular momentum*. The reader is referred to a text or reference in physics or engineering mechanics (such as appear in the reference list at the end of the chapter) for the derivations and detailed discussions of these fundamental principles.

Methodology

As mentioned above, accident reconstructions are *a posteriori* investigations and analyses of accidents or events for the purpose of trying to determine how and what occurred during the event consistent with the available data and governing natural laws. In this sense, accident reconstructions are similar to human postmortem examinations or autopsies. Accident reconstruction is defined as the scientific process of analyzing an accident utilizing the available data and the appropriate natural or physical laws. The availability of data and physical evidence in combination with the reconstructionist's knowledge and understanding of scientific/engineering principles is the critical factor in determining the quality, accuracy, and reliability of the reconstruction.

The basic philosophy of accident reconstruction is the one in which the analyst/reconstructionist generally works backward in time and space. Detailed documentation of the accident scene, to the extent possible, is critical for a rigorous, quantitative accident reconstruction. The starting point for the analysis is the final rest position(s) of the vehicle(s), accident debris, and sometimes victims/bodies in the cases of pedestrian impacts and occupant ejections from vehicles. From this known information/data, the reconstruction proceeds backward in time and space to the point(s) of impact, and if sufficient physical evidence exists, to points in time and space prior to the impact(s). The natural laws utilized will be Newton's laws of motion. In addition, the reconstructionist must use physical evidence left at the scene as a result of the accident, as well as relevant vehicle data and scene geometry. Typically, this information includes vehicle weights (masses) inclusive of the vehicle occupants, miscellaneous items and cargo, location(s) of the point(s) of impact on the roadway and on the vehicle(s), distances and trajectories moved postimpact, tire marks, gouge and scratch marks on the roadway, vehicle crush damage and damage locations, paint transfer, debris locations, body contact marks, location of blood and other body tissue, etc. The evidence left at the scene, or within (or on) a vehicle, or on a body is sometimes referred to as the witness mark. Typically, it is the police or presiding law enforcement agency, who are the first responders to an accident scene and they are in responsible charge of documenting, photographing, and measuring/mapping the accident scene and preserving the available evidence. Obviously, the quantity and quality of the available data will affect the accuracy and reliability of the reconstruction, and proper documentation of the scene is critical to ensuring a high-quality, scientifically reliable accident reconstruction. It is better if the scene and vehicles are inspected as soon as possible after an accident. Not only do witness marks and transfer evidence tend to degrade/disappear with time, but also vehicles can be repaired, or disposed of, and the scene may change. Often, the reconstructionist becomes involved in an investigation long after the accident occurred and may not have the opportunity to examine the scene and vehicles. He/she will then have to rely upon the previously documented evidence including, but not necessarily limited to, measurements and photographs. High-quality photographs contain a large amount of useful information, some of which can be quantitatively extracted utilizing photogrammetric techniques, that is, the science of making measurements from photographs. Examples of information obtained from photographs include the character(s) and relevant length(s) and geometric characteristic(s) of tire marks, pothole depths, as well as vehicle crush patterns and dimensions. It is noted that not all tire marks are skid marks, and a proper identification of the type of tire mark is necessary to properly characterize the motion and/or acceleration/deceleration of the vehicle that left the tire mark in question. In addition to dimensional information and data being obtained from photographs, photographs can often be digitally enhanced—not merely enlarged—to reveal significant details such as sharp edges of potholes, tire defects paint transfer evidence, tire tracks, and so on. Although actual physical inspections are always best, good-quality photographs may also play a key role in formulating reliable and supportable scientific opinions concerning the happening of an accident, or catastrophic event/failure.

The main objective of an engineering accident reconstruction is to obtain quantitative answers and this requires making calculations based upon Newton's laws and the available data. This often involves the calculation of vehicle impact speeds and/or the change in velocity ($\Delta\overline{V}$) of a vehicle for the purpose of assessing the severity and injury-producing potential of an impact and/or accident sequence. The correlation between $\Delta\overline{V}$ and the injury-producing potential of an accident is discussed in greater detail below when considering the biomechanical and occupant kinematic responses to an impact. Calculations can be performed by hand, computer, or both. Numerous accident reconstruction computer programs are commercially available, and it is not the purpose of this chapter to discuss these programs. For a more detailed discussion, the interested reader is directed to the chapter appearing herein on Computer Methods in Accident Reconstruction. Suffice it to say that computer programs are based on Newton's laws, as they must be, and many also contain special algorithms and techniques particular to automobile accident reconstruction. As a word of caution, it is noted that some programs are so user-friendly that they can be used by individuals lacking in an adequate educational background and/or sufficient experience to obtain results that may not be valid for a particular accident. It is also noted for completeness that accident reconstruction programs should not be confused with *simulation* programs. As opposed to reconstruction programs, which utilize the governing laws of physics and available scene data to work backward from the final rest positions of the vehicles to the conditions at impact, and then before impact if sufficient data are available, simulation programs, which contain many assumptions and limitations, proceed forward in time and space commencing before, or at, impact and then continue proceeding to the postimpact and rest positions of the vehicles. Solutions obtained utilizing these (simulation) programs are very sensitive to the initial conditions and assumptions made by the reconstructionist and are not unique, that is, several answers can be obtained for the same problem.

Accident Reconstruction

Often, Newton's laws are cast in other forms for ease in performing calculations and solving a problem. As discussed above, alternate forms of the Newton's laws are the work-energy principle and impulse-momentum principle. It is emphasized that these principles are not separate laws of physics but are derived from the Newton's second law and are mathematically integrated statements of the second law. Ignoring ground reaction forces at the tires, which are typically small compared to the equal and oppositely directed collision forces acting on the vehicles (Newton's third law) at the moment of impact, the linear impulse-momentum principle is very useful in analyzing collision problems because the total (impulsive) external force acting on the system of colliding vehicles is zero at impact. This is equivalent to stating that the linear momentum of the vehicle system immediately before (or at) impact is conserved and leads to the well-known principle of conservation of linear momentum. It is emphasized that the principle of conservation of linear momentum is a vector statement and directions, therefore, must be properly accounted for. It is also worth noting that although linear momentum is conserved at impact, the kinetic energy of the system of vehicles, in general, is not (for inelastic collisions). That is, energy is permanently dissipated in crushing/deforming the vehicles. However, as illustrated below, if sufficient scene information is available, it is not necessary to use energy methods to solve a problem.

To illustrate the methodology and calculation procedures discussed above, consider the common case of a right angle (90°) intersection collision between two vehicles (**Figure 1**). As shown, vehicle 1 proceeds into an intersection from left to right (in the positive x-direction), whereas vehicle 2 proceeds into the intersection traveling in the positive y-direction. Subscripts 1 and 2 are used to denote quantities for vehicles 1 and 2, respectively. A collision occurred, and the vehicles came to rest in the positions shown. In this illustration, each vehicle skidded before impact (shown in bold by the lines behind and leading up to the vehicles at impact and denoted by the lengths d_1 and d_2). Skidding means that the wheels locked up due to driver braking and the vehicles then slid, not rolled, to the point of impact (POI). It is parenthetically noted that most modern motor vehicles are equipped with antilock braking systems (ABS), which prevent full wheel lockup during a hard brake application, and oftentimes do not leave discernable physical evidence of braking even during a hard brake application. However, in this illustration, it is assumed that both vehicles did skid/slide before impact. Investigation at the scene determined the POI to be as shown. Location of the POI on the roadway is typically determined by abrupt changes in the direction of the skid marks, collision scrub marks from the tires, and/or gouges in the roadway surface made by vehicle components and/or portions of the vehicle being displaced

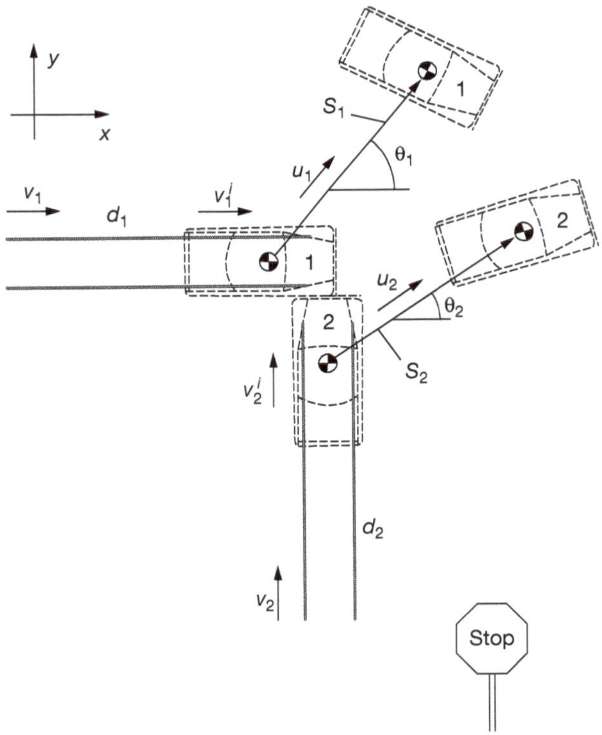

Figure 1 Intersection collision.

downward into forceful contact with the roadway surface. The location of solid vehicle debris on the roadway is, in general, not a good indicator of the POI because it travels when it is dislodged from the vehicle and does not fall straight down to the roadway surface and stop. However, fluid spatter, resulting from the sudden release of high-pressure fluid as oftentimes occurs when a radiator bursts at impact, indeed may be a good indicator of the point/area of impact as the explosive release of radiator fluid occurs very quickly. However, the reader is cautioned when identifying a POI through the use of fluid evidence that high-pressure fluid spatter patterns not be confused with fluid debris left at an accident scene by slowly draining/discharging fluids that take time to deposit on the roadway surface, as such evidence will, in general, not be a good indicator of the POI. Impact speeds are designated with the superscript *i* and postimpact speeds upon vehicle departure from the POI are designated by u_1 and u_2. The distances moved by the centers of gravity (or mass) of the vehicles from impact to their final rest positions are denoted by the distances s_1 and s_2, while the angles measured from the centers of gravity are also shown in the figure. The data shown in **Figure 1** are very complete and representative of an excellent on-scene investigation. In this case, the goal of reconstruction is to determine the speeds of the vehicles at impact ($v_1{}^{i}$ and $v_2{}^{i}$) and at commencement of preimpact skidding (v_1 and v_2). Although this example is relatively simple, it does illustrate many of the principles used in accident reconstruction.

All of the basic equations utilized to solve this problem are derived from Newton's second law. As the philosophy is to work backward from the points of rest to the POI, the postimpact speeds u_1 and u_2 of the centers of gravity of the vehicles immediately after impact, that is, at the point of vehicle separation, are determined. The reasonable assumption is made that the vehicles did not simply roll to their rest positions, but that the vehicles were displaced laterally and may have also rotated. This means that the tires were also moving laterally and essentially sliding over the road surface. For simplicity, the vehicles are considered point masses, and their rotations from impact to their final rest positions are not explicitly considered in this example. Noting that the centers of gravity of the vehicles moved for distances s_1 and s_2 from impact to final rest, using Newton's second law, it can be shown that the postimpact speeds are given by

$$u_1 = \sqrt{2\mu g s_1}$$

$$u_2 = \sqrt{2\mu g s_2}$$

where μ is the coefficient of friction (or drag factor) between the tires and roadway surface and g is the acceleration due to gravity. At the earth's surface, g is approximately equal to 9.8 m s^{-2} (32.2 ft s^{-2}). The coefficient of friction μ is a nondimensional parameter, which is a measure of the relative slip resistance of the two surfaces in contact. It can be measured at the scene immediately following the accident before conditions change, which is preferred as long as it is done correctly, or ranges can be obtained from published references for tires on various roadway surface materials under a variety of conditions in the absence of measurements. For the example under consideration, it is assumed that μ was determined to be 0.65, which represents a typical value for a road surface in an average condition.

The principle of conservation of linear momentum (which derives from Newton's second law) states

$$m_1\bar{v}_1 + m_2\bar{v}_2{}^{i} = m_1\bar{u}_1 + m_2\bar{u}_2$$

where m_1 and m_2 are the respective vehicle masses and bars over the velocities are used to denote vectors.

Considering the components of linear momentum in mutually perpendicular *x*- and *y*-directions, the following are obtained.

$$x: \quad m_1 v_1 = m_1 u_1 \cos\theta_1 + m_2 u_2 \cos\theta_2$$

or

$$v_1{}^{i} = u_1 \cos\theta_1 + \frac{m_2}{m_1} u_2 \cos\theta_2$$

$$y: \quad m_2 v_2{}^{i} = m_1 u_1 \sin\theta_1 + m_2 u_2 \sin\theta_2$$

or

$$v_2{}^{i} = \frac{m_1 u_1}{m_2} \sin\theta_1 + u_2 \sin\theta_2$$

Note that in this example, since the collision occurred at a right angle, application of the principle of conservation of linear momentum in the *x*- and *y*-directions gives rise to two decoupled (independent) equations for the impact speeds, without the necessity of solving two simultaneous equations.

Since the vehicles skidded to impact, again using kinematic relations derived from Newton's second law, it can be shown that the vehicle speeds at the commencement of their respective preimpact skidding are given by

$$v_1 = \sqrt{2\mu g d_1 + (v_1{}^{i})^2}$$

$$v_2 = \sqrt{2\mu g d_2 + (v_2{}^{i})^2}$$

To obtain a numerical solution, consider the following set of input data. Vehicle 1: $m_1 = 1750$ kg (weight = 3860 lb), $s_1 = 6.0$ m (19.7 ft), $d_1 = 26.0$ m (85.3 ft), $\theta_1 = 48°$; vehicle 2: $m_2 = 2100$ kg (weight = 4630 lb), $s_2 = 4.6$ m (15.1 ft), $d_2 = 7.6$ m (24.9 ft), $\theta_2 = 36°$. Substituting these data into the above equations yields

$$u_1 = 8.7 \text{ m s}^{-1} \; (31.3 \text{ km h}^{-1}) \; (19.5 \text{ mph})$$
$$u_2 = 7.6 \text{ m s}^{-1} \; (27.4 \text{ km h}^{-1}) \; (17.0 \text{ mph})$$
$$v_1{}^{i} = 13.2 \text{ m s}^{-1} \; (47.5 \text{ km h}^{-1}) \; (29.5 \text{ mph})$$
$$v_2{}^{i} = 9.9 \text{ m s}^{-1} \; (35.6 \text{ km h}^{-1}) \; (22.1 \text{ mph})$$
$$v_1 = 22.5 \text{ m s}^{-1} \; (81.0 \text{ km h}^{-1}) \; (50.3 \text{ mph})$$
$$v_2 = 14.0 \text{ m s}^{-1} \; (50.4 \text{ km h}^{-1}) \; (31.3 \text{ mph})$$

where the answers were first calculated in m s^{-1} and then converted to km h^{-1} and mph.

Hypothetically, the results of the above analysis can be used as follows. Suppose driver 2 claimed that he stopped at the stop sign, located 10 m (32.8 ft) from the POI in the negative y-direction, as shown in **Figure 1**, and further claimed that the reason for the accident is driver 1, who had the right of way, was speeding through the intersection. The speed limit was posted at 80 km h^{-1} (50 mph). Examination of the numbers obtained from this analysis indicates that vehicle 1 was sensibly traveling at the speed limit when the brakes were locked up. Furthermore, vehicle 2 could not have obtained a speed of 50.4 km h^{-1} (31.3 mph) at the start of preimpact skidding if that driver had started from rest at the stop sign. In fact, for a two-wheel drive vehicle, driver 2 likely could not have reached an impact speed of 35.6 km h^{-1} (22.1 mph) if he/she accelerated at a constant rate from rest at the stop sign to the POI even if no preimpact skidding (deceleration) occurred. This has been noted for two reasons. First, driver 2 may claim that the skid marks are not from his vehicle. Second, often there are no preimpact skid marks left by the vehicles, as may be the case for vehicles that do not experience brake lockup, that is, those equipped with ABS, or there may be only partial (or no) brake application at all before impact. In this case, the linear momentum analysis would then be able to reliably determine the impact speeds only. However, based on the results of the analysis just presented, an accident reconstructionist would then opine to a reasonable degree of engineering and scientific certainty that vehicle 2 did not stop at the stop sign before proceeding into the intersection. It is noted for completeness, however, that as more vehicles are equipped with electronic data recorders (EDRs), or "black boxes," accident reconstruction calculations can be checked and supplemented with data obtained from the EDRs. For further discussion on EDRs, the interested reader is referred to the chapter on EDRs appearing herein.

Another quantifiable parameter used in accident reconstruction and biomechanical injury analysis is the change in velocity ($\Delta\overline{V}$) of a vehicle due to an impact. Noting that $\Delta\overline{V}$ is a vector quantity that must account for direction in its computation, it is not simply the change in speed of a vehicle. $\Delta\overline{V}$ is defined as the change in velocity of a vehicle from its impact velocity to its immediate postimpact (separation) velocity and is a quantifiable measure of crash severity and injury-producing potential. As acceleration is defined as the time rate of change of velocity, $\Delta\overline{V}$ is strongly related to, and can be correlated with, the average acceleration of the vehicle during the impact, which typically occurs over a time duration typically expressed in milliseconds. The magnitude and direction of the $\Delta\overline{V}$ vector are significant in determining occupant motion.

To illustrate the proper computation of vehicle $\Delta\overline{V}$'s, the numbers obtained in the previously discussed intersection collision example are used. Again, noting and emphasizing that $\Delta\overline{V}$ is a vector quantity, it is generally incorrect (except for the special case of a one-dimensional analysis) to simply algebraically subtract the speeds (magnitudes of the velocities) before and after impact to obtain $\Delta\overline{V}$. Hence, it is incorrect to calculate $\Delta\overline{V}$ of vehicle 1 as $31.3 - 47.5 = -16.2$ km h^{-1} (-10.1 mph) decrease. The correct calculation requires using the components of the velocity vector in the x- and y-directions as follows.

$$\begin{aligned}\Delta v_x &= u_1 \cos\theta_1 - v_1{}^{i} = 31.3 \cos 48^\circ - 47.5 \\ &= -26.6 \text{ km h}^{-1} \; (-\,16.5 \text{ mph}) \text{ decrease} \\ &\quad \text{(points in negative } x\text{-direction)}\end{aligned}$$

$$\begin{aligned}\Delta v_y &= u_1 \sin\theta_1 - 0 = 31.3 \sin 48^\circ \\ &= 23.3 \text{ km h}^{-1} (-\,14.5 \text{ mph}) \text{ increase} \\ &\quad \text{(points in positive } y\text{-direction)}\end{aligned}$$

Hence, the magnitude of the $\Delta\overline{V}$ vector for vehicle 1 is:

$$\begin{aligned}\Delta V_1 &= \sqrt{(\Delta V_x)^2 + (\Delta V_y)^2} \\ &= \sqrt{(-26.6)^2 + (23.3)^2} \\ &= 35.4 \text{ km h}^{-1} \; (22.0 \text{ mph})\end{aligned}$$

Similarly, for vehicle 2, it would be incorrect to calculate $\Delta\overline{V}$ as $27.4 - 35.6 = -8.2$ km h^{-1} (-5.1 mph) decrease. The proper calculation for vehicle 2 follows using the velocity components.

$$\begin{aligned}\Delta V_x &= u_2 \cos\theta_2 - 0 = 27.4 \cos 36^\circ \\ &= 22.2 \text{ km h}^{-1} \; (13.8 \text{ mph}) \text{ increase} \\ &\quad \text{(points in positive } x\text{-direction)}\end{aligned}$$

$$\begin{aligned}\Delta V_y &= u_2 \sin\theta_2 - v_2{}^{i} = 27.4 \sin 36^\circ - 35.6 \\ &= -19.5 \text{ km h}^{-1} (-\,12.1 \text{ mph}) \text{ decrease} \\ &\quad \text{(points in negative } y\text{-direction)}\end{aligned}$$

with the magnitude of the $\Delta\overline{V}$ vector for vehicle 2 being

$$\begin{aligned}\Delta V_2 &= \sqrt{(22.2)^2 + (-19.5)^2} \\ &= 29.5 \text{ km h}^{-1} (18.4 \text{ mph})\end{aligned}$$

Note that in this case, the incorrect calculations seriously underestimate the magnitudes of the $\Delta\overline{V}$ vectors and say nothing about their directions. $\Delta\overline{V}$ (and its vector components) for each vehicle is shown in **Figure 2**. As shown, the $\Delta\overline{V}$ values are those for the centers of gravity of the vehicles, which are not the same as those for the occupants, particularly when vehicle rotations are considered and there is relative motion between the occupants and the vehicles. Furthermore, by Newton's third law, the impact forces acting on the vehicles must be equal and opposite and it is these impact forces which determine the vehicle accelerations by Newton's second law. Hence, the vehicles' $\Delta\overline{V}$'s have to be oppositely directed, as shown, although their magnitudes will generally be unequal since the vehicles have different masses. It is also noted for completeness that in the calculations above, the $\Delta\overline{V}$'s were computed and expressed in terms of the global, or scene fixed, coordinate system. Additionally, it is noted that since vehicles' $\Delta\overline{V}$'s are often calculated to study occupant kinematic motion and to evaluate the injury-producing potential of an accident, it is often convenient to express $\Delta\overline{V}$ vectors in terms of vehicle-fixed coordinate systems. Once properly calculated in any well-defined coordinate system, a straightforward application of coordinate transformation rules can be used to express the $\Delta\overline{V}$ vector (and its components) in any other well-defined coordinate system.

Before closing the calculation section, it is important to mention another concept often utilized in accident reconstruction analyses. This is the concept of obtaining vehicle speeds from crush damage. It is noted immediately that only estimates to the $\Delta\overline{V}$ can be obtained from an idealized crush analysis and not individual vehicle speeds, in general, although there are special cases where vehicle speeds can be obtained, if the investigator fully understands how the $\Delta\overline{V}$'s were calculated and how they should be combined or utilized, such as a collinear collision with one vehicle known to be stopped at the time it was impacted. Another fairly common special case is that where a vehicle is essentially stopped at the POI by the impact, as may occur when a vehicle collides with a wall at a moderate-to-high speed, where the elastic rebound can be ignored. In such a case, the damage-based vehicle $\Delta\overline{V}$ calculation for the subject vehicle will also yield the approximate magnitude of the impact speed of the vehicle as (neglecting the relatively low rebound velocity) the immediate postimpact speed can be approximated to be zero.

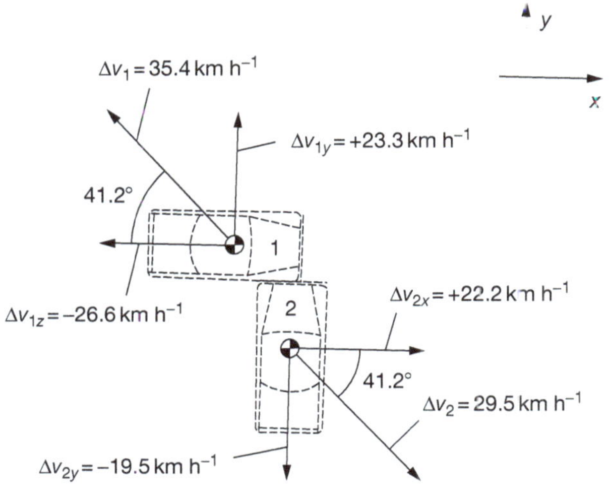

Figure 2 Δv.

Crush damage calculations are based on an idealized energy dissipation algorithm, which is utilized in many computer programs. This algorithm contains assumptions on vehicle structural behavior, is based on an idealized crush pattern, and also depends on certain measured properties of the vehicles known as stiffness coefficients, which are obtained from crash testing of vehicles. However, the damage algorithm is not a law of physics and its limitations and assumptions must be understood by accident reconstructionists before making damage-based calculations and drawing conclusions from them.

It is well known that significant discrepancies can arise when the results of actual crash tests of vehicles are compared to the idealized energy dissipation crush analyses. If the data permit, a momentum calculation, which does not require any assumptions on vehicle damage or crush, is preferable to a damage-only calculation. Note that in the example previously given, it was not necessary to use any aspect of vehicle crush or energy considerations in order to reconstruct the vehicle speeds. However, this should not be interpreted to mean that crush calculations are unimportant and should never be made. Sometimes, there is insufficient scene data available with which to perform a momentum analysis in order to determine speeds. In this situation, a crush analysis based on actual field measurements or on photogrammetric determinations of crush can provide useful information. In addition, a skilled and knowledgeable reconstructionist can often obtain bounds on $\Delta\overline{V}$'s, which may be significant in reconstructing an accident. Care and caution must always be used when interpreting the results of crush calculations.

Occupant Kinematics and Related Concepts

The previous section dealt only with the reconstruction of vehicle speeds (velocities), and no attention was paid to the occupants. Occupant kinematics is the term used to describe the motion of occupants relative to the vehicle in which they are riding, and not the absolute motion of the occupant with respect to a fixed frame of reference. An occupant kinematics analysis may be a significant part of an accident reconstruction if an injury analysis is necessary. The complete determination of an occupant's motion in a crashing vehicle can be the most difficult part of an accident reconstruction and analysis.

The design purpose of vehicle seat belts is to restrain occupant motion in order to minimize or prevent occupant contact with interior vehicle structure. However, depending on both the magnitude and direction of $\Delta\overline{V}$, that is, the crash severity, occupants can still impact vehicle structures, which will affect their occupant kinematics. In addition, it is the contact between the occupant and the vehicle structure, the so-called "second collision," that can cause serious injury or death in an accident. The second collision causes an occupant to rapidly decelerate, and by Newton's second law, thus have high direct contact forces imposed upon his/her body. Even properly worn and designed seat belts cannot prevent an occupant from impacting interior structure in all types of crashes. However, there is a dramatic difference in the protection afforded by different types of seat belt systems. For example, combination lap–shoulder belts are superior to lap-only belts. Lap (only) belts do not prevent upper torso jackknifing, which can lead to serious injuries, even death, in relatively minor crashes. "Lap-belt syndrome" injuries, such as abdominal injuries and tension fractures (chance fractures) of the lumbar spine, can occur in frontal crashes with lap-belted (only) occupants, typically seated in the rear of the vehicle. In a given accident, these same injuries would likely not have occurred if the vehicle manufacturer had provided lap–shoulder belts in the rear. Fortunately, rear-seat lap-only belts are rapidly vanishing from properly designed and manufactured vehicles, and now all new vehicles to be sold in the United States must be manufactured with lap–shoulder seat belts at all rear seat and front outboard occupant seating positions. It is also worthwhile to review a few occupant kinematics scenarios for restrained and unrestrained occupants. In general, an occupant will tend to initially move toward the impact, that is, toward the direction which is opposite of the vehicle $\Delta\overline{V}$ and the resultant vehicle acceleration. This is why care and caution was used in the illustrative example where $\Delta\overline{V}$ was properly calculated for both vehicles. Because of the directions of the vehicle $\Delta\overline{V}$'s (**Figure 2**), occupants in vehicle 1 will tend to move forward and to their right (relative to vehicle 1), whereas occupants in vehicle 2 will tend to move forward and to their left (relative to vehicle 2). In a purely frontal vehicle crash, occupants will tend to move forward relative to the vehicle. If unrestrained, front-seat occupants can impact the steering wheel, dashboard, and windshield, whereas rear-seat occupants can impact the backs of the front seats. After the initial motion into, or toward, interior vehicle structure, the occupants will rebound or be deflected and/or impact each other and otherwise undergo complex motions. Seat belts can limit the occupant excursions in the vehicle but, depending on $\Delta\overline{V}$ of the vehicle, may not prevent interior vehicle contacts. In a vehicle, which is impacted in the rear, occupants will first tend to move backward into their seats (relative to their vehicle) and potentially ramp up the seat backs, followed by a rebound phase toward interior vehicle front structure. Unrestrained occupants can move forward into interior structure, whereas restrained occupants will not rebound as far. However, before rebounding, an occupant's head and neck may extend or hyperextend over the seat back and/or head restraint leading to vertebral column injuries typically located in the cervical or high thoracic spine, although depending on the severity of the impact and the physical condition of the occupant, lumbar injuries cannot be ruled out. Occupant anthropometry is significant and tall people, greater than the 95th percentile in height, *may* be at high risk in vehicles where the head restraint cannot be adjusted high enough or may not even be adjustable at all. Although it is not discussed in detail here, low-$\Delta\overline{V}$ rear-end accidents also have the potential for causing cervical and lumbar spine injuries (such as bulging or herniated intervertebral discs), but these typically have a delayed onset. Another occurrence in rear impacts is the failure of seat backs, even in relatively minor crashes. When this occurs, occupants can tumble into the rear of a vehicle (back seat or cargo area) resulting in serious injuries. It is not commonly appreciated that the seat is an integral part of proper occupant packaging and failure of the seat back, or the entire seat not remaining in place, can lead to serious injury or death for both restrained and unrestrained occupants. Lateral crashes can be particularly devastating because the nearside occupant immediately adjacent to the impact site on the vehicle may be at high risk of serious or fatal injury independent of the design of the vehicle. Initial occupant movement is toward the impact and it is not uncommon for the nearside occupant to physically contact the impacting vehicle or roadside object, such as a pole or a tree. Lap–shoulder belts would generally not be effective in preventing lateral movement in this type of crash. Padding may be effective in minimizing the effect of interior vehicle contacts, and lateral (side) air bags, which are now readily available, will also offer some protection. A restrained far-side occupant in a lateral crash will, in general, derive benefit from a lap–shoulder belt, which will limit occupant excursions toward the far-side lateral impact site. However, an unrestrained far-side occupant can move across the vehicle into interior structures as well as into other occupants, which can cause injuries or death.

Detailed descriptions of occupant kinematics generally require the use of very complex computer programs. These programs are fundamentally based on Newton's laws but also require the human body to be modeled as an engineering structure. In addition, a large amount of vehicle information is required such as physical properties of restraint systems and those interior vehicle occupant compartment (and other) structures that can be contacted by occupants. These include, but are not limited to, padding, door structures, steering wheel, air bags, seat belts, dashboard, and so on. It is also generally necessary to specify a large amount of input information (data) in order to achieve an accurate solution. Even slight changes in the input data can result in significant changes in the output and, hence, care must be exercised when interpreting the results of a complicated occupant kinematics computer-generated solution.

Biomechanics of Injuries

Biomechanics is a compound word which means the application of Newtonian mechanics to biological systems, including but not limited to the human body. The field of biomechanics is vast, with applications beyond those considered herein in an automotive context. For example, to name a few areas, biomechanicians are involved in the design of prostheses, artificial organs, anthropomorphic test devices, bioinstrumentation as well as with the analysis of biological systems on the macroscale and the microscale, such as cellular mechanics. Biomechanics of injuries has already been mentioned several times herein. The remaining discussion provides a few additional concepts, but a complete discussion of injury biomechanics is beyond the scope of this chapter.

Injuries are often correlated with the vehicle $\Delta\overline{V}$ for both restrained and unrestrained occupants. Apart from what has already been mentioned, it should be noted that for unrestrained occupants, $\Delta\overline{V}$ physically corresponds to the velocity of the occupant relative to the vehicle. For example, if a vehicle is traveling at 50 km h^{-1} and crashes into a brick wall, unrestrained occupants will move forward at 50 km h^{-1} relative to the vehicle as the vehicle is rapidly decelerated and brought to a stop by the frontal crash.

Since $\Delta\overline{V}$ is a vector quantity, injury correlations appear in the biomechanics literature as a function of $\Delta\overline{V}$, seating position, and the type of crash (frontal, lateral, rear, etc.). It must be emphasized that the injury correlations are statistical in nature, not absolute, and only attempt to quantify the probability of injury.

Sometimes people walk away from a high-$\Delta\overline{V}$ accident without permanent injuries, whereas others are seriously injured or killed in a low-$\Delta\overline{V}$ accident. Furthermore, the severity of the injury is usually given in terms of the abbreviated injury scale (AIS) published by the Association for the Advancement of Automotive Medicine (AAAM). The AIS is based on anatomical injury at the time of the accident and does not score impairments or disabilities that may result from the injuries. **Table 1** gives the injury severity code as used in the AIS.

Table 1 Injury severity scale

AIS	*Injury severity*
0	None
1	Minor
2	Moderate
3	Serious
4	Severe
5	Critical
6	Maximum (currently untreatable, fatal)

The concept of threshold injury criteria is also important to the understanding of injury biomechanics. Threshold injury criteria, along with human tolerance conditions, refer to those combinations of forces, moments, impact durations, stresses, strains, and so on, which will result in traumatic injury. Caution is in order here since it is not reasonable to expect that a given injury criterion, if it exists at all, will apply across the entire spectrum of the human population. This is because biomaterials (human tissues) have a normal variability range in response to loading, and all people are not equal in their biomechanical response to impact. Human impact response can be, and is, a function of many factors including but not limited to age, gender, preexisting conditions which may or may not be symptomatic, anthropometry (body size and measurements), and so on.

Currently, there are essentially three threshold injury criteria in common use. These are written into the Federal Motor Vehicle Safety Standards (FMVSS), which vehicle manufacturers must comply with by law. The FMVSS is an attempt to require a minimum level of safety and crashworthiness for all automobiles sold in the United States. The injury criteria refer to the head, femur, and chest and are as follows for a front-seat lap–shoulder seat-belted occupant in a vehicle during a 30-mph frontal barrier crash test.

The first, known as the head injury criterion (HIC), requires that a certain mathematical integral expression explicitly dependent on the resultant acceleration at the center of gravity of a test dummy's head in a frontal crash cannot exceed the numerical value of 1000 or else the standard has not been met and the vehicle fails to comply. There are many limitations to the HIC, which are worth noting. The HIC does not even define or discriminate between types of head injuries, which can range from subdural hematomas to diffuse axonal injuries to skull fractures. In addition, the use of a test dummy may be open to question when attempting to assess the effects of injury on live human beings; a live person may survive a HIC significantly greater than 1000 but may be killed in a crash when the HIC is significantly less than 1000.

The second criterion refers to the force measurement made in the femur of a test dummy during a frontal crash. The compressive force in the femur cannot exceed 10.0 kN (2250 lb) or else the standard has not been met and the vehicle fails to comply. This injury criterion, similar to all injury criteria, does not account for the wide and normal range of biomaterial variability among individuals. Compressive fracture loads can be considerably less than 10.0 kN for a significant portion of the population.

The third criterion refers to the resultant acceleration measured at the center of gravity of the thorax (chest) of a human surrogate test dummy. This criterion requires that the acceleration cannot exceed 60 times the acceleration due to gravity (60 g) except for time intervals whose cumulative duration is not more than 3 ms in a test crash. Depending upon the test dummy used, another chest injury criterion, which relates to chest deflection, may also be required. As above, shortcomings in these criteria are that they do not distinguish between types of chest trauma (rib or sternum fractures, transected aorta, etc.) and do not account for normal population variability.

For completeness it is noted that other injury criteria have also been proposed and/or are undergoing intense research investigation. Biomechanics injury research is a vast and complex field which requires the collaborative efforts of engineers, physicians, and other researchers in order to understand the response of the human body to impact. The brief discussion presented herein has only been intended as an introductory exposure to the field. Major advances in understanding the biomechanics of injury can be expected as forensic applications to automobile accident reconstruction become more commonplace.

See also: **Engineering:** Accident Investigation – Determination of Cause; Airbag Related Injuries and Deaths; Electronic Data Recorders (EDRs, Black Boxes).

Further Reading

Accident Reconstruction Technology Collection, 2012. CD-ROM. Society of Automotive Engineers.
Association for the Advancement of Automotive Medicine, 1990. The Abbreviated Injury Scale. Association for the Advancement of Automotive Medicine, Des Plaines, IL.
Batterman, S.C., Batterman, S.D., 1999. Forensic engineering. In: Caplan, Y.H., Frank, R.S. (Eds.), Medicolegal Death Investigation, Treatises in the Forensic Sciences, second ed. The Forensic Sciences Foundation Press, Colorado Springs.
Batterman, S.C., Batterman, S.D., 2002. Forensic Engineering. McGraw-Hill Yearbook of Science and Technology. McGraw-Hill, New York.
Batterman, S.D., Batterman, S.C., 2002. Delta-V, spinal trauma, and the myth of the minimal damage accident. Journal of Whiplash and Realtyed Disorders 1 (1), 41–52. The Howarth Medical Press.
Batterman, S.C., Batterman, S.D., 2004. Forensic Engineering. Legal Medicine, American College of Legal Medicine, sixth ed. Mosby, Inc., Philadelphia, PA.
Batterman, S.D., Batterman, S.C., 2011. Introduction to forensic engineering and accident reconstruction. In: Mozayani, A., Noziglia, A. (Eds.), The Forensic Laboratory Handbook Procedures and Practice, second ed. Humana Press, New York, NY.
Batterman, S.C., 1990. Education of forensic engineers. American Academy of Forensic Sciences (Newsletter) 20 (6).
Beer, F.P., Johnston, E.R., 1977. Vector Mechanics for Engineers: Statics and Dynamics. McGraw-Hill, New York, NY.
Bohan, T.L., Damask, A.C. (Eds.), 1995. Forensic Accident Investigation: Motor Vehicles. Michie Butterworth, Charlottesville.
Brach, R.M., 1991. Mechanical Impact Dynamics; Rigid Body Collisions. Wiley, New York.
Campbell, K.L., 1974. Energy basis of collision severity. In: Proceedings of the 3rd International Conference on Occupant Protection, Paper 740565. Society of Automotive Engineers, Troy, MI.
Carper, K.L., 1989. Forensic Engineering. Elsevier, Amsterdam.
Damask, A.C., Damask, J.N., 1990. Injury Causation Analyses: Case Studies and Data Sources. Michie, Warrendale, PA.
Evans, F.G., 1973. Mechanical Properties of Bone. Charles C Thomas, Springfield, IL.
Greenwood, D.T., 1977. Classical Dynamics. Prentice-Hall, Inc, Englewood Cliffs, NJ.
Hyzer, W.G., 1989. Forensic photogrammetry. In: Carper, K.L. (Ed.), Forensic Engineering. Elsevier, Amsterdam.
Limpert, R., 1984. Motor Vehicle Accident Reconstruction and Cause Analysis, second ed. Michie, Tokyo.
McElhaney, J.H., Roberts, V.L., Hilyard, J.F., 1976. Handbook of Human Tolerance. Japan Auto Research Institute, Tokyo.
Mills, P.J., Hobbs, C.A., 1984. The probability of injury to car occupants in frontal and side impacts. In: Proceedings of 28th Stapp Car Crash Conference. Society of Automotive Engineers, Warrendale, PA, pp. 223–235.
Nahum, A.N., Melvin, J., 1985. The Biomechanics of Trauma. Appleton-Century-Crofts, New York.
Newman, J.A., 1980. Head injury criteria in automotive crash testing. In: Proceedings of the 24th Stapp Car Crash Conference. Society of Automotive Engineers, Warrendale, PA, pp. 703–747.
Pike, J.A., 1990. Automotive Safety; Anatomy, Injury, Testing and Regulation. Society of Automotive Engineers, Warrendale, PA.
Smith, R.A., Noga, J.T., 1982. Accuracy and Sensitivity of CRASH, Paper 821169. Society of Automotive Engineers, Warrendale, PA.
Society of Auto Engineers, 1986. Human Tolerance to Impact Conditions as Related to Motor Vehicle Design, SAE J885. Society of Automotive Engineers, Warrendale, PA.
Society of Automotive Engineers, 1993. Biomechanics of Impact Injury and Injury Tolerances of the Head–Neck Complex, SAE PT-43. Society of Automotive Engineers, Warrendale, PA.
Society of Automotive Engineers, 1994. Biomechanics of Impact Injury and Injury Tolerances of the Thorax-Shoulder Complex, SAE PT-45. Society of Automotive Engineers, Warrendale, PA.
Stapp Car Crash Conference Proceedings, 1998. CD-ROM. Society of Automotive Engineers.
Thorpe, J.F., Middendorf, W.H., 1979. What Every Engineer Should Know About Product Liability. Marcel Dekker, New York.
US Dept of Transportation, 1986. CRASH 3 Technical Manual. NHTSA, Cambridge, MA.
Weinstein, A.S., Twerski, A.D., Pichler, H.R., Donaher, W.A., 1978. Products Liability and the Reasonably Safe Product: A Guide for Management, Design and Marketing. Wiley, New York.
Yamada, H., 1973. Strength of Biological Materials. Krieger, Huntington, NY.
Yeh, H., Abrams, J.I., 1960. Principles of Mechanics of Solids and Fluids; Particle and Rigid Body Mechanics. McGraw-Hill, New York.

Key Terms

Acceleration, Accident reconstruction, Analogy, Biomechanics of injury, Classification, Comparison, Crime, Crush damage, Delta-v, Epistemology, Evidence, Force, Forensic, Forensic engineering, Kirk, Linear momentum, Locard, Mass, Mechanics, Method, Newton's laws, Occupant kinematics, Paradigm, Science, Set, Speed, Taxon, Taxonomy, Vector, Velocity.

Review Questions

1. If the "basic unit of forensic science is the trace," how does this fit within forensic engineering's conceptual framework? What would be a "physical remnant" for engineering?
2. What are the three levels that Locard's principle needs to address?
3. Besides Locard's principle, what else do Crispino and Houck consider to be a "native" forensic principle?
4. What are the nonnative principles that forensic science uses? Give an example of each one in action.
5. What is the difference between uniqueness and individualization?
6. How is forensic engineering defined?
7. Why are not there degree programs in forensic engineering?
8. What are the major functions of a forensic engineer?
9. What other disciplines must forensic engineering deal with?
10. What types of cases would a forensic engineer work on?
11. Batterman states that laws are vectors statements and not scalar ones. What does that mean?
12. What are the three principles useful for solving dynamics problems?
13. What does POI stand for?
14. What does biomechanics mean?
15. What does AIS stand for?
16. What are the three threshold injury criteria?
17. What is HIC?
18. Why are good-quality images important in accident reconstruction?
19. What kinds of transfer, a la Locard, occur in forensic engineering?
20. Beyond proving that a product or material was defective, what else is necessary for the forensic engineer to determine in injury cases?

Discussion Questions

1. How is forensic engineering like the rest of forensic science? How is it different? What aspects of nonforensic engineering does it still embrace? Does forensic engineering "fit" within the wider discipline of forensic science?
2. Investigations are of necessity *a posteriori*, or after the fact. What limits does this place on forensic engineering investigations and what they can say?
3. Why would the complete determination of an occupant's motion in a crashing vehicle be the most difficult part of an accident reconstruction?
4. Is forensic engineering a separate discipline from engineering? If so, how? If not, does "forensic engineering" actually exist?
5. What kind of "proxy data" do forensic engineers work with? How is their proxy data different from other forensic sciences?

Additional Readings

Brando, F., Iannitelli, A., Cao, L., Malsch, E.A., Panariello, G., Abruzzo, J., Pinto, M.J., 2013. Forensic information modeling: a new forensic tool. Civil Engineering Magazine Archive 83 (1), 48–53.
Lacasse, S., 2016. Forensic geotechnical engineering theory and practice. In: Forensic Geotechnical Engineering. Springer, India, pp. 17–37.
Poulos, H.G., 2016. A framework for forensic foundation engineering. In Forensic Geotechnical Engineering. Springer, India, pp. 1–15.
Reznik, Y.M., 2013. Geochemistry as a valuable tool in forensic engineering. Journal of Performance of Constructed Facilities 27 (5), 666–669.

Section 2. Investigations

> Engineers are not superhuman. They make mistakes in their assumptions, in their calculations, in their conclusions. That they make mistakes is forgivable; that they catch them is imperative. Thus, it is the essence of modern engineering not only to be able to check one's own work but also to have one's work checked and to be able to check the work of others. — Henry Petroski

A good mantra for life and work is, "Mistakes are alright, as long as they are not repeated." This could be engineering's motto. Yet mistakes do happen, communications are missed, designs fall short, and materials are misused. Those mistakes or missteps are baked into the objects forensic engineers examine, providing, in a sense, the history of the object, its design, and manufacture. "[O]bjects created in the past are the only historical occurrences that continue to exist in the present"[1] and forensic engineers deconstruct the past by examining the object's failure, revealing its history as well as its flaws. History may occur in time and space but reconstructions take place in structure and process.

> It's here that the methods of historians and scientists—at least those scientists for whom reproducibility cannot take place in the laboratory—roughly coincide. For historians too start with surviving structures, whether they be archives, artifacts or even memories. They then deduce the processes that produced them. Like geologists and paleontologists, they must allow for the fact that most sources from the past don't survive, and that most daily events don't even generate a survivable record in the first place.[2]

This construction/deconstruction concept may be unique to engineering as a forensic science: They are the only ones who both build their evidence and examine it when it fails.

Collection and Chain of Evidence

F Poole, Forensic Services Group, New South Wales Police Force, Parramatta, NSW, Australia

Glossary

Denaturing of DNA The process by which double-stranded DNA unwinds and separates into single-stranded strands through the breaking of hydrogen bonds between the bases.
DNA amplification A process called polymerase chain reaction. A DNA testing procedure that mimics the cell's ability to replicate DNA, essentially copying it a millionfold.
Electrostatic lifting device A device consisting of a high-voltage supply used with a special conductive lifting film to transfer a dry origin footwear impression electrostatically from a surface to a film.
Latent evidence Evidence at a crime scene that cannot be seen with the naked eye. Examples might be a blood stain that was bleached out, semen stains that cannot be seen without special lighting, or a fingerprint that cannot be seen without powdering.
Probative value Evidence which is sufficiently useful to prove something important in a trial.
Trace evidence Evidence such as hairs, fibers, and residue as well as other microscopic evidence that may not be visible to the naked eye.

[1] Prown, J., 1982. Mind in matter: an introduction to material culture theory and method. Winterthur Portfolio. 17 (1), 1–19. Quote appears on page 3.

[2] Gaddis, J.L., 2002. The Landscape of History. Oxford University Press Oxford, UK. Quote appears on page 41.

Introduction

The role of crime scene investigators at major scenes and incidents is to identify and recover physical and trace evidence that has some probative value in the crime. Crime scenes may involve single or multiple locations and these are referred to as the primary and secondary scenes. For example, the primary scene may be where a person has been killed; however, secondary scenes may result when the body has been moved to a different location, and/or the weapon is found at another location. All locations involved form part of the overall crime scene and the evidence collected from each scene helps solve the puzzle piece by piece.

Forensic evidence collected must be relevant to the crime, not be at risk of being misplaced, lost, or tampered with, not be at risk of contamination from other sources and remain intact throughout the process of collection and analysis. For this reason, the preservation of forensic evidence commences at the time of arrival of the first police responder at the crime scene. At this time, an immediate assessment must be made and action taken to protect any fragile evidence at risk of loss or damage. From this point on, crime scene investigators must be meticulous in their approach to the collection of evidence and maintain an unbroken chain of custody from the time of collection, to the laboratory for analysis, and through to the presentation of the evidence in a court of law.

Scene Examination

Forensic evidence can appear in many forms, including physical, trace, and latent evidence. In the twentieth century, forensic scientist Edmond Locard advanced the theory on which most forensic science is founded today, and that is that "every contact leaves a trace." Locard's principle of exchange is based on the theory that any contact made between two objects will result in transfer of material from one to the other. Therefore, using this principle, items of evidence that have probative value can be determined by assessing areas where the offender, victim, and items within the scene have potentially come into contact with one another.

Initial information received upon arrival at the scene and updates from investigating police will help indicate what is, or could be, potential evidence. The following are considered at the scene or during subsequent investigations: the area the offender used to enter and exit the scene; areas within the scene that have been disturbed; anything that the offender has potentially handled or come into contact with; anything that has been left behind by the offender, that is, a weapon.

Successful identification, recovery, and collection of evidence are dependent upon a methodical and diligent approach to scene examinations. The analysis of any evidence should take into consideration the possible destruction of other trace evidence. The development of a crime scene plan ensures that systematic searches are conducted, and appropriate light sources and chemical enhancement techniques are used by applying least destructive to most destructive methodology. Sequencing the order in which the examinations and/or analysis is conducted is paramount to preserving the integrity of all available evidence types.

Evidence Collection

Once all identified evidence has been recorded in situ by way of notes, sketch plans, photography, etc., appropriate techniques are used to collect the evidence. To prevent damage or contamination of the sample, systematic collection/sampling techniques are used. By following a collection sequence, the potential evidence on the item is maximized while providing the highest level of protection to that evidence. For example, a gun used in the commission of a crime would first need to be fingerprinted, then swabbed for DNA, then submitted for test firing. If this sequence is not followed, one or more evidence types would be lost.

Collection techniques are many and varied and include methods such as swabbing, swatching, lifting, vacuum sweeping, picking, and casting. These techniques are selected according to the type of evidence, the substrate the evidence is located upon, and the physical properties of the evidence itself.

For example, swabbing is the preferred sampling method for biological stains, for example, blood, semen, and saliva; however, if a confirmatory test indicates either semen or saliva on an item of clothing, the preferred method of collection is to cut out a swatch of the stained fabric. Trace or contact DNA can be collected using either a swab or a tape lift. Swabbing involves rubbing a moistened swab over the stain, which rehydrates the cells attached to the area, making them easier to retrieve through the rubbing action of the swab. Tape-lifting removes trace or contact DNA from the item using sterile or ultraviolet (UV)-treated sticky tape. Both techniques allow a large surface area to be sampled, which concentrates any DNA present to a smaller, analyzable sample.

Tape lifts may be used to collect not only trace material such as fingerprints and trace DNA but also macroscopic material such as hairs and fibers. The adhesive side of the tape is repeatedly and firmly patted or rolled over the item causing loosely adhering trace evidence to stick to the tape. The collected lifts are placed onto a transparent backing such as clear plastic sheeting or glass slides which protects the evidence against contamination and permits samples to be easily viewed and removed for further examination or comparison. Hand-picking of hairs, fibers, vegetation on clothing, etc., is another acceptable method of recovery using disposable tweezers. Gel lifts are useful for collecting shoeprints in dust. Similarly, electrostatic dust-lifting kits are also commonly used for lifting shoe marks off many surfaces.

Vacuuming is another technique used to collect material from inaccessible areas or where there is a large amount of scattered material. There are several different types of vacuum cleaners used for this purpose. One such vacuum cleaner has a one-use only attachment head that the material is collected into. Once collected, the attachment is removed and sealed in an evidence bag and the contents are later examined in the laboratory. Another type of vacuum cleaner has changeable filters; however, the filter and catchment area of the vacuum cleaner must be changed and rigorously cleaned between each vacuuming to avoid contamination. This method of collection should be used subsequent to other collection techniques as it is indiscriminate and may result in the collection of a large amount of extraneous material.

Casting is a technique used for collection of three-dimensional impressions such as shoe marks in soil or a jemmy mark in a doorframe. This technique involves filling of a three-dimensional footwear impression with material such as Dental Stone that sets, capturing the characteristics that were left in that impression by the footwear. Mikrosil is a casting material used to capture three-dimensional tool mark impressions.

When collecting forensic evidence, it is absolutely essential to properly preserve each item so that the evidence on the item is not damaged or degraded in any way, which would limit its potential evidentiary value. For example, all bloodstained clothing items need to be air dried before packaging. When the garment is fully dried, it should be packaged in a paper bag, as paper allows for the circulation of air. If the garment is packaged in plastic, the garment would become moldy which may result in the denaturing of the DNA.

All items of evidence need to be handled with the utmost of care to prevent contamination. Contamination of evidence can occur in many ways such as handling of evidence without gloves and coughing or sneezing near evidence items destined for DNA analysis. Control measures are put in place to minimize potential contamination including the wearing of personal protection equipment (PPE) such as disposable overalls, gloves, masks, overshoes, and goggles. PPE also provides investigators with a means of self-protection from body fluids that may cause hepatitis or human immunodeficiency virus-related diseases. Another control measure is to regularly change disposable gloves between the collection of evidence items which will reduce the potential for cross-contamination from item to item.

Control Samples

When samples are collected from a crime scene, it is also necessary to collect a control sample of the same material. This control sample is a portion of undamaged material from the same general area as where the crime scene sample is collected. A comparative analysis is conducted between the control sample and the crime scene sample which determines the original components of the material allowing for any introduced substances in the crime scene sample to be identified. For example, during arson investigations, control carpet samples are often collected from preserved areas such as underneath wardrobes for comparison against carpet samples collected from suspected areas of fire origin. Once the composition of the carpet samples are established, any flammable liquids that may have been used to accelerate the fire, such as petrol, can be identified.

An infamous forensic case in the 1970s involved the conviction and subsequent exoneration of Edward Charles Splatt for the murder of a woman in her home. Among other trace evidence, fibers were recovered from the bedsheet on the victim's bed. The three fibers collected were gray in color, like those on Splatt's trousers. Years later, a Royal Commission into the case concluded that the fibers found on the bedsheet had been selectively collected to include only gray fibers and were not representative of Splatt's trousers. This evidence, in part led to the exoneration of Splatt some 13 years later.

This case demonstrates the criticality of collecting a representative sample of all evidence found at a crime scene, just as it is equally important to collect a control sample that is completely representative of every component it comprises. Materials such as patterned fabrics may appear to be homogeneous in composition but may actually be heterogeneous and contain different colored fibers, or in the case of a blended fabric, different fiber types such as polyester and cotton. It is essential that the complete range of both dyed fibers and fiber types are captured in the control sample to ensure accurate comparison with crime scene samples.

Chain of Custody

Evidence management is critical to the outcome of criminal prosecutions. Chain of custody is the process of validating how any piece of evidence has been gathered, tracked, and protected on its way to a court of law. Proving chain of custody is necessary to affirm that the evidence has not been tampered with, changed, or substituted. It is essential for police, scientists, and other specialists involved in the case, to demonstrate that a chain of custody exists on every occasion the item is transferred from one person to the next to show with absolute confidence that there is no possibility of misidentification or adulteration of the evidence. Should the defense counsel question the chain of custody for any piece of evidence, a documented path of continuity can prove that the item presented at court is in fact the same item collected from the crime scene. Where there is any discrepancy about where an item has gone to or who has had possession of the item throughout the process, the judge may rule that the chain of custody has been broken and the item may not be admitted into evidence.

Chain of custody requires three different questions to be answered to verify the accuracy and reliability of court testimony by police and forensic scientists: First, that the piece of evidence in question is actually what it is reported to be; second, that a continuous trail of possession by each individual handling the item can be demonstrated from the time it was collected until the time it was presented in Court; and third, that each person who had possession of the item can state that it essentially remained in the same condition from the moment he or she received it, to the moment he or she released it.

Key factors in proving a chain of custody need to commence at the crime scene itself. When items of evidence are located at a scene by the crime scene investigator, an identification marker displaying a unique number or letter is placed next to each item of evidence and recorded in situ by way of photographs, notes, and sketch plans. This procedure of marking the items for identification and accurately recording the details of each item is the first step in the chain of custody tracking process.

After the evidence has been recorded, appropriate packaging is essential to preserve the integrity of the evidence. All evidence bags or containers must be sealed with tamper-evident tape and appropriately labeled. Tamper-evident tape is designed to pull apart or shatter when the seal of the packaging is broken and is very difficult, if not impossible, to remove all telltale signs of this tape from the packaging. If the sealing tape is not tamper-evident tape, a signature and date across the tape and the bag is required as another control measure of tampering. Details of the evidence item and its collection are recorded on either preprinted label templates on the evidence bags or on adhesive labels, which are affixed to the front of the bag or container. The label information includes the description of the item, the location, date, time of collection, name of the person collecting it, and a unique forensic case number. These labels, whether preprinted templates or affixed as adhesive labels, also have a chain of custody section that must be signed and dated whenever the evidence bag or container is handed by one person to another.

In some jurisdictions, once all the items of evidence have been packaged and sealed, a forensic exhibit list is compiled by the crime scene investigator that comprises an itemized list of all evidence to be handed over to the officer in charge (OIC) of the case. The OIC will sign the bottom of the exhibit list providing continuity that he or she has taken possession of the forensic evidence items.

Once the items of evidence are transported from a crime scene to a police station or laboratory, the chain of custody procedures must be strictly adhered to. Continuity, storage, and security of the evidence are of utmost importance in the chain of custody process. All evidence items must be entered into an exhibit book or electronic case management system and secured in a locked exhibit room while awaiting further analysis. It is imperative that exhibit storage areas have restricted access that is regularly monitored and audited. Access to searching suites, examination areas, and drying cabinets within laboratories often have security controls in place limiting access to authorized personnel only.

When evidence items are removed from exhibit rooms, the items must be either manually signed out of the exhibit book or electronically transferred using a computerized case management system. Electronic transfers can be completed manually on the system or via a bar code label, where the bar code is scanned and movement of the item is confirmed with a personal pin number. This technology has improved the accountability and accuracy of tracking evidence movements within laboratories.

When a sealed evidence bag is opened for further examination or laboratory analysis, the person doing so must sign the chain of custody details that form part of the label. The crime scene investigator or laboratory analyst will thoroughly document their examination or analysis for each evidence item and record any findings or results. Upon completion of the examination or analysis, the evidence item must be resealed inside the evidence bag or container and the chain of custody details on the label must be updated to reflect that the bag or container has been resealed by that person.

When external laboratories receive evidence for analysis, another unique laboratory number will be allocated, first to track the items through that laboratory system and, second, to ensure that each item of evidence can be identifiable by the laboratory analyst in Court.

By following the chain of custody rule to the letter, the case can proceed with the knowledge and confidence that the movement of all forensic evidence can be precisely tracked from crime scene to courtroom. This rigorous process ensures that forensic evidence critical to the case is admissible into judicial proceedings allowing a judge and jury to impartially reach a fair and just conclusion based on the forensic evidence before them.

See also: **Investigations:** Contamination; Evidence Collection at Fire Scenes; Major Incident Scene Management; Packaging; Preservation; Recording.

Further Reading

Brown, M., Wilson, P., Whelan, J., 1992. Justice and Nightmares: Successes and Failures of Forensic Science in Australia and New Zealand. New South Wales University Press, Sydney.

Fisher, B., 2004. Techniques in Crime Scene Investigation, seventh ed. CRC Press, Taylor & Francis Group, Boca Raton, FL.

Geberth, V.J., 2006. Practical Homicide Investigation – Tactics, Procedures, and Forensic Techniques, fourth ed. CRC Press, Taylor & Francis Group, Boca Raton, FL.

Saferstein, R., 2003. Criminalistics: An Introduction to Forensic Science, eighth ed. Prentice Hall, Upper Saddle River, NJ.

Saferstein, R., 2008. Forensic Science: From the Crime Scene to the Crime Lab. Prentice Hall., Upper Saddle River, NJ.
Stauffer, E., Bonfanti, M.S., 2006. Forensic Investigation of Stolen-Recovered and Other Crime-Related Vehicles. Elsevier, Burlington, MA.
Swanson, C.R., Chamelin, N.C., Territo, L., Taylor, R.W., 2009. Criminal Investigation, tenth ed. McGraw Hill, New York.

Relevant Website

http://www.crime-scene-investigator.net—Crime-Scene-Investigation.

Accident Investigation—Determination of Cause

H Steffan, Graz University of Technology, Graz, Austria

Introduction

Car traffic has increased dramatically. Although additional traffic regulations such as speed limits or other restrictions were imposed to reduce accidents, crashes can be seen every day. Many result in vehicle damages or injuries of occupants or other people involved. Claims, resulting from these accidents, often have to be settled in court.

In most countries, special "accident reconstruction experts" assist the judges by reconstructing the accident situations based on information such as eyewitnesses, car deformations, or tire marks. In many countries, these experts are either specially trained police officers or members of the road authorities. In other places, the reconstructions are performed by independent specialists. The necessity to understand the accident for court decisions initiated specific accident research.

Meanwhile, accidents are also reconstructed for many other reasons. Car manufacturers perform in-depth studies of real accidents to both learn more about accident mechanisms and study the potential and the effectiveness of new safety devices in cars. Also for improved road planning, accidents are often reconstructed and analyzed.

There are several questions which must be answered, when reconstructing a car accident. Today's accident reconstruction tries to provide full information about the movement of all involved vehicles or other involved persons or objects from the point of first visual contact to their rest positions. Time history of velocities, positions, as well as crash-related data such as velocity changes, deformation energies, or passenger loads must be analyzed. In addition, prevention analyses are often included. These analyses allow one to determine the conditions necessary to prevent the accident.

There is one general rule in any accident reconstruction: A detailed collection of scene data forms the basis for a good accident reconstruction. Some of the most important scene data can be summarized as follows:

- Vehicle or human rest positions
- Road marks
- Vehicle damages and marks
- Personal injuries

Due to the increased number of cars with antilocking brakes, fewer tire marks, which formed the most objective basis, can be seen and thus one major basis for the reconstruction is lost. On the other hand, the increased performance of personal computers made it possible to use highly sophisticated reconstruction or simulation algorithms to study the accidents.

In recent years, several software products have been developed for the reconstruction of vehicle accidents. Some of them are designed to just calculate vehicle velocities from kinematic relations for specific accident types. Others, containing full vehicle simulators, allow one to simulate the car motion during the accident starting from the point of reaction to the end position for any kind of accident. Due to the availability of powerful PCs, it is now even possible to visualize the three-dimensional (3D) car motion on the screen during the calculation in real time. 3D geometrical information regarding road marks can be imported from drawing programs through defined interfaces such as DXF or from photographs as scanned Bitmaps, and all-important road, car, and tire characteristics can be taken into account as well as the reactions of the driver.

Targets

It is important, when formulating the physical and mathematical model of the car to take into account the fact that many parameters are not well known when reconstructing the accident. Especially, certain conditions of the tires such as their age or air pressure, the condition of the shock absorbers or the steering system are often not well documented. The stiffness of the car body is only known for well-defined crash conditions and not for the specific accident situation. Therefore, it is important to find a compromise between accuracy and the amount of input data necessary to perform the calculation. On the other hand, the user should be able to take into account all parameters which are documented and known to influence the driving and crash behavior. Based on this knowledge, simplified models have to be used, which guarantee that the basic driving behavior is predicted correctly. All necessary input parameters must be defined on the bases of physical parameters, and a simple possibility to vary the parameters guarantees that the expert can cross-check their influence on the simulation result.

The automotive industry uses several simulation models to learn about the driving behavior of their vehicles. These industrial vehicle dynamics simulation programs are optimized to predict the driving behavior under well-defined initial and boundary conditions. As a result, these models require

 http://dx.doi.org/10.1016/B978-0-12-802718-9.00006-5

many input parameters. For the reconstruction of a vehicle accident, such detailed knowledge, especially regarding the suspension, the tires, and the road conditions, is normally not available. In addition, the steering as well as the degree of braking is often not known. It is thus difficult to use these simulation models for the reconstruction of vehicle accidents.

Regarding collision models, a similar problem exists. Several programs (mainly Finite Element based) exist, which allow the calculation of the deformations for well-defined collision conditions. To get a good agreement with real impacts, they require a very detailed knowledge of the vehicle structure and a very powerful computer. Some 10 000 000 degrees of freedom are required to model one vehicle properly. In addition, every simulation has to be calibrated with similar crash tests. This is the major reason why Finite Element programs are only used in a few specific cases for accident reconstruction.

Several computer programs have been developed especially for the reconstruction of vehicle accidents. They allow the calculation of the vehicle motion and collisions based on various physical models. PC-CRASH is one of these programs with a worldwide distribution. It uses a kinetic time forward simulation of 3D vehicle dynamics and combines it with several collision models. So, the accidents can be reconstructed starting from the conflict point to the end position for all involved cars simultaneously. The reconstruction is performed in an interactive graphical environment, which allows a sketch or photograph of the accident scene to underlay the reconstruction. For an effective presentation of the results, 3D animations can be created directly from the calculated results.

Accident Analysis

Postcrash Movement

In the case of a conventional accident analysis, the reconstruction is normally started from the rest positions of the involved vehicles backwards to the collision position. To determine the postcrash velocity of the vehicles, an average deceleration has to be estimated. In many cases, where no tire marks are available, a straight movement of the vehicle is assumed to determine the distance of postcrash travel. Depending on road situation and involved vehicle, an average deceleration level is assumed which typically lies in the range of 0.1–10 m s^{-2}. The postcrash velocity can then be calculated according to the formula

$$v = \sqrt{2as + v_0^2} \qquad [1]$$

where v represents the postcrash velocity (m s^{-1}), a the average deceleration (m s^{-2}), and s the postcrash travel of the vehicles center of gravity (m).

This is the way to determine the postcrash velocity in a conventional method. The major problem within this method is to estimate the vehicles' average deceleration during a complicated postcrash movement. In this phase, the vehicle may have been rolling or sliding, and this makes a huge influence on the amount of energy dissipated. To overcome this problem, vehicle simulators are often used to compare the vehicles' movement with marks found on the road or on the vehicles. Thus, the postcrash velocity can be determined more accurately.

Collision Model

During the collision, the contact forces between the two involved vehicles vary over time. These forces depend on the local vehicle structures, the deformation velocity, the contact situation, and several other parameters. As these dependencies are highly nonlinear and very difficult to formulate, the treatment of these parameters through their integral values has proven to be more efficient (impulse based).

Modern Finite Element programs allow one to investigate the time variation of the contact forces. However, these programs require enormous calculation times and a huge amount of modeling. Simplified models such as the "CRASH algorithm" have proven to produce large errors under several conditions. In many cases, insufficient knowledge regarding the structural deformation behavior is available to estimate the proper parameters.

Therefore, "crash hypotheses," which only compare the velocity conditions of both vehicles before and after the impact, have been used with great success for the reconstruction of accidents. As modern programs provide both impulse-based and force-based models, they will both be discussed.

Material Properties in the Contact Area

During the contact phase, large forces may occur, which cause deformations to one or both collision partners. These deformations may remain fully or partially after the impact (plastic deformations) or they may fully recover (elastic deformations). Through the definition of the parameter k, the amount of elasticity for a crash can be defined. This parameter k is only valid for a whole crash situation and not for one of the partners. Therefore, one crash partner may include a high degree of elasticity when impacting with partner A and may include a high degree of plasticity when impacting with partner B (**Figure 1**).

Through a sample of a straight central impact, the parameter k can be explained easily:

To ensure that the two partners may collide, the velocity of partner 1 must be higher than that of partner 2. In phase 1

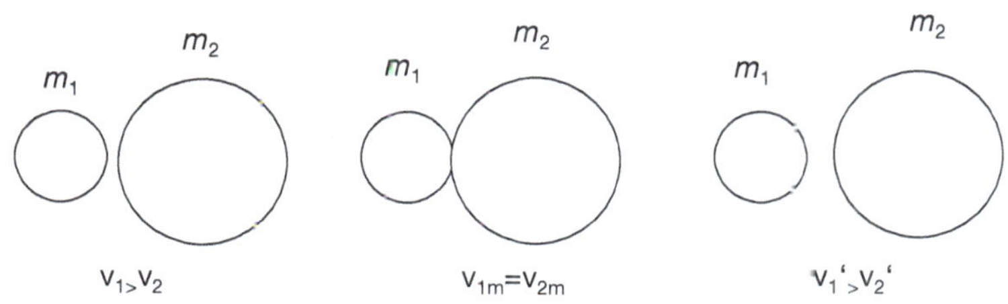

Figure 1 Deformation to one or more collision partners.

contact forces act, which reduce the velocity of mass 1 and increase the speed of mass 2. They are equivalent in size but with opposite direction. The relation between acceleration and contact force is calculated from

$$ma = F \quad [2]$$

where m defines the mass, a the acceleration, and F the contact force.

The end of phase 1 is defined when both partners do have equal velocities. It is called "compression" phase. In phase 2, the forces reduce again until the two masses separate. This phase 2 is also called "restitution."

In this case, the coefficient of restitution is defined as

$$k = -\frac{\Delta v'}{\Delta v} = -\frac{v_2' - v_1'}{v_2 - v_1} \quad 0 < k < 1$$

where $k = 0$ represents a fully plastic impact and $k = 1$ a fully elastic impact.

The term in the numerator $\Delta v'$ is referred to as the relative separating velocity of the two vehicles after impact and the term in the denominator Δv their relative closing velocity before impact.

Eccentric Impacts

Figure 2 shows a sample of a measurement of the two acceleration components for an eccentric impact. The compression impulse S_C and restitution impulse S_R are now defined as

$$S_C = \int_{t_0}^{t_m} F dt \quad [3]$$

$$S_R = \int_{t_m}^{t_1} F dt \quad [4]$$

where t_0 is the time of first contact, t_m defines the time of identical velocity for both vehicles at the point of impact, and t_1 describes the time of separation.

In the case of a generic eccentric impact, the "coefficient of restitution" is now defined as a ratio between restitution impulse and compression impulse.

$$\varepsilon = \frac{S_R}{S_C} \quad [5]$$

The total exchanged impulse is calculated from

$$S = S_C + S_R = S_C^*(1 + \varepsilon) \quad [6]$$

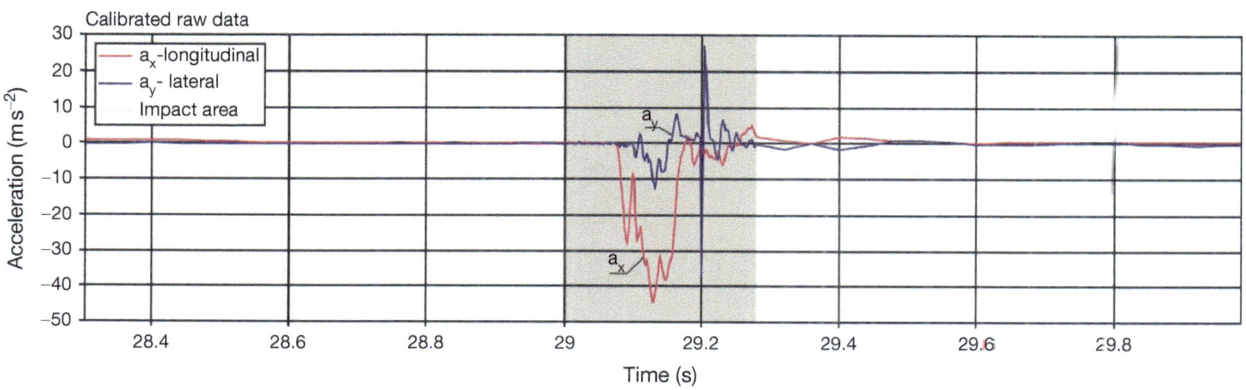

Figure 2 Measured accelerations.

In a full impact, the velocities of both vehicles at the impulse point must be identical at the end of the compression phase.

For simplicity, the impact model is only derived here in 2D. In some programs such as PC-CRASH, the models are also extended to 3D. They are identical except that all velocities, the momentum, and angular momentum are defined in 3D coordinates, and all three components of the velocity and rotational velocity are taken into account.

As can be seen in **Figure 3**, a local coordinate system is defined which originates at the "impulse point" P. The components of the relative velocity for both vehicles at the impulse point can be calculated from

$$V_{1t} = v_{s1t} + \omega_{1z} n_1 \qquad [7]$$

$$V_{1n} = v_{s1n} + \omega_{1z} t_1 \qquad [8]$$

where V_{1t} defines the velocity component of the impulse point for vehicle 1 in direction t and V_{1n} in direction n.

So, the components of the relative velocity for both vehicles at the impulse point can be calculated from

$$V_t = V_{1t} - V_{2t} \qquad [9]$$

$$V_n = V_{1n} - V_{2n} \qquad [10]$$

In addition, the balance of momentum can be formulated for both vehicles:

$$m_1 \left(v'_{s1t} - v_{s1t}\right) = T \qquad [11]$$

$$m_1 \left(v'_{s1n} - v_{s1n}\right) = N \qquad [12]$$

$$m_2 \left(v'_{s2t} - v_{s2t}\right) = -T \qquad [13]$$

$$m_2 \left(v'_{s2n} - v_{s2n}\right) = -N \qquad [14]$$

The balance of angular momentum can be formulated as follows:

$$I_{1z}\left(\omega'_{1z} - \omega_{1z}\right) = Tn_1 - Nt_1 \qquad [15]$$

$$I_{2z}\left(\omega'_{2z} - \omega_{2z}\right) = Tn_2 - Nt_2 \qquad [16]$$

When combining these equations, the change of the relative velocity for both vehicles at the impulse point can be calculated from

$$V'_t = V_t + c_1 T - c_3 N \qquad [17]$$

$$V'_n = V_n - c_3 T - c_2 N \qquad [18]$$

with

$$c_1 = \frac{1}{m_1} + \frac{1}{m_2} + \frac{n_1^2}{I_{1z}} + \frac{n_2^2}{I_{2z}} \qquad [19]$$

$$c_2 = \frac{1}{m_1} + \frac{1}{m_2} + \frac{t_1^2}{I_{1z}} + \frac{t_2^2}{I_{2z}} \qquad [20]$$

$$c_3 = \frac{t_1 n_1}{I_{1z}} + \frac{t_2 n_2}{I_{2z}} \qquad [21]$$

To be able to solve these equations and to calculate the postimpact velocities and rotations, two additional assumptions have to be made.

According to Kudlich and Slibar, two different kinds of impacts are treated.

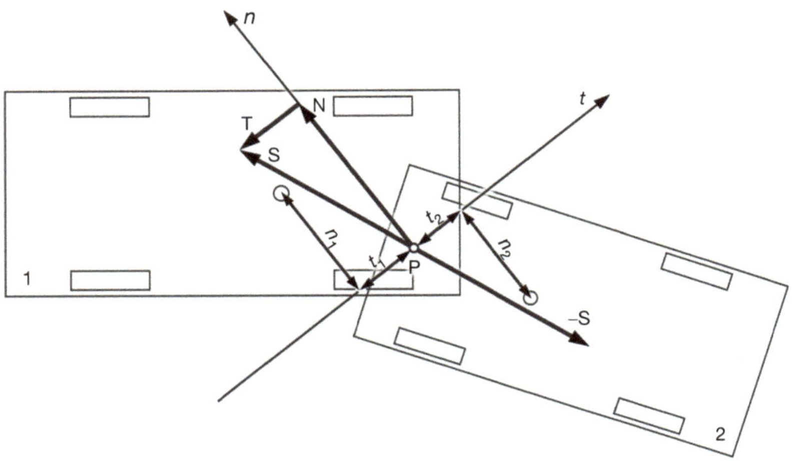

Figure 3 Impact configuration.

The Full Impact

In the case of a full impact, two additional assumptions are made:

1. No relative movement between both vehicles can be found in the impulse point at the end of the compression phase.

$$T_c = \frac{V_n c_3 + V_t c_2}{c_3^2 - c_1 c_2} \quad [22]$$

$$N_c = \frac{V_n c_1 + V_t c_3}{c_3^2 - c_1 c_2} \quad [23]$$

2. The average between compression and restitution impulse is defined by the coefficient of restitution, which is defined according to eqn. [6].
3. So, the components of the total impulse can be calculated from

$$T = T_c(1 + \varepsilon) \quad [24]$$

$$N = N_c(1 + \varepsilon) \quad [25]$$

These equations are sufficient to calculate all postimpact velocity conditions for both involved vehicles in the case of a full impact.

The Sliding Impact

In certain collisions, the two vehicles will never reach identical velocities in the impulse point during the impact. In such a case, a contact plane has to be defined, along which the two vehicles slide. The impulse point must coincide with this plane. For such a situation, the following two assumptions are made:

1. No relative movement between both vehicles can be found in the impulse point at the end of the compression phase in direction normal to the contact plane. So, N_c can be calculated from eqn. [24].
2. The direction of the impulse is limited by a friction (μ). This value defines the friction between the two impacting vehicles:

$$T = \mu N \quad [26]$$

3. The average between compression and restitution impulse is again defined by the "coefficient of restitution" according to eqn. [6], and T and N can again be calculated from eqns. [25] and [26].

The coefficient of restitution, which can be handled as input or output parameter, is easy to define. It generally lies in the range between 0 and 0.3. The higher the deformations of the vehicles, the lower the coefficient of restitution. Only for very small deformations, values higher than 0.3 are possible.

In principle, coefficient of restitution k can range from -1 to 1, with negative values being characteristic of the so-called penetrating impact, in which the point of common velocity will not be reached. From a numerical viewpoint, this can be considered by adding a negative $\Delta v'$. It is typical of a bullet shooting through a plank but extremely rare in the case of vehicle accidents.

Energy Equivalent Speed

As modern cars are mostly designed to absorb energy during the collision, the amount of damage found on the vehicles can also be used to determine the precrash velocity. For most vehicles, crash tests are performed, which are also published. They show the vehicle's deformation after an impact with a well-defined barrier. **Figure 4** shows the deformations of a certain car type in the frontal area once when impacting a rigid barrier with a speed of 15 km h^{-1} and once with a speed of 48 km h^{-1}. Significant differences can so be found in the amount of damage imposed on the vehicle.

This knowledge can also be used for the accident reconstruction.

When comparing the deformations found on the vehicle to those of the reference tests, the amount of energy absorbed by the vehicle due to the impact can be estimated. As deformation energies are not a very handsome quantity, they are recalculated into velocities and called "energy equivalent speed" (EES). EES is so defined according to eqn. [28]:

$$\text{EES} = \sqrt{\frac{2E_{\text{Def}}}{m}} \quad [27]$$

where E_{Def} defines the energy of vehicle residual deformation and m the vehicle mass.

Thus, the conservation of energy can be used in addition:

$$E'_{\text{kin}1} + E'_{\text{kin}2} = E_{\text{kin}1} + E_{\text{kin}2} - E_{\text{Def}1} - E_{\text{Def}2} \quad [28]$$

where $E_{\text{kin}i}$ represents the kinetic energy of vehicle i before and $E'_{\text{kin}i}$ after the impact.

In the USA, the concept of EBS (equivalent barrier speed) is often used. Considering a vehicle deformed in an accident, EBS is the velocity at which an identical vehicle would have to collide with a stiff and nondeformable barrier to obtain identical permanent deformation. On the other hand, EES covers only the part of velocity that is necessary to get plastic deformation, neglecting the energy of elasticity E_E, absorbed without leaving any residual.

$$\frac{m \times \text{EBS}^2}{2} = \frac{m \times \text{EES}^2}{2} + E_E \quad [29]$$

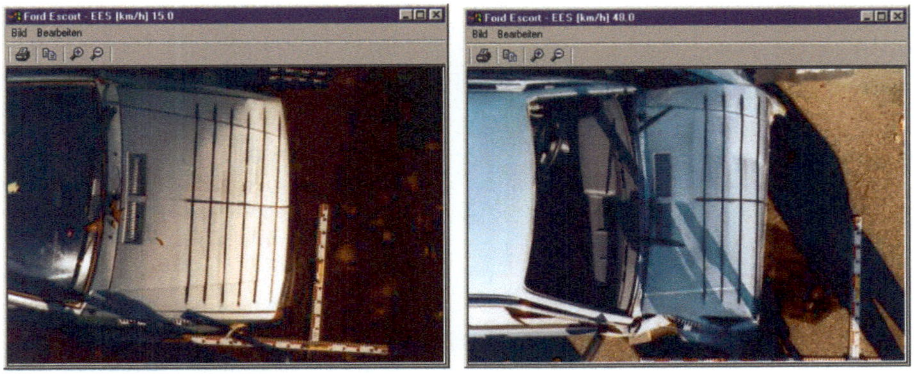

Figure 4 Vehicle damage after frontal impact with 15 and 48 km h^{-1}.

The dependence combining the two parameters can be easily derived using the coefficient of restitution k:

$$\text{EES} = \text{EBS}\sqrt{1 - k^2} \qquad [30]$$

Both EES and EBS values of a deformed vehicle can be obtained using crash test data collected in EES catalogs or databases meeting the CRASH3 standards.

Stiffness-Based and Mesh-Based Impact Models

The impact can be also simulated by means of two other models: stiffness based or mesh based. In the former, one vehicle uniform solid is converted into several hyper ellipsoids (shown in **Figure 5(a)**), whereas in the latter one the mesh structure of vehicle external shell is applied (**Figure 5(b)**). The shell is described by means of a 3D mesh of a structure analogous to the finite elements method. The idea was used to detect constraints' interpenetration during impact, determination of forces' operation, and constraints' displacements into new positions, according to the superimposed constraints. Some simplifications in relation to the FEM make it possible to increase the efficiency of calculations, with satisfactory accuracy preserved.

What is characteristic to both models is that during simulation of vehicle deformation phase impact force is calculated as a function of time.

The difference between the Kudlich–Slibar model and the stiffness-based (also mesh-based) model is illustrated in **Figure 6** which shows an impact of a car against a solid wall at 55 km h^{-1}. The following denotation has been adopted: a_{KS}, v_{KS} (acceleration and velocity, respectively, in the Kudlich–Slibar model) and a_F, v_F (acceleration and velocity, respectively, in stiffness-based model). During impact simulation with the Kudlich–Slibar model, velocity change $v - v'$ occurs as a single step within an infinite short time step $\Delta t = 0$. Deceleration diagram a_{KS} in **Figure 6** refers to vehicle motion state immediately before and immediately after the impact, neglecting the infinitely small instant of the impact itself. Since in the sample case the car was neither braked nor accelerated, $a_{KS} = 0$.

In the Kudlich–Slibar model, the time history for deceleration is not necessary to calculate the postimpact velocity. As a result, neither curve for impact force is determined. In the stiffness-based model, on the other hand, velocity v_F and acceleration a_F during deformation are calculated as functions of time, which means determination of time history of impact force.

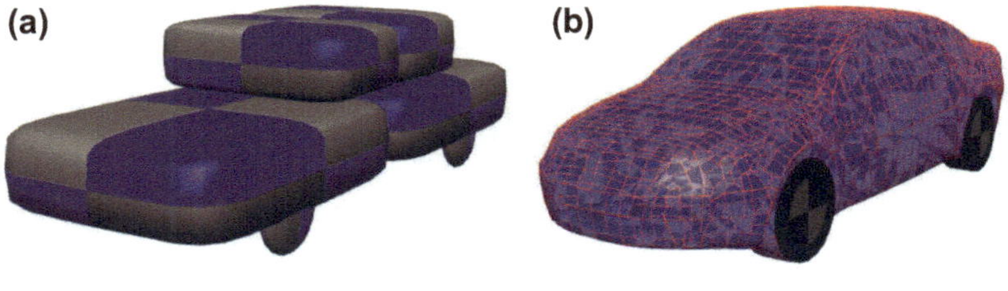

Figure 5 Vehicle structures used in (a) stiffness-based impact model and (b) mesh-based impact model.

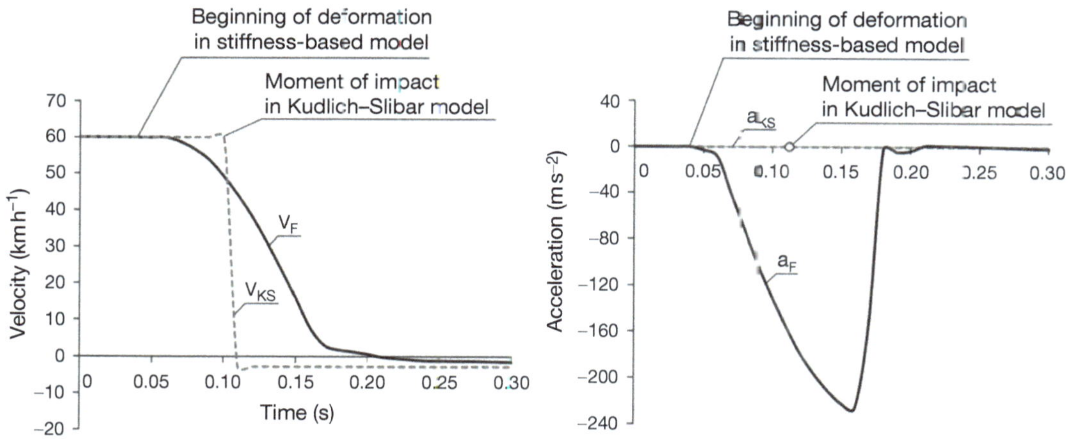

Figure 6 Accelerations and velocities during a car impact against a stiff wall: aKS and vKS—acceleration and velocity, respectively, in the Kudlich–Slibar model; aF and vF—acceleration and velocity, respectively in the stiffness-based model.

Precrash Movement

As the impact speed in many cases is not the relevant speed, precrash analysis is also of great importance. They allow one to determine where the drivers or other involved persons reacted and when conflicts were created. In the reaction point, the velocities, relative positions, and visibilities can then be investigated. Through this calculation, the cause of the accident and the failures of the involved persons or vehicles can be identified. There are several types of failures or their combinations, which cause accidents: The vehicle velocity not being adapted to traffic situation and insufficient attention of the driver are only some of the situations causing accidents. Whenever accidents are investigated, the situation has to be analyzed from the reaction point. Only, in this way, the real cause can be found.

The Driver's Reaction

The point of reaction can be determined through various methods, depending on the accident situation.

In principle, the time from the reaction to the collision position normally consists of three major phases:

1. reaction time,
2. lag, and
3. action time.

Reaction Time

The reaction time defines the necessary time to identify the situation, decide the kind of reaction, and start the action through activation of certain muscles. The reaction time depends on many parameters, such as age, tiredness, and also on parameters which are difficult to estimate, for example, an eye blink may enlarge the reaction time by approximately 0.2 s. One aspect, which is also of greatest importance, is the visibility of the object. In the case of low contrasts or small light differences, reaction time may increase dramatically. This is the reason why so many pedestrian accidents occur during nighttime. In addition, there is a significant difference if the resulting action has to be done by the arms or by the legs. Due to the longer distance between brain and leg, an action being performed with the leg will require a significantly longer reaction time.

Typical reaction times are between 0.5 and 1.5 s. Racing drivers are known to react during the race within a time of 0.3 s.

Lag

The lag is defined through the amount of time required by the technical system to an act. For the steering system, a lag does not exist. However, for the brake system, a lag between 0.1 and 0.3 can be found. This lag is mainly caused by the amount of time necessary to push the brake pedal to such a level that the full brake force is applied to the wheels. One technical solution to reduce this time is the so-called brake assistant, which automatically applies the full brake force, when the brake pedal is pushed very fast

Action Time

The action time defines the time when the action (mainly braking or steering) is active.

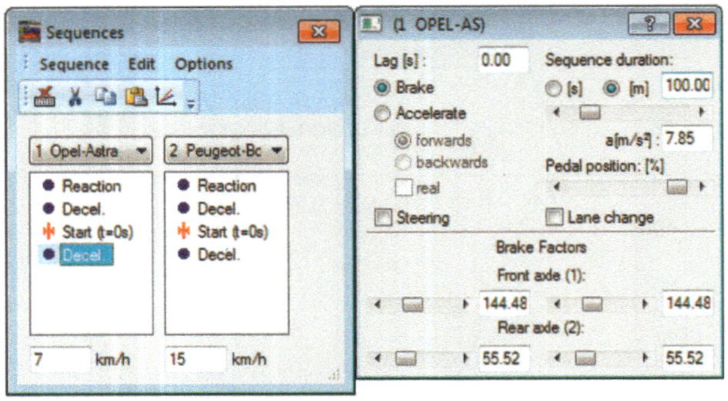

Figure 7 Definition of driver actions.

Prevention Analysis

When the point of reaction was found, the so-called prevention analysis must be performed. These calculations are used to determine a fictive scenario where the accident would have not occurred. In this case, parameters such as initial velocity, shorter reaction time, better visibility, or higher friction on the road are varied. These analyses are used not only to determine the influence of the driver's failure but often also to change the road design to prevent similar accidents.

Sample Case

The following sample case demonstrates the analysis of a vehicle–vehicle accident, mainly caused by one driver overrunning a "Stop" sign.

The first step when simulating a new accident is the identification of all involved vehicles. Modern simulation programs allow one to access various databases containing all necessary geometrical and mass data.

In a second step, the involved cars can be moved to the "average" collision position. Here, their correct overlapping must be taken into account.

To define the friction conditions and driver's actions, the so-called sequences can be given. They can be defined in a very simple and flexible way. The different steering, brake, and acceleration actions can be defined by listing the actions. The values for the individual actions can then be given. The validity of one sequence can be limited by definition of a time interval or a travel distance of the vehicle's center of gravity. The brake or acceleration forces can be given for every wheel independently. This is important to simulate wheels locked up through the accident.

Changes within the friction due to oil or wet areas can be defined by identifying the individual areas. The corresponding friction coefficient must then be specified (**Figure 7**).

After these definitions have been given, the impact can be calculated and the postcrash movement will be simulated automatically. As a result, the movement of the involved vehicles including their wheel traces can be seen on the screen. It can then be compared to a previously created DXF drawing of the scenario. As an alternative, a scanned Bitmap can be underlaid. So, the preimpact velocities can be varied as well as all other impact parameters. PC-CRASH as one of the mayor software tools also provides an optimization tool, which allows one to vary the parameters automatically. Through the definition of target functions, the most plausible solution will be found immediately.

The following figures show the movement of two vehicles after a 90° impact in steps of 200 ms.

A dry asphalt surfaced road was assumed.

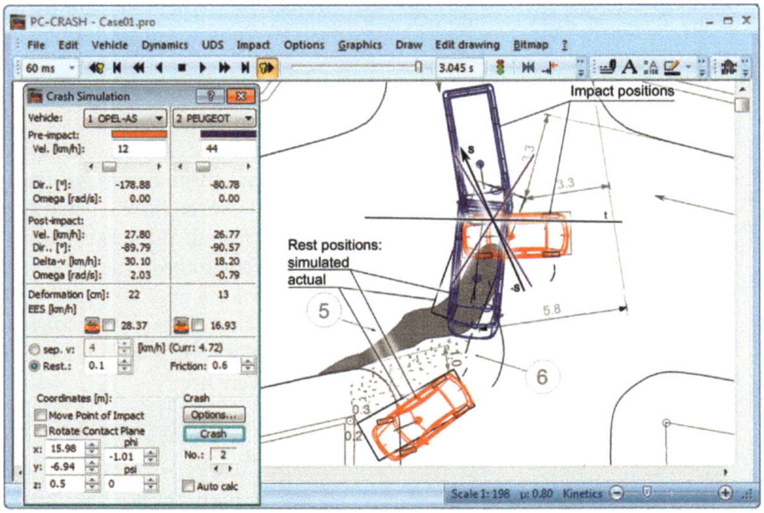

Figure 8 Definition of impact values including results.

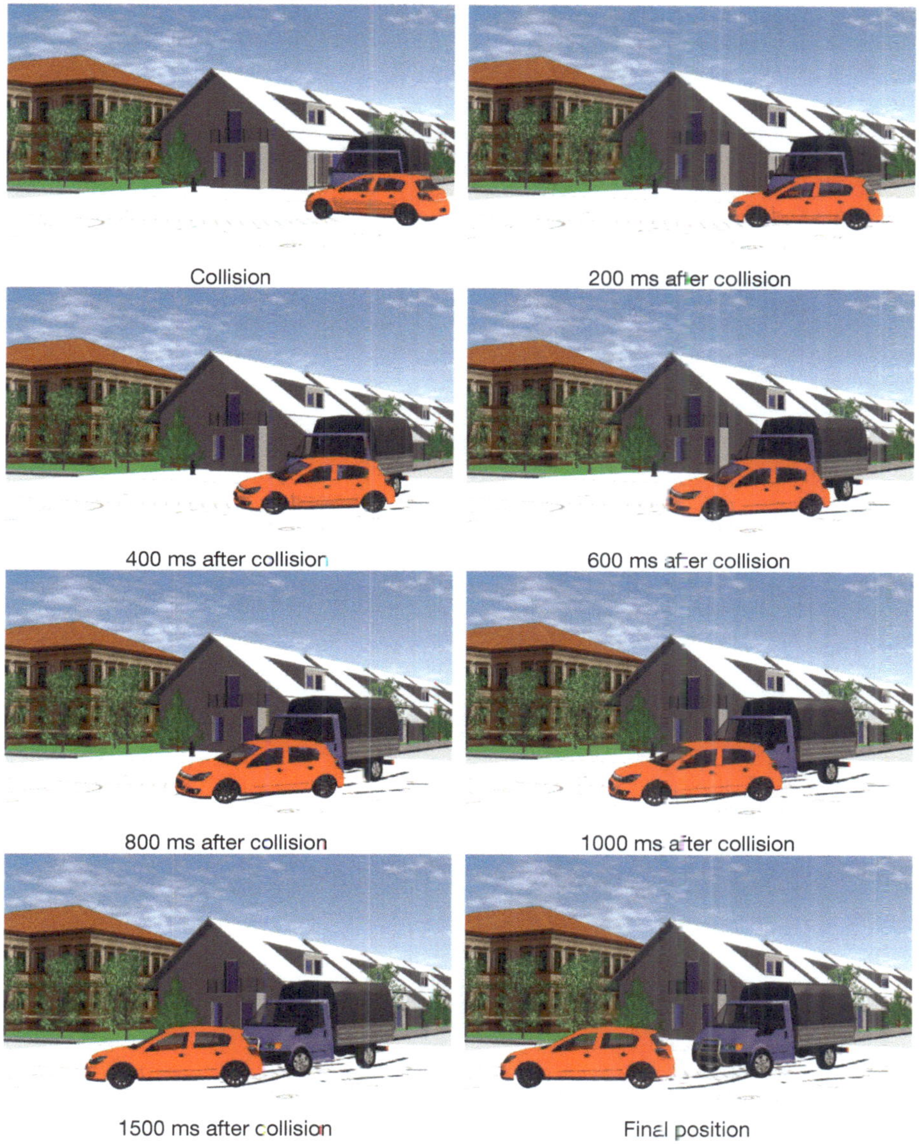

Figure 9 Vehicle movement after impact.

Results

The major fault, in this case, was on the side of the driver of the personal car, as he should have stopped before the crossing (**Figures 8 and 9**).

However, due to the fact that there were brake traces drawn from the blue truck before the impact, which had a length of 7 m, the initial speed of the truck could be determined to be 60 km h^{-1}. The truck driver reacted 1.7 s before the collision. In the case of driving at a speed of 48 km h^{-1} or lower, the accident would have been prevented, as the truck would have stopped before the impact. For these calculations, a reaction time of 1 s and a brake lag of 0.2 s was assumed.

If the driver of the truck would have been driving at a speed of 50 km h^{-1}, the truck would have reached the collision position 2.4 s after he had reacted. This means the crossing personal car would have continued to drive for 0.7 s and thus had passed an additional distance of approximately 2.3 m. This distance would have not been satisfactory to prevent the accident. So, the accident would have also occurred at an initial speed of 50 km h^{-1} in the reaction point, but impact speed would have been reduced from 44 km h^{-1} to approximately 15 km h^{-1}.

Summary

This article outlines the basic method to determine the cause of an accident. In addition to conventional reconstruction, modern numerical methods are outlined. All three phases of an accident, precollision, collision, and postcollision are discussed. Prevention analysis is performed within a sample case.

See also: **Engineering:** Airbag Related Injuries and Deaths; Forensic Engineering/Accident Reconstruction/Biomechanics of Injury/Philosophy, Basic Theory, and Fundamentals; Human Factors Investigation and Analysis of Accidents and Incidents.

Further Readings

Brach, R.M., 1991. Mechanical Impact Dynamics – Rigid Body Collisions. Wiley, New York.

Campbell, K.L., 1965. Energy Basis for Collision Severity. SAE 740565.

CRASH3 Technical Manual, 1986. Accident Investigation Division. NCSA, NHTSA.

Day, T.D., 1995. An Overview of the HVE Vehicle Model. SAE 950308.

Day, T.D., Hargens, R.L., 1982. An Overview of the Way EDCRASH Computes Delta-V. Paper No. 820045. SAE, Warrendale.

Gnadler, R., 1971. Das Fahrverhalten von Kraftfahrzeugen bei instationärer Kurvenfahrt mit verschiedener Anordnung der Hauptträgheitsachsen und der Rollachse. University of Karlsruhe, Karlsruhe (Dissertation).

Kudlich, H., 1966. Beitrag zur Mechanik des Kraftfahreug-Verkehrsunfalls (Dissertation, TU-Wien).

Lugner, P., Lorenz, R., Schindler, E., 1984. The connexion of theoretical simulation and experiments in passenger car dynamic. In: Hedrick, H.J. (Ed.), The Dynamics of Vehicles on Roads and on Tracks. Swets & Zeitlinger, Lisse, pp. 317–330.

McHenry, R.R., February 1976. Extensions and Refinements of the CRASH Computer Program. Part I. Analytical Reconstruction of Highway Accidents. DOT-HS-5–01124.

McHenry, R.R., Segal, D.J., DeLeys, N.J., October 1967. Computer simulation of single vehicle accidents. In: Stapp Car Crash Conference. Anaheim, CA.

Rill, G., 1994. Simulation von Kraftfahrzeugen. Vieweg & Sohn Verlagsgesellschaft mbH, Braunschweig/Wiesbaden.

Slibar, A., 1966. Die mechanischen Grundsätze des Stoßvorganges freier und geführter Körper und ihre Anwendung auf den Stoßvorgang von Fahrzeugen. Archiv für Unfallforschung 2 (1), 31.

Steffan, H., 1993. PC-CRASH, a simulation program for car accidents. In: ISATA; 26th International Symposium on Automotive Technology and Automation, Aachen.

Steffan, H., 1999. PC-CRASH 6.0: A Simulation Program for Car Accidents. Users and Technical Manual, DSD, Linz, Austria.

Steffan, H., Moser, A., 1996. The Collision and Trajectory Models of PC-crash. SAE 960886.

Steffan, H., Moser, A., 1998. The Trailer Simulation Model of PC-crash. SAE 980372.

Human Factors Investigation and Analysis of Accidents and Incidents

SA Shappell, Embry-Riddle Aeronautical University, Daytona Beach, FL, USA
DA Wiegmann, University of Wisconsin, Madison, WI, USA

Glossary

Decision errors These errors represent intentional actions that proceed as intended, but the plan proves inadequate or unsuitable for the situation. Usually, they are related to knowledge, problem solving, and procedural errors, such as improper procedure, exceeded ability, and poor decision.

Error Normal accepted behavior that fails to meet the desired outcome.

Human Factors In 2000, the International Ergonomics Association defined ergonomics (human factors) as "the scientific discipline concerned with the understanding of interactions among humans and other elements of a system, and the profession that applies theory, principles, data, and other methods to design in order to optimize human well-being and overall system performance."

Perceptual errors These errors can occur when the operator has perceived features of the environment inaccurately (sensory degradation); the misperception could be visual, auditory, olfactory, etc. Mostly they are related to detection, awareness, and understanding errors, such as distance misjudgments and visual illusion.

Skill-based errors These errors represent unintended/unconscious behaviors of operators, which usually occur in the operator's execution of a highly practiced task relating to procedure or a routine. In general, they are related to attention failures, memory lapses, and inadequate techniques, such as an omitted step in procedure, an omitted checklist item, and poor technique.

Violation Refers to the willful disregard of accepted practices and norms within a field of endeavor. The keyword is willful; that is, the individual knew what the accepted practice was but chose to ignore it.

The need to manage human error comes as no great revelation to anyone involved in operations where the consequences of failure are dire. Truth be told, however, the "battle cry" that human error is associated with some 60–80% of all accidents/incidents in complex, high-risk systems has become passé in many organizations. The reason behind this attitude is that human error statistics have not changed appreciably in more than half a century. While safety professionals can all agree that something "must" be done to reduce errors, a growing number of executives in the boardroom are becoming skeptical of the idea that something actually "can" be done.

So where does one start when managing human error? Most quality and safety professionals are very familiar with the traditional system safety approach illustrated in **Figure** 1. While there are many variants to the approach, most involve the following components: collecting relevant data, identifying and assessing hazards, identifying/developing interventions, assessing intervention feasibility, implementing intervention, and system monitoring/program evaluation. Ideally, this is a dynamic process involving the real-time identification of threats, deployment of interventions, and, hopefully, improvements in the process.

Historically, this traditional approach to system safety has been effective in addressing mechanical and engineering problems within a variety of operational contexts; however, it has been facilitated by the development of a set of engineering tools and techniques for implementing each step. Not surprisingly, given the success of system safety in the engineering world, many quality/safety professionals have been quick to adopt this same approach when attempting to manage complex human factors issues.

Unfortunately, the requisite tools and techniques for employing system safety concepts within human factors have been largely ineffectual or nonexistent. As a result, it has been virtually impossible to get beyond the first step in the process—data collection. In fact, when an accident or incident does happen, the most common response is to simply collect more data. However, merely gathering more data about the occurrence of errors is "not" the solution. In fact, most quality and safety departments are already swimming in data, particularly

 http://dx.doi.org/10.1016/B978-0-12-802718-9.00007-7

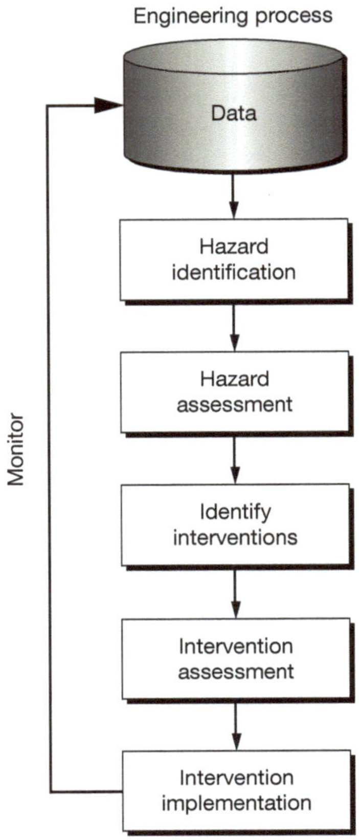

Figure 1 The safety management process.

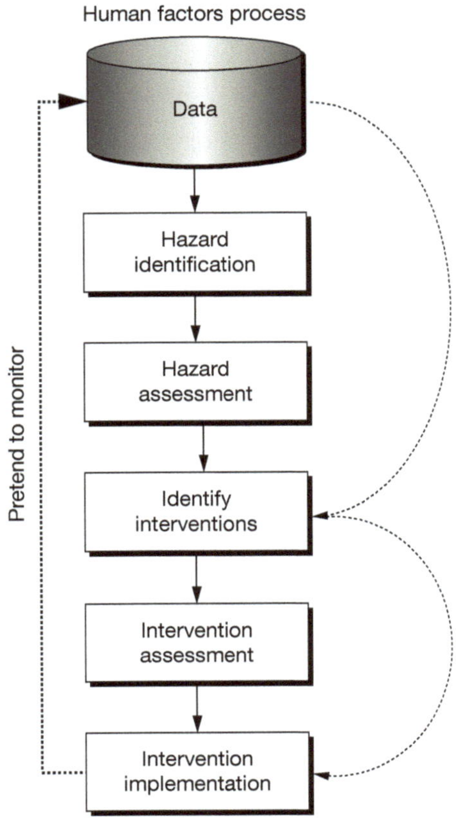

Figure 2 Lack of effective human factors tools results in key steps in the safety process being skipped.

given the recent advances in technology that have increased the amount of information available.

As a result, those tasked with solving the human error dilemma have become frustrated with just watching the "data bucket" fill up while having no means to interpret the information contained within the bucket. After all, implementing intervention programs that improve safety is the ultimate goal, and one should not confuse data collection with interventions.

Rather than systematically analyzing the aggregate data for significant trends, the modus operandi of many organizations has been to simply "cherry pick" the hazards identified in the most recent incident or focus on select "high-profile" events that have captured everyone's attention, thereby bypassing the next two critical steps in the system safety process and jumping right on to identifying interventions (see **Figure 2**). However, these may have no real bearing on the more significant systemic hazards.

To make matters worse, there are few tools for systematically generating effective intervention programs that target specific forms of human error. Consequently, quality/safety professionals have been forced to bypass two more crucial steps in the safety management process—generating and evaluating intervention strategies (see **Figure 2**). Indeed, the typical process for ginning up error prevention programs is to use simple intuition, expert opinion, or "pop psychology." Unfortunately, such an approach often results in people pushing their pet projects, or the person with the highest authority having the last say on what gets done. Hence, this entire system engineering approach to human error management, although conceivably a good idea, has met with little success.

Over the last decade, human factors psychologists and engineers have been working on ways to improve the human factors system safety process by developing human factors tools and techniques. The remainder of this chapter describes one of the tools born out of aviation accident investigation to facilitate error identification and intervention development within the context of forensic science.

HFACS—A Human Factors Investigative Tool

It is generally accepted that human error is not the isolated act of a particularly unlucky individual. Rather, it is the result of a chain of events that often culminate in a less than desired outcome. This is not particularly new, as industry has embraced a sequential theory of human error for nearly a century. However, some would argue that it was not until James Reason published his "Swiss cheese" model of human error in 1990 that industry truly began to examine human error in a systematic manner.

Reason's approach to error causation is based on the assumption that there are fundamental elements of all organizations that must work together if safe and efficient operations are to occur. Taken together, these elements comprise a "productive system" as depicted in **Figure 3**. For example, forensic analysis can be viewed as a complex productive system whose "product" is the identification and classification of forensic evidence. These so-called "productive activities," in turn, require the effective integration of human and mechanical elements within the system, including, among other things, an effective scientist–laboratory interface to ensure optimal performance.

Even before productive activities can occur, certain "preconditions," such as reliable and well-maintained equipment and a well-trained and professional workforce, need to exist. After all, most analysts work within a highly structured organization that requires effective middle management and careful supervision. Furthermore, such management and supervision is needed across numerous departments within the larger organization. But even the best managers need guidance, personnel, and money to perform their duties effectively. This support comes from decision makers who work even further up the chain of command, charged with the responsibility of setting goals, and managing the available resources of the organization.

In most organizations, this highly structured and orchestrated system functions well. However, what about those rare occasions when the wheels do come off? According to Reason, accidents occur when there are "breakdowns" in the interactions among the components involved in the production process. These failures degrade the quality of the system making it more vulnerable, and hence more susceptible, to erroneous conclusions. As illustrated in **Figure 3**, these errors can be depicted as "holes" within the different layers of the system, which transform what was once a productive system into a failed or broken-down process. Given the image of Swiss cheese that this illustration generates, the theory is often referred to as the "Swiss cheese" model of error causation.

So, what are those "holes in the cheese" or failure points in the system? Drawing upon Reason's concept of latent and

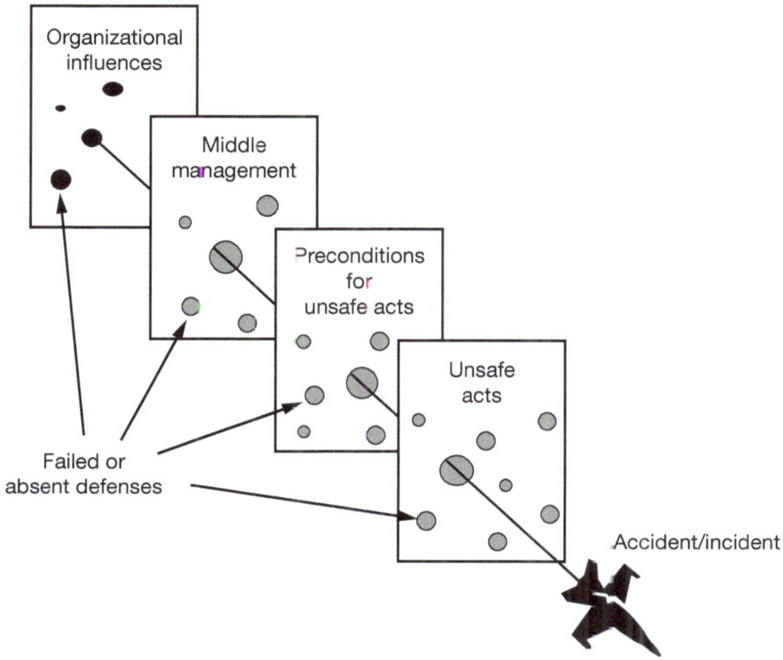

Figure 3 Components of a productive system.

active failures, the Human Factors Analysis and Classification System (HFACS) describes human error at each of four levels: (1) the actions of the operators (e.g., bench-level scientists and field investigators in forensics), (2) the preconditions for those actions (i.e., the conditions that influence human behavior), (3) the middle management (i.e., the individuals whose role it is to assign work), and (4) the organization itself.

In other words, the HFACS framework goes beyond the simple identification of "who did what wrong" to provide a clear understanding of the reasons "why" the error occurred in the first place. In this way, errors are viewed as consequences of system failures and/or symptoms of deeper systemic problems, and not simply as the fault of the examiner whose fate was tied to being at the "pointy end of the spear."

HFACS was originally developed and used as an aviation accident investigation and analysis tool for the U.S. Navy and Marine Corps. Since then, the framework has been adopted by a number of other aviation, transportation, and high-risk industries. More recently, HFACS has been modified by the Expert Working Group on Human Factors in Latent Print Analysis chartered by the National Institute of Standards and Technology for use in forensic science. The brief discussion that follows is a product of that effort.

Analyst Actions

The actions of the forensic analyst can be loosely classified as either errors or violations (see **Figure 4**). While both can occur within most settings, they differ markedly when the rules and regulations of the scientific field are considered; that is, while errors represent normal accepted behavior that fails to meet the desired outcome, violations refer to the "willful" disregard of accepted practices within the field. It is within these two overarching categories that HFACS describes three types of errors (decision, skill-based, and perceptual) and types of violations.

Decision errors

Successful decision making depends upon three fundamental ingredients: (1) information—that is, information (e.g., fingerprints, DNA) must be of high quality and other variables such as the surface from which the evidence was obtained must be known; (2) knowledge—analysts must have the necessary knowledge, training, and background to reliably assess the information provided; and (3) experience—there is ample evidence in the literature that suggests that experience also contributes to decision making. Ultimately, when information, knowledge, or experience is lacking or degraded, errors can occur. Often referred to as honest mistakes, errors typically manifest as poorly executed procedures, improper choices, or simply the misinterpretation and/or misuse of relevant information (see **Table 1**).

Skill-based errors

In contrast to decision making, skill-based behavior occurs with little or no conscious thought. For instance, little thought goes into turning one's steering wheel in an automobile or typing on a keyboard. Likewise, in latent fingerprint examination, most analysts orient latent prints for examination with little or no conscious thought. In other words, they do not have to think about it—they just do it. The difficulty with these highly practiced and seemingly automatic behaviors is that they are particularly susceptible to attention and/or memory failures. Unfortunately, even the technique or "skill" with which we do a task can lead to erroneous conclusions (see **Table 1**).

Perceptual errors

While decision and skill-based errors are the most common error forms seen in occupational settings, perceptual errors may also impact the forensic laboratory. No less important, these "perceiving" errors arise when sensory input is degraded, or "unusual." For example, latent fingerprint analysis is a decidedly visual comparison task that requires optimal visual conditions to insure accuracy and reliability. Consequently, laboratory environmental factors such as the intensity, type, and direction of lighting can influence what an analyst perceives as key features of a print. When confronted with less than ideal conditions, an analyst may perceive that there is more or less information in a given print than there actually is (see **Table 1**).

Violations

Although there are many ways to distinguish between types of violations, two forms have been identified based on their etiology. The first, routine violations tend to be habitual by nature and are often enabled by a system of supervision and/or management that tolerates such departures from the rules. Often referred to as "bending the rules," the classic example is that of the individual who drives his/her automobile consistently 5–10 mph faster than allowed by law. While clearly against the law, the behavior is, in effect, sanctioned by local authorities (police) who often will not enforce the law until more excessive speeds are observed. What is more, speeding on the interstate is something that most people do, and often it becomes a "routine" part of the individual's behavior.

In contrast, driving at 120 mph in a 55-mph zone would be considered an exceptional violation. It is important to note, that while most exceptional violations tend to be particularly dangerous, they are not considered "exceptional" because of their extreme nature. Rather, they are regarded as exceptional because they are neither typical of the individual(s) nor condoned by management. In essence, they are "exceptions" to normal behavior.

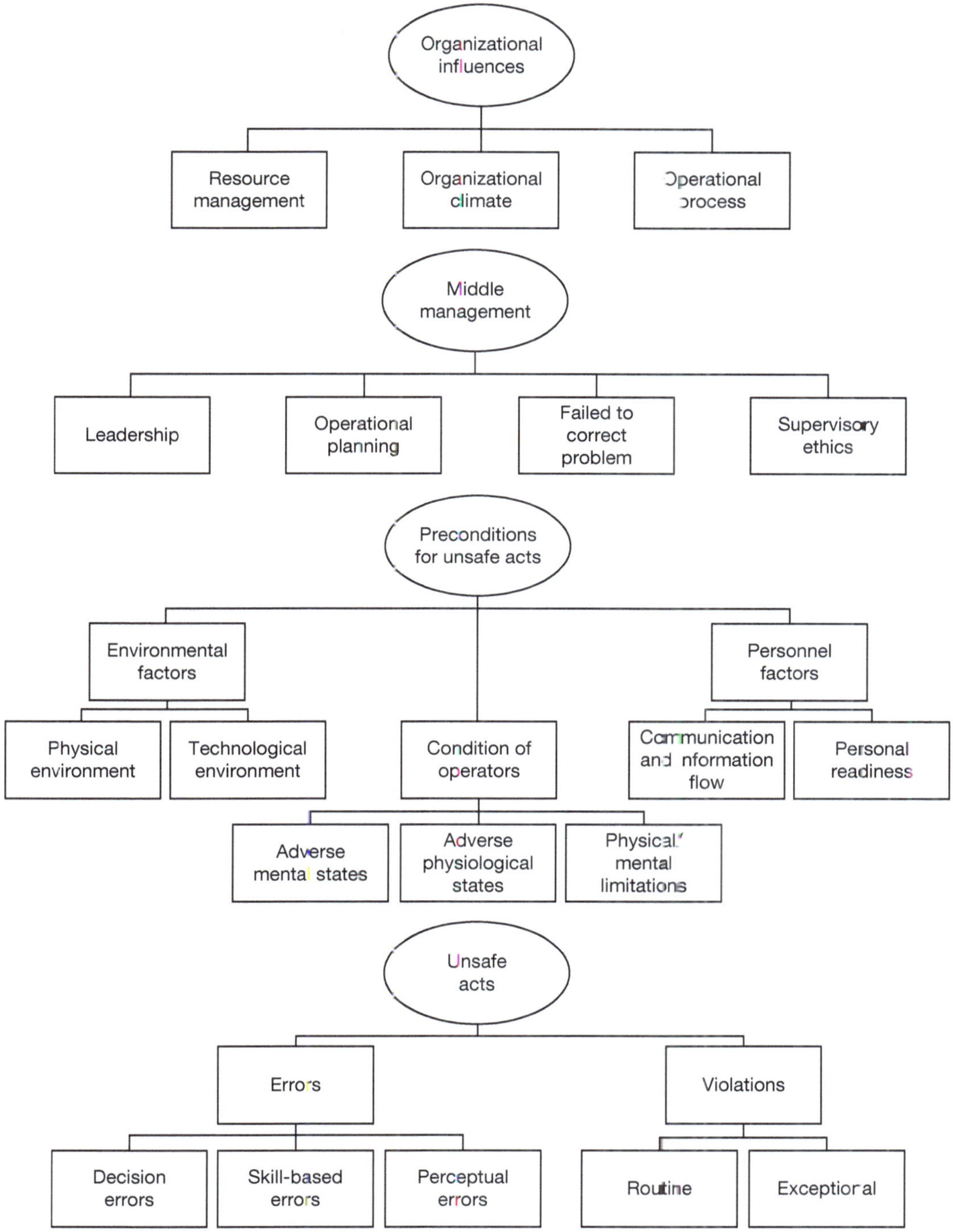

Figure 4 The Human Factors Analysis and Classification System (HFACS).

Table 1 Selected examples of analyst actions

Errors	*Violations*
Decision errors • Improper orientation of data • Print improperly deemed unsuitable • Printer improperly deemed suitable • Improper anatomical source of data • Poor target group selection • Hurried examination • Thorough analysis not conducted • Incomplete search • Failure to search all exemplars • Misprioritized level of effort • Failure to use available technology • Conclusion exceeded ability of examiner • Insufficient data to support conclusion • Failure to recognize exemplars are insufficient/inadequate *Skill-based errors* • Poor search technique • Incomplete comparison due to failure of memory • Incomplete comparison due to attention failure • Failure to recognize distortion (i.e., variation in appearance) • Failure to properly weigh the significance of the data • Failure to find target • Inability to reach a conclusion • Misinterpreted the data • Distraction *Perceptual errors* • Failure to perceive due to poor contrast • Visual illusion in print	• Intentionally issue an unsupportable conclusion • Failure to follow SOPs • Refusal to use available technology • Consistently failed to mark a difficult print • Consistently conduct an incomplete search • Failed to maintain objectivity

Note: This is not a complete listing.

Unfortunately, depending on the setting and conditions an individual is faced with, what may be a routine violation in one organization or situation may be an exceptional violation in another. For instance, driving 5–10 mph over the speed limit on a US interstate may be condoned by local authorities and is done by most people, while driving that same 5–10 mph over the posted speed limit in a school zone certainly would not be condoned nor does everyone do it. Likewise, depending on the given laboratory procedure, a violation may be perceived as routine (everyone does it) or exceptional (rarely occurs). Often, the actual causal factor (e.g., failure to follow standard operating procedures (SOPs)) is the same in both categories, only the circumstances surrounding the violation are different. Consequently, all violations are simply included within one category as shown in **Table 1**.

Preconditions Analyst Actions

Simply focusing on errors, however, is like focusing on a medical patient's symptoms without understanding the underlying disease state that caused it. As such, investigators must dig deeper into the preconditions for why errors occurred. Within HFACS, seven causal categories within three major subdivisions are described (see **Figure 4** for an illustration and **Table 1** for examples within the forensic sciences) (**Table 2**).

Adverse mental states

It is well known that human performance is affected by the mental state of the individual. Attitudes such as frustration, anger, and even misplaced motivation have all led to decision and skill-based errors when not recognized and addressed by individuals. Likewise, cognitive states such as inattention, distraction, and task overload have all resulted in a variety of errors. Indeed, one of the more insidious adverse mental states is fatigue and stress associated with many occupations. Consider, for example, a forensic analyst who is called to a crime scene in the middle of the night and is then expected to show up the next day to attend to his or her normal caseload. Can management reasonably expect performance to remain at optimal levels? Likewise, the stress associated with working

Table 2 Selected examples of preconditions of unsafe acts

Condition of analyst	*Personnel factors*	*Environmental factors*
Adverse mental states	*Communication and information flow*	*Physical environment*
• Overconfidence	• Failure to convey adequate information	• Workspace design
• Lack of confidence	• Failure to use all available resources	• Inadequate lighting
• Distraction	• Lack of teamwork	• Inadequate ventilation
• Mental fatigue	• Confusing/conflicting directions/demands	• Noise interference
• Apprehension with reaching a conclusion	• Inadequate communication with management	• Excessive heat or cold
• Preoccupation with personal problems	• Inadequate communication between shifts	• Clutter
• Task overload	• Lack of communication	*Technological environment*
• Frustration	• Standard terminology not used	• Inadequate/outdated
• Channelized attention	• Lack of report writing skills/poor documentation	• Software or equipment
• Complacency	*Fitness for duty*	• Defective equipment
• Misplaced motivation	• Lack of sleep	• Defective tools
• Stress	• Self-medicating with prescription or over-the-counter medications	• Inadequate information technology
• Expectancy	• Use of illicit drugs and alcohol	• Confusing manuals
• Boredom	• Suffering from hangover	
• Anger	Inadequate nutrition	
• Peer pressure	• Reporting for work in ill-health	
• Haste		
• Bias		
• Compromised integrity/ethics		
Adverse physiological states		
• Medical illness		
• Impairment due to drugs or alcohol		
• Physical fatigue		
• Eyestrain		
Physical/mental limitations		
• Performance limiting physical condition		
• Visual limitations		
• Dyslexia		
• Color vision deficiency		
• Chronic psychological disorder		
• Incompatible intelligence/aptitude		
• Limited experience/proficiency		

Note: This is not a complete listing.

through large backlogs with limited resources or unrealistic quotas and turn-around times set by well-meaning, but uninformed, managers can lead to errors.

Adverse physiological states

Equally important, however, are those adverse physiological states that affect performance. The traditional work environment for many forensic scientists includes a standard desk, a magnifier or two, general lighting, and a variety of specialized tools with the analyst bent over a desk sifting through evidence to draw their conclusions. The strain on the neck and back can be considerable and may impact their performance. Likewise, many comparisons today are performed utilizing computer monitors, which may expose the analyst to monitor glare, resulting in headache or eyestrain. Medical illness is also known to have an impact on performance; yet, people sometimes go to work with illnesses, such as a cold, migraines, or the flu, or worse, they are self-medicating with remedies that can adversely influence their decision-making ability.

Physical and/or mental limitations

Physical/mental limitations include those instances when an individual simply may not possess the necessary aptitude, physical ability, or skill to conduct an examination adequately. After all, just as everyone cannot play linebacker for his or her favorite professional football team or be a concert pianist, not everyone has the aptitude or physical attributes necessary to be a forensic examiner. Moreover, physical and mental aptitudes change as we age. Left unabated, normal aging processes can adversely influence performance.

Communication and information flow

Communication and information are essential to case management. The analyst must effectively work with a variety of analysts, managements, investigators, and attorneys. Indeed, coordination among multiple "players" (i.e., analysts, detectives, and the district attorney) is essential in case management. Poor communication will drastically reduce gathering of the necessary information that may affect casework. Case documentation is another mode of communication critical during technical review and for archival purposes.

Fitness for duty

Individuals, must, by necessity, ensure that they are adequately prepared for work. Consequently, this category was created to account for those instances when examiners do not get enough rest, report to work when ill, or take medications that may impair their ability to perform their duties.

Physical environment

The term physical environment refers to things such as heat and lighting toxins in the workplace, all of which are well known to impact human performance.

Technological environment

The technological environment that individuals often find themselves in can also have a tremendous impact on their performance. The term technological environment encompasses a variety of issues including the design of equipment and controls, display/interface characteristics, checklist design, and automation, to name a few. One needs to look no further than the design of automobiles to see poor design influencing performance. After all, who has not tried to turn on the lights in a rental car only to see the wipers engage instead? Similar design flaws in the laboratory can be catastrophic. Whether one is having difficulty operating a DNA sequencer or dealing with flawed software, how one designs equipment can have a dramatic influence on analyses and interpretation of complex results.

Middle Management

Ultimately, analysts and those at the front lines of forensic science are responsible for their actions. However, in some instances, they are the unwitting recipients of a variety of latent failures attributable to those who manage them. To account for these threats to quality, the overarching category of "supervision" was created within which four categories (leadership, operational planning, failure to correct known problems, and supervisory ethics) are included (**Table 3**).

Table 3 Selected examples of middle management

Inadequate supervision
- Lack of appropriate incentives
- Inadequate performance measurement evaluation
- Unrealistic expectations
- Insufficient initial or ongoing training
- Inadequate assessment of required skills
- Failure to track job qualifications/skills
- Lack of coaching/training on skills
- Change introduced without training
- No measurement of training effectiveness
- Inadequate leadership job knowledge
- Inadequate monitoring of work
- Personality conflicts between manager and employee
- Failed to provide professional guidance/oversight
- Failed to set a proper example

Operational planning
- Unclear/conflicting assignment of responsibility
- Inadequate communication of policy, procedure, practices, and/or guidelines
- Improper/insufficient delegation of work
- Setting objectives, goals, or standards that conflict
- Improper/insufficient delegation of work
- Setting objectives, goals, or standards that conflict
- Failure to provide adequate rest breaks
- Inadequate matching of individual qualifications and job/team requirements
- Unrealistic deadlines or quotas
- Inadequate documentation

Failed to correct a known problem
- Failure to correct known workplace hazard
- Improper performance is rewarded or tolerated
- Failure to correct inappropriate behavior (hostile work environment)
- Failure to initiate corrective actions
- Failure to correct known reportable problem
- Failure to update/revise operator guidelines, policies, and procedures

Supervisory ethics
- Management implied haste
- Undue influence toward a desired outcome
- Violation of operating guidelines, policies, and procedures by a manager
- Manager encourages bending of the rules
- Enabling excessive risk taking
- Failure to enforce rules and regulations
- Fraudulent documentation

Note: This is not a complete listing.

Leadership

At a minimum, supervisors must provide the analysts in their charge the opportunity to succeed. Toward these ends, it is expected that effective leaders will provide analysts with adequate training, professional guidance, oversight, and operational leadership, and that all will be managed appropriately. When this is not the case, examiners can become isolated, thereby increasing the likelihood that errors will occur. For example, it is not difficult to imagine a situation where a manager becomes so overwhelmed with answering e-mail, attending telecons, preparing countless presentations, etc., that they fail to provide sufficient oversight of their analysts. In the end, the employees are left with little guidance and oversight, as the "phantom manager" is lost in an electronic world of corporate demands. In contrast, some supervisors take oversight to extremes, becoming overcontrolling and more concerned with minute details rather than the accuracy of the work. In effect, by micromanaging their analysts, decision making is often delayed, information flow between employees is often restricted, and overall confidence and efficiency can be diminished.

Operational planning

The risks associated with failures of management and supervisors come in many forms. For example, occasionally the operational tempo and/or schedule are planned such that individuals are put at unacceptable risk and, ultimately, performance is adversely affected. As such, the category of operational planning was created to account for all aspects of improper or inappropriate scheduling practices, which may involve such issues as, rest breaks, workload, deadlines, etc. Consider, for example, the supervisor who assigns a complex case to a junior or less-experienced analyst rather than someone with more experience who may be more capable. In some ways, the supervisor may "set up" the analyst for failure. Likewise, it is not uncommon for supervisors and management to continue to "go to the well" and overload the high performer while ignoring those who are less capable or difficult to work with. In either event, the top performers are often burdened with additional work and in some circumstances may be prone to error when they would otherwise perform optimally.

Failure to correct known problems

The remaining two categories of unsafe supervision, the failure to correct known problems and supervisory ethics, are similar, yet considered separately within HFACS. The failure to correct known problems refers to those instances when deficiencies among analysts, equipment, training, or other related safety areas are "known" to the supervisor, yet are allowed to continue uncorrected. For example, the failure to consistently correct or discipline inappropriate behavior fosters an unsafe atmosphere but is not considered a violation if no specific rules or regulations were broken. Similarly, there are times when optimal equipment is not purchased or necessary repairs are not done on existing equipment, yet the mission can still be carried out—just not as efficiently or without undue burden on analysts.

Supervisory ethics

This category is reserved for those instances when supervisors willfully disregard existing rules and regulations. While everyone would agree that putting undue influence on forensic analysts toward a desired outcome is clearly an example of poor supervisory ethics, more subtle ethical violations can result in adverse outcomes in the forensic laboratory. For instance, permitting unqualified forensic analysts to do casework can lead to errors that may go unnoticed. Likewise, encouraging the bending of rules and procedures in the interest of reducing backlogs is a classic failure of supervisory ethics.

Table 4 Selected examples of organizational influences

Resource management
• Selection of personnel
• Inadequate evaluation, promotion, upgrade policies
• Inadequate hiring, firing, and promotion policies
• Staffing/manning issues
• Inadequate matching of qualifications for job
• Use of inadequate contractor/vendor
• Ineffective cost cutting
• Unfunded directives
• Lack of logistical support
• Failure to correct known design flaws
• Purchasing unsuitable equipment/parts
Organizational climate
• Inaccessibility/visibility of upper management
• Inadequate formal accountability for actions
• Unclear/conflicting assignment of responsibility
• Dysfunctional organizational culture
• Organizational values not clearly defined or articulated
Organizational process
• Work/production schedules that produce bad decisions
• Overextending resources
• Unrealistic quotas
• Organizational induced time pressure
• Lack of appropriate standards, policies, or guidelines
• Procedures not integrated into training process
• Quality process not adequately implemented/maintained
• Inadequate/inappropriate performance measures
• Failure to maintain accreditation

Note: This is not a complete listing.

Organizational Influences

Where decisions and practices by laboratory supervisors and managers can adversely impact forensic analyst performance, fallible decisions of upper-level management may directly affect the supervisors and the personnel they manage. Unfortunately, these organizational influences often go unnoticed or unreported by even the best-intentioned quality assurance personnel. Within this tier, the HFACS framework describes three organizational failures: (1) resource management, (2) organizational climate, and (3) operational processes (Table 4).

Resource management

This category refers to the management, allocation, and maintenance of organizational resources, including human resource management (selection, training, and staffing), monetary budgets, and equipment design (ergonomic specifications). In general, corporate decisions about how such resources should be managed center around two distinct objectives—the quality of the analyses and on-time, cost-effective operations. In times of prosperity, both objectives can be easily balanced and satisfied. However, there may also be times of fiscal austerity that demand some give and take between the two. Unfortunately, history tells us that quality is often the loser in such battles, as quality improvements and training are often the first to be cut in organizations experiencing financial difficulties.

Organizational climate

The concept of an organization's climate has been described in many ways; however, here it refers to a broad class of organizational variables that influence worker performance. For example, one tell-tale sign of an organization's climate is its structure, as reflected in the chain of command, delegation of authority and responsibility, communication channels, and formal accountability for actions. Likewise, an organization's policies and culture are also good indicators of its climate. Consequently, when policies are ill-defined, adversarial, or conflicting, or when they are supplanted by unofficial rules and values, confusion abounds, and safety suffers within an organization.

Operational process

Finally, operational process refers to formal processes (operational tempo, time pressures, production quotas, etc.), procedures (performance standards, objectives, etc.), and oversight within the organization. Poor upper-level management and decisions concerning each of these organizational factors can also have a negative, albeit indirect, effect on operator performance and system safety.

Using HFACS to Identify and Address Threats to Quality and Safety

As described earlier, identifying what went wrong without fully understanding why it happened can lead to ineffective interventions/mitigation strategies. Consider, for example, a forensic analyst who inadvertently forgets to document a key step in the analysis, leading to an erroneous conclusion. On the surface, management might view such an error as unforgivable and resort to some form of admonishment or worse. However, without knowing "why" the analyst forgot to document the key step, one cannot recommend a practical intervention apart from one that may appease those looking for some degree of accountability.

On the other hand, had a full human factors analysis been conducted, management might have uncovered the following:

1. *The analyst had been working overtime for several consecutive days (operational planning) in an attempt to relieve an extensive backlog of cases that had accumulated leading to fatigue (adverse mental state) and stress (adverse mental state).*
2. *The analyst had recently welcomed their first baby into the family and was still not sleeping through the night while off-duty (fitness for duty).*
3. *With the holidays fast approaching, the analyst was busy making holiday travel plans including airline reservations, with a newborn and conflicting work schedules, thereby compounding the stress he was undergoing (adverse mental states).*
4. *Due to budget cuts (resource management), the staff was reduced by 20% resulting in an increase in workload (adverse mental state) without necessary support.*
5. *In an effort to reduce the workforce, management chose to offer the employees early retirement as an incentive to retire, resulting in the more experienced analysts retiring early (resource management).*
6. *The analyst who forgot to document the key step in the analysis was relatively new, with little experience in the field (physical/mental limitation) and would not ordinarily be tasked with such a complex and high-profile case (operational planning).*
7. *Because of the reduced staffing and increased workload (resource management), the analyst was not provided necessary training (leadership) and was given minimal oversight (leadership) during the investigative process.*

Arguably, such a scenario would likely change management's view of the necessary steps to prevent this sort of event from happening again. Moreover, punishment for such a transgression clearly would not have prevented the error from occurring in the future. Indeed, it is well known that one cannot punish one's way out of an error. If anything, punishment drives reporting underground and hampers safety and quality.

What makes frameworks such as HFACS effective is that each of the causal categories is based upon years of human factors research and has very specific interventions associated with it. For example, if one is trying to address decision errors, effort should be aimed at ensuring that useful and complete information is available, and that the analyst has acquired the knowledge to deal with the information through training and the requisite experience to know when he or she may need additional training or assistance. Likewise, if one were trying to address skill-based errors, efforts should be aimed at reducing distractions, ensuring that critical items are addressed through either checklists or other memory aids and that the initial and recurrent training is sufficient to ensure the skill and proficiency of the analyst.

Closing Thoughts

The field of human factors accident investigation and quality management is a rich and growing discipline with a history dating back decades. Recently, with the integration of tools such as HFACS into traditional system safety models, management in general and forensic management in particular are better able to identify and address heretofore unknown human causal factors. Of note, however, while special emphasis was placed on HFACS in this short chapter, there are many other tools and techniques available to those interested in human factors accident investigation.

The interested reader is encouraged to reach out to organizations such as the Human Factors and Ergonomics Society at www.hfes.org for more information.

See also: **Engineering:** Railroad Accident Investigation and Reconstruction.

Further Reading

Australian Transport Safety Bureau, 2007. Human Factors Analysis of Australian Aviation Accidents and Comparison with the United States. ATSB Transport Safety Investigation Report, B2004/0321.

Baysari, M.T., Mecintosh, A.S., et al., 2008. Understanding the human factors contribution to railway accidents and incidents in Australia. Accident Analysis and Prevention 40, 1750–1757.

Beaubien, J.M., Baker, D.P., 2002. A review of selected aviation human factors taxonomies, accident/incident reporting systems and data collection tools. The International Journal of Applied Aviation Studies 2, 11–36.

Celik, M., Cebi, S., 2009. Analytical HFACS for investigating human errors in shipping accidents. Accident Analysis and Prevention 41, 66–75.

Dambieer, M., Hinkelbein, J., 2006. Analysis of 2004 German general aviation aircraft accidents according to the HFACS model. Air Medical Journal 25 (6), 265–269.

Dekker, S., 2006. The Field Guide to Understanding Human Error. Ashgate Publishing Company, Aldershot, Great Britain.

Dismukes, K., 2009. Human Error in Aviation. Ashgate Publishing Company, Aldershot, Great Britain.

ElBardissi, A.W., Wiegmann, D.A., et al., 2007. Application of the human factors analysis and classification system methodology to the cardiovascular surgery operating room. The Annals of Thoracic Surgery 83, 1412–1419.

Gaur, D., 2005. Human factors analysis and classification system applied to civil aircraft accidents in India. Aviation, Space, and Environmental Medicine 76 (5), 501–505.

Heinrich, H.W., Petersen, D., Roos, N., 1980. Industrial Accident Prevention: A Safety Management Approach, fifth ed. McGraw-Hill, New York.

Iden, R., Shappell, S., 2006. A human error analysis of U.S. fatal highway crashes 1990–2004. In: Proceedings of the 50th Annual Meeting of the Human Factors and Ergonomics Society, San Francisco, CA.

Jennings, J., 2008. Human factors analysis and classification: applying the Department of Defense system during combat operations in Iraq. Professional Safety 44–51.

Krulak, D.C., 2004. Human factors in maintenance: impact on aircraft mishap frequency and severity. Aviation, Space, and Environmental Medicine 75 (5), 429–432.

Li, W.-C., Harris, D., 2005. HFACS analysis of ROC Air Force aviation accidents: reliability analysis and cross-cultural comparison. The International Journal of Applied Aviation Studies 5 (1), 65–72.

Li, W.-C., Harris, D., 2006. Pilot error and its relationship with higher organizational levels: HFACS analysis of 523 accidents. Aviation, Space, and Environmental Medicine 77 (10), 1056–1061.

Li, W.-C., Harris, D., et al., 2008. Routes to failure: analysis of 41 civil aviation accidents from the Republic of China using the human factors analysis and classification system. Accident Analysis and Prevention 40 (2), 426–434.

Paletz, S., Bearman, C., Orasanu, J., Holbrook, J., 2009. Socializing the human factors analysis and classification system: incorporating social psychological phenomena into a human factors error classification system. Human Factors 51 (4), 435–445.

Pape, A.M., Wiegmann, D.A., et al., 2001. Air Traffic Control (ATC) related accidents and incidents: a human factors analysis. In: 11th International Symposium on Aviation Psychology, Columbus, OH, the Ohio State University.

Reason, J., 1990. Human Error. Cambridge University Press, New York.

Reason, J., 1997. Managing the Risks of Organizational Accidents. Ashgate Publishing Company, Aldershot, Great Britain.

Reinach, S., Viale, A., 2006. Application of a human error framework to conduct train accident/incident investigations. Accident Analysis and Prevention 38, 396–406.

Shappell, S., Wiegmann, D., 2009. Developing a methodology for assessing safety programs targeting human error in aviation. The International Journal of Aviation Psychology 19, 252–269.

Tvaryanas, A.P., Thompson, W.T., et al., 2006. Human factors in remotely piloted aircraft operations: HFACS analysis of 221 mishaps over 10 years. Aviation, Space, and Environmental Medicine 77 (7), 724–732.

Wiegmann, D., Shappell, S., 2003. A Human Error Approach to Aviation Accident Analysis: The Human Factors Analysis and Classification System. Ashgate Publishing Company, Aldershot, Great Britain.

Woods, D., Dekker, S., Cook, R., Johannesen, L., Sarter, N., 2010. Behind Human Error. Ashgate Publishing Company, Aldershot, Great Britain.

Major Incident Scene Management

J Horswell, Approved Forensics Sendirian Berhad, Selangor, Malaysia

This chapter is reproduced from the previous edition, volume 1, pp. 428–432, © 2000, Elsevier Ltd., with an updated Further Reading section supplied by the Editor.

Background

There are three critical operational stages of a criminal investigation. These are as follows:

- control and coordination of the criminal investigation;
- the criminal investigation;
- the forensic investigation.

Normally the coordination of a major criminal investigation is delegated to a senior investigating officer; likewise a complex forensic investigation must be coordinated by a senior forensic investigator. To manage major incident scenes and multisited crime scenes the crime scene coordinator/manager must understand forensic science and criminal investigation. The crime scene manager must surround him- or herself with competent, skilled and qualified forensic investigators, not prima donnas.

This chapter only deals with the management of major incident scenes from a forensic investigator's point of view.

Scene Control and Coordination

Without proper control and coordination, information may not reach the crime scene investigator. This may lead to his or her efforts being aimless, and the leads uncovered may never be passed on to investigators for follow-up action. This is most important when the crime scene is large and there are several crime scene investigators present processing the scene, or where there are secondary scenes away from the primary scene. There must be a flow of information back and forth between the senior investigator and crime scene investigator. This is one of the functions of the crime scene manager.

Approach to Crime Scene Investigation

The examination of a crime scene and subsequent collection of potential physical evidence require special skills, knowledge, and aptitude. The manner in which a crime scene examination is conducted may be a critical factor in determining the success of an investigation. The proper examination of a crime scene requires a disciplined approach and systematic application of the various observation, recording, and collection techniques, as well as an in-depth knowledge of forensic science.

Examining a crime scene is often a demanding task, and in many instances requires physical and mental stamina as well as team leader/member skills.

Forensic science has become a powerful aid to criminal investigations, with courts placing much emphasis on the results. Accordingly, the manner in which evidence is collected, the observations made, and the results of tests and comparisons conducted are vigorously examined by the courts.

A systematic approach to crime scene investigation will ensure the following:

- good coordination between investigation and crime scene examination teams;
- an efficient, effective, and thorough examination;
- less fatigue;
- orderly recording and collection of potential evidence;
- effective observations and deductions.

Initial Assessment

Before attending the crime scene it is important to obtain the best possible assessment of the circumstances relating to the crime. It is also important to receive a briefing from the senior investigating officer who has been appointed to conduct the investigation. From a forensic viewpoint, a crime scene coordinator should be appointed. This person will remain the focal point of contact between all the players who will subsequently become involved in the forensic investigation. This person should be a senior crime scene investigator who will be responsible for chairing all subsequent meetings with investigating police and for the coordination of all aspects of the forensic investigation. This includes the allocation of human resources to multiple scenes.

Homicide will be discussed here as the model crime scene as this is the most serious crime against the person.

Forensic investigators are in the same position as investigating officers. They need answers to the same questions: Who? What? When? How? Where? Why? Some of these questions can be answered at the homicide scene.

 http://dx.doi.org/10.1016/B978-0-12-802718-9.00008-9

Who?

- Who is the deceased?
- Who reported finding the deceased?
- Who saw or heard anything?

What?

- What happened?
- What crime, if any, has been committed?
- What were the actions of the deceased?
- What were the actions of others involved?
- What are the witnesses saying?
- What injuries, marks, clothing, and personal property are on the deceased?
- What is the estimated time of death?
- What possible weapons were used?
- What cause of death information can be gleaned from the deceased?
- What was the manner of death?

When?

- When was the deceased discovered?
- When was the deceased last seen alive?
- When were the police notified?

How?

- How did the deceased get to the death scene?
- How long has passed between injury and death?
- How did the deceased sustain the injuries?

Where?

- Where was the body discovered?
- Where did the death occur?
- Where was the deceased last seen?
- Where did the injury/ies occur?
- Where were the witnesses during the incident?

Why?

- Why was the deceased at this location?
- Why was a crime committed?
- Why was this type of weapon used?
- Why did this death occur at this time?
- Why was the deceased found at this time?

The more detail that can be obtained about what happened, the easier it is to determine what resources are required in examining the scene and the significance which is to be placed on certain aspects of the evidence.

Scene Security

An initial assessment of the crime scene will be made by the first officers attending the scene. The scene will be secured by them to an extent based on the information available at the time. The crime scene manager, who will normally be a senior member of the crime scene staff, must attend the scene at the earliest possible opportunity to take charge of the management of the crime scene. He or she will usually be accompanied by a crime scene investigator or a team of crime scene investigators who will undertake the crime scene examination. The size of the crime scene/s will dictate the amount of resources allocated to the particular incident. It is imperative that the crime scene manager has the authority to allocate the amount of resources required.

Once the crime scene is handed over to the crime scene manager, a reassessment of the scene security should be made to ensure that it is adequate. There should be a formal protocol for the handing over of a crime scene. This ensures control and the maintenance of the scene's chain of custody.

It is an essential element of any prosecution where forensic evidence is involved, to prove the security of the scene and that it was maintained throughout the subsequent examination/investigation. Therefore the objectives of securing the crime scene are as follows:

- To prevent evidence being destroyed or contaminated.
- To ensure security of information; generally information is only released to the media by a media liaison officer or the senior investigating officer.
- To ensure chain of custody of the scene is maintained as is necessary with any item of potential evidence.
- To remove from the scene all unnecessary persons including police officers and the media. It must be remembered that the more people present, the greater the potential for contamination and destruction of evidence. Large numbers of persons present will also inhibit the proper processing of a crime scene.
- To ensure that all evidence has been recorded and recovered. This may include securing the scene until the results of the postmortem or scientific analysis are at hand.

There are a variety of methods for securing the crime scene, for example:

- posting guards;
- rope or printed tape cordons;
- the strategic placing of vehicles;
- the use of markers, flags, signs;
- locking rooms or areas within buildings or using the external walls of a building as the barrier;
- establishing safe walk areas (common approach path) with tape or purpose-built raised stepping plates.

Occupational Health and Safety

The well-being of the crime scene investigator/s is the primary responsibility of the crime scene manager. He or she must be aware of fatigue and well-being of the investigators. Appropriate protective clothing and equipment should be made available. Breaks should be organized for the forensic investigators and refreshments should be on hand during those breaks. Scene guards should also be part of the crime scene operation, regardless of the area they originate from. There should be an area designated where food and drink can be taken, equipment can be stored, and rubbish can be accumulated.

All personnel on-site should be briefed regarding the following:

- safety hazards;
- smoking and eating;
- the location of critical areas;
- the use of telephones and toilets.

Systematic Collection of Potential Evidence

After the general survey of the crime scene, the sequence in which evidence is to be collected and areas searched should be apparent. Priority should be given to the following:

- any items that are in danger of being destroyed by wind, rain, vehicles, animals, tides, and human movement;
- the collection of any evidence that will enable access to a deceased or any critical area of the crime scene, such as along entry and exit paths;
- those critical areas of the crime scene that may render most evidence or, once processed, enable the removal of a body, or the remainder of the examination to begin;
- any area that may give a quick indication as to the identity of the offender/s;
- any areas that, when processed, will permit the release of scene guards and other resources;
- a general examination of the remainder of the crime scene for potential evidence.

Systematic and Sequential Approach to the Search and Recovery of Potential Evidence

In establishing the manner and sequence of collecting potential evidence, consideration must be given to both, the possible destruction of evidence and to the approach which will yield the best result in terms of useful information. Consultation with other specialists, such as forensic scientists and forensic pathologists, as to the sequence and method of collection may be necessary to ensure the best result; however, this may not always be possible at the scene.

Some examples of collection procedures are as follows:

- Macroscopic evidence should be collected from an area before it is powdered for fingerprints.
- Bloodstains and forensic evidence should be collected from an area before searching for fingerprints.
- Sweepings from the floor around a safe need to be collected before the magna brush is used.
- Polished floors need to be examined first with oblique lighting to locate latent shoemarks/footprints.
- Visible fibers, hairs, and other trace material should be collected from an area before applying general collection techniques, such as tapelifts, sweeping, and vacuuming.
- Tapelift areas of interest before removing deceased persons, collecting sheets and blankets.

In searching critical areas, a search conducted in a clockwise or anticlockwise direction from a fixed point, or conducting a line strip search, makes for a systematic examination. A systematic approach reduces fatigue and ensures a more comprehensive search by minimizing the chance of missing potentially valuable evidentiary material.

Larger objects should be examined before smaller objects and all items should be packaged and labeled at the time of collection.

Examination Records

In order to conduct a thorough, systematic crime scene investigation a proforma should be developed for each activity. These will not be enlarged on here as each jurisdiction will have its own subtle differences, but a list of subject headings for each category of examination is given below. These should be pre-prepared checksheets that will provide the examiner with comprehensive notes taken during the examination; these proforma records should be available for the following:

- crime scene log—activities undertaken at the scene, including movements in and out of the scene;
- formal handover of the crime scene;
- list of environmental conditions at the crime scene;
- list of activities and observations at the crime scene;
- exhibit list;
- rough sketch of the crime scene;
- photographs taken at the scene;
- list of specialists attending and the number of times they were involved in the examination.

Ongoing Case Management

Once the scene work is completed, the emphasis changes to the coordination of further examinations and the communication and flow of information of the results from forensic examinations to investigators, and from investigators to forensic examiners. If it is not practical for the crime scene coordinator to chair further case management meetings, another senior crime scene investigator may be nominated to maintain that contact and coordination of the ongoing case management.

Summary

Management of major or minor crime is a matter of seizing control of the crime scene and then the coordination of resource management, along with a systematic approach to processing the scene.

Major crime scenes vary in size and complexity. Some may require many crime scene investigators; others which are uncomplicated may require only one or two crime scene investigators.

Overall, scene management and the maintenance of a two-way communication flow are the two essential ingredients to successful scene management. Regular case management meetings must be held to keep all stakeholders abreast of the latest available information. These must be recorded in the case notes.

See also: **Chemistry/Trace/Forensic Geosciences:** Crime Scene Considerations; **Investigations:** Crime Scene Analysis and Reconstruction; Recording; Recovery of Human Remains; **Management/Quality in Forensic Science:** Risk Management.

Further Reading

Fisher, B., Fisher, D., 2003. Techniques of Crime Scene Investigation, seventh ed. CRC Press, Boca Raton, FL.
Fisher Barry, A.J., 1993. Techniques of Crime Scene Investigation, fifth ed. CRC Press, Springfield, IL.
Goddard, K.W., 1977. Crime Scene Investigation. Reston, Reston, VA.
Horswell, J., 1993. Crime scene investigation. In: Freckelton, Selby (Eds.), Expert Evidence. Law Book Company, North Ryde, NSW.
Kirk, P., 1974. Crime Investigation, second ed. Wiley, Chichester.
Svensson, A., Wendel, O., 1974. Techniques of Crime Scene Investigation, second ed. Elsevier, New York.

Crime Scene Analysis and Reconstruction

WJ Chisum, (Retired) President of California Association of Criminalists and American Society of Crime Lab Directors, Elk Grove, CA, USA

Glossary

A Clue It is a physical item that, due to its environment and its relationship to the incident in question, yields information about that incident.

Bias It is a preference or an inclination, especially one that inhibits impartial judgment.

Critical thinking It is the intellectually disciplined process of actively and skillfully conceptualizing, applying, analyzing, synthesizing, and/or evaluating information gathered from, or generated by, observation, experience, reflection, reasoning, or communication, as a guide to belief and action. In its exemplary form, it is based on universal intellectual values that transcend subject matter divisions: clarity, accuracy, precision, consistency, relevance, sound evidence, good reasons, depth, breadth and fairness.

Emergency responses Concentrate on what society expects in a life-threatening situation. First, the scene must be secure from further threats by ensuring that the offender is no longer present; second, is the medical preservation of life; third, is the investigative phase which has as one goal the preservation of evidence.

Evidence dynamics It is any influence that adds, changes, relocates, obscures, contaminates, or obliterates physical evidence, regardless of intent.

Staging It is when evidence has been purposefully altered by the offender to mislead authorities and/or redirect the investigation.

> Physical evidence cannot be wrong; it cannot be perjured; it cannot be wholly absent. Only in its interpretation can there be error. Professor Paul L. Kirk.

Crime Analysis and Reconstruction

Crime analysis has been defined in several different ways depending upon the role the administrator, investigator, or forensic scientist has, which is defining it. This chapter will discuss crime analysis as a basis for crime reconstruction using the physical evidence.

Crime analysis starts within the confines of yellow barrier tape under the watchful eye of law enforcement officers and investigators and continues throughout the documentation and collection of the evidence. It is a broad inquiry that seeks to establish the record of the physical evidence produced during an event or a series of related events. It is ultimately the unified result of aggregated crime scene examination and processing efforts; forensic laboratory examinations and analyses; medicolegal examinations and analyses; and collateral victim–suspect history. Crime scene analysis is used to establish the facts; the forensic scientist then uses those facts to determine how the crime proceeded.

Crime scene analysis depends upon objective observations without premature interpretation that may be influenced by otherwise harmless bias. Bias is a natural component of human thought processes and takes many forms. To be able to recognize and minimize bias is a prerequisite for reconstruction. The reconstructionist that does not go to the crime scene is impaired by the biases of those selecting the evidence to be analyzed.

For the purpose of reconstruction, evidence is anything that supports or refutes anyone's theory about any of the phases that occur in the precommission, commission, and postcommission of a crime.

Phases of a Crime

Crimes have certain behavioral processes or phases in common, of course, not all of the phases will be present in every crime. The following are the phases that may occur:

Fantasy: The person thinks about the crime, whether it is getting money and what can be done with it or the rape or murder of a person. Thinking about these crimes is fantasizing. They may write about their thoughts, even on the Internet, or draw sketches giving information regarding their reasons for the crime, they are thinking about doing.

Planning: When the person decides to actually commit the crime they may discuss it with others in person, on the telephone, or on the Internet. They may even prepare a kit

 http://dx.doi.org/10.1016/B978-0-12-802718-9.00009-0

for committing the crime or purchase a gun. Not all crimes are planned in advance, sometimes they are spontaneous, occurring as the opportunity presents itself.

Approach: This may take different forms in the approach to a victim; anything from a sudden attack to a subtle con as used in the movie "The Silence of the Lambs" to get the girl into the van.

Control: The perpetrator has control of the situation. This may be in the form of transportation, such as inside a vehicle, has tied the victim, has locked the doors, or in a nonviolent crime, has gained entry without setting off an alarm.

Aggression: The perpetrator now starts the attack. This is the primary criminal act such as robbery, rape, assault, homicide, or other crime.

Response: The victim responds to the act of aggression. The victim may fight, breaking the control of the perpetrator, taking control of the situation themselves such as obtaining a weapon or by talking the perpetrator out of further acts, or attacking the perpetrator with violence. The victim may also panic, resulting in running away or assuming a fetal position.

Disposal/display: The perpetrator may continue aggressive acts on the victim resulting in "overkill" or in necrophilia. They may try to cover up the crime by dismemberment and/or disposal. They may pose the victim to satisfy their fantasy or for further "embarrassment" to the victim when they are found.

Staging: The perpetrator may rearrange the scene or plant something that shifts the blame away from him/her, to implicate someone else or to make a homicide appear to be a suicide or accidental death.

Flight: The perpetrator leaves the scene, perhaps taking a souvenir or trophy.

Alibi: The perpetrator devises a story to show that he/she could not be responsible.

Fantasy: The perpetrator fantasizes about what happened and then starts thinking about how to do it again (the serial criminal). They review the actions and think how they could improve on them and start the cycle over again.

Crime analysis should also include an evaluation of the victim. Why was this victim chosen? Why was this location chosen? These questions may or may not be answerable from the physical evidence at the scene. Some will require that the victim and witnesses be interviewed by the investigators. However, no statements should be accepted unless consistent with the physical evidence.

These phases should be kept in mind as the crime scene is analyzed. Later, when they have been noted, they should provide a basis for reconstruction and discovery of less-important evidence or lead to other analyses. Are there items of evidence that contribute to one or more of these phases? Do not assume for instance that the blood pattern on the wall is cast off from a weapon produced in the aggression phase. The victim's response may have been to grab the wound, then move their hand rapidly casting off the accumulated blood. The crime scene analysis needs to describe the weapons, wounds, locations, the scene, and the background of the victim before bloodstains are interpreted. The analysis of the crime scene should give the investigators information that can be used in interviewing witnesses and potential suspects. They need to be aware of the scene characteristics, physical dimensions, initial phases of emergency response and ensure that the subjects being interviewed correlate their stories with those facts.

Evidence Dynamics

One must enter each scene with an open mind, have no assumptions regarding the types of evidence that one may encounter. Assumptions are made though one may not realize one is making it. For example, one receives a call to respond to a crime scene and one is given information about the case. If one has seen cases of this type before, one knows what to expect at the scene. One tends to work the scene as one had done in the past. One sees what one expects to see. This attitude can result in serious impediments to crime scene analysis and to the investigation. For instance, if one is being sent to a death scene, one should not be told that it is a suicide. It could be staged to appear to be a suicide but may really be a homicide. As stated by Hans Gross "we must forget the past and work each case based only on what we see." This was a warning against one of the various forms of bias that is encountered.

One must realize, too, that the scene is not necessarily the same as it was immediately after the crime happened. The scene may have been changed by the perpetrator, victim, or the witnesses or by time and the environment. It has definitely been changed by the first two phases of emergency response. The first officers at the scene have the primary duty of protecting life: first their own, then the lives of others, and the duty to apprehend the perpetrator. The protection of evidence has low priority. Preservation of fingerprints on a closet door knob are not a concern, and should not be, when the officer opens the closet to see if the perpetrator may be hiding therein.

The second phase of emergency response is the life-saving efforts of medical technicians or others giving CPR or other life-saving measures. Evidence is frequently destroyed, altered, or created by these activities. Investigators, technicians, and forensic scientists who respond to crime scenes must be cognizant of these changes.

The first officer must be interviewed before the crime scene is entered. Find out what has been changed by the first responders. If the officers moved a weapon or lifted the body to search for one, the evidence will be altered. Giving CPR to a victim will cause blood patterns that can be confusing. Such actions can cause misinterpretation of the evidence in the crime scene analysis. Time will be wasted analyzing the bloodstain patterns caused by the actions of the EMTs transporting the victim. Interview someone who is familiar with the scene area to see what has been changed by the actions of the offender(s) and the victim(s).

The victim is a part of the scene. Movement of the body may change bloodstain patterns. The transportation of the body to the morgue in a body bag will change and destroy pattern and trace evidence. All these changes have to be anticipated. Some are predictable but others are not, such as insect or animal damage to the corpse.

The autopsy constitutes a secondary scene. The crime scene analyst will have a different background than the medical examiner (hopefully a board certified forensic pathologist) and, therefore, will look for different things. Teamwork and open communication by the analyst, investigative personnel, and the pathologist will result in maximum information from autopsy. In some jurisdictions, the doctor does not allow outsiders in the autopsy. Yet the doctors should have the scene described to them and need to communicate with the investigators in some manner as the autopsy is being conducted.

The Reconstructionist needs all the information possible in order to develop the who, what, where, when, how, and sometimes the why of the crime. The crime scene must be thoroughly documented. New technology based on the "total station" and high-definition digital photo integration producing three-dimensional holograms of the crime scene are possible, but must be supplemented with sketches and photos on the computer. It has been said that "any reconstruction can only be as good as the information provided." New instruments provide more detailed information but isolation of certain photos may be necessary to see relationships that build a reconstruction.

Role of Physical Evidence

There are several different classifications used by forensic scientists to classify the evidence. Some classify based on the types of analyses that will be undertaken such as trace, biology, chemistry, etc. Others by the types of crimes, homicide, assault, burglary, etc. The reconstructionist benefits from recognizing the role, the physical evidence plays in the reconstruction. The use of this classification has the analyst starting the reconstruction process when the evidence is first recognized.

Evidence may be transitory, relational, or functional.

- Transitory evidence is something that does not remain, it could be an odor, water which evaporates, ice, a burning cigarette, clothing position of the victim, temperature etc. If transitory evidence is not noted upon arrival at the scene, it will be lost.
- Relational evidence has a relationship to other items at the scene and/or a relationship to the crime. Essentially it is clue evidence. A pistol, bullet, casing, powder, and trajectory are related to each other.
- Functional evidence is how something works. It can be mechanical, chemical, electrical, electronic, or biological.

The evidence may be further subdivided into the role it plays in the crime:

- *Action* evidence establishes direction and velocity. It shows what the actors did or what was done to them.
- *Location* evidence establishes the positions of the actors or where some action took place.
- *Directional* evidence establishes the direction something was going or the direction it came from.
- *Sequencing* evidence establishes the order of occurrence of events.
- *Limiting* evidence establishes the crime scene.
- *Ownership* evidence establishes the owner of the items.
- *Implied* or derived evidence is the absence of something.
- *Temporal* evidence establishes the duration of an event, distance, and when a crime occurred.
- *Psychological* evidence is something at the crime scene that is added, removed, or arranged in a particular way to satisfy a need of the perpetrator.

The above classification scheme is part of the crime analysis. By listing the items of evidence and assigning them the roles that they play in the crime, new relationships and logical conclusions can emerge which will aid in the formation of a reconstruction.

Reconstruction—Historical Perspective

To reconstruct a crime, or any event, one must be able to see the effects and infer the cause. People observe the cause and then see the effects, but most do not do the reverse.

> In solving a problem of this sort, the grand thing is to be able to reason backward. That is a very useful accomplishment, and a very easy one, but people do not practice it much. In the every-day affairs of life it is more useful to reason forward, and so the other comes to be neglected. There are 50 who can reason synthetically for one who can reason analytically (Doyle, 1887).

Sherlock Holmes was the epitome of the "scientific detective." He conducted experiments and had collections of various materials so that he could identify not only what they were but, in many instances, where they came from. He also used scientific thinking to solve cases. Sherlock Holms was fictional; Arthur Conan Doyle wrote the stories. He apparently based the methods on his mentor Dr Joseph E. Bell.

Did Doyle come up with reconstruction on his own or was it part of the investigative processes at the time? The stories show how clues are used to "think backward" and reconstruct the incidents. "Even the forensic scientist himself owes his existence and attitudes a good deal more to Conan Doyle, and a good deal less to Newton and Darwin, than he is usually prepared to admit" (Kind, 1977).

Scientific reconstruction of crimes was practiced in the 1930s. Henry Rhodes stated that "from one point of view it might be said that the technique of reconstruction is a modern development in the detection of crime." Its object was to decide how the crime was probably committed and the order of events. "Scientific instruments and expert diagnosis make possible increasingly precise technique of reconstruction. It is a necessary part of the study of any modern crime concerning which there is an element of doubt" (Rhodes, 1933).

While an element of doubt seems to be a questionable criteria for needing to reconstruct, forensic scientists seldom get to see those cases where there is no element of doubt. They do not get the case unless someone wants them to develop some information about it. "Information can only be received where there is doubt; and doubt implies the existence of alternatives—where choice, selection, and discrimination is called for."

The "Wizard of Berkeley," Edward O. Heinrich was a practicing reconstructionist in the 1930s–40s. He stated that his method of criminal investigation was:

> Crime analysis is an orderly procedure. It is precise and it follows always the same questions that I ask myself. Let us consider what they are:
>
> Precisely *what* happened?
> Precisely *when* did it happen?
> Precisely *where* did it happen?
> *Why* did it happen? *Who* did it?

The average investigator seems to give immediate attention to the *why* and *who* but he takes what happened for granted. We must analyze the method of the crime before we can analyze its purpose and look for the criminal.

> I have always held that the criminal virtually labels every crime he commits. My procedure is to reconstruct the crime by visualizing the habits and actions of the criminal. I do this by using the debris that the criminal leaves behind and relocating it with respect to the criminal episode. So I make the natural sciences interpret what I have observed (Bloch, 1958).

Heinrich did not teach classes on Forensic Science even though he was a Professor of Criminology at UC Berkeley. That role fell to Paul L. Kirk. Professor Kirk was also interested in the how, when, what, and where, more than he was in the why and who of crimes. He instilled a questioning attitude in his students that resulted in many of them becoming leaders in the forensic field.

Who Does Reconstruction

A well-documented scene is the primary requisite for a reconstruction. Any evidence not collected may result in an incomplete or, worse, an erroneous reconstruction. Therefore, any bias that influence the recording of the crime scene or the collection of the evidence will cause errors. Unfortunately, the forensic scientists who respond to crime scenes are diminishing. Evidence technicians, for the most part, do not have the same approach. Too many are told what to do and what to collect. There are also some who are very competent and do try to eliminate any bias from their crime scene work and reconstruct the events of a crime. However, they lack the science that allows them to interpret the crime laboratory findings. This too, is not a new situation, O'Hara and Osterburg described the laboratory analyses as "a routine affair which may be entrusted to an ordinary technician," and stressed "the importance of field work for the working criminalist" (O'Hara and Osterburg, 1949).

Detectives frequently do preliminary reconstructions. They then select evidence that supports their theory of the crime. Unfortunately, the physical evidence has not been analyzed and the detectives are basing their reconstructions on experience. "The use of one's experience as a knowledge base for developing reconstruction hypotheses and theories, or for inferring the cause from an examination of the effects, is commonplace. It is consequently important to bear in mind that not all experience is equal, not all experience is sufficiently instructive, and not everyone learns from his or her experiences" (Chisum and Turvey, 2007). The purely "experience-based reconstructionist" may give examples of his conclusions to demonstrate how he reconstructs but often will not be able to show the logic and science behind his methods. The absence of such "long division" in his work is in effect an absence of science. In addition, if an arrest is made or a suspect is identified, then the detective concentrates on proving his theory and that this suspect is guilty.

If there is no forensic scientist looking at the case, attorneys will form a reconstruction to present their evidence to the jury or court. The DA has conviction of the crime as the primary goal. The reconstruction they make will make the crime heinous and the defendant's role in it as significant as possible. The defense will take the opposite view and try to present a reconstruction that eliminates any possibility that the defendant is responsible for the crime. Attorney's have no scientific basis for their reconstructions, just that they support their side of the case. The reconstruction should not be left until the final moment (court) before it is presented.

Methods of Reconstruction

The reconstruction of a crime is to reconstruct several small things first. For example, in a shooting case: the bullet bearing the victim's DNA is compared to test bullets; the weapon is fired (with the same ammunition) at various distances to determine the muzzle to target distance from the gun powder residue; the angle through the body; the location of blood spatters and their nature; and the location of any casings from the weapon are used to establish the position of the shooter relative to the victim. This is utilizing the scientific information in reconstruction of a shooting (a singular event or event segment not the entire crime). Each hypothesis is now tested against the stories told by the shooter, the victim, if alive, and any witnesses. It is also tested against any hypotheses the police or prosecutors may have. The individual laboratory tests that were done are used to strike down any hypothesis that has a conflict with the scientifically developed information. This is the primary use of reconstruction, *premise negation*.

The hypotheses that were proposed by the laboratory analyses are that the DNA match on the bullet shows that the bullet struck the victim; that the blood spatter pattern has significance; the powder patterns do give the distance; and the casings eject in a certain manner. These are the things that are tested in the experimental phase. The results are, provided they are done correctly, as scientifically valid as possible. There may be other evidence that will shed more light on what happened. However, one is not looking to establish a theory of reconstruction; one is eliminating the stories that are being told about the case. This is the primary use of reconstruction and builds a rapport with the investigator. One uses the physical evidence to develop what are called facts that are used to test the theories or stories about the incident.

One method is to take all the stories that have been presented. List them across the top of a table as "theory 1," "theory 2,"... Then list the physical evidence derived information in column 1. Mark the chart for each item of information with a "C" for consistent, "X" for excluded, "I" for inconclusive. This works well when witnesses tell multiple versions of events or when there is a disagreement between investigators. It is a simple way of using reconstruction, if all the versions are excluded, refine the elements that are consistent to see if another theory can be developed. This is essentially the same as the "Storyboard" concept that Rynearson uses in his "play" analogy process.

Brian Parker stated that "the goal of the crime lab is to prove the investigator wrong." This philosophy may sound wrong to many, but it is the basis of tests. If one adopts this philosophy then one will be eliminating much of the bias that is inherent in almost all endeavors we undertake. But, one must extend that philosophy to all theories that one is presented with, even those of colleagues.

Bevel and Gardner have developed an event-sequencing methodology for reconstructing crimes. They define the overall crime as an incident, events are major aspect of the occurrence, and specific actions as event segments.

A crime scene analysis then is reverse engineered. We use evidence and data to define event segments. Since this reconstruction method attempts to define the crime or incident by establishing these events and event segments, we call the process event analysis. The steps taken to create a crime scene reconstruction using event analysis are as follows:

- collect data,
- establish specific event segments (time snapshots),
- establish which event segments are related to one another,
- sequence those related segments, establishing a flow for that event,
- consider all possible sequences, auditing the evidence when necessary to resolve contradictions,
- based on the event segment sequence, establish final order of the events themselves
- flowchart the entire incident and validate the sequence.

One of the criticisms of this methodology was that it was quite similar to the multievent sequencing methodology developed by Gustav Benner many years ago. The two have got together and shared information. This has resulted in Mr Benner making suggestions as how his flowcharts and methods could be adapted for crime scene reconstruction. This produces a nice charted sequence of events. However, the sequence cannot be developed without information (not data–data are used to develop information). The relationships that exist between items are not always obvious nor are the "event segments."

Event segments must also be studied with scientific methods. Each event segment is a hypothesis and will have alternatives. These alternatives must be eliminated so that the event segment becomes the most probable. Failure to follow the process with all inferences (hypotheses) will not result in a meaningful reconstruction.

There are several approaches to understanding relationships between items of evidence. Critical thinking is needed which means using imagination, logical, and scientific reasoning processes. It also entails "thinking outside the box," after all the incident one is reconstructing is something that occurred "outside the box" whether one defines the "box" as the law or the norms for society.

Faye Springer of Sacramento uses a 3 × 5 card for each item of evidence. She lists what is known about that item on the card. Then she spreads the cards and associates them as she sees things in common. This is essentially a manual technique of *mind-mapping* which also has proved to be a useful technique.

The author uses "The Brain®" software. The Brain® gives one the ability to see relationships and to do event analysis on the computer. It helps to see data, develop information, and put it in a logical order. The full version (as opposed to the free version) allows photos, files and Internet sites to be incorporated into the matrix. These methods can also use the classification scheme for evidence (directional, temporal, etc.) as given previously.

If there is a group, then *brainstorming* is a technique whereby several minds are working on the problem. This has proved to be a good technique when done properly. There are instructions on how to brainstorm on the Internet.

Joe Rynearson proposed that a crime scene is like a play.

> The commission of a crime can be likened to a movie or a play. There are actors, the stage, backdrops, acts, and scenes, as well as dialogue and actions revolving around a script which has a defined plot. Each scene also includes "stage directions," which give additional information as to where actors should stand, move, and how they should act. If we understand the stage (the crime scene), then we have a chance to learn what actions may coincide with them. A crime scene reconstruction may portray specific actions and perhaps even place them in sequence; however, it is unlikely that a continuous "flow of actions" (like the script of the play) can ever be achieved. Though some investigators would represent a crime reconstruction as a continuous portrayal of the entire "plot and script of the play," this is not possible. The very best that can actually be achieved is to develop the separate (and brief) "stage directions" for each scene of the play and of *ALL* of the logical alternatives to connect them together.

To summarize, the forensic scientist studies the reports, photos, and diagrams. They use logical inferences to come up with hypotheses for all kinds of causes for the effects seen. They then eliminate hypotheses until they have eliminated all the alternatives but one. When they "have eliminated the impossible then whatever remains, however improbable, must be the truth" (Doyle, The Sign of Four, 1890).

See also: **Behavioral:** Criminal Profiling; **Foundations:** Forensic Intelligence; History of Forensic Sciences; Overview and Meaning of Identification/Individualization; Principles of Forensic Science; Semiotics, Heuristics, and Inferences Used by Forensic Scientists; Statistical Interpretation of Evidence: Bayesian Analysis; The Frequentist Approach to Forensic Evidence Interpretation; **Investigations:** Collection and Chain of Evidence.

Further Reading

Bevel, T., Gardner, R., 2008. Bloodstain Pattern Analysis: With an Introduction to Crime Scene Reconstruction, third ed. CRC Press, Boca Raton.

Cherry, C., 1957. On Human Communication. MIT Press Classic, Cambridge.

Chisum, W., Turvey, B., 2007. Crime Reconstruction. Academic Press, San Francisco.

Doyle, A., 1887. A Study in Scarlet. Berton's Christmas Annual, London.

Doyle, A., 1890. The Sign of Four. Doubleday, London.

James, S., Nordby, J., 2005. Forensic Science: An Introduction to Scientific and Investigative Techniques, second ed. CRC Press, Boca Raton.

Kind, S., 1977. The Scientific Investigation of Crime. Forensic Science Services, Ltd, Harrogate.

Kirk, P., 1953. Crime Investigation: Physical Evidence and the Police Laboratory. Interscience, New York.

O'Hara, C., Osterburg, J., 1949. An Introduction to Criminalistics. Indiana University Press, Bloomington. www.forensic.demon.co.uk.

Springer, F., 1993. Private Communication.

Parker, B., 1971. Private Communication.

Rhodes, H., 1933. Clues and Crime. John Murray, London.

Rynearson, J., 1997. Evidence and Crime Scene Reconstruction, fifth ed. National Crime Investigation and Training, Redding, CA.

Turvey, B., 2008. Criminal Profiling: An Introduction to Behavioral Evidence Analysis, third ed. Elsevier Academic Press, San Francisco.

Key Terms

Accident analysis, Accident investigation, Admissibility, Casting, Chain of evidence, Collision analysis, Collision model, Continuity, Control, Crash test, Crime phases, Crime reconstruction, Crime scene investigator, Crime sequencing, Criminal investigation, Critical thinking, DNA, Emergency response, Energy equivalent speed, Event analysis, Evidence, Evidence dynamics, HFACS, Human error, Human factors, Impacts, Incident investigation, Initial assessment, Internal assessment, Multi-event sequencing, Numerical methods, Passenger loads, Personal injuries, Physical evidence, Possession, Premise negation, Prevention analysis, Quality management, Sample, Scene control and coordination, Sequence, Staging, System safety, Tape lift, Tire marks, Vehicle damages.

Review Questions

1. What is a chain of custody?
2. Why is a chain of custody important in forensic engineering?
3. If the chain of custody is "broken," what are the consequences?
4. What are the questions that must be answered in an accident reconstruction?
5. What are some of the reasons for reconstructing accidents?
6. Why are Finite Element programs used only in a small number of accident reconstructions?
7. How can the point of driver reaction be determined?

8. What proportion of accidents and incidents are accounted for by human error? Why is this figure somewhat misleading?
9. What is HFACS?
10. What is the "Swiss Cheese" model of error causation?
11. How can an organization influence forensic analyst performance?
12. Why is a formal system of accident investigation necessary?
13. What is a "major incident scene"?
14. What are the challenges of scene control and coordination of a major scene, especially a large one?
15. What is the purpose of scene security given that a scene is controlled?
16. According to Chisum, what phases may occur in a crime?
17. What is the difference between transitory, relational, and functional evidence?
18. Why is the primary use of reconstruction "premise negation"?
19. Why is avoiding bias and having an open mind important in a forensic engineering case?
20. Why is experience alone not enough to conduct accident reconstruction?

Discussion Questions

1. What factors could limit the reconstruction of an accident? How could those factors be mitigated?
2. Why is collecting more data after a safety incident or accident not helpful?
3. How is the "Swiss Cheese" model used? Why is it an important concept in error investigations? Does it apply in all instances? Why or why not?
4. What would be the purpose of a "formal handover" of a major crime scene? Why would this be important? How could this facilitate ongoing case management?
5. How is a crime scene like a play? Does this analogy work? Why or why not?

Additional Readings

Houck, M., Crispino, F., McAdam, T., 2012. Science of Crime Scenes. Elseiver.

Kassin, S.M., Dror, I.E., Kukucka, J., 2013. The forensic confirmation bias: problems, perspectives, and proposed solutions. Journal of Applied Research in Memory and Cognition 2 (1), 42–52.

Peng, Y., Yang, J., Deck, C., Otte, D., Willinger, R., 2014. Development of head injury risk functions based on real-world accident reconstruction. International Journal of Crashworthiness 19 (2), 105–114.

Püschel, K., 2015. Forensic medical aspects of traffic accident reconstruction. Journal of Local and Global Heath Science 2015 (2), 94.

Struble, D.E., 2013. Automotive Accident Reconstruction: Practices and Principles. CRC Press.

Section 3. Transportation Investigations

> Any design, whether it's for a ship or an airplane, must be done in anticipation of potential failures.—Henry Petroski

Planes, trains, and automobiles, as the saying goes. Humans have relied on motorized transportation for over a century.[1] With vehicles come not only mechanical failures but also human ones; mass transit leads to mass fatalities. Much like a fire scene, the scene of a transit incident can be difficult to sort out (as well as emotionally troubling) due to damage to the vehicles and the surrounding environment.

The administrative and technical model for these types of investigations is a unique body called the National Transportation Safety Board (NTSB). The NTSB, established in 1967, conducts independent investigations of all civil aviation accidents in the United States and major accidents in the other modes of transportation. The NTSB is independent of government agencies related to transportation, such as the Department of Transportation, or affiliated organizations, including the Federal Aviation Administration. The NTSB has no regulatory or enforcement powers. The NTSB's analysis of factual information and its determination of probable cause cannot be entered as evidence in a court of law: This is by design to keep it above the fray of individual cases. If a transportation accident is determined to be a criminal act, the FBI takes over the investigation and the NTSB can advise.[2]

Forensic engineering is intimately involved with digital evidence, given the increasing prevalence of data recorders ("black boxes," although the transportation industry almost never uses this term and also, they are bright orange in color) in airplanes, locomotives, and passenger vehicles The marriage of digital information and physical reconstruction yields highly precise answers but massive amounts of data. Keeping up with modern materials, technical design, and "big data" presents a formidable challenge to forensic engineering.

Railroad Accident Investigation and Reconstruction

RW Halstead, IronWood Technologies, Inc., Syracuse, NY, USA

Introduction

This chapter provides a broad introduction to the field of railroad accident investigation and reconstruction. The term "investigation" denotes the manner in which the scene of a railroad-related accident is secured and relevant forensic data are collected and preserved, whereas "reconstruction" denotes the application of scientific principles and analysis to the data collected so as to determine with the greatest degree of precision possible the following:

1. The sequence of factors that led up to the accident.
2. Whether the accident could have been prevented or its severity reduced through improvements in operating procedures, engineering design, or maintenance practices.

The "investigator" and "reconstructionist" may be, and often are, the same person. The final railroad accident reconstruction product may be used by governmental agencies contemplating additional regulatory activity or evaluating the

[1] Although the first stationary gasoline engine, developed by Carl Benz, ran for the first time on New Year's Eve 1879 (www.daimler.com), motorized vehicles did not become accessible to the populace until the early twentieth century.

[2] www.ntsb.gov.

effectiveness of existing regulations. It may be used by railroad companies to improve their internal engineering and operating procedures and may be used to prosecute or defend criminal or civil legal proceedings.

Railroad accidents, as many other kinds of accidents, are rarely the product of a single factor or event, but most often result from a sequence of events, the removal of any one of which precludes the accident from occurring. Thus, it is incumbent on the railroad accident investigator and reconstructionist to cast a wide net for potential causal events, for any one of which might prove crucial to an understanding of an accident event and its future prevention.

While this chapter is modeled on North-American (primarily United States) operating and engineering practice, the investigative and reconstruction techniques set forth herein have broad application to the science of railroad accident reconstruction worldwide. As it is not possible to cover the entire breadth of the field of railroad accident investigation and reconstruction in any depth in this chapter alone, this chapter instead offers detailed information on a number of specific areas of interest in the field, while inviting the reader to pursue related information via supplemental sources.

Types of Railroad Accidents

Railroad accident investigators and reconstructionists may be called upon to examine a variety of railroad-related accidents, including but not limited to

- collisions between two or more trains (head-on, rear-end, sideswipe),
- derailments,
- collisions between trains and vehicles or pedestrians,
- platform gap injuries,
- electrocution (railroad employee or civilian), and
- railroad employee injuries (United States: Federal Employer's Liability Act).

Factors that may have contributed, in whole or in part, to the accident may include

- excessive speed by train crew or motorist;
- inadequate sight distance for train crews and/or motorists at level crossings around vegetation or standing equipment on adjacent tracks;
- operating rules noncompliance by train crews;
- unfamiliarity of train crew with the section of track over which they were operating leading up to the accident;
- inadequate design and/or maintenance of track, level crossing surface, signaling systems for trains or motorists, or safety appliances on locomotive or cars;
- failure to thoroughly test signal system revisions or upgrades before placing in service;
- fatigue—train crew and/or motorist;
- impairment by drugs and/or alcohol;
- inability of motorist to accurately judge closing speed of train;
- impatience of the motorist at level crossings;
- improper operation of traffic signals at traffic intersections interconnected with active warning devices at level crossings; and
- failure to conduct safety briefing before performing task and/or failure to adequately supervise the performance of a potentially hazardous task.

Event Recorders

As time goes on and railroad signal and train control systems become more sophisticated, more and more systems are incorporating event-recording capability, many of which are useful to the accident reconstructionist.

Among the first sources of data typically pursued for preservation and analysis in a railroad accident, investigation is from the locomotives involved in the accident. In the United States, every leading locomotive (or cab car, if the train is being shoved by a locomotive at the rear) that exceeds 48.3 km h^{-1} (30 min h^{-1}) must be equipped with a functioning event recorder that must record at least the most recent 48 h of operation (49 CFR § 229.135a). Other countries may have their own event recorder installation and data retention requirements. The recorder may record data on a tape-based medium or in digital memory. Data extraction is usually performed through removal of a tape, memory module, or download of data to a laptop computer. Nonradar-based event recorders require the accurate measurement of the diameter of the wheelset from which the recorder is receiving data in order to facilitate data analysis.

Data from the locomotive event recorder are often encrypted and must be processed by the event recorder manufacturer's proprietary software for analysis. The date, time, and exact location of the locomotive at the time of data extraction should be noted, and a chain of custody document should be established and maintained as the data are handed off from person to person to the point of final analysis.

Locomotive event recorders typically sample their input channels and record data records in either a second-by-second manner or "on demand" (only when one or more data input channels change state). Data input channels typically include time, speed, distance, throttle and braking status, and wayside signal aspects. Newer digital recorders accept more input channels than do the older tape-based units and may include horn and headlight status and safety-critical messages from the train control system.

It is important to note that as the locomotive travels down the track, the (nonradar-based) event recorder samples its inputs and records data at its designed interval, but the recorder itself has no idea of where it is located along the track at the time at which each specific data sample is taken. In order to determine the train's location at specific times in the event recorder data, at least one specific event recorder data record must be correlated with a specific location along the track. This is commonly done by associating the point on the track at which the locomotive came to rest with the first event recorder data record that shows a speed of zero. Once correlated in this manner, the location along the track corresponding to any event in the event recorder data (brakes first applied, horn first sounded, etc.) can be quickly determined by adding or subtracting their corresponding distance values in the data.

Authentication of locomotive event recorder data can be accomplished by comparing the progression of the data over several kilometers or several tens of kilometers of territory before the collision point. Recorded speeds are compared to areas of known speed restrictions, horn soundings are compared to known level crossing locations, and stops (event records displaying a speed of zero) are compared with locations of known station stops to verify that the offered data most likely pertain to the subject accident and that the recorded measurement of wheel size is reasonably accurate.

Increasingly, locomotive event recorders are being equipped with the ability to record GPS (Global Positioning System) coordinates corresponding to each individual data record. When present, GPS data offer a means of independent verification of the analysis detailed above.

If more than one locomotive or leading cab control car is involved in the accident, data from all available event recorders should be obtained. The resulting data sets can be cross-compared to ensure data integrity and verify the accuracy of the wheel diameter measurements used for data interpretation.

Additional data useful in a railroad accident investigation may be extracted from the following:

- Train dispatcher's computer logs—recording all dispatcher control inputs and resulting status indications from field signaling equipment, typically located in the main dispatching center.
- Telephone and radio audio recordings of voice transmissions from dispatcher to train, train to dispatcher, and train to train.
- Signal system computer logs—typically stored in the field by solid-state interlocking controllers or equivalent and/or by computer-controlled level crossing control equipment.
- Nearby defect detectors—"hot box" (overheated wheel bearings), flat wheel, dragging equipment, or excessive dimension (high/wide). Detectors often record not only the existence of a defect, but also the speed of the train at that instant for comparison with corresponding locomotive event recorder data.
- Automatic equipment identifier locations—scanners read magnetic tags on the side of cars as the train passes—give a listing of each of the cars in the train (and their order) and the speed of the train as it passed through the scanner.
- Cellular telephone tower records—when obtained from the telephone company (in the United States, commonly referred to as "call detail records" and "cell tower" and "switch" records), they can be used to determine whether or not the train operator's cell phone was in use at a critical portion of the approach to the accident location. "Smart phones" that internally record GPS positioning data can also be used to position the train and/or motorist at critical points in the approach to the accident, particularly in cases in which locomotive event recorder data do not exist or are unusable.
- Locomotive video—although not required in the United States (yet), more and more major railroads are equipping their locomotives with on-board video cameras. The cameras record to a computer hard drive and are usually interfaced with the locomotive's existing event recorder.
- Surveillance video from surrounding cameras mounted on residential, commercial, or governmental structures that may have captured part or all of the accident scenes on video. Retention periods are typically short and timely data recovery is essential.
- Motor vehicles—most modern airbag systems record precrash data related to the driver's speed, use of brakes and seat belts, etc., in modules commonly referred to as an EDR (event data recorder) or SDM (sensing diagnostic module) that can be downloaded from the vehicle to a laptop computer through the use of specialized hardware and software.
- Motor vehicles—General Motors Corp. OnStar system or equivalent, if equipped, offers an independent record of the vehicle's GPS position as a function of time.

Once data are secured from all possible sources, it must be synchronized in order to assemble a comprehensive picture of the events leading up to the accident. To do this, a data source that uses the GPS time base is selected (if available), and all other data sources are synchronized to it by identifying one or more common events that appear across multiple event data sources, taking into account network delay and update frequency for each data source.

After all available event data have been accurately synchronized, a more complete picture of the events that led up to the accident begins to emerge. Movement of the respective vehicles can be plotted with respect to time, recorded precollision use of horn and braking can be evaluated, and the operation of the signal system can be recreated with a reasonable degree of accuracy. The synchronized event data can then be combined with physical evidence collected at the scene in a final analysis of the accident event.

Procedural Steps to Follow in Railroad Accident Investigation and Reconstruction

While there are many different kinds of railroad accidents, certain investigative procedures tend to be applicable to almost all. The generalized list given below represents typical investigative activities and is not intended to supersede more specific procedures that may already be in force in the locale in which the accident has occurred.

Collision With Motor Vehicle or Pedestrian

1. If the train is still at the scene, observe a demonstration of the leading locomotive or cab car's horn, headlight, ditch lights (if equipped), and bell.
2. Download the event recorder on every locomotive and cab car on the train. Note the exact time that the data were downloaded.
3. Download the video recorder on all equipped locomotives and cab cars.
4. Measure the distance from the point of impact to the front of the leading locomotive or cab car at rest. This distance value will be required to analyze the locomotive event recorder data. Mark the location of the front of the leading locomotive at rest on the side of the rail with a paint stick or equivalent, ensuring that that point will be preserved even after the train is moved.
5. Perform or observe testing of the active warning devices (flashers, gates, etc.) at the level crossing, if so equipped. Ensure that all governmentally required tests that apply to level crossing warning devices are performed, along with all additional tests required and customarily performed by the railroad entity that has primary maintenance and testing responsibility for the warning devices. Ensure that the results of all such testing comport with those expected for a nondefective, properly functioning system.
6. Download and preserve data from all computer-controlled warning device control equipment or auxiliary event recorders connected thereto.
7. Examine the pavement surface in the vicinity of the crossing—photograph, measure, and diagram all pavement gouges and tire skid marks attributable to the accident vehicle. Observe the condition of the pavement surface at the track for roughness, potholes, and so on.
8. If the level crossing is humped, take measurements sufficient to create a cross-sectional view of the crossing's profile, then compare its shape to generally accepted standards (United States: AASHTO standards).
9. Examine signage that governed the accident vehicle's approach to the collision for completeness, placement and visibility, and reflectivity (if at night).
10. Inspect the scene for the presence of vegetation or other fixed obstruction that may have impaired the motorist's ability to see down the track for approaching trains. If any is found, measure and diagram its relationship to the level crossing.
11. If the level crossing has active warning devices and they are interconnected with traffic signals at an adjacent intersection, perform or observe testing of the combined system to ensure that the traffic signals displayed the proper phases as required by the approach of the train.
12. Map the overall accident scene and debris field. Collect and analyze any debris that may have bearing on the accident, including all crossing signal equipment replaced immediately after the accident.
13. Collect statements from all eyewitnesses and others with potentially useful information.
14. Scan the area for security cameras oriented in the general direction of the level crossing. If any are identified, arrange to obtain the video footage from the time of the accident.
15. Collect the remains of any shattered light bulbs from the vehicle, locomotive, or crossing signals. It may be possible to perform a "stretched filament" analysis on them to determine if they were lit at the instant of impact.
16. If the level crossing is equipped with gates, collect video footage depicting the gates going up and down repeatedly. Back in the office later, the video tape can be run in slow motion and very precise gate operating times can be determined.
17. Take lots of photographs and video footage from a variety of angles and distances from the crossing. Use a 50 mm lens for all photography—it most closely replicates the perspective of the human eye.
18. Secure audio file copies of all radio and telephone footage between the train crew and train dispatcher (if applicable) for the 5–10 min leading up to the collision.
19. Note the weather at the time of the accident, as well as sun angle.
20. If the level crossing was equipped with active warning devices, examine the incoming commercial power feed—determine if commercial power was on or off at the time of the accident.
21. Determine if the motorist/pedestrian/train crew employees were tested for the presence of alcohol or drugs.
22. Determine if cell phones were in the possession of and/or were in use at the time of the accident.

Collisions Between Trains

Although their frequency has greatly decreased over the years through the increased use of advanced technology and operating practice refinement, train collisions still occur in the

United States and around the world. Potential collision causes include the following:

1. Train dispatcher mistakenly authorized two trains to occupy the same block of track simultaneously.
2. One or both train crews incorrectly interpreted track block authority, leading them to enter a section of track for which they had no authorization to occupy.
3. Signal system failed by displaying a "false clear" signal to one or both trains, a signal conveying information that was more favorable than intended. Alternatively, the proper signal was displayed but was misinterpreted by the crew due to sun glare or other missed vision cue.
4. Signal system functioned properly, but train crew failed to observe and/or act on restrictive signal information due to inattention, fatigue, or drug or alcohol impairment.
5. Braking system failure occurred on one or both trains.
6. Excessive speed by one or both trains, leading to an inability to stop short of a restrictive signal as required.

When the collision occurs in territory that is equipped with a signal system, all available signal system event recorders must be downloaded, and their data content must be synchronized and carefully interpreted to determine if any "false clear" signals were displayed to the accident crews. In addition, complete postaccident "service testing" must be performed on the signal system, in which all possible combinations of train routings, including those being conducted at the time of the accident, are simulated and the signal system outputs are compared to those intended.

Derailments

Derailments may be caused by track defects, equipment defects, signaling defects, poor train handling or other human factors, improperly loaded freight, improper train makeup (heaviest cars at rear of train), or any combination of these factors.

Upon arrival at the accident scene, an investigator will typically begin at the location of the wreckage and walk down the track opposite the train's direction of travel, looking along the rails and track structure for the point of derailment (POD). This is the first evidence of a wheel climbing the rail, commonly seen as a prolonged scuff or scrape mark along the top of the rail, followed by spike and tie damage as the wheel flange contacts and begins to ride along the top of the tie plates or ties. While the initial POD is commonly found within 100 m or so of the general wreckage, it may also appear up to several kilometers from the main wreck scene, indicating that derailed equipment was pulled for some distance in a derailed state before it finally "went off exploring."

Once the initial POD is located, the track structure for at least 50–100 m before the POD should be examined in detail for conformance with governmentally required and generally accepted industry standards for track construction and maintenance. In the United States, these minimum requirements can be found in 49 CFR (Code of Federal Regulations) Part 213. Track defects that, depending on their severity, may lead to derailment include

1. broken joint bar between two rails (especially in older rail and at cold temperatures);
2. broken rail (internal defect in steel, especially in older rail and at cold temperatures);
3. excessive cross-level between rails—one rail higher or lower than its adjacent rail can cause car body roll and resulting wheel lift and may also cause harmonic rocking at specific speeds: commonly caused by poorly tamped rail joints;
4. poor track alignment—"wavy" track;
5. track gauge (distance between rails) too wide (causes wheelset "drop-ins") or too narrow (causes wheelset/rail climbovers, rail rollovers);
6. ties deteriorated to the point that they are unable to hold spikes or other fasteners—leading to rail rollovers, particularly the high rail on curves;
7. track switch or turnout to adjacent track not properly adjusted or maintained—"point gap" between switch point and stock rail—allows incoming wheel flanges to "pick the point" and derail;
8. insufficient ballast or rail anchoring to support track or to restrain longitudinal movement under load (often resulting in "heat kinks");
9. insufficient spikes or fasteners to hold the rail in place, allowing rail rollover under load, particularly the high rail on curves; and
10. insufficient rail superelevation on curves for the authorized track speed.

To be considered potentially causal in nature, track defects identified during the postaccident inspection must have existed prederailment and not have been caused by the derailment.

Another potential cause of a derailment is equipment (rolling stock) failure. Common causes of equipment failure that can lead to derailment include

1. broken wheel—may be caused by defective steel or thermal cracking from brake shoe heating;
2. broken axle—defective steel;
3. roller bearing burn off—lack of lubrication, internal flaws, manufacturing defects: check data from nearest hot bearing detector for confirmation;
4. railcar truck (or "bogie") hunting—oscillating back and forth between the two rails as the train progresses, exerting high lateral forces on the rails, most common on straight track with worn rail at 70 kph or higher; and
5. truck (or "bogie") warp—causes wheelsets to move in a canted fashion relative to the track. This accelerates wear

on the track and railcar components and may cause a wheel-lift derailment if sufficiently pronounced. This would be present when bogie components are worn and "out of square" relative to the direction of travel.

The search for an equipment-caused derailment begins at the leading car in the line of derailed cars. Any equipment-related derailment cause will likely be found in this car, but in certain conditions, may be found in the first or second car to the rear of the leading derailed car, but rarely further back than this in the line of derailed cars. Bogies and wheelsets (which are often separated from their respective cars and sometimes partially buried in the ground) should be matched with their corresponding cars, if possible, and examined in detail for any of the above defects. As with track, in order to be considered potentially causal, the defect must have existed prederailment and not have been caused by the derailment. This can often be determined by the presence or absence of corrosion in cracks and fractured steel surfaces of wheels, axles, and bogie components. Freshly cracked or fractured surfaces will not exhibit corrosion on the fractured surfaces. Partial corrosion (corrosion up to a point on the fractured surface) often indicates that the crack existed for some time before it finally sheared completely. The services of a metallurgist should be secured to complete the analysis whenever such evidence is present.

Another potential derailment cause, although much rarer than any we have discussed thus far, may come into play when a train is passing over a turnout (switches trains from one track to another) that is power-operated (as opposed to hand-operated) and controlled remotely by a train dispatcher. Signal track circuitry at the location detects the presence of a train and electrically locks the switch, preventing it from changing position, for the duration of time that the train is passing over it. If this signal track circuitry fails in some fashion, it may be possible for the power-operated turnout to change position as the train is passing over it, causing a derailment.

Derailments sometimes result from improper train makeup and/or train handling. The placement of a block of heavier cars near or at the rear of a string of empty or lighter cars can produce "slack run-ins" if the train brakes are not applied with the necessary finesse. The heavier cars will tend to bunch toward the leading end of the train, derailing lighter cars between them and the locomotives. This effect is especially pronounced on curves. A locomotive engineer's failure to control slack action in a long train, particularly in an area of undulating grades, can produce the same effect.

Human Factors Considerations—Level Crossing Accidents

No analysis of level crossing accidents would be complete without consideration of the human factors such as performance of motorists as they approach crossings. This subset of human factors research has been studied extensively over the past 40 years, culminating in hundreds of research reports, many of which are available on the Internet. As such, the field, as it relates to level crossings, is far too voluminous to cover here in the level of detail that it merits. We can, however, explore some of the high points and whet the reader's appetite for further investigation on the subject.

In general, motorists are imperfect decision-makers and often underestimate the speed at which trains approach crossings, leading them to attempt movements over the track, which they have insufficient time to complete safely. The perception/reaction time (PRT) required for a motorist to perceive a hazard and begin to take corrective action varies with expectancy. The less expected a hazard is or the more complex the environment from which hazard clues must be drawn, the longer the PRT. In most cases, an assumed PRT of between 1.5 and 2.5 s is typical for level crossing accident reconstruction, unless a more detailed value derived from actual study or reenactment at the subject crossing is available.

The motorist's prior observations of one or more "false activations" of active warning devices at level crossings (instances in which the active warning devices falsely indicate the approach of a train, indicating that it is not safe to proceed over the crossing when in fact it is) tend to reduce the credibility of crossing warning devices in general, prompting them to attempt passages over crossings in violation of activated flashers and gates. Studies have shown that motorists tend to assign a decreased credibility to activated warning devices at a particular crossing even if their observations of "false activations" occurred at other crossings.

Frequent false activations of level crossing active warning devices have many potential causes, principal among them being poor maintenance on the part of the railroad company. Failure to maintain clean, mud-free track ballast (the crushed stone that the track rests on) within the approach to the crossing is the most common cause of false activations. This condition worsens over time, as repeated rail traffic grinds portions of the crushed stone into dust that then combines with water (rain, melted snow and ice, road salt applied over level crossings in winter for the benefit of highway traffic) to form an electrically conductive slurry bridging the two rails. This electrically conductive slurry interferes with the crossing signal equipment's ability to reliably detect approaching trains. The equipment is designed to recognize the onset of this unsafe condition and activate the warning devices continuously until ballast conditions improve (rain stops and ballast dries out) or the equipment is repaired or readjusted by signal maintenance personnel. This unsafe condition is easily preventable by careful and proactive attention to track and crossing signal maintenance.

Motorists often exhibit impatience with activated level crossing warning devices, particularly if they have been held for long periods of time, with or without the passage of a train, by such warning devices in the past. This impatience may prompt them to attempt unsafe maneuvers around activated flashers and lowered gates. The higher the degree of reliability with which the active warning devices at crossings operate, the greater their credibility is and the fewer violations of activated warning devices by motorists that are likely to occur.

See also: **Engineering:** Accident Investigation—Determination of Cause; Human Factors Investigation and Analysis of Accidents and Incidents.

Further Reading

British Columbia Institute of Technology, School of Transportation, 2006. Course RAIL-0107: Rail Accident Scene Investigation, Course Notes.

Halstead, R.W., 2006. In: Railroad Collision Investigation, Presentation to the Pennsylvania State Police Reconstruction Seminar – Course Notes (Available on Request).

Halstead, R.W., 2011. In: Railroad Accident Investigation and Reconstruction, Presentations to the California Accident Reconstruction Society – Course Notes (Available on Request).

Loumiet, J.R., Jungbauer, W., 2005. Train Accident Reconstruction and FELA and Railroad Litigation, fourth ed. Lawyers & Judges Publishing Co.

Relevant Websites

www.apta.com—U.S. American Public Transit Association.

www.arema.org—U.S. American Railway Engineering and Maintenance-of-Way Association.

http://www.ite.org/bookstore—U.S. Federal Highway Administration Railroad-Highway Grade Crossing Handbook – Revised second ed. (2007).

http://mutcd.fhwa.dot.gov—U.S. Federal Manual on Uniform Traffic Control Devices.

www.fra.dot.gov—U.S. Federal Railroad Administration.

www.fta.dot.gov—U.S. Federal Transit Administration.

www.ntsb.gov—U.S. National Transportation Safety Board.

Aircraft Flight Data Recorders

VL Grose, US National Transportation Safety Board, Washington, DC, USA

Glossary

Adversary truth Pursuit of truth in the law involving a process of confrontation between opposing parties that eliminates voluntary disclosure of uncertainties, must be certain, known firsthand, and excludes hearsay, rumors, and guesses.

Air Line Pilots Association (ALPA) Representing more than 53 000 pilots of 37 US and Canadian airlines, it is chartered by the AFL-CIO and the Canadian Labour Congress and is a member of the International Federation of Air Line Pilot Associations.

Analog information Continuous phenomena with no quantized or discrete variances whose recorded result is subject to interpretation as to meaning and significance.

Bureaud'Enquêtesetd'Analysespourla sécurité de l'av-iationcivile (BEA) The French authority responsible for safety investigations into accidents or incidents in civil aviation.

Coalition of Airline Pilots Associations (CAPA) A trade association comprised of over 28 000 professional pilots in five individual member unions.

Cockpit voice recorder (CVR) An electronic device used to record conversation in the cockpit, radio communications between the cockpit crew and others (including conversation with air traffic control personnel), as well as ambient sounds.

Collision Safety Institute (CSI) An independent traffic collision research, training, and crash consulting organization with a mission to provide state-of-the-art training and technology transfer to both public and private entities involved in the analysis and evaluation of motor vehicle collisions.

Digital information Discrete values of quantized time, frequency, and amplitude that are not subject to interpretation as to content.

Event data recorder (EDR) An electronic device used to record information related to vehicle (automobile, truck, locomotive) performance that can be retrieved and analyzed following an accident.

Federal Aviation Administration (FAA) The national aviation authority of the United States. An agency of the Department of Transportation, it has authority to regulate and oversee all aspects of civil aviation.

Flight data acquisition unit (FDAU) A junction box that receives various discrete, analog, and digital parameters from a number of sensors and avionic systems, then routes them to a flight data recorder (FDR).

Flight data recorder (FDR) An electronic device used to record hundreds of specific aircraft flight performance parameters, including control and actuator positions, engine information—all on a time basis.

National Transportation Safety Board (NTSB) An independent US Federal agency charged with determining the probable cause of transportation accidents, promoting transportation safety, and assisting victims of transportation accidents and their families. It has five board members, each nominated by the President and confirmed by the Senate to serve 5-year terms.

Probable cause Conclusions reached by an investigating body as to the factor or factors that caused an accident.

Proximate cause An event directly related to a loss and legally held to be its cause.

Technical truth Pursuit of truth in scientific work that includes not only what is known but also open admission of uncertainties, assumptions, suppositions, and hunches—all of which are probabilistic that is, never certain and always subject to reversal should new findings disprove them.

Underwater locator beacon (ULB) A device fitted to both CVRs and flight data recorders that, when triggered by water immersion down to depths of 14 000 ft, emits an ultrasonic pulse of 37.5 kHz at once per second.

 http://dx.doi.org/10.1016/B978-0-12-802718-9.00011-9

Introduction

Commercial aircraft crashes arouse wide public interest for several reasons. They generally involve massive, unexpected death and destruction. The flying public easily correlates an airliner crash with its own perceived risk beyond any possible personal control. Even though the probability of a crash per flight is one in every 19 000 years, it is one of the six most feared causes of death. So such disasters appear mysterious—but demand a raison d'etre. Aircraft flight data recorders (FDRs) help to answer that demand.

Pursuit of Truth

Airliner crashes and criminal acts both provoke public concern for their avoidance—focused on what needs to be eliminated or corrected. Crash-free airliners and crime-free society are admirable goals but never likely to be fully achieved.

Approaches to perfection for both goals are similar but not identical. Truth in airliner technology involves traditional scientific freedom to look anywhere at any time under unrestricted rules. In contrast, truth sought under rules of criminal law is bound by restrictions as to its relevance. These two types of truth may lead to contrasting use of recorders.

Adversary Versus Technical Truth

Adversary truth pursued in the law involves a process of confrontation between opposing parties. That eliminates the voluntary disclosure of uncertainties. To be admissible in court, information must be certain and known firsthand. Hearsay, rumors, and guesses are not allowed.

Technical truth sought in scientific work includes not only what is known, but also open admission of all uncertainties. So assumptions, suppositions, and hunches have a large role in science. In contrast to public perception, science is probabilistic—never certain and always subject to reversal and should new findings disprove it. **Figure 1** illustrates these two types of truth.

These two truths are not competitive because they have different truth objectives. However, digital recordings may play contrasting societal roles in providing information. Law enforcement's use of recorded automobile data (e.g., velocity and location) does not enjoy the same public endorsement that recorded aircraft data used in postcrash analysis does.

Immediately after an airliner catastrophe, the public expects someone to determine not only what happened but also why it happened. In the United States, the National Transportation Safety Board (NTSB) pursues both.

That process consists of nine steps:

- On-scene fact finding
- Collection of physical evidence

Two truths
Vernon L. Grose, DSc

Adversary truth Only what has been elicited as *known*	Technical truth What is *known* plus full disclosure of all *uncertainties*
Fault-finding, blame assessment	Accident-preventing
Protection of deficiencies	Openness to deficiencies
Reaction to events	Foreseeability
Post-facto review	"A priori" projection
Involuntary, screened data	Voluntary exposure
First-hand facts	Hunches, possibilities

Figure 1 Two truths.

- Interrogation of witnesses
- Conduct of public hearings for sworn expert testimony
- Analysis of all obtained evidence (including FDR readouts)
- Formulation of conclusions
- Determination of causation
- Creation of recommendations to preclude a similar accident
- Summary and publication of the determinative process

This evidentiary pursuit can often require a year. All nine steps revolve around history—dealing with what has already been transpired but was unplanned and unforeseen. Most require human judgment that is subjective rather than objective.

Compounding that search is the likely disintegration of the aircraft and its contents because of their velocity at impact and release of onboard energy sources that frequently results in wide distribution of wreckage.

What is being sought? Accurate knowledge of something that can never be fully replicated. That airliner crash knowledge deficiency likewise exists in criminal situations that forensic sciences attempt to solve. Therefore, pursuit of what, how, and why of airliner crashes—utilizing FDR data—is explored for its application to forensic sciences.

Desired Knowledge

Knowledge about past events—regardless of nature—is always limited. Airline disasters suffer this limitation in several ways.

First, a limit exists because there are seldom any surviving on-scene witnesses. So knowledge can only be inferred by deduction.

Second, even if witnesses survive, humans are not accurate observers. They have limited sensory capacity at best.

Eyewitnesses of any accident are rarely unanimous in recalling what happened.

Third, human recollection of major disasters often atrophies over time as testimonies disagree. Therefore, individual witness memory is quite unreliable.

Fourth, there can be motivational conflict in obtaining truth of what has transpired. On one hand, desire for revenge, emotional resolution, justice, future prevention, or historical accuracy stimulates intensity to pursue such knowledge. But on the other hand, witnesses of tragedies often desire to suppress or purge painful recollection of them.

Despite these limitations, postevent knowledge is essential to understanding how both airline crashes and major criminal acts occur.

Seeking Causation

A force for obtaining knowledge of past events is the belief that they were caused. This emanates from the philosophical concept of causality—where one event is believed to produce a responding subsequent event. An airliner crash thereby becomes an outcome of an initial cause or causes that enabled the disaster to occur.

Single Versus Multiple Causes

Seldom are catastrophes due to a singular cause. The NTSB, initially part of the Department of Transportation (DOT), was separated in 1974 from DOT by Congress as an independent agency and charged with determining the probable cause(s) of aircraft accidents. However, for many years thereafter, controversy and political pressure within the NTSB forced it to promote singularity of causation.

On one hand, advocates of adversary truth favored a sharp, singular focus on defining fault that would facilitate legal action following a crash. Underlying this position was blame assessment but thereby subordinating loss prevention (the prime NTSB objective).

On the other hand, proponents of technical truth insisted that causality is never singular. Being forced to discard the known causes (or list them only as contributory to a single cause in order to select only one) needlessly fragmented the causality search.

Testimony before the US Senate on October 22, 1985, resolved this controversy in favor of the scientific position that retains pursuit of causation plurality and assures focus on maximizing safety.

Approximate Versus Proximate Cause

There are two types of causation in the law: cause-in-fact and proximate or legal cause. Cause-in-fact says, "But for this, the accident would not have occurred..." while proximate cause is an event directly related to a loss and held to be its cause.

Given that aircraft crashes have several causes, the purpose of seeking causation is to (1) define, (2) isolate, and (3) remove or control all factors believed to have contributed to the crash.

The US Air Commerce Act uses "probable cause" to define aircraft accident causation. This term stops short of "proximate cause" to remove any stigma of blameworthiness. This distinction allows investigators to freely discuss anything related to an accident without fear of legal action.

This contrast between proximate and probable cause may likely affect digital recording in forensic science. Beyond that distinction, two contrasting—negative and positive—concepts regarding accident causation in **Figure 2** may also influence forensic data recording.

First, the public believes that most airliner crashes somehow occur in spite of all countermeasures incorporated to prevent them—much like a snake mysteriously slithering with diabolical intent through a maze of preventive measures.

A second view—proposed by Dr Nestor B. Kowalsky—is that an aircraft crash is a positive, successful event—inevitable when preceded by a series of committed or omitted acts. It is created by decisions. Thereby, FDRs identify some of the dominoes to be eliminated to preclude future crashes.

Role of Evidence in Establishing Truth

Knowledge of truth about the past—whether pursued by adversary or technical means—ultimately becomes classified as evidence. It has different meanings in the law (adversary truth) and in airliner crash investigation (technical truth).

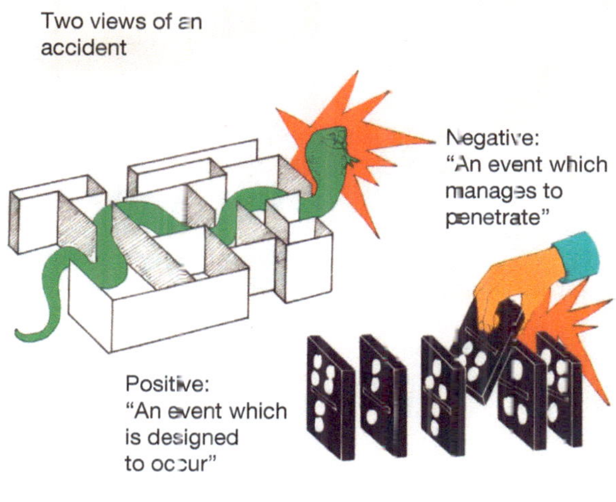

Figure 2 Contrasting views of causation.

Evidence produced by FDRs is generally objective (free from human manipulation) and can be interpreted by experts in science, engineering, psychology, sociology, management, and operations.

Digital Versus Analog Evidence

The FDR evidence is either digital or analog. Digital information consists of discrete values of quantized time, frequency, and amplitude. Those signals are not subject to interpretation as to content (though there may be interpretive disagreement regarding raw data conversion into engineering units). Analog information consists of continuous phenomena with no quantized or discrete variances. Its recorded output is therefore subject to interpretation as to meaning and significance.

Multiengine, turbine-powered aircraft carrying over 10 passengers must have an FDR. A cockpit voice recorder (CVR) is required for multiengine aircraft with two pilots, which carry over six passengers. FDR inputs are digital while CVR inputs are analog. When both recorders are recovered and their outputs can be correlated on a common time line, unique evidence of what has transpired is available.

Validity of FDR digital data—being virtually free of interpretation—is universal, giving it objectivity over CVR analog evidence. CVRs record more than cockpit conversation. All audio data in the cockpit, including switch clicks, wind noise, control equipment operation, and ambient acoustics such as engine pitch are available. However, translating, interpreting, analyzing, and affirming recorded conversation requires excellent linguistic and cultural skills.

Aircraft Flight Data Recorder History

The genesis of FDRs was a series of mysterious crashes in 1953 of the world's first jet airliner—the British "Comet." The Comet revolutionized commercial flight in altitude, speed, quietness, and freedom from vibration—until several of them crashed inexplicably.

Cause of Comet disasters was not soon determined. Dr David Warren (1925–2010), a principal research scientist for the Aeronautical Research Laboratories in Melbourne, Australia, was engaged to investigate the crashes.

Dr Warren early proposed recording flight crew voices in the cockpit. Initially, his idea raised little interest. He built an experimental device to store 4 h of speech and flight instrument readings. He then published a report "A Device for Assisting Investigation into Aircraft Accidents." But it took 5 years for the device to be accepted.

Warren's first recorder was produced in 1958. It was about the size of an adult hand but was rejected by Australian aviation experts who said it had "little immediate direct use in civil aircraft." Pilots joined opposition by calling it "a spy flying alongside." Warren then took it to England where their aviation experts endorsed it—urging its installation in all British civil aircraft.

Before long, aircraft flight recorders were adopted—for both cockpit sounds and aircraft performance—by 1964, the US Federal Aviation Administration (FAA) mandated them in commercial aircraft.

Black Boxes

News media early began to call FDRs "black boxes." This designator may be traced to its use in technology as "a device, system, or object viewed solely for its input, output, and transfer characteristics without concern about its internal workings." In that sense, how black boxes function may be unknown—even unnecessary to be understood but only to be useful—and thereby opaque or black.

However, FDRs are never black in color. In stark contrast, they are painted International Orange or Yellow (per US Federal Standard 595) to increase their visibility for recovery from amidst wreckage.

Cockpit Voice Recorder Hostility

From the beginning, pilots opposed recorders in a cockpit based on their reduction of the pilot's personal privacy.

No other profession endures the surveillance, oversight, and potential for second guessing—particularly after death—that CVRs represent. Voice recordings cannot fully provide rationale for past human behavior, especially under stressful circumstances. CVRs personify George Orwell's "1984" Big Brother syndrome to pilots.

On the other hand, airline passengers totally surrender their lives to pilots when the door closes, and the aircraft departs the terminal. This unusual transfer of life-and-death responsibility to someone unknown seems to justify unusual accountability and oversight.

Beyond this pilot–passenger trust relationship is the larger issue of maintaining airline safety. Obtaining recorded cockpit audio following a fatal crash enhances prevention of future disasters—invaluable in advancing flight safety.

Thus, tension between pilot freedom and airline passenger dependence on the pilot is likely to continue.

Evolution of Recording Processes

Flight recording technology has progressed in parallel with commercial aircraft development. The first generation of recorders—installed in early jet airliners such as B-707, DC-8, and Caravelle—recorded five analog aircraft parameters (heading, altitude, airspeed, vertical accelerations, and time).

To survive high impact velocity and sustained postimpact fire, the readouts of five parameters were embossed onto a metal foil (incanol steel) used only once (i.e., not recorded on a repeatable loop). Although the recording foil was nearly indestructible, crash survival of these recorders was a serious problem. By 1965, the original 100 g (100 times the accelerative force of impact by simply being dropped) requirement was upgraded to 1000 g.

A second FDR generation emerged after aircraft performance data proved insufficient for determining accident causation. Magnetic tape technology (recording data in digital format) was adopted for FDRs, replacing metal foil. These upgraded FDRs covered additional parameters such as engines, flight controls, and flaps. Use of magnetic tape, however, required even more complex fire and crash protection. This second class of FDRs was installed in B-747, DC-10, L-1011, and A300 wide-body aircraft during the late 1960s and early 1970s.

Based on Dr David Warren's earlier work, advanced audio recording technology enabled capture of a wider range of cockpit sounds, including crew conversation, air traffic control communications, and aircraft noises. By 1965, all US commercial airlines carrying over 20 passengers had a CVR retaining the latest 30 min of cockpit sounds.

A third generation of solid-state recorders was available by 1990. These recorders store data in semiconductor memories or integrated circuits—replacing electromechanical techniques. One of its advantages is that such memory does not require scheduled maintenance or overhaul. At present, solid-state recorders provide vastly expanded data while also saving airline operational costs when the aircraft is in service. Stored flight data in these recorders can be readily evaluated for aircraft flight performance or whether a device requires maintenance.

Adaptation of solid-state recorders for the CVR occurred later than for the FDR—primarily because more memory capacity was required for audio data. In 1992, a 30-min CVR became available. By 1995, a 2-h CVR was introduced

Most current FDRs receive inputs via specific data frames from a flight data acquisition unit that records many flight parameters—including control and actuator positions, engine information, and time of day. A minimum of 88 parameters are currently required (only 29 were required until 2002), but some systems today monitor even thousands of variables. Most parameters are recorded a few times per second, but some data are stored at a much higher frequency. FDRs record a minimum time frame of 25 h of data in a continuous loop.

Recorder Survivability

Two simultaneous and violent forces are unleashed on recorders when an aircraft crashes—high impact velocity and sustained high-temperature fire. As technically sophisticated as a recorder may be, it is useless unless it survives those forces generated by the very event it is designed to record. Therefore, recorder survival is fundamental.

First, if the aircraft is integral when impacting the ground, recorders must survive not only that initial blow but also a possible secondary impact if they are launched by explosion of jet fuel. Early recorders were installed in the forward avionics bay of the aircraft and required to survive a 100 g impact. After several crashes, recorders failed to survive. At present, the acceleration requirement has been raised from 100 g to the current 3400 g for 6.5 ms.

The second survival factor for recorders is fire generated by Jet A or Jet A-1 fuel with open air burning temperatures of 1000 °C. This fuel can ignite other aircraft materials surrounding the recorders, resulting in even higher ambient temperatures. Damage by fire is dependent on not only temperature but also time at temperature. So, at present, recorders must survive 1100 °C for 1 h.

Additional recorder requirements exist for penetration resistance, static crush, deep-sea pressure, sea water immersion, and fluid immersion.

Installed Location of Flight Data Recorders

After increasing recorder impact requirement to 1000 g, the FAA also required relocation of both CVR and FDR recorders to the rear of the aircraft where survival was believed to be higher. **Figure 3** shows the position of both recorders and their recording pickup instrumentation.

Relocation rationale was that most crashes occur while the aircraft is moving forward and considerable energy is absorbed before the tail impacts the ground. Further, as the aircraft breaks apart, the tail is often separated from remaining wreckage and fuel sources in the wings. Of course, this reasoning does not apply when the aircraft disintegrates in the air before hitting ground or water, for example, following an airborne explosion.

Underwater Locator Beacons

Both CVR and FDR have an underwater locator beacon (ULB) to assist in locating them following an overwater airliner crash. The ULB—called a "pinger"—is activated once it is immersed in water. Therefore, it never emits a signal following crashes on land. The pinger transmits a 37.5-kHz acoustical signal detected only with a special receiver. It can transmit from depths down to 14 000 ft.

Recorder manufacturers install pingers as part of their individual recorder models. The airline owning and installing a recorder is required to replace its pinger batteries to keep them operational. Required pinger life is 6 years.

Government Regulation

The Air Line Pilots Association International (ALPA), founded in 1931, currently represents 59 000 pilots of 39 US and Canadian airlines. The ALPA was an early and vocal opponent of the CVR.

Figure 3 Flight data recording and storage.

An initial ALPA–NTSB compromise stipulated (1) 30-min duration of retained recorded information, (2) flight crew erasure of recording once on the ground, and (3) disclosure of recorded information only for accident investigation. CVR acceptance by airline pilots is thereby viewed as a reasonable, needed component of aircraft crash investigation.

However, political pressure occasionally arises to expand CVR use beyond accident investigation to disciplinary enforcement of pilot behavior that violates the "sterile cockpit" rule. That rule prohibits, below 10 000 ft, any cockpit conversation not directly related to flight operation.

The source of such pressure is generally public reaction—stimulated by dramatic news media coverage—to obtained CVR recordings that reveal bizarre or irrelevant chatter or behavior in the cockpit.

In February 2010, the US Senate Pilot Professionalism Act (S.3048) was introduced "to improve air safety by authorizing the limited use by air carriers of information collected through CVRs and flight data recorders, to prohibit tampering with such devices, and for other purposes."

Both the ALPA and the Coalition of Airline Pilots Associations (CAPA) representing over 28 000 pilots objected seriously to this bill because it could allow airlines to administer discipline—even terminate or require a pilot to pass a proficiency check in a simulator—based on CVR and FDR findings.

Ultimately, S.3048 was not adopted.

Public Access to Voice Recordings

Interest in listening to final minutes of a fatal aircraft crash occasionally arises. On some occasions, families of deceased pilots have demanded to listen to those final moments of life. News media also seek recordings to satisfy public interest after crashes. Increasing electronic capability and exposure via the Internet have made recordings available—although a large percentage of alleged recordings are created artificially.

Historically, the NTSB has exercised tight discipline in permitting cockpit voice recordings to be heard only by parties engaged in translating and converting conversation to written format. The recorders are the property of the involved airline. Once a transcription of the CVR has been officially documented, the recorder is released back to its airline owner who is responsible for whatever public access is subsequently obtained.

The US Congress has imposed reasonable restrictions on the use of CVR materials. It exempts CVR data from Freedom of Information Act requests. In the Transportation Safety Act, Congress limited public disclosure of CVR data involving aircraft crash investigations. However, that Act allows the NTSB to publicize any transcript segment believed relevant to a crash—provided it holds a public hearing or if it is discussed along with other factual reports at the same time. The Board is also allowed to refer to CVR recordings in its safety recommendations.

A court may allow discovery of a CVR recording or transcript without prior public release of the information. If the CVR transcript is publicly released, the transcript becomes discoverable. Each CVR audio transcription released to the public by the NTSB carries the warning shown in **Figure 4**.

The Bureau d'Enquetes et d'Analyses (BEA) is the French equivalent to NTSB. Citing Article 14 of the European Regulation of October 20, 2010, as its policy standard governing public release of CVR transcripts, it came into effect on December 2, 2010, following the crash of Air France Flight 447.

Current Recorder State-of-the-Art

Recorder technology continues to expand. At periodic intervals, the FAA upgrades performance requirements as shown in **Figure 5**. These revisions result from technological, operational, and financial forces that demand increasing knowledge of truth about the past to advance airline safety.

Image Recorders

Superiority of visual and aural images over oral recordings to obtain knowledge of the past is undeniable. Therefore, video technology increases to be considered to augment CVR output. Cockpit video recorders are not yet required in commercial airliners, but they are installed in smaller aircraft and helicopters.

Cultural Repercussion of Recorders

Two airliner crashes are summarized to illustrate the significant role recorders have played in international, political, cultural, and business controversies, thereby highlighting the influence that recorders can contribute to ascertaining truth.

NTSB warning on cockpit recording transcription

- The reader of this report is cautioned that the transcription of a CVR tape is not a precise science but is the best product possible from an NTSB group investigative effort.
- The transcript, or parts thereof, if taken out of context, could be misleading.
- The attached CVR transcript should be viewed as an accident investigation tool to be used in conjunction with other evidence gathered during the investigation.
- Conclusions or interpretations should not be made using the transcript as the sole source of information

Figure 4 CVR transcript protection. CVR, cockpit voice recorder; NTSB, National Transportation Safety Board.

EgyptAir Flight 990

On October 31, 1999 the Boeing 767 en route from JFK airport in New York to Cairo crashed into the Atlantic Ocean about 60 miles south of Nantucket Island, Massachusetts, killing all 217 people onboard. As the crash occurred in international waters, the Egyptian Civil Aviation Authority (ECAA) had investigation responsibility. But since ECAA lacked adequate technical resources, the Egyptian government asked the NTSB to conduct the investigation.

Within 2 weeks NTSB proposed transferring the investigation to the US Federal Bureau of Investigation (FBI) since evidence suggested that (1) a criminal act had taken place and (2) the crash was intentional rather than accidental. The Egyptians rejected this proposal. So the NTSB continued to lead the investigation.

As evidence of the deliberate crash increased, the Egyptian government reversed itself, and the ECAA began its own investigation. The two investigations reached conflicting conclusions. NTSB concluded that the cause was a deliberate action of the Egyptair Relief First Officer (RFO). ECAA found the cause to be the mechanical failure of the airplane's elevator control system.

Both CVR and FDR provided determinative evidence that (1) the RFO had committed suicide by deliberately diving the aircraft into the ocean and (2) no mechanical failure scenario could have produced the aircraft movements recorded by the FDR.

An international firestorm erupted. Egypt's state-owned *Al Ahram Al Misai* called RFO Al-Batouti a "martyr." The Islamist *Al Shaab* newspaper even accused US officials of secretly recovering the FDR, reprogramming it, and throwing it back into the water to be publicly recovered.

Given strong Egyptian cultural aversion to suicide, the possibility of a common NTSB–ECAA agreement on probable cause evaporated. The NTSB probable cause, however, avoided discussing suicide. The airplane's departure from normal cruise flight and subsequent impact with the Atlantic Ocean was a result of the RFO's flight control inputs. The reason for the RFO's actions was not determined.

Air France Flight 447

On June 1, 2009, the Airbus A330-200 aircraft crashed in the Atlantic Ocean en route from Rio de Janeiro to Paris, killing 228 persons. After 30 months of investigation, BEA had not determined probable cause—as CVR and FDR were not recovered for 23 months. CVR and FDR unavailability allowed widespread and conflicting speculation—overshadowed by several factors seldom acknowledged publicly.

FAA proposed revisions of aircraft recorders 24 February 2005	
Cockpit voice recorders (CVRs)	
Proposed	Current
Two hours recording time	15–30 min recording time; valuable communications can be taped over
No magnetic tape recorders	Magnetic tape OK; can sustain damage in crash
10-minute independent backup power supply	Recorder stops if aircraft electrical power fails
Standardized recording start: Begins when pilots start checklist	Variable requirements for recording start
Mounted in box separate from FDR (except helicopters)	Separate box is FAA policy but not a regulation
No single electrical failure can disable both CVR and FDR	Both CVR and FDR can stop working if aircraft electrical power fails
Flight data recorders (FDRs)	
Proposed	Current
Measurement of control surface (rudder, ailerons, elevators, etc.) movements every 0.0625 s	Measurements every 0.25 or 0.5 s; may not permit accurate reconstruction of movements in all circumstances
Measurement of pilot inputs on control wheel, control column, rudder pedals (airplanes aircraft) every 0.0625 s	Measurements every 1 s; may not permit accurate reconstruction of forces in all circumstances
Measurement of cockpit controls and flight control surfaces (helicopters) 4X per second	Measurements 2X per second; may not permit accurate reconstruction of movements in all circumstances
Mounted in box separate from CVR (except helicopters)	Separate box is FAA policy but not a regulation
No single electrical failure can disable both CVR and FDR	Both CVR and FDR can stop working if aircraft electrical power fails

Figure 5 Aircraft recorder updating. FAA, US Federal Aviation Administration.

First, intensive worldwide competition in the commercial aircraft market between Airbus and Boeing meant that US markets were pitted against those in Europe—even free enterprise (Boeing) challenging government enterprise (Airbus).

Second, these two aircraft giants employ contrasting approaches to the pilot–aircraft interface. To summarize, Boeing designs aircraft control around the pilot as primary, augmented by automation while Airbus reverses those roles—automation prime, pilot as secondary. While the Airbus approach eases the piloting load, its automation is more complex and less intuitive.

Third, where Boeing has continued to use the traditional control yoke in front of the pilot, Airbus employs a control sidestick beside the pilot. Advocates for both approaches abound but controversy remains—primarily because the "feel" or feedback to pilot varies considerably.

Air France announced 3 days after the crash that an onboard monitoring system had transmitted—before crashing—a 4-min series of electronic messages concerning failures and warnings in navigation, autoflight, flight controls, and cabin air-conditioning. These sketchy data—while insufficient for cause determination—ignited extensive but unresolved prognosis about both crew and aircraft performance that persisted for over 2 years.

On June 7, 2009, the AF447 vertical stabilizer was recovered (confirming crash location), so initial search for the recorders ensued. Three more independent recovery attempts occurred before May 2, 2011, when both CVR and FDR were found in about 13 000 ft of water. Amazingly, data on both recorders were recovered.

During the years since the crash, recorders have remained the prime focal point. Their recovery, however, failed to resolve its probable cause even after nearly 3 years likely because of two unique factors—the responsibility for investigation and anticipated legal culpability. In contrast to NTSB investigations in the United States, the French government opened two independent and simultaneous investigations of AF447.

Technical investigation

BEA was charged with the investigation since the aircraft was of French registration and crashed over international waters. Representatives from Brazil, Germany, UK, and United States were directly involved. In addition, observers were appointed from China, Croatia, Hungary, Ireland, Italy, Lebanon, Morocco, Norway, South Korea, Russia, South Africa, and Switzerland because their citizens were onboard.

Criminal investigation

French law requires—for any accident involving loss of life (but implying no presumption of foul play)—that manslaughter be charged against Air France and Airbus. This investigation has been overseen by the Gendarmerie nationale, which conducted it through its aerial transportation division (Gendarmerie des transports aériens or GTA) and its forensic research institute (Institut de Recherche Criminelle de la Gendarmerie Nationale or FR).

Expanded Employment of Recorders

As data recorders gain acceptance, aviation is not the only transportation mode utilizing them. In 1993, the Federal Railroad Administration (FRA) required event data recorder (EDRs) in locomotives.

An EDR is also being installed in automobiles to record vehicle information immediately before and/or during a serious crash. Police and crash investigators download EDR data to learn what happened to the vehicle and how its safety features are performed. In some cases, those data help establish culpability.

Most EDRs are built into a vehicle's airbag control module and record data on airbag deployment. Some vehicles also record precrash data, such as engine throttle and vehicle speed from the engine control module. Certain airbag and engine control modules store only diagnostic trouble codes and whether there was a signal to deploy supplemental restraint systems (i.e., airbags and belt tensioners).

Built-in vehicle modules are not considered to be an EDR. Therefore, they are not governed by Federal regulations, as are EDRs that record vehicle speed before a crash or speed change during impact.

In summary, there is no way that David Warren could have foreseen how aircraft flight recorders would impact both civil and criminal litigation. They also affected economics of the expert witness field. The Collision Safety Institute reports that expert testimony based on EDR data has increased both criminal convictions and death-penalty sentences, as well as high-dollar verdicts in civil cases.

See also: **Engineering:** Analysis of Digital Evidence; Human Factors Investigation and Analysis of Accidents and Incidents; **Forensic Medicine/Clinical:** Airplane Crashes and Other Mass Disasters; **Investigations:** Explosions; Major Incident Scene Management; Recording; **Legal:** History of the Law's Reception of Forensic Science.

Further Reading

Craig, M.W., 2008. Thinking Outside the Black Box: How an Electronic Security Device Became a Police Informant. http://works.bepress.com/mary_craig/1.

Grose, V.L., 1987a. Managing Risk. Prentice Hall, Englewood Cliffs, NJ.

Grose, V.L., 1987b. Commercial Airline 'Margin of Safety'. Presidential Aviation Safety Commission. Aviation Safety Commission report ISBN:13: 9780160032875.

Grose, V.L., 1987c. Coping with Boredom in the Cockpit Before It's Too Late. Presidential Aviation Safety Commission. Aviation Safety Commission report ISBN:13: 9780160032875.

Grose, V.L., 1995a. Your next airline flight: worth the risk? Risk Management 42 (4), 47–56.

Grose, V.L., 1995b. Technology's impact on human risk. Proceedings of the Marine Safety Council 52, 57–63.

Air-Bag-Related Injuries and Deaths

WS Smock, University of Louisville, Louisville, KY, USA

This chapter is reproduced from the previous edition, volume 1, pp. 1–9, © 2000, Elsevier Ltd.

Introduction

During the past decade the motoring public had been shocked to learn that air bags, a life-saving device promoted by the automotive industry, can also induce severe and fatal injuries. Over the last 10 years in the United States, nearly 200 men, women, and children have been fatally injured by deploying air bags. Thousands more have sustained serious nonfatal injuries, including cervical spine fractures, closed head injuries, and multiple fractures and amputations of digits and hands. Ironically, the vast majority of these serious and fatal injuries occurred in low and moderate speed collisions in which little or no injury would have been otherwise expected.

Historical Context

The first air bag patents were filed in 1952. Ford and General Motors began experimenting with these early prototypes in the late 1950s. Based on documents from the automotive industry, it was apparent, even as early as 1962, that deploying air bags had the potential to induce serious and fatal injuries, particularly to children. These early documents include analyses of tests conducted by Ford at their automotive safety research office in the late 1960s. The tests demonstrated that there was sufficient force associated with air bag deployment to traumatically eject a child from a vehicle. Ford also noted the amputation of a steel-hinged arm from a dummy, secondary to the explosive force of deployment. When testing involved the use of animal models, the list of severe and fatal injuries grew. Cardiac rupture, hepatic rupture, splenic rupture, aortic and vena cava transection, atlanto-occipital dislocation, cervical spine fractures, severe closed head injury, and decapitation were observed.

Testing in the 1980s by the automotive manufacturers continued to demonstrate the risk of injury induction. One study conducted and reported by General Motors indicated "many of the exposures were at loading severities beyond the level representing an estimate of nearly 100% risk of severe injury." These laboratory tests were completed and the knowledge available to the entire automotive industry well *before* air bags were placed in American-made vehicles.

With 40 years of research behind us and all of the resultant data before us, it is apparent that steering wheel and dash-mounted air bags, devices designed to protect occupants in high-speed frontal collisions, can also maim and kill.

Automotive Industry

During the period of 1960–1990, the automobile industry embraced an extensive research and development program regarding air bags. This testing involved anthropomorphic dummies, anesthetized swine and baboons, and even human volunteers. It was recognized early on that occupants in the path of the deploying air bag, or who were positioned close to the air bag at the moment of deployment, were at a very significant risk of receiving a severe or fatal injury. In January 1970, at a meeting of engineers from General Motors, Ford, and Chrysler, this information was discussed. As a result, a Chrysler engineer wrote: "This is a very serious problem that must be resolved for the inflatable restraint. Having a child directly in front of the bag when it inflates could prove fatal."

General Motors tested air bags on baboons at Wayne State University during the early 1970s. The research indicated that "if the head is in the path of the deploying air bag, it is concluded that injury is likely to occur in the form of brain or neck injury to a child."

Testing by Ford in the 1970s revealed that individuals involved in collisions at less than 32 kph (20 mph) experienced more severe injuries and loading forces in vehicles equipped *with* an air bag than those without one. Ford also noted that there was "overwhelming evidence that air bags may be dangerous for small children." In 1972, a Ford engineer wrote the following warning, which was never placed in a vehicle: "The right front seat should be used only by persons who are more than five feet [1.52 m] tall and are in sound health. Smaller persons and those who are aged or infirm, should be seated and belted in the rear seat."

In a series of tests conducted by General Motors in the 1980s, anesthetized swine were placed with their chests in close proximity to the air bag module. The tests revealed that when the swine's thorax was impacted by the force of the deploying air bag and the air bag module, significant thoracic and abdominal injuries were sustained. These injuries in one case included 17 rib fractures; 2 cardiac perforations; a splenic laceration; a liver hematoma, and death within 30 min. The proximity of the occupant to the deploying bag and module cover were the pivotal contributory factors.

Human Injuries

The serious and life-threatening injuries that were originally observed in the industry's laboratories using animal models began to be observed in humans on US roadways in the 1990s. The first six driver air-bag-related deaths which were investigated by the government and the automotive industry revealed that the majority of these victims were women of short stature. It was also noted that the fatal injuries could be incurred even if the occupant was restrained by a lap and chest belt. The injuries initially seen included massive head injuries with diffuse axonal injury, subdural hematomas, and skull fractures. Additional injuries evaluated included cervical spine fracture, cardiac perforation, pulmonary contusions, and multiple rib fractures.

Sodium azide is the explosive propellant used to initiate the deployment cycle in most air bag designs in use today (**Figure 1**). When sodium azide is ignited, the deploying air bag explodes toward the occupant at speeds of up to 336 kph (210 mph). An air bag system has two components, either of which may induce injuries: the canvas-covered air bag itself and the air bag module cover (**Figure 2**). Injuries incurred during deployment are relevant to the component inflicting them.

Obviously, the types of injuries which result from impact with the canvas air bag are different from those which result from impact with its module cover. There are three phases to air bag deployment: "punch out," "catapult," and "bag slap."

Figure 1 The air bag is transported as an explosive material. Sodium azide is the explosive propellant used in the majority of air bag modules.

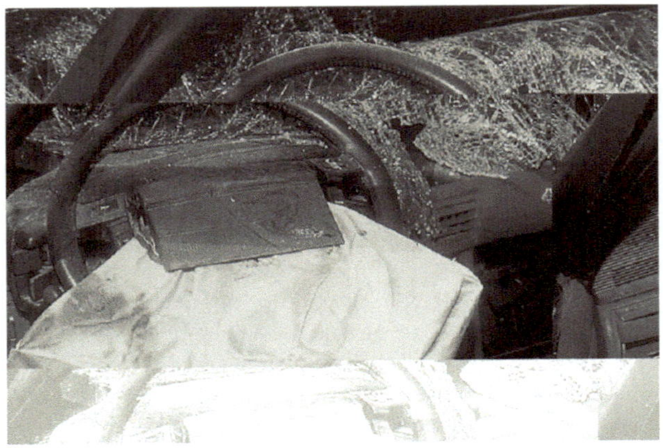

Figure 2 The injury-producing components of the air bag system are the air bag and the module cover which overlies it.

Injuries can be inflicted at any point during the deployment process:

- *Punch out*: This is the initial stage of deployment. If the bag makes contact at this stage, the following injuries can result: atlanto-occipital dislocation, cervical spine fracture with brain stem transection, cardiac, liver and splenic lacerations, diffuse axonal injuries, subdural and epidural hematomas, and decapitation.
- *Catapult*: This is the midstage of deployment when rapidly inflating bag "catapults" or drives the head and neck rearward. This occurs with sufficient energy to rupture blood vessels, ligaments, and fracture cervical vertebrae. The neck injuries occur as the result of cervical spine hyperextension.
- *Bag slap*: This is the final stage of deployment which occurs at the bag's peak excursion. Appropriately named, this happens when the canvas bag's fabric may "slap" the occupant's face, resulting in injuries to the eye and epithelium.

The air bag module covers are located in the steering wheel on the driver's side and in the dashboard panel on the passenger side. As the bag deploys, the module cover is also propelled outward at speeds of up to 336 kph (210 mph). Most steering wheel designs house the horn within the air bag module compartment. Hand and arm injuries observed in individuals whose extremities were in contact with the module at the moment of its rupture include degloving, fracture dislocation, fracture dislocation, and amputations (partial and complete of digits and forearms). If the module cover makes contact with an occupant's face, head or neck, skull fractures and severe or fatal head injuries, and decapitations have also been observed. The driver's side cover is generally made with a rubberized plastic type of material, while the passenger side may have a metal housing. Contact with either type can prove fatal.

Specific Injury Patterns

Ocular

The eye is extremely vulnerable to air-bag-induced injury. These injuries range from corneal abrasions from contact with the air bag, chemical burns from contact with unburned sodium azide, to retinal detachment and globe rupture from the blunt force trauma of the expanding bag (**Figure 3**). The wearing of eyeglasses in some cases has proven to be of benefit, as it offers a degree of barrier protection between the eye and the deploying bag.

Face and Head

The most common injury associated with air bag deployment is that of facial abrasion. The abrasions result from a sliding contact between the bag and the face (**Figure 4**). The injuries are not chemical "burns" but deep abrasions.

Cranial and Intracranial

When acceleration forces are applied to the cranial vault, a variety of traumatic injuries to the brain and surrounding structures will result. These include subdural hematomas, cortical contusions, atlanto-occipital dislocations, skull fractures, and brain stem transections. Cranial injuries may result from contact with either the deploying of bag or module cover (**Figure 5**).

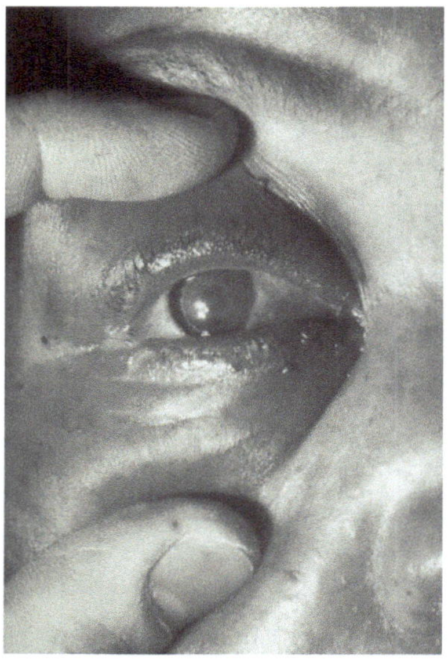

Figure 3 This patient sustained a severe corneal abrasion secondary to the membrane forces associated with air bag deployment.

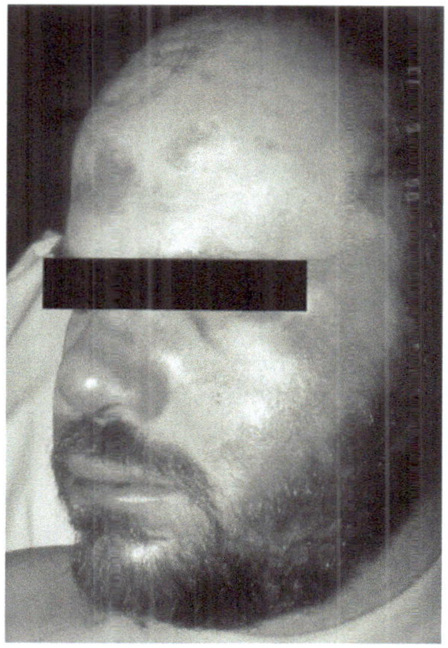

Figure 4 Abrasions to the cheeks, forehead, and nose are the most common injury associated with air bag deployment.

Cervical Spine

The blow from an air bag or module cover which produces a rapid and violent hyperextension of the cervical spine of the driver or passenger will have significant consequences for the cervical vertebrae. Injuries commonly seen as a result of hyperextension include atlanto-occipital dislocation, comminuted fractures of one or more upper cervical vertebrae rupture of the anterior and posterior longitudinal spinal ligaments, and cervical spine disarticulation with transection of the cervical cord. The majority of these injuries are associated with the upper cervical vertebrae, although lower cervical vertebrae injuries have been observed.

Figure 5 Cranial or facial contact with the deploying bag.

Extremities

The upper extremities are very vulnerable to traumatic injury from the deploying bag and its module cover. When an individual's hand or forearm is on or near the module cover at the moment of deployment, the occupant can expect to sustain multiple fractures, and/or tissue degloving or amputation of fingers, hand, or forearm (**Figure 6**). The horn-button-within-the-module cover design significantly increases the risk of injury to the occupant's upper extremities at the moment of deployment. Many of these upper extremity injuries are associated with an occupant's attempt to blow the horn, the button of which is located within the module cover. Forces from air bag deployment may be transmitted to the hand, wrist, or forearm and may even involve the humerus. It is not unusual to see significantly comminuted fractures involving the wrist, forearm, elbow, and distal humerus (**Figures 7 and 8**).

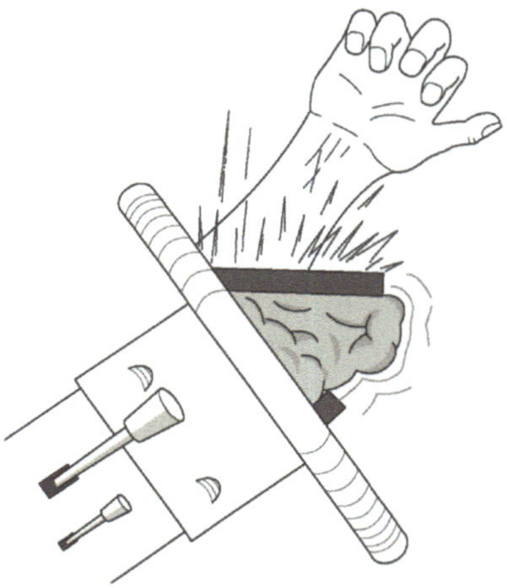

Figure 6 When the forearm is located in a horizontal fashion.

The vehicles whose module covers are of a higher mass have the propensity to inflict more severe injuries. Some of the worst offenders are module covers located on the passenger side, which may have a soft coating of plastic on the exterior but have an underlying piece of rigid metal (**Figure 9(b)**). The placement of hands on the passenger side dashboard, in a bracing maneuver, has resulted in the traumatic amputation of hands and forearms (**Figure 9(a)**).

Respiratory

The by-products of combustion as well as other inert materials within the air bag may produce a white cloud within the vehicle. Many occupants have thought that this indicated a vehicle fire. This whitish material is principally cornstarch, talc, and the by-products of sodium azide combustion. There may be a small percentage of unburnt sodium azide present within this powder as well. Inhalation of these materials can result in a chemical pneumonitis and the induction of asthma-type symptoms. These by-products may also cause a chemical irritation of open wounds and a basic (high pH) burn to the eyes.

Sample Cases

Case 1

A 35-year-old female, 1.57 m (5′ 2″) and 50 kg (110 lb.), was the restrained driver in a 1991 Ford Taurus (**Figure 10**). The vehicle's front bumper grazed a guardrail, which resulted in deployment of the driver's air bag. The patient sustained an immediate respiratory arrest, with subsequent declaration of brain death 12 h later.

A postmortem examination was conducted and revealed the following injuries: significant midface trauma with bilateral epistaxis; corneal abrasions; contusions of the chest; left

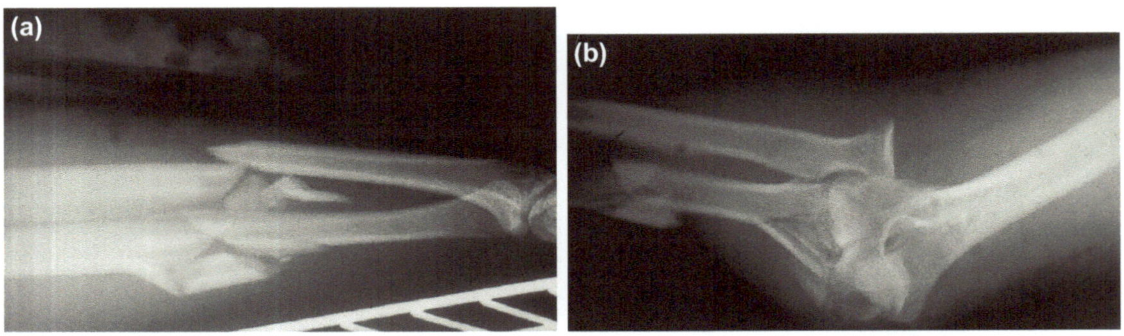

Figure 7 When the hand, wrist, or forearm is on or near the module cover at the moment of deployment, for example when blowing the horn, significant fractures, degloving injuries, and amputations will result. (a) This open-comminuted bending fracture of the radius and ulna was the result of contact with the module cover. (b) This patient sustained a comminuted fracture of the proximal and midshaft ulna as well as a radial head dislocation from impact with the module cover.

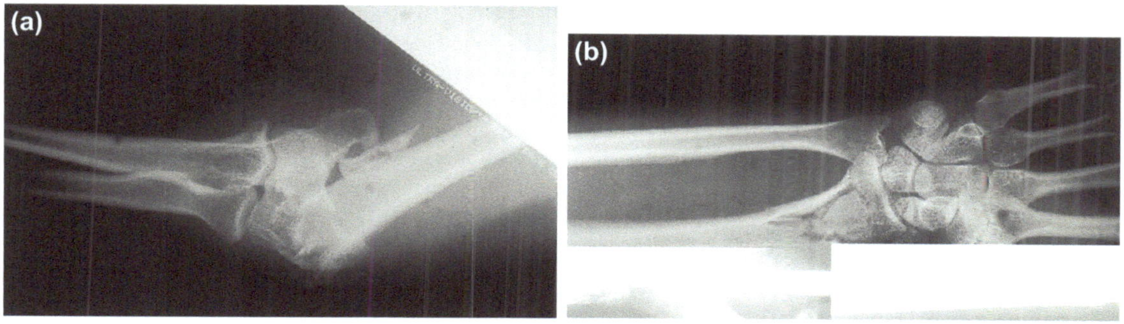

Figure 8 (a) and (b) These severely comminuted fractures resulted when this individual was blowing her horn. She sustained a frontal impact of approximately 15 kph, which resulted in air bag activation. The horn activation button was located within the module cover, which explains why the patient had her forearms over the module cover at the moment of deployment.

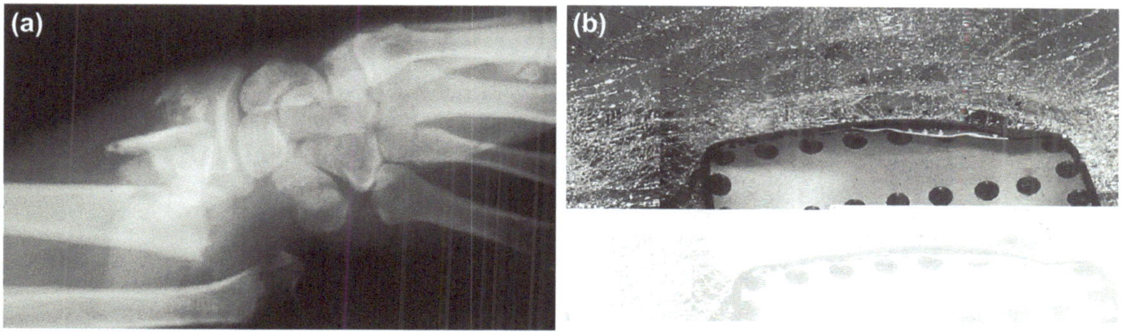

Figure 9 (a) This partially-amputated wrist was the result of placement of the front seat passenger's hand on the dashboard at the moment of deployment. (b) The underlying structure of the module cover was metal.

Figure 10 (a) and (b) A 1991 Ford Taurus was involved in a very minor glancing collision between the front bumper and a guardrail. There was no underlying frame, fender, or structural damage.

subdural hematoma (overlying the frontal and parietal regions); subarachnoid hemorrhage, and severe cerebral edema.

Examination of the exterior of the vehicle revealed very minor damage which was limited to the front bumper. The examination of the interior revealed a tear on the left lower portion of the module cover (**Figure 11**). This tear was the result of contact between the left side of the victim's face and the deploying module (**Figures 5 and 11**).

Case 2

A 4-year-old male was the lap belt-restrained occupant of the front right seat of a 1995 GEO Metro (**Figure 12**). The shoulder portion of the harness was placed underneath the right arm. The patient was catapulted from the front passenger seat to the rear of the vehicle. The patient was found pulseless and apneic on arrival of emergency medical services.

A postmortem examination was conducted and revealed the following injuries: an atlanto-occipital dislocation with brain stem transection; large subdural hemorrhages; multiple rib fractures with underlying pulmonary contusions; liver and splenic lacerations; clavicular fracture, and significant facial abrasions underlying the mandible bilaterally and on the right cheek (**Figure 13(a)**). Examination of the abdomen also revealed a lap belt contusion below the umbilicus (**Figure 13(b)**).

Examination of the vehicle revealed front end damage consistent with a change of velocity of less than 15 kph (9 mph). There was no damage to the driver or passenger compartments. Examination of the passenger bag revealed the presence of blood and tissue transfer. The patient's injuries resulted from blunt force trauma to the chest and abdomen as well as a hyperextension injury of the neck with a rapid rearward acceleration.

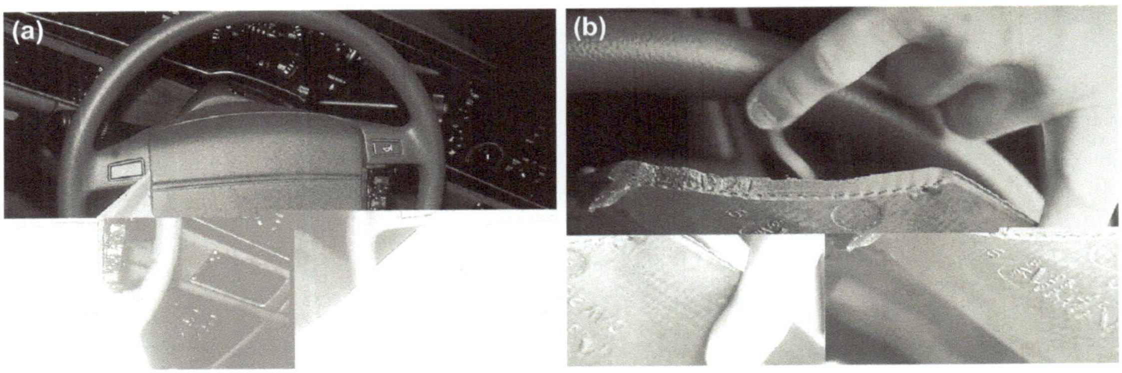

Figure 11 (a) The left lower corner of the module cover exhibited tearing from contact with the left side of the driver's face. (b) Close-up of the module cover reveals the presence of a torn area. There is also an area which indicates this piece was removed by a sharp implement.

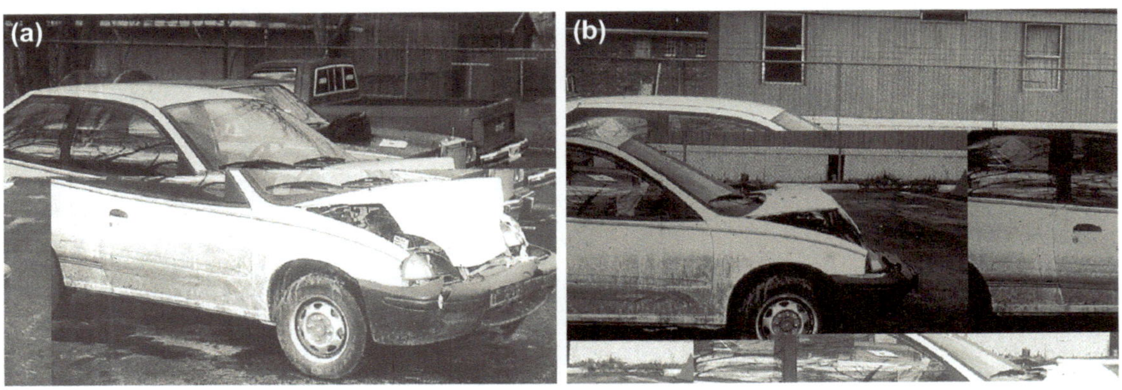

Figure 12 (a) and (b) A 1995 GEO Metro with minor front-end damage consistent with an impact speed of less than 15 kph (9 mph).

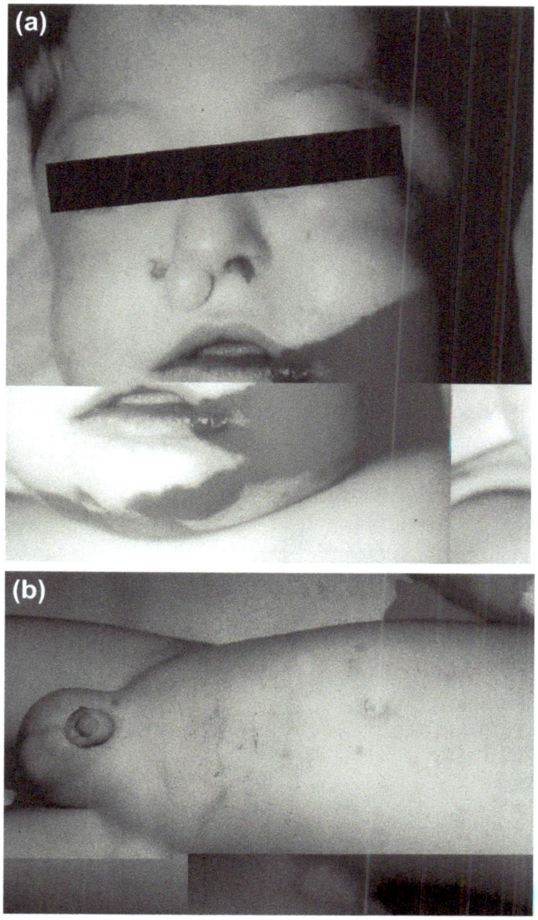

Figure 13 (a) This fatally injured 4-year-old male exhibits significant facial abrasion, overlying the mandible as well as abrasion on the left cheek. This was the result from contacts with the upwardly-deploying air bag. (b) Examination of the patient's abdomen reveals a horizontally-oriented contusion consistent with a lap belt.

Case 3

A 35-year-old female was the front seat passenger in a 1995 Nissan Altima. The patient was in a lap-shoulder belt, with the passenger seat in the most rearward position. The vehicle was stopped in a line of traffic, waiting for a traffic signal to turn, when it was hit from behind by a vehicle traveling at the speed of approximately 5 kph (3 mph). The rear impact pushed the Nissan forward into the trailer hitch of the truck in front. This resulted in air bag deployment. The air bag impacted the patient's left eye. Examination of the face revealed significant periorbital trauma. There were abrasions on the forehead as well as on the cheek and chin. Examination of the eye revealed the presence of chemosis, a hyphema, and a retinal detachment (**Figure 14**).

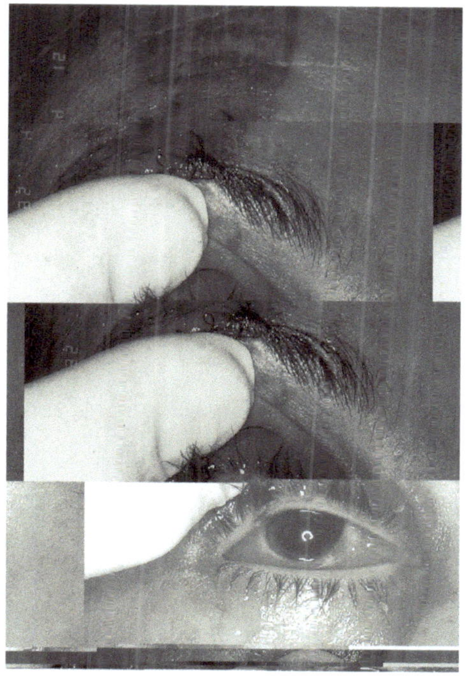

Figure 14 This 1.72 m (5′ 8″) restrained patient suffered permanent retinal detachment secondary to air bag contact.

Examination of the vehicle revealed a 5 × 5 cm dent in the right front bumper (**Figure 15**). There was no significant rear end damage. Examination of the bag revealed transfer of makeup and blood to the bag.

Forensics of Air Bag Injuries

Locard's principle regarding the transfer of physical evidence between two impacting objects is dramatically evidenced in the case of air bags and air-bag-induced injuries. The transfer of evidence to the air bag itself may take various forms. Blood and epithelial tissue transfer is, of course, common but transfer of makeup, including lipstick, rouge and mascara, to the air bag is also seen (**Figure 16**). Analysis of the blood spatter pattern on the bag may assist the investigator in determining the position of the occupant and the configuration of the steering wheel at the moment of air bag deployment

Examination of the air bag module cover may reveal the presence of trace evidence. Depending on the design of the module cover, there may actually be tearing or bending of the module cover, indicative of contact with an occupant's more rigid (bony) surface: face or forearm. Scuff-type marks on the module cover indicate contact with an object, frequently the forearm (**Figure 17**). Fabric imprints may also be seen on close inspection (**Table 1**).

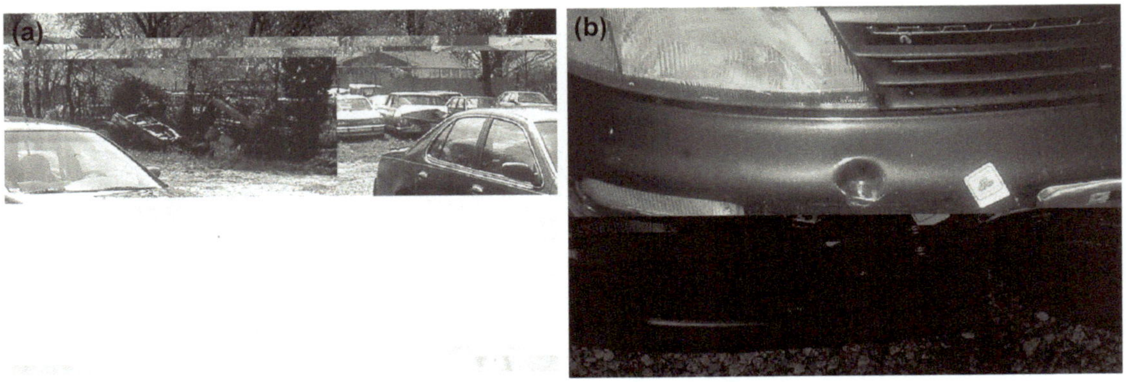

Figure 15 (a) and (b) A 1995 Nissan Altima. The only damage to the vehicle is a 5 × 5 cm indentation in the front bumper from impact with the ball of a trailer hitch.

Figure 16 Close examination of an air bag may reveal a multitude of transferred evidence. This evidence will include hair, blood, epithelial tissue, and facial makeup.

Figure 17 Close inspection of a module cover may reveal the presence of scuffing and fabric imprints.

Table 1 Trace evidence

Air bag
Blood
Makeup
Hair
Tissue
Module cover
Tearing of cover material
Scuff marking
Fabric imprint
Tissue
Hair

Summary

In the United States, the deployment of air bag has been responsible for nearly 200 deaths and thousands of severe injuries. It is clear that the forces of deployment are not insignificant and must be respected by the vehicle occupant. A review of the literature indicates that the serious and fatal injuries, which were once produced in the laboratory setting, are now being observed in "real-world" collisions. Many clinicians may not be aware of the injury-producing power of the deploying air bag and must be informed of the patterns associated with air-bag-induced injuries. The motoring public must also be informed and warned of the dangers of air bag deployment, just as the automotive industry was over 30 years ago.

See also: **Engineering:** Accident Investigation – Determination of Cause; Forensic Engineering/Accident Reconstruction/Biomechanics of Injury/Philosophy, Basic Theory, and Fundamentals.

Further Reading

Horsch, J., Lau, I., Andrzegkad, et al., 1990. Assessment of Air Bag Deployment Loads. SAE 902324. Society of Automotive Engineers, Detroit.

Huelke, D.F., Moore, J.L., Compton, T.W., et al., 1994. Upper Extremity Injuries Related to Air Bag Deployments. SAE 940716. Society of Automotive Engineers, Detroit.

Huelke, D.F., Moore, J.L., Ostrom, M., 1992. Air bag injuries and accident protection. Journal of Trauma 33, 894–897.

Landcaster, G.I., Defrance, J.H., Borrusso, J., 1993. Air bag associated rupture of the right atrium. New England Journal of Medicine: 328, 358.

Lau, I., Horsch, J.D., Viano, D.C., et al., 1993. Mechanism of injury from air bag deployment loads. Accident Analysis and Prevention 25, 29.

Mertz, H.J., Weber, D.A., 1982. Interpretations of Impact Response to a Three Year Old Child Dummy Relative to Child Injury Potential. SAE 826048. Society of Automotive Engineers, Detroit.

Mertz, H.J., Driscoll, G.P., Lennox, J.B., et al., 1982. Response of animals exposed to deployment of various passenger and portable restraint system concepts for a variety of collision severities and animal positions. In: National Highway and Transportation Safety Administration. 9th International Technical Conference on Experimental Safety Vehicles, pp. 352–368 Kyoto, Japan, November 1982.

Parents for Safer Air Bags, 1997. The Air Bag Crisis: Causes and Solutions.

Prasad, P., Daniel, R.P., 1984. A Biomechanical Analysis of Head, Neck and Torso Injuries to Child Surrogates Due to Sudden Torso Acceleration. SAE 841656. Society of Automotive Engineers, Detroit.

Smock, W.S., 1992. Traumatic avulsion of the first digit, secondary to air bag deployment. In: Proceedings 36. Association for the Advancement of Automotive Medicine, p. 444. Des Plains, IL.

Smock, W.S., Nichols, G.N., 1995. Air bag module cover injures. Journal of Trauma, Injury, Infection and Critical Care 38, 489–493.

Viano, D.C., Warner, C.V. 1976. Thoracic Impact Response of Live Porcine Subjects. CvSAE 760823. Society for Automotive Engineers, Detroit, pp. 733–765.

Electronic Data Recorders (EDRs, Black Boxes)

W Rosenbluth, Automotive Systems Analysis, Inc., Reston, VA, USA

Abbreviations

Accel	Accelerator
ECU	Electronic control unit
EDR	Event data recorder
VUT	Vehicle under test

Glossary

A/D Analog/digital (converter used in D/A processes).

ABS Antilock braking system, which prevents wheel lockup on mixed-friction road surfaces, thereby enhancing vehicle stability during braking.

AE Algorithm enable, also called algorithm wake-up, the acceleration threshold value (*x* time) which triggers the ECU software to start considering whether the cumulative crash pulse can be predicted to reach a value for which one or more passive restraints should be deployed.

ALDL Assembly Line Diagnostic Link, a historic term created by G.M., referring to a common diagnostic connector on a vehicle in the era before the common SAE J1962 (OBD-II) connector was in use. However, this term is now common in automotive jargon.

Algorithm wake-up Also called algorithm enable, the acceleration threshold value (*x* time) which triggers the ECU software to start considering whether the cumulative crash pulse can be predicted to reach a value for which one or more passive restraints should be deployed.

APS Accelerator position sensor, a sensor of the accelerator pedal which presents a position-analog voltage to an engine electronic control computer (engine control ECU).

CAN Controller area network (data bus), an advanced and robust intravehicle communications media for data transfer between variations on-vehicle system control ECUs.

D/A Data acquisition, a system whereby various vehicle parameters can be recorded on a common timeline for use in later analyzing vehicle performance.

DAWN Data acquisition with vehicle networks (HEM Data Corp.).

Delta-V The (abrupt) change in velocity during a particular event (such as would be caused by an impact event). This is distinguished from the approach velocity (vehicle velocity just before an impact event).

DLC Diagnostic Link Connector, see ALDL.

DPID Data packet identification number, usually for multiple data values (per SAE J 2190, p. 4).

DUT Device under test, refers to a specific component which is the subject of a particular test step.

ECU Electronic control unit. A typical on-board automotive computer assembly with a dedicated functional purpose.

EDR Event data recorder. Typically, a control ECU which incorporates nonvolatile memory (EEPROM) which records vehicle and system parameters in the event of a significant event (such as a crash which causes airbag deployment).

EEPROM Electrically erasable read-only memory (typical NVRAM implementation). Memory which saves its contents via a physical charge-storage property so that it retains its data even if removed from system power. Thus when system power is removed (as in a crash event), the EEPROM memory contents are saved.

ESC Electronic stability control, a system which can modulate engine output and/or apply individual wheel braking when it senses vehicle acceleration anomalies. Usually implemented in conjunction with ABS functions.

J1962 SAE J1962 connector. Refers to SAE J1962. The standard diagnostic port on all US domestic sold vehicles since 1996. Also referred to as the OBD-II connector (on-board diagnostics, version II) and the ALDL (Assembly Line Diagnostic Link).

KAM Keep alive memory, random-access memory (or a subportion thereof) which is forced to retain its data while energized with system power. When system power is removed, the KAM memory contents are lost.

 http://dx.doi.org/10.1016/B978-0-12-802718-9.00013-2

ms Millisecond, 1/1000 s.

NVRAM Nonvolatile random access memory (see EEPROM).

OBD-II On-board diagnostics connector, version II, a now Federally mandated common diagnostic connector for use in diagnosing on-board systems DTCs (originally purposed for emissions-related DTCs).

PCB Printed circuit board, insulating substrate with plated wires that form interconnects between various integrated circuit components and discrete components (resistors, capacitors, diodes, etc.).

PCM Powertrain control module. The engine electronic control computer (engine control ECU).

PID Parameter identification number, usually for a single data value (per SAE J 2190, p. 4).

RAM Random-access memory, memory in an ECU which is writable to save in-process values during the ongoing function of a control ECU. Such memory typically operates only when system power is applied and thus loses its contents when the vehicle key is turned off, or in a significant crash event.

ROM Read-only memory, memory in an ECU which has fixed contents (such as a control program and/or default calibration constants). Such memory is fixed and not writable by the normal production ECU.

SLOT SLOT factors are defined by SAE J2178-2 to be scaling, limit, offset, and transfer function specifications that allow hexadecimal encoded engineering data to be interpreted or mapped into engineering units such as DTCs, RPM, mph, lbf, psi, seconds, volts, amps, Gs, etc.

SRS Supplemental restraints system. The passive restraints system in a vehicle, which operates without operator intervention, typically controlling frontal airbags, side airbags, curtain airbags, and seatbelt pretensioners.

Thrott Throttle. Typically in EFI (electronic fuel injection) vehicles, an air valve. Fuel is adjusted by the PCM (engine computer, engine control module, ECM) to achieve an optimum mixture for the air input.

TPS Throttle position sensor, a sensor on the throttle body of an engine which presents a position-analog voltage to an engine electronic control computer.

VSS Vehicle speed sensor (usually on the transmission output shaft). Usually a reluctance or Hall effect device which presents a voltage pulse train whose frequency is directly proportional to the rotational speed of the transmission output shaft (and thus directly related to the speed of the vehicle). A reluctance device is a wire coil in a magnetic circuit which produces a voltage output as the magnetic field changes. A Hall effect device is a semiconductor element which produces a resistive change in its properties in response to a transverse magnetic field, thus producing a voltage difference when a constant current is flowing in that device.

WSS Wheel speed sensor (associated with each wheel, used for ABS/ESC functions). Usually a reluctance or Hall effect device which presents a voltage pulse train whose frequency is directly proportional to the rotational speed of the wheel on which it is mounted and thus can be used for comparison to other wheels having similar sensors, thereby allowing an ABS ECU to determine if there is excessive braking slip on any particular wheel.

Background of Electronic Data in Ground Vehicles

Superceding a long tradition of analog and mechanical technology for control of vehicle functional systems, current ground vehicle control technology is dominated by pervasive use of electronic controllers. Electronic control is achieved by using small-package, in-vehicle-mounted, dedicated, real-time computers (electronic control units, ECUs) to accomplish the logic and magnitude control of vehicle system parameters. Such controllers allow for increased calibration accuracy and enhanced diagnostics and serviceability when compared to their traditional analog forebears. ECU memory is used to save diagnostic information which often includes diagnostic trouble codes (DTCs) and information about the status of inputs (representing vehicle parameters) to that ECU at historical moments in the operating life of that ECU. When the ECU memory is used to save data triggered by a crash event, and that data can be later accessed for investigation purposes, that ECU is often referred to as an event data recorder (EDR).

Typical ECUs incorporate a controller (computing) device, memory, sensor input circuits, and actuator power driver circuits. (Often computational, memory, and analog/digital-conversion functions are integrated into a single integrated circuit device. Such devices are identified as a microcontroller unit (MCU) device.) ECU (MCU) memory can be broadly characterized by three categories.

1. ROM (read-only memory)—a fast read-only technology which has fixed contents and is typically used to store application program code and base (default) calibration data.

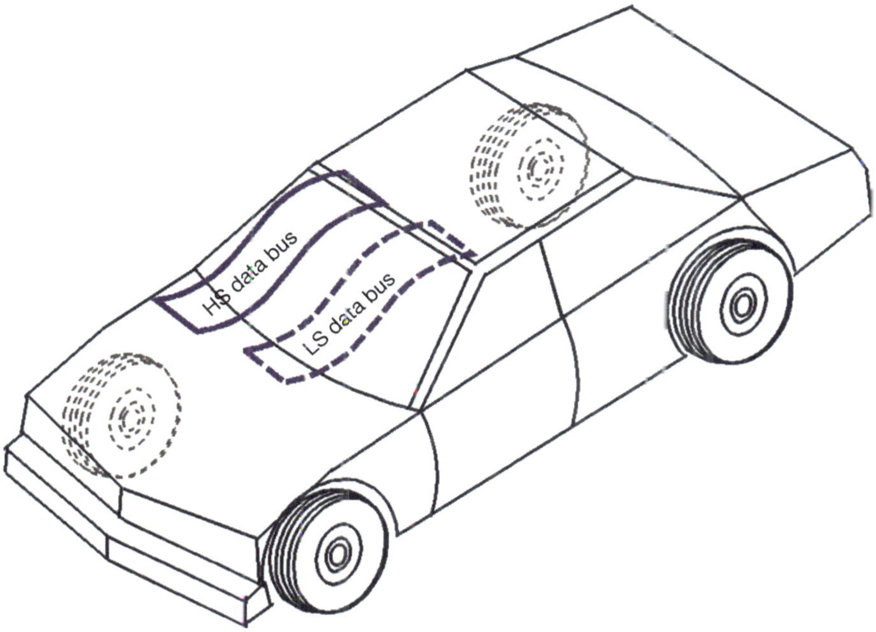

Figure 1 A generic wireframe vehicle with a dual data bus implementation, a high-speed bus (HS data bus) and a low-speed data bus (LS data bus).

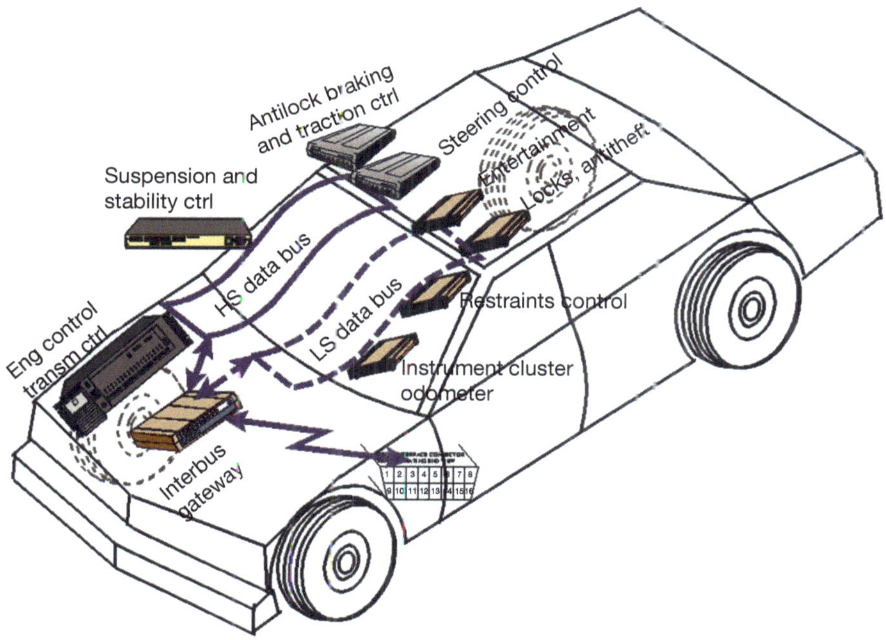

Figure 2 Vehicle operating system controllers requiring real-time intercommunication are grouped on the high-speed bus (HS data bus) and other systems controllers not requiring real-time intercommunication are grouped on the low-speed bus (LS data bus). (Although individual LS data bus controllers certainly act independently in real time. For instance, the restraints controller must autonomously act within 50 ms of the inception of a crash event to calculate the potential magnitude of such a crash event and then deploy passive restraints if appropriate. However, once it has acted with respect to restraints deployment, the need to communicate with other controllers can take 1000 ms or more.) This figure also shows how interbus communications are typically accomplished by using an interbus communications controller, often called gateway.

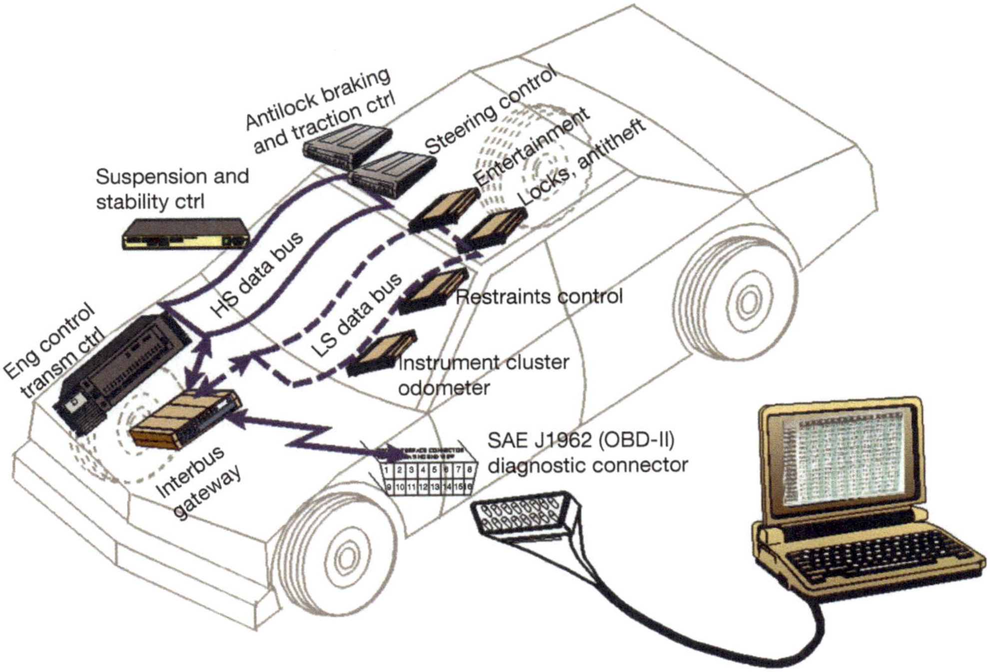

Figure 3 Vehicle diagnostics may be accomplished using a standard diagnostic data port directly accessing the gateway, which then allows indirect access to both data buses. Laptop-based diagnostics are shown in the illustration, but other small specialized computer terminals called scanners or code scanners can also be used in this manner. The standard diagnostic port is defined by SAE J1962 and has been mandated on US-sale vehicles since 1996. OBD-II, on-board diagnostics connector, version II.

2. RAM (random-access memory)—a fast read–write technology which is used as a calculation scratchpad for the algorithms in the computing element.
3. NVRAM (nonvolatile random-access memory)—a usually slower read–write technology which is used to save update or adaptive calibration data and diagnostic information. NVRAM can be battery dependent (KAM, keep alive memory, usually a part of RAM), or it can be a different technology which retains data even when the ECU is disconnected from a vehicle electrical system (EEPROM, electrically erasable programmable read-only memory). (EEPROM data retention times are often specified for periods of 10–20 years.)

Diagnostic information almost always includes DTCs and often includes information about the status of inputs (representing vehicle parameters) to that ECU at historical moments in the operating life of that ECU. (A common use of KAM is the storage of freeze frame data in powertrain control module (PCM) ECUs to assist in the diagnosis of emissions-related PCM DTCs.)

Current vehicle systems now typically relying on electronic control mechanisms, with the inherent ability to save selected data in NVRAM, include electronic fuel injection, engine emissions feedback, antilock braking, traction control, stability control, antitheft, occupant safety protection, and rollover protection. These various ECUs are invariably communicating with each other on one or more internal vehicle data bus(es). To give the reader an immediate reference, a series of stepped-complexity symbolic representations of distributed control ECUs and vehicle diagnostic data bus(es) are shown in **Figures 1–4** Since some, or all, of the ECUs may contain NVRAM data, there can be crash-related event data in more than one ECU, and this is clearly shown in **Figure 4**.

Retrieving ECU NVRAM Data for Use in Crash Investigations

A Word about the Data

EDR data exist as binary bits stored in the nonvolatile memory of a particular ECU. Such data must be accessed via electronic data

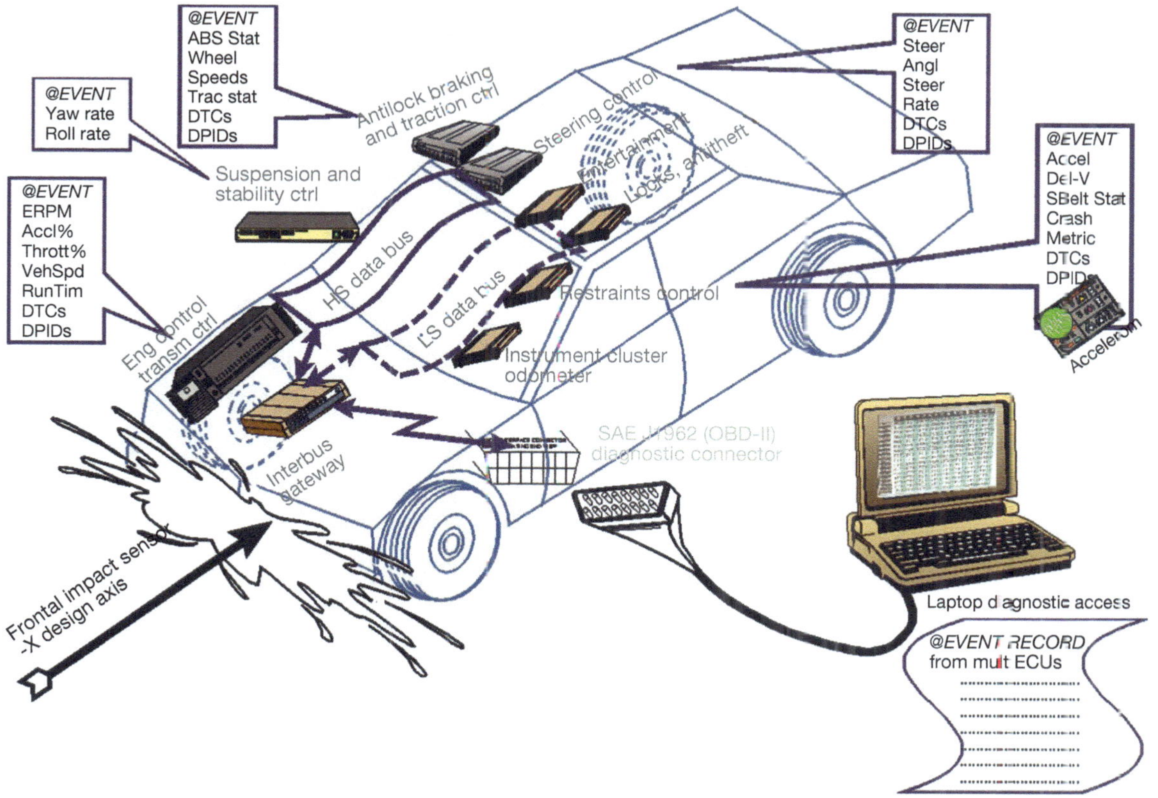

Figure 4 Vehicle operating parameters may be stored in different electronic control unit (ECU) system controllers and how these parameters, if captured at an event, may be accumulated to form an EVENT RECORD which may be retrieved via a specialized laptop access. If one ECU is designated as a central point to save the various parameter data in its nonvolatile memory, that ECU is typically designated as the event data recorder. *DTC*, diagnostic trouble codes; *DPID*, data packet identification number.

retrieval methods. Additionally, since binary bit representations can be long and are nearly impossible to assimilate, such data are normally represented in its most economical human readable form, hexadecimal notation. Hexadecimal notation is a base 16 arithmetic notation, with characters (0–9, A–F) corresponding to decimal values of 0–15 of the original binary value. Hexadecimal notation is most often represented with a leading dollar sign, "$." Hexadecimal notation typically considers "byte" quantities which consist of 8 bits of data. Each byte consists of two ½ bytes, each containing a hexadecimal character representing 4 bits (a "nibble"). For example, a binary value of "1001" = $9 = 9 decimal, and a binary value of "1110" = $E = 14 decimal. Just as in decimal arithmetic, when the maximum digit count is achieved ($F = 15 dec for hexadecimal), the next count is a carry to the left column. The first carry in decimal arithmetic (the next count after 9) gives a representation of 10, the decimal quantity, 10. The first carry in hexadecimal arithmetic (the next count after $F) gives a hexadecimal representation of $10; the decimal quantity, 16. As a last example, the binary value "0011 0101," represented as hexadecimal byte = $35, stands for a decimal quantity of 53 (3 × 16 + 5).

Post-crash-event investigations involving those vehicles now typically utilize such data, as translated into standard engineering units, and such investigations now commonly reference data stored in various EDRs in both bullet- and target-involved vehicles.

Retrieving the Data

The most common method to access EDR data on passenger vehicles and light trucks is via a standard vehicle network interface, the SAE J1962 port (a/k/a the OBD-II port (on-board diagnostics connector, version II), the DLC or the ALDL (Assembly Line Diagnostic Link)) as shown in **Figure 3**.

However, other, equally valid, methods can include direct umbilical-to-ECU interrogation and direct umbilical-to-EEPROM interrogation. If the objective of EDR data retrieval is its use in a litigation process, with either method, a key consideration is the reliability and consistency of output data and the preservation of the source data for other investigators.

With respect to passenger vehicles and light trucks, there are public and proprietary retrieval tools which can retrieve EDR data and present that data both as hexadecimal bytes and as translated engineering values. The most common data retrieval tool, specifically designed for impact-data retrieval from select supplemental restraints system ECUs and from select engine ECUs, is the Bosch/Vetronix CDR® (crash data retrieval tool) and this tool is employed by most vehicle crash investigators. The CDR operates using a serial data interface to the vehicle diagnostic network (see **Figure 2**), and the CDR capabilities are available for specific vehicle application sets. There are CDR options for data retrieval from the J1962 port (OBD-II connector) or from an ECU directly, both employing a serial data interface.

While the CDR presents an orderly and tested means of EDR data retrieval, there are many vehicle models with NVRAM-incorporated ECUs (EDRs) not covered in the CDR application set. Many investigators, when confronted with non-CDR-covered vehicle do not pursue EDR data retrieval. Many manufacturers and EDR suppliers can access such non-CDR-coverage data with proprietary data retrieval tools; however, there is little, if any, public access to such proprietary tools. Additionally, when manufacturers and EDR suppliers provide data to other parties, that data can be seriously incomplete (or incompletely translated). For succeeding discussions, vehicles contained within the CDR application set are referred to as CDR vehicles, and vehicles outside of the CDR application set are hereafter referred to as non-CDR vehicles.

There are three options to retrieve EDR/ECU data.

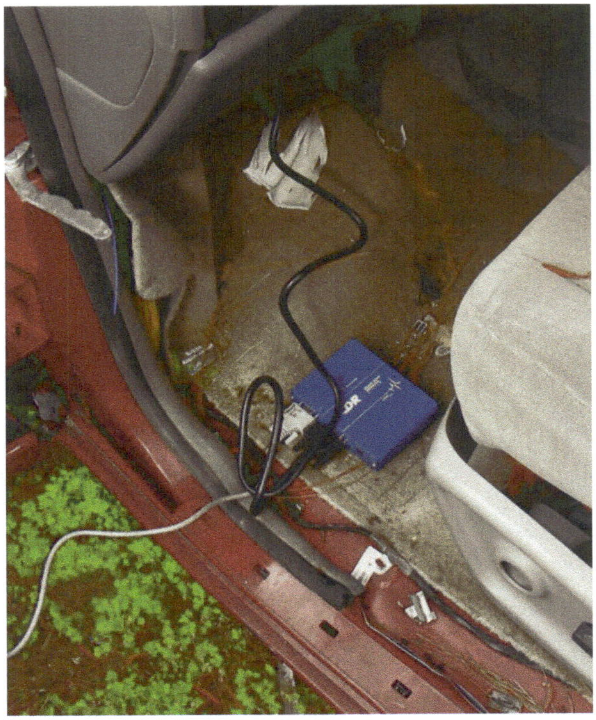

Figure 5 A typical CDR (crash data retrieval tool) connection to the vehicle J1962 data port (aka the OBD-II (on-board diagnostics connector, version II) port or the ALDL (Assembly Line Diagnostic Link) port).

Method 1. Via a vehicle serial data link communication path to the vehicle diagnostic communication port

Method 1 serial communications operate through the vehicle diagnostic connector (SAE J1962, OBD-II, connector) as described above. Often, the data retrieved via this method are DPID (data packet identification number) data, which is a representation of the actual raw NVRAM data, but not the actual NVRAM data itself. For direct access of the actual NVRAM data, there is usually a special mode and security feature incorporated into the serial access protocol—and this prevents general public access to the actual raw stored event data (**Figures 5–7**).

Method 2. Via a serial data link communication path connected via a direct umbilical cable to the target ECU

Alternative serial communications operate via a direct umbilical cable to the EDR/ECU itself (also incorporating the security access feature). The direct-ECU umbilical method is invariably nonforensically neutral and can be nonconforming with ASTM E2493-07. This is because direct connections to the EDR/ECU generally present an incomplete system environment to the EDR/ECU and will often lead to test-induced data changes (e.g., generating DTCs for nonexisting sensors, squibs, etc.). Such violations of ASTM E2493-07 practices, when the investigation involves a litigation context, can lead to conditions considered as "data spoliation" in that context (**Figure 8**).

Method 3. Via a direct umbilical to the EEPROM component on the printed circuit board within the EDR assembly

This method requires disassembly of the ECU to access the printed circuit board, cleaning the EEPROM component contacts with a solvent, and operates by using a direct electrical connection of the EEPROM device and an EEPROM reader. This process retrieves the raw binary data in hexadecimal notation format. This process, while far more tedious and time-consuming than methods 1 and 2, overcomes the serial security access controls (by avoiding it entirely), retrieves the actual raw NVRAM binary data (in hexadecimal

BOSCH

Hexadecimal Data

Data that the vehicle manufacturer has specified for data retrieval is shown in the hexadecimal data section of the CDR report. The hexadecimal data section of the CDR report may contain data that is not translated by the CDR program. The control module contains additional data that is not retrievable by the CDR system.

$01	B2	3C	00	E6	64	69
$02	00	00	00	E0	18	00
$03	00	00	00	00	00	00
$04	70	88	40	00	00	00
$05	3F	00	00	06	C0	E2
$06	3F	00	00	06	C0	E2
$07	00	00	C0	00	22	00
$08	00	08	08	50	50	00
$09	04	64	04	64	7F	00
$0A	00	00	00	00	00	00
$0B	08	F6	00	00	FE	00
$12	FF	7F	00	84	40	00
$17	03	03	03	02	00	00
$18	03	03	00	00	00	00
$19	00	00	00	00	00	00
$1A	00	00	00	00	00	00
$1B	00	00	00	00	00	00
$1C	4F	4F	00	4F	00	01
$1D	32	C1	00	00	00	00
$1E	00	00	00	00	00	00
$1F	0C	0D	31	32	5C	5C
$20	20	93	03	00	00	00
$21	7F	7F	7E	7F	7F	7E
$22	01	00	00	00	00	00
$23	00	FF	FF	1E	59	00
$24	00	FF	1E	59	F9	00
$25	82	C0	00	00	00	00
$26	41	4F	4F	4F	4F	00
$27	00	01	01	01	01	01
$29	26	26	26	26	26	00
$2A	46	46	46	46	46	00
$2B	00	00	00	00	00	80
$2C	01	01	00	01	00	40
$2F	00	00	00	00	00	80
$30	00	00	00	00	00	80
$36	8C	00	00	00	00	00
$37	00	A5	00	00	00	00
$38	00	00	00	00	00	00
$39	00	FF	FF	1E	58	00
$3A	FF	1E	58	00	00	00
$3B	FF	FF	FF	FF	FF	FF
$3C	FF	FF	FF	FF	FF	FF

Figure 6 Virtual hexadecimal data (i.e., DPID (data packet identification number) data) as retrieved by the CDR (crash data retrieval tool).

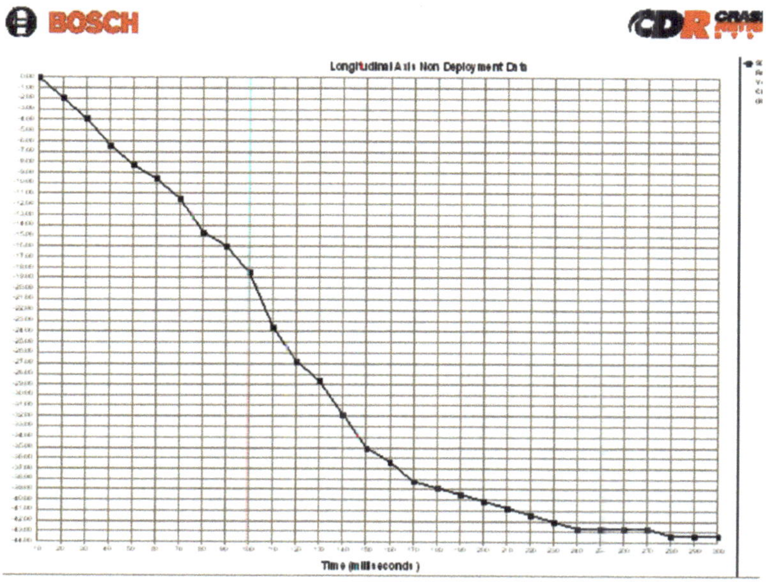

Figure 7 Typical integrated longitudinal Delta-V data as produced by the CDR (crash data retrieval tool) postprocessing software which translates the retrieved DPID (data packet identification number) data.

Figure 8 A data retrieval using a CDR (crash data retrieval tool) direct connection to a supplemental restraints system electronic control unit.

representation), and also avoids the problem of test-induced data changes. Thus this process (as is method 1) is forensically neutral and is in conformity with ASTM E2493-07. As this methodology works independently of serial data communications, it is especially valuable for non-CDR vehicles. It has been demonstrated for retrieval of crash event data in EDRs on vehicles from domestic, European, and Asian manufacturers (**Figures 9 and 10**).

The Use of ECU NVRAM Data in Crash Investigations

Post-crash-event crash investigations involving those vehicles now typically utilize such EDR engineering data, and such investigations now commonly reference data stored in various EDRs in both bullet- and target-involved vehicles. It is incumbent upon the investigator wishing to introduce such data as "evidence" in the litigation process to be able to confirm to the court that such "evidence" is stated with reasonable engineering certainty and that it meets certain clear tests for such admission. These tests are typically lumped under the umbrella terms, "Frye criteria" or "Daubert criteria." These tests require that the retrieval, resultant translation, and interpretation of EDR data must:

1. Be founded on an underlying methodology of scientific tests, documentation of test results (charts, graphs, etc.), which have been reviewed by peers who have confirmed the reliability and acceptance of test methods. This can include refereed publication of tests and methods.
2. Be repeatable by any peer scientist/engineer using the conditions and/or data recorded in the subject tests. This means that other investigators must be able to repeat the subject tests and derive the same test results from the EDR at issue.
3. Have a reasonable error rate. This means that the error rate for the method employed is consistent and that it is accepted by other practitioners in the field.
4. Be consistent with industry or commonly accepted standards that define or direct the tests and results postulated. This means that the procedures (methodology) must be compatible with applicable sections of ASTM E860, ASTM E1188, and ASTM E2493.
5. Pass a test of relevance and linkage to the facts at issue in the subject litigation. Obviously, the essence of the tests used must be consistent with the conditions of the accident under investigation, and the results presented must be relevant to the issues under consideration by the trier of fact (the jury).

Figure 9 Electronic control unit disassembly and the direct electrically erasable programmable read-only memory data retrieval process.

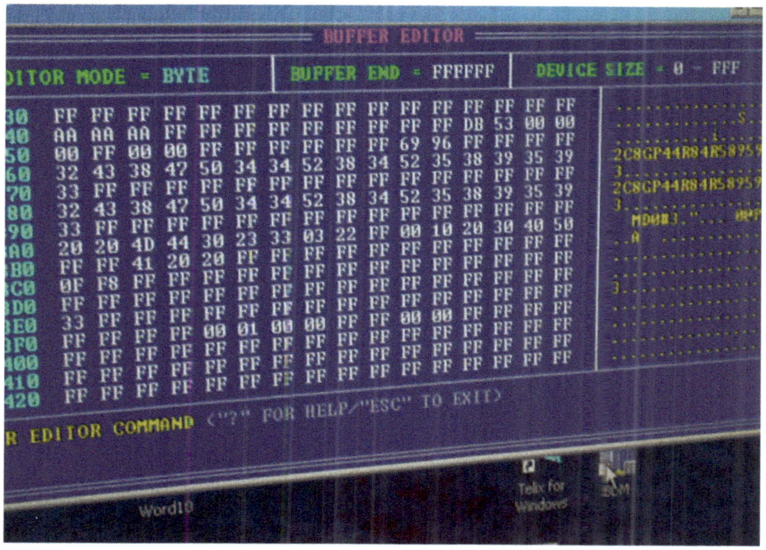

Figure 10 A portion of the raw hexadecimal data retrieved through a typical direct electrically erasable programmable read-only memory (EEPROM) data retrieval process. Note that in this case, the immediate ASCII translation shows the vehicle VIN as written to the EEPROM. (The VIN was correct for the donor vehicle.)

Acknowledgment

The author wishes to acknowledge Mr Fred Chandler, Chandler & Sons Automotive, Herndon, VA, who contributed to many aspects of the testing and raw data retrieval work described herein.

See also: **Engineering:** Analysis of Digital Evidence; Forensic Engineering/Accident Reconstruction/Biomechanics of Injury/Philosophy, Basic Theory, and Fundamentals; Human Factors Investigation and Analysis of Accidents and Incidents; Investigation and Analysis of Structural Collapses.

Further Reading

Note: It is hoped that this introduction to EDRs encourages the reader to continue his or her research on the subject. To assist with that, several references for further reading are presented below.

American Society for Testing and Materials E1188–95, 1995. Standard Practice for Collection and Preservation of Information and Physical Items by a Technical Investigator. ASTM International, West Conshohocken, PA, pp. 19428–22959.

American Society for Testing and Materials E860–97, 1997. Standard Practice for Examining and Testing Items that Are or May Become Involved in Litigation. ASTM International, West Conshohocken, PA, pp. 19428–22959.

American Society for Testing and Materials E2493–07, 2007. Standard Guide for the Collection of Non-volatile Memory Data in Evidentiary Vehicle Electronic Control Units. ASTM International, West Conshohocken, PA, pp. 19428–22959.

Bosch/Vetronix. CDR[W] Data Retrieval Tool Vehicle Coverage Information, Can Be Found at Vetronix Corporation 2030 Alameda Padre Serra, Santa Barbara, CA 93103–1716, USA, 1 800 321-4889. http://www.boschdiagnostics.com/testequipment/diagnostics/cdr/Pages/CDRHome.aspx.

Chidester, A., Hinch, J. Roston, T., 2001. Real world experiences with event data recorders. In: Proceedings of the 17th International Technical Conference on the Enhanced Safety of Vehicles (ESV) Conference, Amsterdam, The Netherlands, 4–7 June 2001. Washington, DC: National Highway Traffic Safety Administration, DOT HS 809 220. Paper Number 247, 11 pgs.

Chidester, A., Hinch, J. Mercer, T., Schultz, K., 1999. Recording automotive crash event data. In: National Highway Traffic Safety Administration Paper, International Symposium on Transportation Recorders, Arlington, VA, USA, May 3–5, 1999.

Gabler, H.C., Hampton, C.E., Hinch, J., 2004. Crash Severity: A Comparison of Event Data Recorder Measurements with Accident Reconstruction Estimates. Rowan University, Glassboro, NJ. National Highway Traffic Safety Administration Washington, DC, 8 pp. Accident reconstruction 2004, Warrendale, SAE, 2004, pp. 81–88; Report No. SAE 2004–01–1194; UMTRI-98091.

Gabler, H., Hinch, J., Steiner, J., 2008. Event Data Recorders – A Decade of Innovation. SAE International, Warrendale, PA, ISBN 978-0-7680-2066-3. Product Code: PT-139.

Ishikawa, H., Takubo, N., Oga, R., et al., 2009. Study on Pre-crash and Post-Crash Information Recorded in Electronic Control Units (ECUs) Including Event Data Recorders. National Research Institute of Police Science, Japan. Paper Number 09–0375.

Kowalick, T., 2004. Fatal Exit: The Automotive Black Box Debate. Wiley.

Murthy, S.K., Satish, C.A., n.d. Diagnostics for Automobiles – A Snapshot. Dearborn Electronics, Bengaluru. http://www.deindia.com/images/downloads/whitepapers/Diganostcs_for_Automobiles-A_Snapshot.pdf.

Niehoff, P., Gabler, H., Brophy, J., Chidester, A., Hinch, J., Ragland, C., 2005. Evaluation of Event Data Recorders in Full Systems Crash Tests. National Highway Traffic Safety Administration, United States. Paper No: 05–0271.

Reust, T.J., 2004. The Accuracy of Speed Captured by Commercial Vehicle Event Data Recorders. Accident Science, Newhall, CA, 8 pp. Accident reconstruction 2004, Warrendale, SAE, 2004, pp. 115–122; Report No. SAE 2004–01–1199; UMTRI-98091.

Riling, J., 1995. Sensing and Diagnostic Module for Airbags. SAE Technical Paper, 952682. http://dx.doi.org/10.4271/952682.

Rosenbluth, W., 2001. Investigation and Interpretation of Black Box Data in Automobiles: A Guide to the Concepts and Formats of Computer Data in Vehicle Safety and Control Systems. The American Society for Testing and Materials (ASTM) and the Society of Automotive Engineers (SAE).

Rosenbluth, W., 2009. Black Box Data from Accident Vehicles, Methods for Retrieval, Translation and Interpretation, ASTM Monograph 5. ASTM International, West Conshocken, PA.

Rosenbluth, W., 2010. Collecting EDR Data for Crash Investigations, Forensic Magazine. Vicon Publishing, Amherst, NH.

Ruth, R., Daily, J., 2011. Accuracy of Event Data Recorder in 2010 Ford Flex during Steady State and Braking Conditions. Detroit, MI: SAE 2011-01-0812.

Ruth, R., West, O., Nasrallah, H., 2009. Accuracy of Selected 2008 Ford Restraint Control Module Event Data Recorders. Detroit, MI: SAE_2009-01-884.

Takubo, N., Ishikawa, H., Kato, K., et al., 2009. Study on Characteristics of Event Data Recorders in Japan; Analysis of J-NCAP and Thirteen Crash Tests. SAE Technical Paper 2009-01-0883. http://dx.doi.org/10.4271/2009-01-0883.

Gaasbeck, Van, Stephen, 01 February 2007. How to challenge black box data. Trial Magazine.

Relevant Websites

http://www.cdr-system.com—Crash Data Group: Crash Data Retrieval Tool.

Analog Tachograph Chart Analysis

RF Lambourn, Transport Research Laboratory, Wokingham, UK

Glossary

Odometer A device that displays the total distance traveled by a vehicle since its manufacture, usually incorporated in its speedometer.

Rolling road A set of rollers on which the wheels of a vehicle may be placed in order to test its speedometer.

Tachograph Device that records speed and distance traveled by a vehicle, primarily for the purpose of regulating a driver's hours of work and also of use in determining speeds etc. in incidents.

Tachograph chart A stiff paper chart, usually circular, on which traces recording speeds, distances, and modes of work are scribed.

Introduction

The tachograph is a device fitted to motor vehicles, which makes a recording of the speed traveled and distance driven, together with details of the driver's periods of work and rest. Although found in vehicles worldwide, it is particularly used in the countries of the European Union (EU), where it is a requirement in most larger goods and passenger-carrying vehicles. Analog tachographs make this recording on a paper chart. However, in the EU, these devices are obsolescent, with vehicles registered from 2006 having to be fitted with digital tachographs that store the data electronically.

A note on digital tachographs will be found at the end of this chapter, but otherwise it is concerned only with analog tachographs and lays particular emphasis on the instrument as it is specified in EEC Regulations. However, the principles of chart analysis and the use of tachograph data described here are applicable to all types of analog device. In any case, the user of tachograph data must be aware of the characteristics of the particular make and model from which it comes, to appreciate the amount of information available and its limitations. The range of models is large, and this chapter cannot therefore cover every detail of every instrument that may be encountered.

The Forensic Use of Tachograph Data

Tachograph data have two main forensic applications:

1. in road accident investigation, to determine the speed at and immediately before an incident, the rate of braking, and the manner of driving;
2. in general criminal investigation, to find the route traveled and the time when a vehicle was at a particular location.

The Tachograph Chart

Figure 1(a) and (b) shows typical paper tachograph charts and illustrate the information provided by the instrument. The shape and design are not specified by the EEC Regulations, but the chart pattern shown here has become the standard for the industry. Other designs do exist, but the instruments that require them are obsolete.

The chart is a disk (123 mm in diameter) coated with a material that blackens under pressure. In the center is an area where the driver writes his name, the starting and finishing places for his day's driving, the date, vehicle number and odometer readings.

Printed around the edge of the chart, and repeated further in, is the timescale of 24 h. Between these is the field where the speed is recorded, usually with a maximum of 125 km h^{-1}. Further in is the work-mode field, where the driver indicates whether he is driving, engaged on other work, or resting. The indication is made by either the track in which the recording line runs (**Figure 1(a)**) or the thickness of the line (**Figure 1(b)**). It is sometimes possible to extract useful information from the mode-of-work line when the vehicle has been moving very slowly.

The innermost recording is of distance, shown by a zigzag line. One stroke of this line is made during 5 km of travel; a complete V indicates 10 km of movement, whereas other distances will create partial strokes.

 http://dx.doi.org/10.1016/B978-0-12-802718-9.00014-4

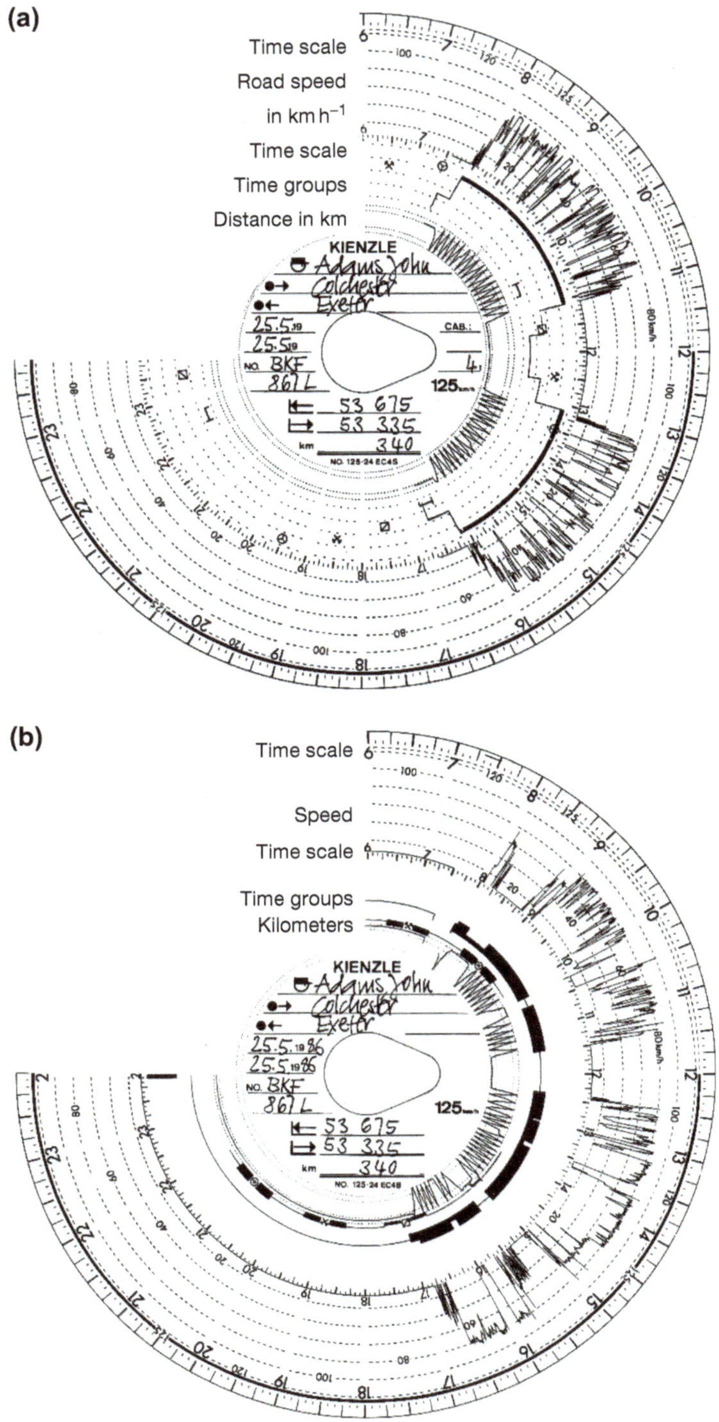

Figure 1 Tachograph chart showing recordings with (a) "manual" time group recordings and (b) "automatic" time group recordings (Continental Automotive Trading UK Ltd.).

Figure 2 The face of a Veeder Root model 8400 tachograph.

Figure 3 The Kienzle MTCO1324 modular tachograph. The lower part is the drawer onto which the chart is placed.

The Tachograph Instrument

Older models of tachograph combine their function with that of a speedometer in a single unit and are therefore designed to be mounted on a vehicle's dashboard in front of the driver. **Figure 2** shows such an instrument. More recent models, which are called "modular," have an appearance akin to a car radio or CD player and are located elsewhere in the cab, while being linked to a conventional speedometer. **Figure 3** shows an example of a modular tachograph.

Externally the older instruments (as in **Figure 2**) have the usual speedometer features—a speed dial and an odometer display—together with a clock and knobs with which the driver and a colleague can indicate their modes of work. (Most instruments hold two charts to cater for there being two drivers crewing the vehicle.)

Figure 4 An opened Kienzle model 1318 tachograph showing the insertion of a chart.

In these instruments, the chart usually lies behind the face, which is hinged to open downward to allow insertion and removal. **Figure 4** shows an open instrument with a chart about to be inserted. The spindle on which it is mounted rotates once in 24 h. When the instrument is closed, the chart bears against three styluses that move up and down to make the recordings. The spindle and the hole in the center of the chart are oval to ensure the correct orientation of the chart with respect to time.

With the modular instrument in **Figure 3**, the charts are placed on a tray where, when it is pushed into the body of the instrument, they are brought into contact with the recording styluses. **Figure 5** shows a modular instrument with the tray open and a chart in place.

Tachographs may operate either mechanically or electronically. Mechanical types, in which a rotating cable from the vehicle gearbox drives a cup-and-magnet system, are now rarely seen. Typical models are the VDO Kienzle TCO1311 and the Veeder Root 1111. Electronic types receive a train of pulses from a transducer at the gearbox, the frequency of which is translated as speed. There are many models: examples of the unitary type being Motometer EGK100, Jaeger G.50, VDO Kienzle TCO1314 and TCO1318, and Veeder Root 8300 and 8400; and of the modular type being Kienzle MTCO1324 and Stoneridge (Veeder Root) 2400.

The Principles of Chart Analysis

Figure 6 shows a schematic tachograph trace and the quantities to be measured in an analysis. The speed trace is thought of as a series of points connected by straight lines. The speed at each point (v_1, v_2, v_3) is measured, as are the time intervals between them (t_1, t_2, t_3). This provides a plot of speed against time, which can then be integrated to yield a plot of speed against

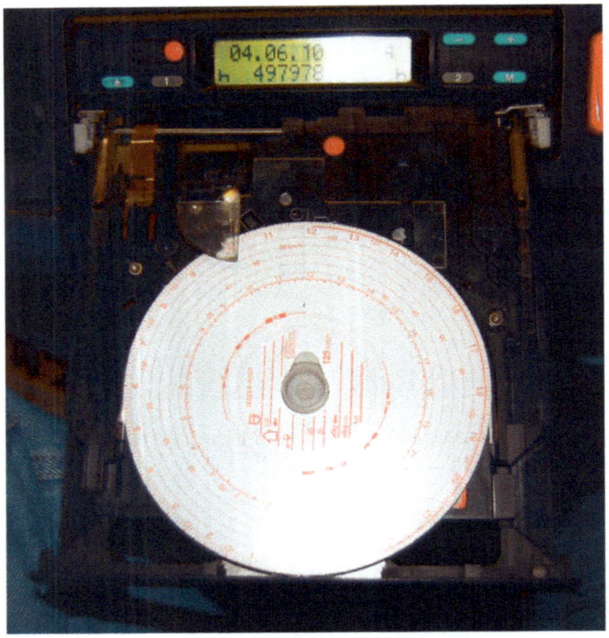

Figure 5 An opened Kienzle MTCO1324 tachograph with a chart placed on the drawer.

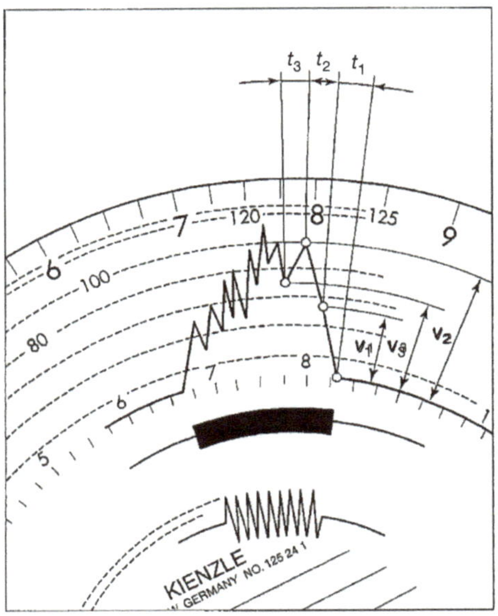

Figure 6 Schematic speed trace showing time intervals and speeds.

distance. **Figure 7** shows a real tachograph trace, **Table 1** shows the tabulated analysis, and **Figure 8** shows the resulting plot. (This trace and analysis is used for the case example at the end of this chapter.)

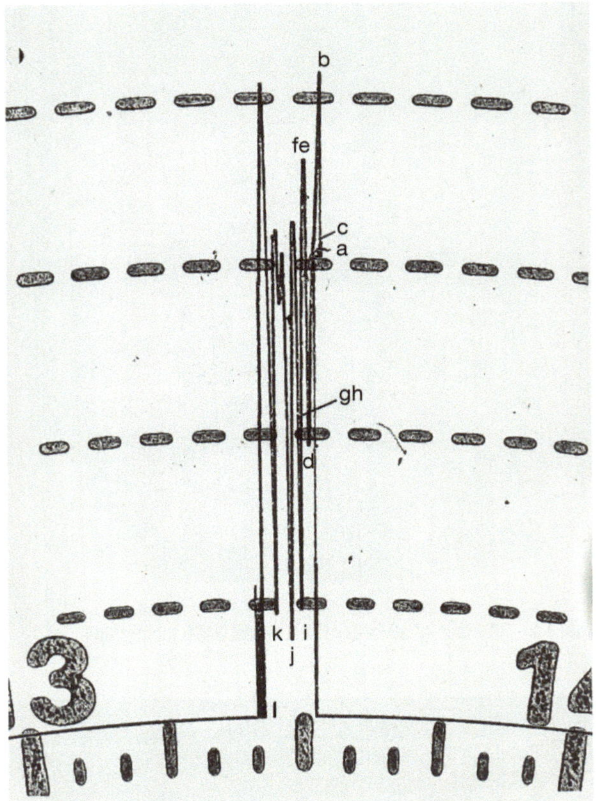

Figure 7 A real tachograph speed trace with analysis points.

There are various approaches to making these measurements. Taking the speed from the chart is relatively straightforward and can be done from a photographic enlargement or with a traveling microscope. Allowance has to be made for the accuracy of the speed recordings. Measurement of the time intervals is, however, not at all easy. Because the chart rotates

Table 1 Data from tachograph chart shown in **Figure 7**: the analysis points *a*, ..., *i* run backward from the end of the driving

Point	*Speed (km h^{-1})*	*Time (s)*	*Distance (m)*
a	–	0	0
b	81	0	0
c	61	32	630
d	38	67	1110
e	71	83	1355
f	71	87	1435
g	40	111	1805
h	39	116	1860
i	19	122	1905

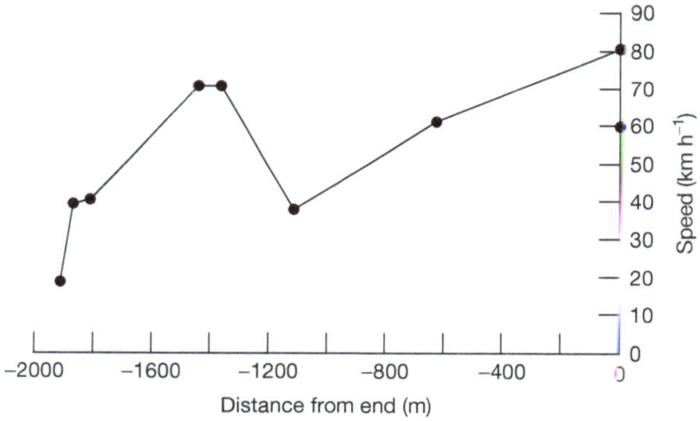

Figure 8 Plot of speed data from **Figure 7** against distance.

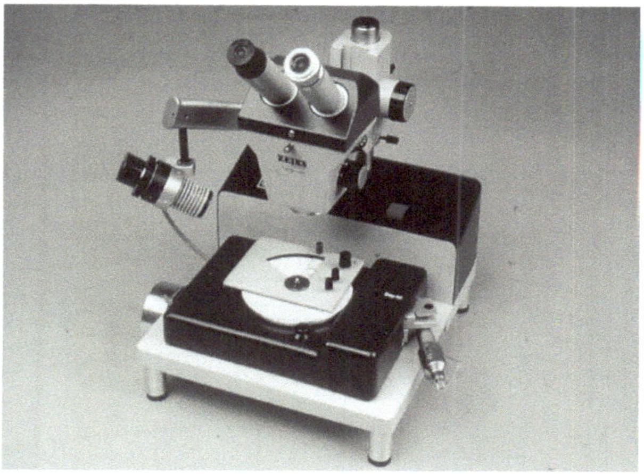

Figure 9 The Kienzle microscope with computer-linking modifications.

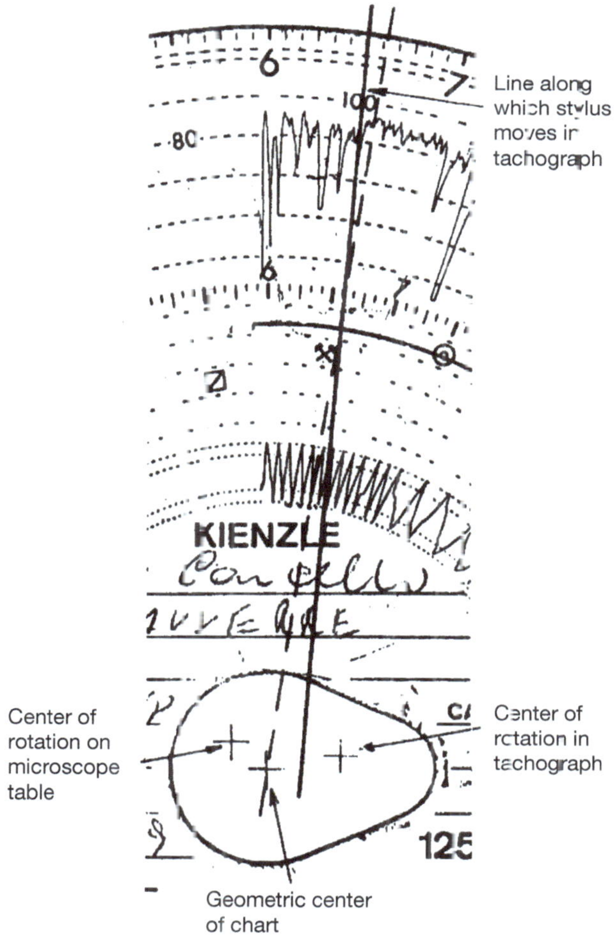

Figure 10 The need for an adjustable cursor when analyzing a chart.

just once in 24 h, 1 min occupies only 0.25°, and in 1 s it turns through only 15 s of arc. It is usual to mount the chart beneath a microscope on a rotating table that is turned by a micrometer.

Figure 9 shows one such microscope, as produced by VDO Kienzle. A glass plate etched with a vertical cursor line, 3 μm wide, lies over the chart, and this combines with a horizontal line in one eyepiece to form a "crosswire" in the field of view. Points for measurement are brought successively under the crosswire by rotating the chart table and moving it backward and forward, and transducers are attached to these two movements. Signals from the transducers, which are interpreted as time and speed, are fed to a microcomputer, where listings of these quantities and graphs of speed versus time and distance are produced.

It is very important to recognize, when measuring time intervals, that the line along which the recording stylus moved was unlikely to have passed exactly either through the center of rotation of the chart when mounted on the microscope or when mounted in the tachograph instrument, or through the geometric center of the chart itself. This is illustrated in an exaggerated form in **Figure 10**. If the vertical cursor is not arranged such that it lies along the line on which the stylus moved, any time intervals measured between points at different speeds will be in error. In the VDO Kienzle microscope, the glass plate carrying the cursor line can be swiveled about a point at the outer edge of the chart, and this adjustment is used to set the line in what the analyst judges to be the correct position.

The Accuracy of the Speed Record

The EEC Regulations require that having been installed and calibrated, the analog tachograph record may not have an error in its speed figures greater than ±6 km h^{-1}. The analyst may be content to use this as a probable error figure, but the author's experience is that although the error is generally not greater than ±3 km h^{-1}, there are occasional cases where it is substantially more than ±6 km h^{-1}. It is therefore desirable that the calibration in each case be checked, and various methods have been devised for this.

Calibration Checks on the Tachograph Installation

The most straightforward method of checking the accuracy of a tachograph system is to employ a "rolling road" of the sort used when tachographs are installed in vehicles. A chart is placed on the instrument, and the vehicle is then effectively driven through a range of known speeds. The chart will then show how each speed is recorded, from which the amount of error is found.

Some agencies find the rolling road method quite inconvenient, and instead use a method that requires no special facilities and that can be employed in any open space. The procedure is in two stages: first, the signal sent from the vehicle's gearbox is measured (either as pulses per kilometer in electronic systems or turns per kilometer in mechanical ones) in one of two ways, and the second, the way the tachograph instrument responds to a known set of input signals is found.

The first of the two ways of measuring the signal is called the 20-m track and is one that may actually be used by some installers. It is fully described in manuals produced by tachograph manufacturers for the guidance of their agents. It is appropriate for vehicles that can be moved slowly in a straight line; thus, they need not be fully mobile, and may have, for example, accident damage to the steering gear or the engine coolant system. The test can even be performed by towing the vehicle.

In this method, the drive to the tachograph is either disconnected and a turn or pulse counter is attached to it, or, with later electronic models, a pulse counter is plugged into a socket in the head itself. The vehicle itself is then driven slowly on a straight and level surface for 20 m, the distance being measured exactly. From the number of pulses and the distance, the rate per meter is calculated: this is usually referred to as the w figure.

The second way is called the static method and is appropriate for vehicles that cannot be moved, either through accident damage or through restriction of space. In this method, a turn or pulse counter is attached in the same way as in the 20-m track method. One of the driven wheels of the vehicle is then lifted off the ground, and its circumference is measured with a tape; later, the other driven wheels are also measured. The raised wheel is then rotated manually 10 times and the resulting turn or pulse reading is noted. From these data, the w figure can be calculated, remembering to take account of the effect of the axle differential, which averages the rotation of the wheels at each end: thus, with one wheel being stationary and the other wheel being turned, 10 turns of the single wheel is equivalent to five turns of the two wheels together and therefore of the whole axle.

If the accident damage is so great that neither of these methods can be used, then the investigator may have to resort to examining gear trains and counting gear teeth.

Having determined w, the response of the tachograph instrument itself is found by driving it in steps through its speed range with a sequence of signals generated by a specially constructed test unit. These units are produced by the tachograph manufacturers for the use of their installation agents and display either the mechanical rotation rate or the electronic pulse rate (in units of min^{-1}). A new tachograph chart is put in the instrument, and a test chart similar to that in **Figure 11** is produced.

The sequence of tests in **Figure 11** is given as follows:

1. The instrument is taken up to its maximum speed (140 km h^{-1} in this example) and then very quickly back to

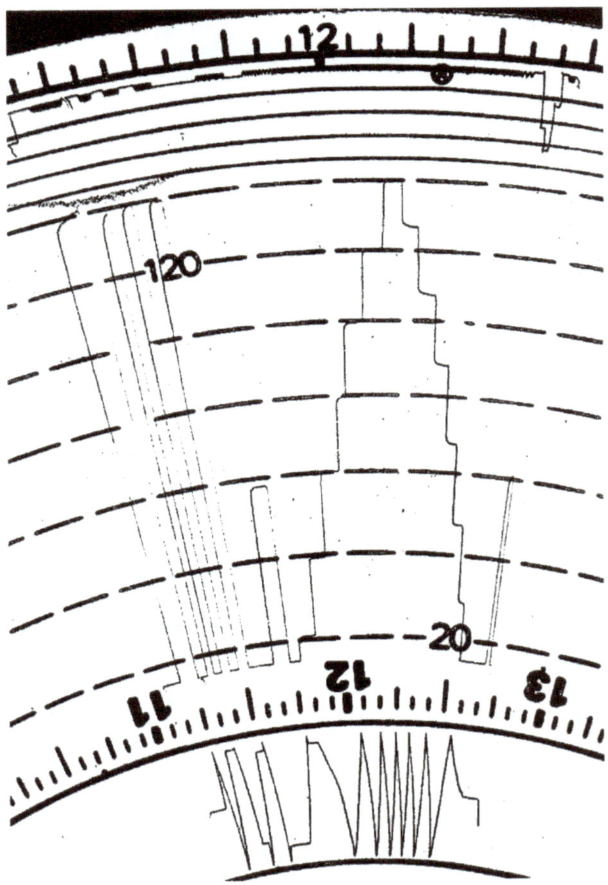

Figure 11 A typical head test chart.

zero to check the straightness of the recording stylus movement.

2. This is repeated three times.
3. Next, the instrument is taken to an indicated speed of 60 km h^{-1}: at this speed the input figure (in min^{-1}) is known as the characteristic coefficient or *k* figure and should be, in a correctly calibrated system, approximately equal to the *w* figure of the vehicle.
4. The speed is taken from zero up to the maximum in 20 km h^{-1} steps, holding each step for 2 min, and then back down to zero at the same speed between the steps (130 km h^{-1}, 110 km h^{-1}, etc.).
5. Finally, the instrument is run at 60 km h^{-1} for as long as it takes 1 km to be recorded on its odometer, and the actual number of turns or pulses delivered is noted: this should equal the κ figure. During the last test, the activity mode switch is also changed through its four settings.

The information from part (4) of the test can now be used, together with the *w* figure, to construct a graph of how the recording on the chart at any point relates to the true speed. This can be done manually or by using the software associated with the Kienzle computer-linked microscope.

The "Route-Trace" Method of Calibration

An alternative to the direct methods of calibration is the so-called route-trace method. This can only be used when the route the vehicle has taken is known and is particularly appropriate when the vehicle is unavailable for testing or has been so seriously damaged as to make the other methods impossible. The tachograph record can also be used to find the route that a vehicle has taken (see Route Tracing below). The calibration method is an inversion of this procedure and briefly involves attempting to match the recorded pattern of driving to the known route. A system that is in calibration will give an immediate match, whereas one that is out of calibration will require an adjustment of its speed by some factor before the match is found. Thus, the calibration method is one of finding what this adjustment factor is.

Problems of Interpretation

Once the cursor on the analyzing microscope has been correctly aligned and the errors, if any, in the speed record have been determined, the chart can be analyzed. However, the limitations of the recording and the individual instruments must be recognized.

Time Intervals

The clock in most tachograph instruments advances the chart in 1-s step, and it is, therefore, not possible in principle to measure the time at a particular point on the chart with a precision greater than ±1 s. When measuring the interval between two points, the precision can therefore never be greater than ±2 s. In any case, the edge of the speed-recording trace is insufficiently smooth for these steps to be resolved. With a good trace, a confidence of ±2 s can probably be assumed, but in some instruments an unsteadiness will be evident, which means that wider uncertainty must be accepted. It is found that although some manufacturers' instruments yield consistently good traces, other manufacturers produce tachographs that are invariably poor.

Figure 12 shows an example of a poor trace. The drive train from the clock to the chart is somewhat loose, and allowing a generally unsteady speed trace as the stylus moves up and down, it permits the stylus, as it falls after rising to a maximum, to track back along the groove it has made on the chart instead of drawing a new line. This effectively drives the clock backward

Figure 12 An unsteady trace from a Jaeger mechanical instrument.

for a few seconds (20 s being typical) until the slack is taken up, and only then the stylus will start to follow a new track.

In some tachographs, severe steering or swerving may cause the recording trace to deviate, and a likely example appears in **Figure 13**, where the period of perfectly constant speed, which can be seen just before the final descent of the trace to zero speed, was probably caused by a swerving action that made to chart shift slightly within the instrument.

Impacts to the Vehicle

A severe impact to the vehicle will generally produce a disturbance of the tachograph recording traces (**Figures 7 and 13**). When such a disturbance is apparent on the speed trace, it is obviously an indication of the speed of the vehicle at the moment of impact and is an item of the information easily read from the chart. However, some caution is needed before interpreting it. The disturbance was properly caused by a shock to the tachograph itself, and its magnitude therefore depends on the proximity of the impact to the instrument. A collision with a pedestrian, where the pedestrian meets the vehicle very close to the instrument panel, can cause a clear disturbance, whereas a severe impact from another vehicle to the rear of an articulated lorry may have little or no effect on the recording.

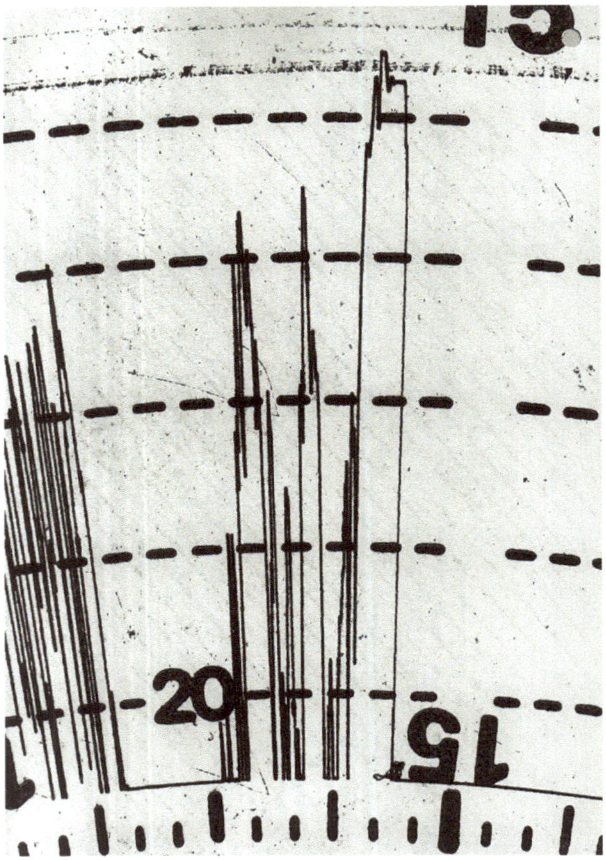

Figure 13 Chart showing lock-up followed by impact at zero speed.

If heavy braking is taking place at the time of the impact, the speed at which the disturbance appears may be affected by other factors.

Finally, a severe impact will often derange the instrument for at least a few seconds, such that little or no reliance can be placed on any features that may immediately follow it.

Response of the Instrument

The EEC Regulations require tachograph instruments to be able to follow "acceleration changes" of $2\ \mathrm{m\,s^{-2}}$. Since during braking the deceleration of a vehicle may well be as much as $7\ \mathrm{m\,s^{-2}}$, there is the possibility that the speed-recording stylus may not be able to keep up with the changing speed. This does not, in fact, appear to be a problem in Kienzle, Motometer, Jaeger, and mechanical Veeder Root tachographs, but there is a difficulty with older electronic Veeder Root instruments.

Earlier Veeder Root 1200 and 1400 series tachographs only just met the $2\ \mathrm{m\,s^{-2}}$ requirement; early 8300 series models had a response of $2.5\ \mathrm{m\,s^{-2}}$, whereas later 8300 series and all 8400 series models have a response of $5.7\ \mathrm{m\,s^{-2}}$. The consequence of a response of 2 or $2.5\ \mathrm{m\,s^{-2}}$ is that during heavy braking, the speed being recorded at any moment will be greater than the true speed, and any event recorded during the braking, for example an impact, will be shown at too high a speed. It also means that it is not possible for an analyst to give an opinion as to whether a particular piece of braking was at a normal or an emergency rate.

A response of $5.7\ \mathrm{m\,s^{-2}}$, however, is much closer to the braking rate, which can be expected of a lorry in an emergency, and will generally allow a reasonable speed for an impact to be recorded.

An even higher response rate leads to a difficulty where there has been a wheel slip during braking.

Tire Slip Effects

The tachograph records the speed of rotation of the wheels to which it is connected, almost always the driven wheels of the vehicle. Therefore, if for some reason those wheels slip, that is, rotate at a speed that does not correspond to the speed of the vehicle over the ground, then the recording will be in error.

Wheel slip particularly occurs during heavy braking, notably when the driven wheels lock. An instrument with a fast response will react to wheel locking as though the vehicle had instantaneously come to a halt: the stylus will fall very quickly to its base position even though the vehicle is still traveling forward, and an impact may even appear to have been at zero

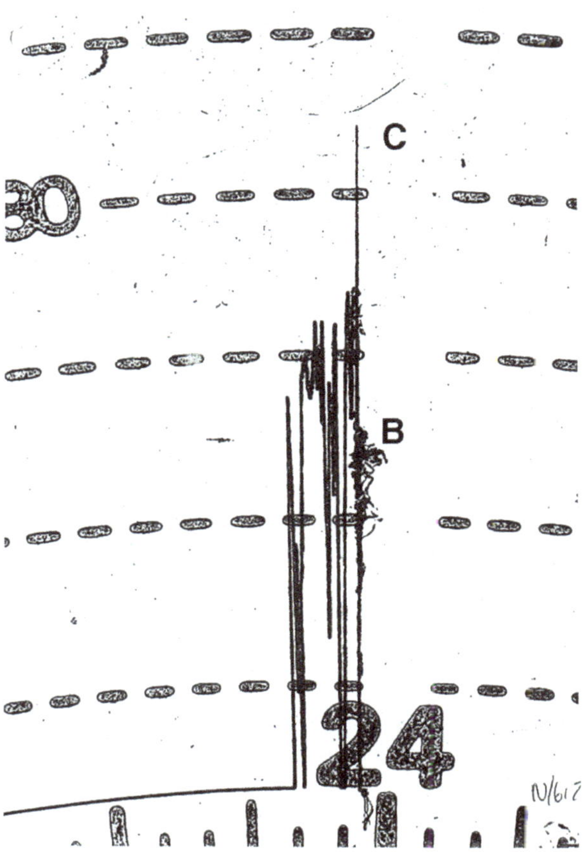

Figure 14 Spin up in rollover.

speed. An example is shown in **Figure 13**, where, because the vehicle has skidded, the prominent impact feature will certainly be at a level much lower than the true speed. The operation of an antilock braking system will also cause the speed of the wheels to be less than the road speed of the vehicle.

The opposite can occur if one of the driven wheels lifts off the road, allowing it to spin. This typically occurs as a vehicle starts to roll onto its side. In **Figure 14**, the speed at C would appear to be that at which the accident has occurred. In fact, the correct speed is at B, from which the trace has jumped to C, as the wheels on one side of the lorry lifted from the road.

Low-Speed Behavior

Incidents at low speed cause particular problems in the interpretation of tachograph data because all instruments have a level below which they do not record speed. In all current models, this is about 5 km h^{-1}, although in some older instruments it is somewhat higher, up to 16 km h^{-1}. Therefore, it can neither be said, for example, whether a vehicle that has slowed before entering a main road actually stopped or merely reduced its speed to, say, 5 km h^{-1} before proceeding nor may it be possible to say whether or not during an apparent stationary period the vehicle was moved a small distance.

Some information can, however, be gained from the mode-of-work trace. Movement of the vehicle generates a thicker line, made by an oscillation, or an increased oscillation, of the recording stylus (see the examples in **Figure 1**). A small gap in the thick line can be taken as a brief stationary period. The threshold for this thick line depends on the model. Mechanical instruments will produce a noticeable movement in the stylus only when the vehicle has moved for about 20 m or more. Electronic instruments have a critical speed at which the line is made. The mode trace in Kienzle 1318 and 1319 instruments, however, operates during driving in the same way as the zigzag distance trace, making one stroke over a distance of 50 m.

Falsifications and Diagnostic Signals

Tachographs have always been prone to attempts by drivers to falsify the recordings, to show either lower speeds or shorter hours of work. Most falsifications are readily apparent on careful examination, for example, by comparing the apparent average speed during a journey with the recorded distance and the time taken, or by comparing the total recorded distance for the day's driving with the difference in the handwritten odometer figures. **Figure 15** shows a commonly attempted falsification, where driving has stopped at 1736, the clock has been rewound, and driving has resumed, apparently earlier, at 1623.

The scope for such tampering is greatly increased with electronic instruments, and therefore, various "diagnostic" features are incorporated in them to show when certain irregularities occur. These depend on the individual models of instrument, and a full account of them is beyond the scope of this chapter. However, two examples are as follows.

If the pulse sender is disconnected in an attempt to make it appear that the vehicle is not being driven, the speed stylus will oscillate between zero and, typically, 30 km h^{-1} to make a broadband (**Figure 16**).

If the electrical supply to the instrument is interrupted in an attempt to halt recording for a period, then, on reconnection, the speed stylus will execute a characteristic movement, for example, a full-scale deflection or a dip to zero speed.

Other attempts at falsification, that can be more difficult to detect, include connecting the wires from the pulse sender to earth; reducing the voltage to the tachograph, such that the clock continues to operate but the other electronic systems fail; and connecting certain spare terminals on the back of the instrument together, again to cause some of the electronic functions to fail.

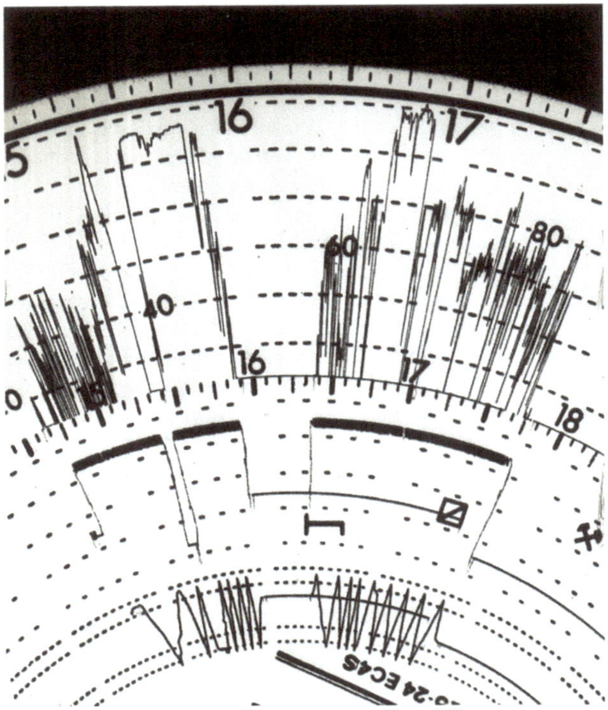

Figure 15 Fraudulent rewinding of tachograph clock. The overlapping traces show that the time was turned back from 1736 to 1623 h.

Figure 16 Broad trace generated when speed transducer is disconnected.

Case Example

Figure 7 is an enlarged section of a tachograph recording made by a tourist bus that came into collision with a pedal cycle. The bus had been on a slip road, approaching a main dual-carriageway road. The pedal cyclist was traveling on the main road and appears to have cut suddenly across the path of the bus. The bus driver braked hard and swerved to his right. **Figure 17** shows the route the bus had taken to the scene of the accident.

The results of the detailed analysis, given in **Table 1** and plotted in **Figure 8**, indicate that (point *i*) about 120 s before the accident the speed of the bus was at a minimum of 19 km h^{-1} and that it was about 1900 m from the scene of the accident. Its speed then increased to a maximum (points *f* and *e*) of 71 km h^{-1} before falling to another minimum (point *d*) of 38 km h^{-1}. It was then about 67 s and 1110 m from the end.

Again the speed increased, reaching its final maximum (point *b*) of 81 km h^{-1} immediately before the collision. There was then heavy braking to a sudden disturbance of the trace (point *a*), which on the chart is at an indicated speed of 60 km h^{-1}. However, the time interval between *a* and *b* is too small to measure, and at the scene, locked wheel skid marks were found from the rear wheels on one side of the vehicle. The rapid fall of the recording trace between the last two points has clearly occurred as a result of the locking of these driven wheels, which accounts for the immeasurably small time interval and also means that the figure of 60 km h^{-1} (because of the characteristics of the particular instrument) is probably significantly less than the speed at which the disturbance occurred.

The disturbance itself consists of a thickening of the trace, caused by a small shock to the stylus, followed by a step to the left (i.e., "backward" in time). The shock would have been caused by the impact at the front of the bus with the pedal cycle, whereas the sudden swerve would have brought about some sideways movement of the components of the tachograph instrument, to create the step to the left.

Route Tracing

The two minima of speed in **Table 1** lie at distances from the end, which, on an accurate map, can be readily identified with a roundabout (point *d*) and an intersection (point *i*). When the analysis is extended back to the start of the driving, two further minima and the start of the journey itself can be matched to a complex road junction and two more roundabouts (points *j*, *k*, and *l* in **Figure 7**). Some very slow driving before point *i* indicates that the bus started its driving from a place just before the roundabout.

In the context of a road accident, the matching of the trace to a known route serves to confirm the accuracy of the analysis process and of the calibration of the tachograph installation. If it had not been possible to check the calibration

Figure 17 Section of Ordnance Survey map showing route of vehicle as determined from the tachograph recording in **Figure 7** (Crown copyright reserved).

independently, this matching, with any necessary adjustment of the speed figures, could have been used as a calibration procedure in itself.

However, if it is imagined that, in different circumstances, the route of the vehicle was unknown, and there was a need for it to be found, it can be seen that working back from the known end point (*a*), it would be possible to do so by comparing distances between speed minima with distances on a map between road junctions. A typical case where this method is applied would be where a vehicle carrying a valuable load is stolen from a known location *X*, driven to an unknown place *Y* and unloaded, and then driven to a third place *Z* and abandoned. From its tachograph record, it may be possible to locate *Y* by tracing back from *Z* or forward from *X*. The success of this method depends on traffic conditions; in heavy traffic, the vehicle may have had to stop so many times at places other than road junctions, where following its route becomes impossible.

Digital Tachographs

EEC Regulations now require that from May 2006, all tachographs fitted to vehicles that fall within their scope be of the digital rather than analog type. Digital tachographs make their recording on a removable "smart card" rather than a paper chart, although a printer within the instrument also provides a summary of some of the data on a paper strip. Digital tachographs have similarities with the event data recorders (EDRs) fitted to some vehicles. Current models of digital tachograph include Siemens VDO DTCO1381, Stoneridge SE5000, and Actia SmarTach L2000.

The specifications to which they all conform require, among other things, that speed is recorded with a resolution of 1 Hz and that the smart card holds the record for the last 24 h of driving (not simply the last 24 h of time). The 1 Hz record of earlier driving is deleted, although simplified data are still held on times and distances traveled and maximum speeds. But manufacturers may choose to exceed the specifications, and, for example, the Siemens VDO DTCO1381 holds a record of the speed at 4 Hz for the *xx* seconds preceding the last stop made by the vehicle.

The reader is referred to EEC Regulation 1360/02, manufacturers' literature and the chapter on EDRs for further information.

See also: **Engineering:** Accident Investigation—Determination of Cause; Electronic Data Recorders (EDRs, Black Boxes).

Further Reading

EEC, 1985. Council regulation (EEC) no. 3821/85. Official Journal of the European Communities L370, 8–21.

EEC, 2002. Council regulation (EEC) no. 1360/02. Official Journal of the European Communities L207, 1–252.

Lambourn, R.F., 1985. The analysis of tachograph charts for road accident investigation. Forensic Science International 28, 181–199.

Lehmann, H., 1992. Die mikroskopische Diagrammscheibenauswertung für die Unfallrekonstruktion (Microscopic analysis of tachograph charts for accident reconstruction). Verkehrsunfall und Fahrzeugtechnik 1, 2–4.

Lowe, D., 1989. The Tachograph, second ed. Kogan Page, London.

Needham, P., 1988. Tachograph chart computer-linked microscopic analysis. Forensic Science International 36, 211–218.

Wach, W., Unarski, J., Duś, A., 2005. Tachograph Chart – Analysis of Intensive Braking Recording. Technical Paper 2005-01-1185. Society of Automotive Engineers, Warrendale, PA.

Traffic Injuries and Deaths

M Graw and J Adamec, Institute of Legal Medicine, Munich, Germany

Road Traffic

Epidemiology

Traffic accidents are a very common cause of injury or even death (one of the leading causes of death in young adults) and as such are of high forensic relevance. The vast majority of deaths and injuries result from road traffic. In 2009, the number of accident victims and injured persons exceeded 33 000 and 2 200 000 in the United States; in the European Union, more than 35 000 persons died and 1 700 000 sustained nonfatal injuries in traffic accidents. WHO reports that approximately 1 300 000 people die each year worldwide and another 20–50 million are injured. According to the WHO statistics, low-income and middle-income countries have significantly higher road traffic fatality rates than high-income countries (the fatality rate in high-income countries is approximately 10 per 100 000 persons, in low-income countries twice as much).

Of the above-reported US fatalities, 72% were car occupants (52% drivers and 20% passengers), 13% motorcyclists, 12% pedestrian, and 2% cyclists; similar percentages have been reported in Europe. According to WHO estimates, the proportion of vulnerable road users among the fatalities is higher in low-income countries, so that globally almost half of those who die in road traffic crashes are pedestrians, cyclists, or riders—motorized two-wheelers. The numbers of fatalities are decreasing in most industrialized countries continuously. In most regions of the world, the epidemic of road traffic injuries is still increasing.

Relevant Injury Types

Accident investigation is a very complex interdisciplinary work aiming at the reconstruction of time history of parameters related to the movement and interactions of vehicles and persons immediately before and during an accident. Injuries constitute very important traces. The routine clinical documentation would typically be insufficient because some minor, and thus for the treatment irrelevant injuries, would potentially not be accounted for. For this reason, a thorough body inspection (in nonfatal injured) or an autopsy by a forensic pathologist is to be performed in order to assess the injuries and collect other traces (small parts from vehicle, etc.) from the body. Care must be taken regarding the anthropometric data—basic body measures as well as the exact locations of all injuries must be ascertained and documented. Apart from descriptions and photographs, computed tomographies (CTs), 3D-scans, and other methods can be employed for injury documentation. In accident investigation, the skin and bone injuries are the most important because they carry the most information about the interaction between the human body and its surroundings. Apart from injuries, an analysis of the clothing is to be performed (though not necessarily by the pathologist).

There are two types of information an injury analysis can offer—injuries enable inferences regarding the injury mechanism, that is, the art of loading, and regarding the amount of loading. Thus, based on injury analysis, the investigator can reason back as to the particular cause. The immediate injury cause is always the force acting in the injured tissue. However, there is not always a clear topographic relationship between the external force acting on the surface of the human body and the injury location; injuries might occur in distant body regions because of force transmission. An example—indirect rib fractures caused by thorax compression—is shown in **Figure 1**.

Skin injuries

Abrasions—superficial skin injuries resulting from tangentially inflicted force (scraping along the body surface). Its direction can be established based on the morphology (bevel descent at the beginning and a tag consisting of accumulated epidermis at the end). The shape of the injury gives clues about the size of the contacting object; sometimes a pattern can be found that reflects some of its morphology.

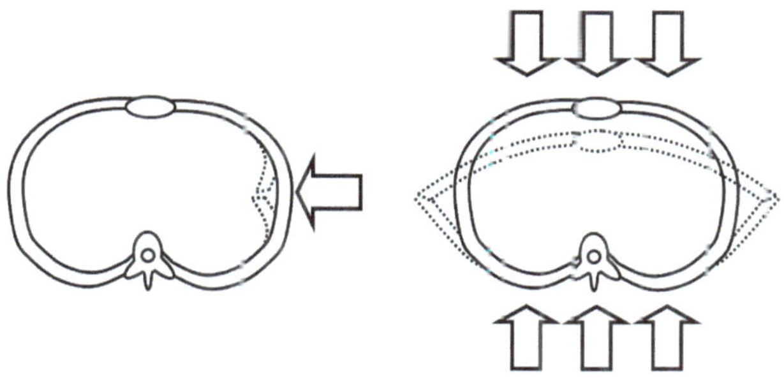

Figure 1 Direct (left) and indirect (right) rib fractures; the former as a result of an interaction with an object in the location of the fracture, the latter as a result of chest compression in the anterior–posterior direction (the ribs brake on the side, i.e., not at the contact location).

 http://dx.doi.org/10.1016/B978-0-12-802718-9.00015-6

Stretch injury—if the skin is stretched beyond its elasticity, injury occurs along the cleavage lines (defined by the natural orientation of collagen fibers). If the skin was stretched as a result of extensive movement of body segments, the injury shows in typical locations over protruding bony structures; such injuries give clues about the segmental movements during the accident. An example is shown in **Figure 2**.

Hematoma—soft tissue bleedings resulting from blood vessel damage. Depending on the size of the blood vessel and the surrounding soft tissue structure, even serious blood loss affecting the hemodynamics might occur. Hematoma shows on the skin typically only after a lag of several hours or days. Its resorption takes up to several weeks and is accompanied by color changes that might help an experienced investigator to narrow down its time of origin. Typically, its shape reflects initially the morphology of the contacting object, and in the course of time, it changes as the blood follows the cleavages in the soft tissue, the muscle fasciae, and so on. Tire profile might leave negative impression on the skin (see **Figure 3**).

Decollement—separation of the subcutis from the underlying fascia as a result of tangential blunt force imposed on the affected body part, accompanied by a formation of a wound cavity that might be filled with blood (see **Figure 4**). This injury type results typically from being ran over by a car tire or from a primary contact between a pedestrian and a car. The overlaying skin is frequently intact; sometimes tire marks can be observed.

Laceration—apart from incisions (cuts), open wounds can result from compression, shearing, and/or tensile force, that is, from blunt force. The edges of the wound might be ragged, bruised, or abraded and thus potentially giving hints as to the contacting object. Sometimes bridging strands (blood vessels, nerves, tissue strands) can be found especially in the deeper layers. The shape of the wound can help to identify the contacting object.

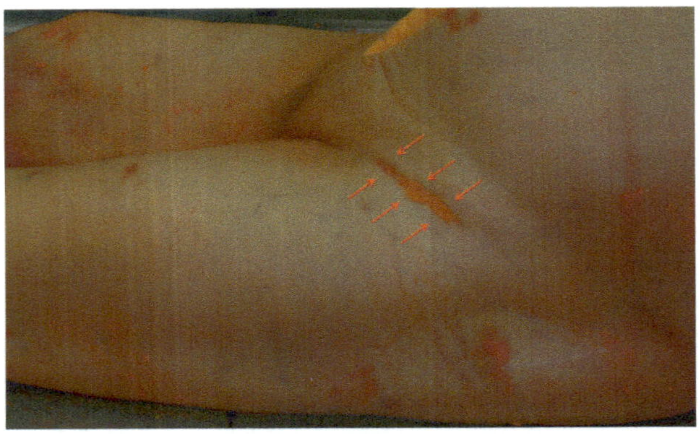

Figure 2 Typical stretch injury of the skin in the groin caused by extensive extension of the left hip (pedestrian accident victim).

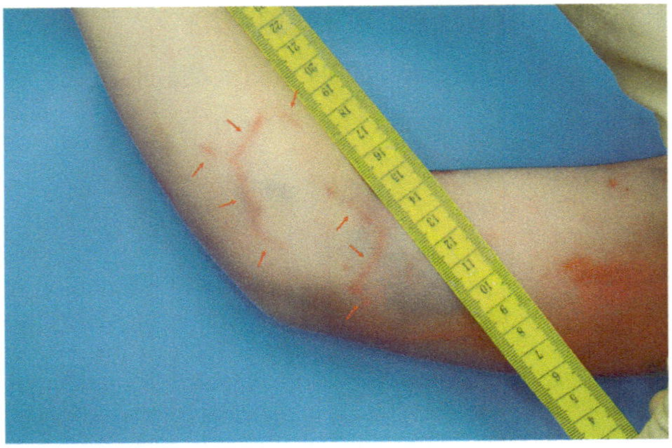

Figure 3 Tire profile marks on the skin of the right arm of an accident victim.

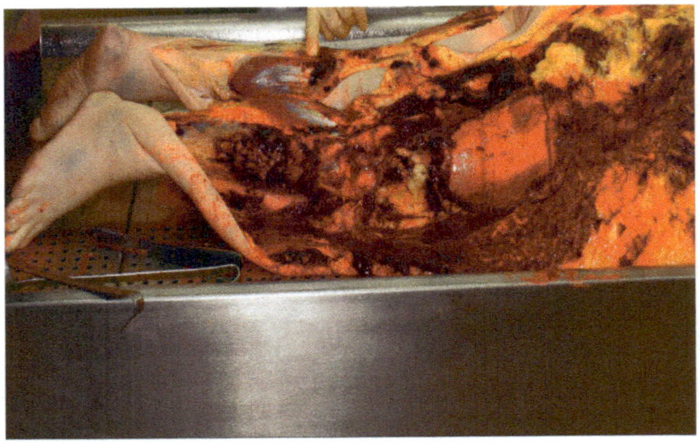

Figure 4 Wound cavity filled with blood developed as a result of direct blunt trauma (primary impact of a pedestrian hit by a passenger car).

Head injuries

Traumatic brain injuries

Traumatic brain injuries (TBIs) are mechanical insults to the brain associated with temporary or permanent impairment of brain function. The distinguishing mark of TBI is an altered state of consciousness.

From a clinical point of view, TBIs are classified by using the Glasgow Coma Scale (GCS) of the patient shortly after the traumatic incident:

Mild TBI: $13 \leq \text{GCS} \leq 15$
Moderate TBI: $9 \leq \text{GCS} \leq 12$
Severe TBI: $\text{GCS} \leq 8$

The GCS quantifies the conscious state of a patient based on his/her eyes (eyes do not open—1 point; eyes open in response to pain—2 points, eyes open in response to voice—3 points, eyes open spontaneously—4 points), verbal (no sound—1 point, incomprehensible sound—2 points, inappropriate words—3 points, confused verbal statements—4 points, normal conversation—5 points), and motor (no movements—1 point, extension in response to pain—2 points, abnormal flexion in response to pain—3 points, flexion/withdrawal in response to pain—4 points, localization of pain—5 points, obeying of commands—6 points) responses.

The traditional and still widely used classification based on the morphology differentiates also three degrees:

1. Commotio cerebri (concussion)—a temporary and reversible impairment of brain function (diagnostic criteria: loss of consciousness, amnesia, nausea/vomit, headache)
2. Contusio cerebri (brain contusion)—a distinct injury of the brain tissue (cerebral contusion, intracranial bleeding)
3. Compressio cerebri (brain compression)—this grade is characterized by an increased intracranial pressure caused by brain swelling and/or intracranial bleeding; since the cranial cavity is not deformable, the only direction the brain can expand is toward the foramen magnum and that is where the brain can be clamped. Secondary bleedings especially in the pons area might follow leading to severe brain damage or even death

Intracranial hemorrhages

Intracranial hemorrhages are bleedings inside of the cranial cavity. According to their localization, they can be divided into following types:

Epidural hematoma—a bleeding between the skull bone and the dura mater. The bleedings are caused by damage to one of the meningeal arteries, typically as a result of skull fracture.

Subdural hematoma—a bleeding between the dura mater and the arachnoidea caused by damage to one or more of the bridging veins, typically in incidents with high head acceleration.

Subarachnoid hematoma—a bleeding between the arachnoidea and the pia mater, caused typically by damage to small brain arteries.

Intracerebral hematoma—bleeding within the brain tissue typically caused by damage to small blood vessels in the convolutions of the brain.

In head impacts with high linear acceleration of the head, focal brain injuries can occur not only directly underneath the impact site, but also on the opposite side of the brain (inertia of the brain and its interaction with the skull being the assumed cause). The hemorrhage areas on the impact side (coup, see **Figure 5**) are typically less pronounced than on the opposite side (contrecoup). High rotational accelerations typically lead to diffuse injuries of the brain tissue (DAI = diffuse axonal injuries).

A striking finding after head trauma are raccoon eyes (unilateral or bilateral)—a periorbital hematoma (bruising around the eyes) Its presence is strongly suggestive of skull base fracture (and not of blunt force acting in the eye region).

Thorax injuries

Heart contusion (contusio cordis)—occurs in blunt chest traumata and can lead to cardiac arrhythmia.

Lung contusion—because of contusions and bruises of lung tissue, the gas exchange might be impaired and/or blood might enter the respiratory tract leading potentially to death through suffocation.

Pneumothorax—if due to an open thorax injury air gets into the pleural cavity (between the lungs and the chest wall), the negative pressure is neutralized and one or both lungs might collapse. Sometimes a ventil mechanism forms that allows air to flow to the pleural cavity but prevents its escape (the so-called tension pneumothorax); the state of the injured might deteriorate rapidly and lead to death within a short period of time. Also an accumulation of blood in the pleural cavity (hemothorax) might lead to breathing and/or heart action impairment. A combination of both (air and blood in the chest cavity) is the hemopneumothorax.

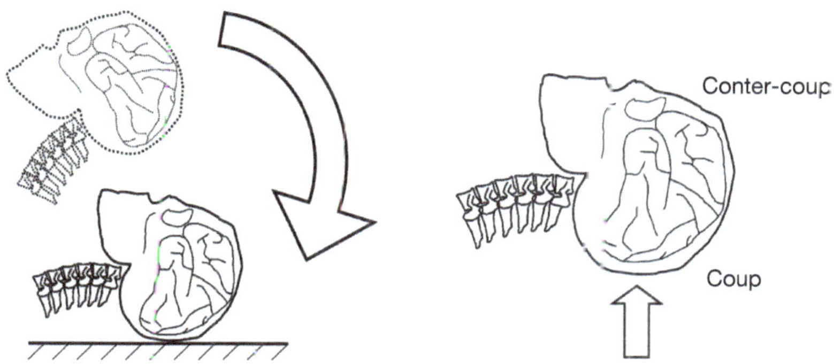

Figure 5 The coup and contercoup injuries.

Flail chest—although simple rib fractures per se do not constitute a relevant danger, serial fractures in various locations can lead to a detachment of a part of the rib cage that moves (due to the pressure within the thorax) opposite to the rest of the chest (paradoxical breathing movement). This constitutes a life-threatening injury. Sharp rib fragments in displaced fractures might cause lung lacerations with (hemo)pneumothorax.

Aortic rupture—tears of the chest aorta are rarely the result of direct force (although sometimes seen in overrun accidents). Typically the aortic rupture is caused by high accelerations (e.g., an impact of a nonbelted driver on the steering wheel).

Diaphragm rupture—can occur in car occupants as well as in pedestrian as a result of an impact. It is a relatively rare finding, but the fatality rate is high.

Pelvic injuries are of high clinical importance because they can be associated with severe bleeding. Fractures of the pelvic ring are typical for side collisions, whereas acetabulum fractures occur primarily due to femoral shaft axial loading, that is, due to force transfer from the knee/foot region in frontal collisions. This mechanism can also lead to posterior dislocations of the hip.

Abdominal injuries

Abdominal injuries result typically from blunt force and can be associated with very high danger to life. Perforations of the digestive system lead to infections and inflammations of the abdominal membrane (peritonitis).

Injuries to the extremities

Fractures of long bones

For accident reconstruction, fractures of long bones, especially of the legs, are of a great importance. Relevant are both the localization and the shape of the fracture. The latter enables to distinguish among torsion fractures, compression fractures due to axial loading of the bone, and bending fractures. All kinds of fractures can be observed in traffic accidents. In bending fractures, the direction of force can be reconstructed—if a bone is bent beyond its tolerance, it starts to fail on the tension side and the fracture lines typically form a wedge pointing in the direction of the acting force (see **Figure 6**). Complicated fracture patterns might be very difficult to interpret.

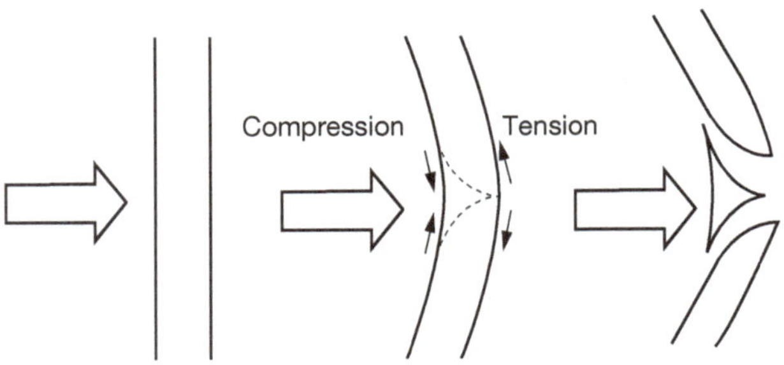

Figure 6 Wedge-shaped bending fracture of long bones, the apex of the wedge showing the direction of the acting force.

Car Occupants

Injuries of the car occupants are frequently the most important traces that enable to reconstruct their position in the car (driver vs passenger). Apart from injuries (among the most important are seat belt injuries—**Figure 7**; injuries caused by pedals, airbag injuries, and injuries from glass fragments—**Figure 8**), a thorough inspection of the car interior (contact locations, possibly with biological traces) must be performed. Since airbags deploy not until the crash, the existence of biological (DNA) traces is strong evidence regarding seating order. Another question that can be clarified by a thorough injury analysis (in combination with other methods) is the seat belt use. In multiphase accidents, the assignment of injuries to individual events (interactions) might be very tricky.

Frontal collisions

Seat belt injuries—occur as a result of the interaction between the restrained occupant and the (typically three-point) belt. In severe crashes, bony injury (ribs, clavicles, and sternum,

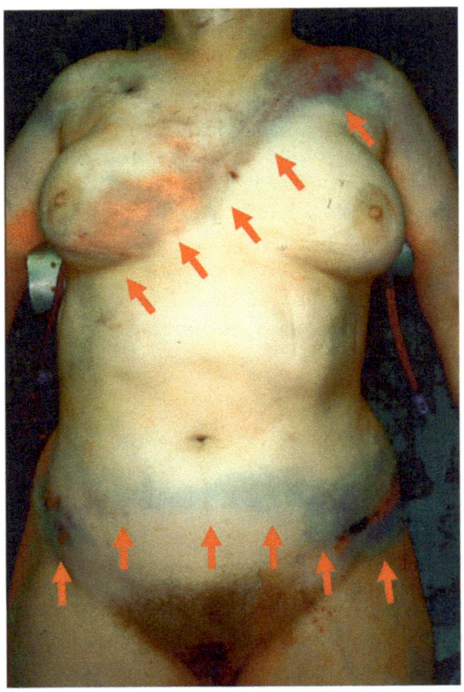

Figure 7 Seat belt injury, car occupant in severe frontal crash; the shoulder belt consistent with the driver position in the car (in left-hand drive vehicles).

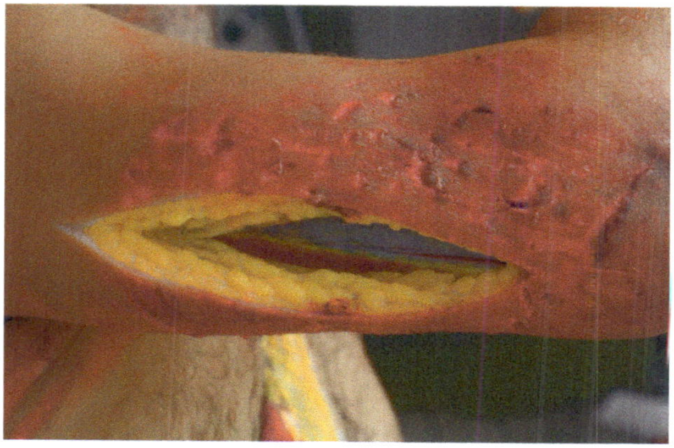

Figure 8 Superficial skin injury caused by shattered glass (right upper arm of a car occupant); the long and deep cut was made by a pathologist.

fractures typically located along the belt contact site) can be found as well as abdominal wall disruptions and mesenteric tears. The typical seat belt sign—a skin abrasion running diagonally across the chest (shoulder belt) and/or across the lower abdomen (lap belt)—can show in an impressive way (see **Figure 7**) or not at all; sometimes, subcutaneous bleeding along the belt line can be found in the autopsy in spite of no visible trauma sign on the body surface. A missing seat belt injury does not necessarily imply that the belt was not used (e.g., the belt might have been rolled out because of precrash movements of the occupant). For the same reason, knee (and/or hip and/or pelvic) injuries as well as head injuries might occur in spite of seat belt use.

Airbag injuries—the interaction with the airbag can lead to face, skull and/or jaw fractures, teeth luxations, eye injuries, burns, or hearing damage. Depending on the hand location at impact, injuries of the hand and/or lower arm might occur (fractures, burns, etc.). Injuries might be caused by eyeglasses, pipes, and other objects between the face and the inflating bag.

Side collisions

With respect to a particular occupant, far-side and near-side (struck side) impact should be distinguished. An impact to the side of the vehicle, where the occupant sits, leads to direct loading via blunt force (the injuries result from cabin intrusion as well as its acceleration). Typically, the impact side of the head, the thorax, and the pelvis are affected. Seat belts are not effective in near side impacts, and so seat belt injuries are not likely to occur. The injuries of occupants on the struck side are typically more severe than on the far side of the vehicle.

In far-side collisions, the torso of the occupants might slip out of the shoulder belt and severe head injuries can occur (interaction of the head with the dashboard, the near-side occupant, or the intruding far-side cabin structures). The upper body of the occupant stays impinged by the belt and the excessive belt loading might lead to abdomen injuries including typical belt signs on the skin.

Rear-end collisions

Low-impact rear-end collisions are being frequently investigated because of whiplash-associated disorders (WADs). Severe crashes might lead to cervical spine injuries (fractures, dislocations) and/or head injuries.

Rollover collisions

The loading imposed on occupants in rollover crashes is variable. Isolated rollovers are rare and belted occupants frequently suffer only minor injuries in spite of extensive car damage. Severe injuries typically occur due to occupant ejection (possible in spite of belt usage if the seat collapses) or due car impact secondary to rollover; in the latter case, the injury patterns resemble those of frontal, side, or rear-end collisions depending on the impact direction. Severe blunt head traumata and/or compression cervical spine injuries occur as a result of roof crush.

Whiplash-Associated Disorders

Nonphysiological movements in the cervical spine region can cause significant disorders ranging from discomfort to neurological symptoms and/or injuries to the bones or ligaments. In Europe, more than one million car occupants report to suffer from WAD after a car crash, the costs are estimated to be between 5 and 10 billion EUR. In the United States, more than one million people are affected every year and the costs are approximately 30 billion dollars. The WADs are typically classified according to the recommendation of the Quebec task force (QTF):

Grade 0	No symptoms
	No physical signs
Grade 1	Symptoms such as pain, stiffness or tenderness
	No physical signs
Grade 2	Symptoms such as in Grade 1
	Musculoskeletal signs such as decreased range of movement and point tenderness
Grade 3	Symptoms such as in Grade 1
	Musculoskeletal signs such as in Grade 2
	Neurological sings such as decreased or absent deep tendon reflexes, muscle weakness, and sensory deficits
Grade 4	Symptoms such as in Grade 1
	Fracture or dislocation or spinal cord injury

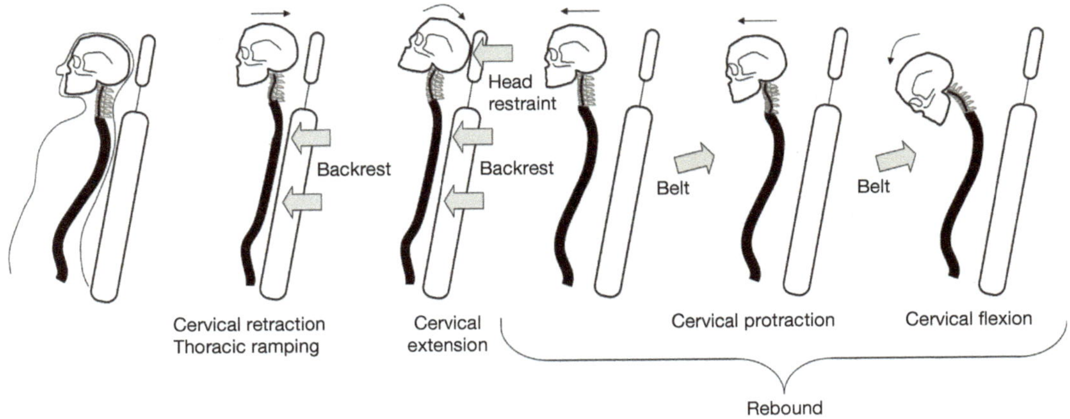

Figure 9 Typical spine kinematics during a rear-end collision. The wide arrows represent the external forces acting on the body, the fine arrows the movement direction.

The exact injury mechanism is not known; microlesions of soft tissue caused by combined translational and rotational movements of spine segments are frequently hypothesized. In traffic accidents, the torso is typically accelerated through the seat; as a result of inertia, a complex relative movement between the head and the torso including both traction and bending of the cervical spine follows. **Figure** 9 shows typical spine kinematics during a rear-end collision—due to a force imposed by the seat back, the thoracic spine flattens and at the junction to the cervical segments, a vertical force component occurs (thoracic ramping); simultaneously, a retraction of the lower cervical spine can be observed. An extension of the cervical spine follows that is limited by the head restraint. Due to contact elasticity, the head as well as the torso typically rebound, which leads to protraction of the cervical spine. The forward movement of the torso is restrained by the seat belt so that a flexion of both upper thoracic and cervical spine follows. As a reference parameter for the amount of load imposed to the passengers, the collision-induced velocity change has been established. Because the collision duration lies typically in a tight range (0.1–0.15 s), the velocity change is proportional to the mean acceleration of the vehicle during the collision. Since the biomechanical tolerance limits were established based on the result of volunteer tests under controlled conditions, care must be taken that the value of the velocity change in the location of the person of interest is reconstructed (due to rotations, the velocity change in different locations of the car varies). The approximate reference values that result from various volunteer studies cannot be generalized as they refer to standardized (lab) situations. The assessment must always take into account the particular situation (especially deviations from the setup of the volunteer studies) and the individual parameters of the person. The most important factors influencing the injury outcome are the posture (especially of the head), the seat (and headrest) adjustment, degenerative or other preexisting alterations, or injuries of the cervical spine, and so on.

Pedestrian Collisions

The interaction between a pedestrian and a car have typically the form of an impact, that is, the pedestrian is hit in an upright position by the car front. Runover collisions (without former impact) are much less common, but they may be of high forensic relevance (drunken pedestrian lying on the road vs camouflage of a homicide vs suicide). The injury outcome is dependent on the pedestrian kinematics, which can typically be divided into the following phases:

1. Primary impact—the contact between the pedestrian body and the car; typically, the lower legs are impacted by the bumper first and the hip and head impacts follow as the pedestrian body wraps around the car (the distance between the road surface and the head impact location measured along the car contour is called WAD = wrap around distance). In box-shaped cars (or in rear-end collision with station wagons, etc.), the impact of the leg, hip, thorax, and head regions can occur simultaneously.
2. Secondary impact—the contact of the pedestrian to the road (impact, sliding, and rolling). The distance between the point of collision and the end position of the pedestrian after the secondary impact is called throw distance; this parameter is frequently used for the collision speed estimation.
3. Tertiary impact—the contact between the pedestrian and the surrounding structures (another car, a tree, also a runover counts as tertiary impact).

The pedestrian kinematics and thus the injury pattern depend to a high degree on the relationship between his/her anthropometry and the geometry of the impacting car. Several different kinematics types can be discerned—wrap kinematics (the pedestrian upper body wraps around the front hood/windshield and after the head impact, he stays on the hood or slides off to the ground), forward projection (the pedestrian is projected forward, typical for box-shaped vehicles or children), fender vault (the pedestrian first wraps around the hood and than exits off the side of the vehicle front), roof vault (the pedestrian wraps around the hood and his legs rotate further upward so that he is lifted), and somersault kinematics (the pedestrian upper body wraps around the car front, after the head impact, the pedestrian is flipped into the air as he continues to rotate). Another collision type is a sideswipe, that is, a movement of a pedestrian along the side of a moving vehicle; in such cases, the pedestrian might twist (rotate about his longitudinal axis) and suffer leg injuries on one side and head injuries on the other side of the body, both resulting from contact to the car.

The most important are the injuries that are able to reconstruct the impact direction—apart from fractures of the long bones of the lower and/or upper leg, the soft tissue findings (contusions, wound caverns, etc.) are of high relevance. Sometimes ankle and/or knee injuries help to reason back as to the impact direction (supination vs pronation vs flexion vs extension in the joint due to impact, etc.). Only very rough estimation of collision speed based on the injuries is possible (and always to be performed also with the knowledge of the car damage); lower-leg fractures can occur even in collisions with a speed of less than 10 km h^{-1}, multiple rib series fractures and/or severe TBIs require typically collision speeds of above 30 km h^{-1}, and amputations of body segments above 80 km h^{-1} (unless the body part gets stuck in a car structure). The assignment of the injuries to individual contacts during a multiphase accident might be very difficult due to overlapping of injury outcomes. Runover leads sometimes to tire profile marks on the skin (that can appear as textile pattern due to clothing), massive soft tissue injuries with décollement is to be expected.

Two-Wheelers

Typically, the most serious injuries of riders are caused by impacts—with the own vehicle (e.g., pelvic ring fractures due to impact on the tank), with other vehicles, road infrastructure (guard railing), and other objects (a tree, etc.). For the accident reconstruction, compression injuries of the hands/lower arms due to handlebar grip are important (e.g., to differentiate the driver and the passenger) as well as superficial injuries morphologically attributable to contacting objects (vehicle parts, etc.).

Helmet injuries—wearing of a helmet at the time of an accident can result in skin abrasions underneath the chin strap or impression mark of parts of the visor. An example of helm injuries is shown in **Figure 10**. In severe crashes, helmets cannot prevent skull fractures and/or brain damage.

In bicycle collisions, severe injuries to the head occur especially if the cyclist does not use helmet and a car is involved. For accident reconstruction, injuries in the upper leg/pelvis region due to the interaction with the bicycle seat might be important (e.g., to clarify whether the cyclist was riding or pushing the bicycle).

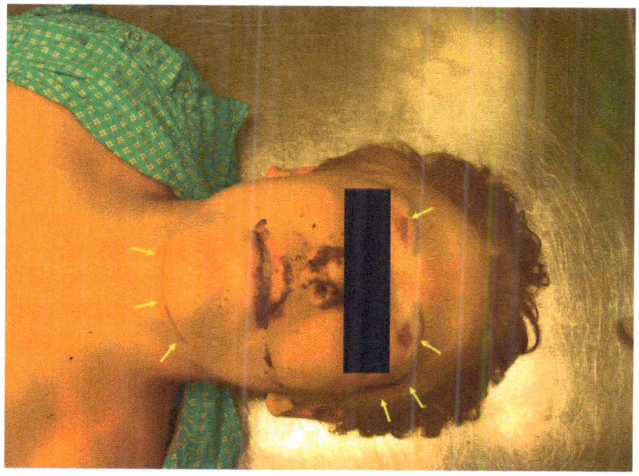

Figure 10 Typical helm injuries (motorcyclist, frontal collision with a passenger car); the helmet was found several meters away from the body, but its damage and the injuries prove it had been worn at the time of accident.

Severe or even lethal injuries result also from a collision between a cyclist and a pedestrian. Fatal injuries are rare and result typically from the secondary fall of the pedestrian with head impact on the road. For accident reconstruction, a characteristic wound on the lower leg of the pedestrian caused by the contact with the front wheel is the most important injury (though typically minor).

Railway

In 2009, 1391 people were killed and another 1350 were seriously injured in railway accidents in the European Union (European railway agency data). The total number of deaths is much higher, because the statistics do not include suicides (2719 in 2009, i.e., more than twice the number of accidental deaths). The number of deaths among railway passengers is very low; within the EU, only 196 persons died in the 3-year period 2007–2009. The US statistics reported 492 deaths and 5264 nonfatal injuries in 2011 (Federal Railroad Administration)

The most important factor determining injury severity in railway accidents is the huge mass of the vehicles. Injuries may be caused by an impact (pedestrian going/running/standing, car occupant), runover (pedestrian lying on the track, with or without contact to the wheels of the train), or dragging along by the train (a person or a whole car). In rare cases, persons fall out of a moving train, also known are deaths and serious injuries resulting from "train surfing" (riding on the outside of a moving train). The differentiation between a suicide and an accident (or a homicide) can be very difficult; in suicide cases, the person lies typically on the abdomen across the rail so that the head or the upper part of the torso are severed by the wheels.

Air Traffic

The number of air accident victims is low and shows irregularities because of the very few events that lead to fatalities—in 2008, there were 170 fatalities reported in the EU, 154 hereof caused by a single major crash. In the United States, there were 30 accidents reported that cost 757 lives in 2009 (Aviation safety network).

In plane crashes, fatal injuries typically result from severe impacts; carbon monoxide or cyanide intoxication is also possible. In crashes into the water, the ultimate cause of death might be drowning.

See also: **Anthropology/Odontology:** Biomechanics of Bone Trauma; **Engineering:** Forensic Engineering/Accident Reconstruction/Biomechanics of Injury/Philosophy, Basic Theory, and Fundamentals; **Forensic Medicine/Causes of Death:** Blunt Injury; **Forensic Medicine/Clinical:** Airplane Crashes and Other Mass Disasters.

Further Reading

Appel, H., Wanderer, S., Meißner, G., et al., 1984. Mechanik und Biomechanik des Unfalls. In: Wagner, H.J. (Ed.), Verkehrsmedizin. Springer, Berlin, p. 480.
Global Status Report on Road Safety, 2009. WHO. www.who.int/violence_injury_prevention (accessed October 2011).
Goldsmith, W., Plunkett, J., 2004. A biomechanical analysis of the causes of traumatic brain injury in infants and children. The American Journal of Forensic Medicine and Pathology 25, 89–100.
Graw, M., König, H.G., 2002. Fatal pedestrian-bicycle collisions. Forensic Science International 126, 241–247.
Naci, H., Chisholm, D., Baker, T.D., 2009. Distribution of road traffic deaths by road user group: a global comparison. Injury Prevention 15, 55–59.
Nahum, A.M., Melvin, J.W. (Eds.), 2002. Accidental Injury: Biomechanics and Prevention. Springer, New York.
Schmitt, K.-U., Niederer, P.F., Walz, F., 2010. Trauma Biomechanics: Accidental Injury in Traffic and Sports. Springer, Heidelberg.
Spitzer, W.O., Skowron, M.L., Salmi, L.R., et al., 1995. Scientific monograph of the Quebec task force on whiplash associated disorders: redefining "whiplash" and its management. Spine 20 (Suppl. 8), 1–73.

Airplane Crashes and Other Mass Disasters

O Peschel, Institut für Rechtsmedizin der Ludwig Maximilians Universität München, München, Germany
W Eisenmenger, University Hospital München, München, Bavaria, Germany

Glossary

Identification Process of assigning a preexisting individual's identity and name to a corpse by different means of forensic sciences.

Mass disaster Man-made or natural incident with more fatalities than the local resources can handle; some sources declare incidents involving more than 100 victims as a mass disaster.

Definition

Generally, a mass fatality is described as a man-made or natural incident with more fatalities than the local resources can handle. As important infrastructures such as rescue and treatment services may be damaged, a certain dependency on the character and locality of the incident is obvious. Events of the last decade show the great variety of scenarios influencing the possibilities and requirements for forensic pathology:

- 2000 Cable car disaster—Kaprun, Austria; 155 victims
- 2001 World Trade Center—USA; 3000 victims
- 2002 Refugee—Ferry Le Joola—Gambia; 1863 victims
- 2003 Explosion of gas mine—Chuandongbei; 191 victims
- 2004 Tsunami—Thailand and Indonesia; 230 000 victims estimated
- 2005 Train crash—Amagasaki, Japan; 107 victims
- 2006 Car ferry—Dhiba, Saudi Arabia; 1026 victims
- 2007 Bomb attack—Bhutto's parade, Pakistan; 135 victims
- 2008 Plane crash, Spanair—Madrid, Spain; 154 victims
- 2009 Air France Flight 447—Rio de Janeiro to Paris, Brazil; 228 victims
- 2010 Earthquake—Haiti; 200 000 victims estimated
- 2011 Tsunami and reactor disaster—Fukushima, Japan; 20 000 victims

The identification of the victims is the main objective of forensic efforts related to mass disasters. Criminalistic reconstruction is very important too, but the consequences of missing identifications seem to be greater for the bereaved than the lack of information concerning the reason of the incident. Knight describes some reasons for identification efforts:

- Ethical and humanitarian need to know which individual has died, especially for the information of surviving relatives.
- To establish the fact of death with respect to particular individuals for official, statistical, and legal purposes.
- To record their identity for administrative and ceremonial purposes with respect to burial or cremation.
- To discharge legal claims and obligations in relation to property, estate, and debts.
- To prove claims for life insurance contracts, survivor's pensions, and other financial matters.
- To allow legal investigations, inquests, and other tribunals, such as those held by coroners, procurators fiscal, medical examiners, judges, and accident inquiries to proceed with accurate information about the identity of the decedent.
- To facilitate police inquiries into overtly criminal or suspicious deaths, as the identity of the deceased person could be a vital factor for investigations.

Comparable to the great variability of the different scenarios listed earlier, there are great differences in the ways of handling identification problems. Information concerning all forensic efforts is not regularly published in the forensic literature or in political statements. Full autopsies were performed in the Kaprun cable car disaster (156 victims) (**Figures 1 and 2**) but not in the Eschede train disaster in 1998 that claimed 101 victims. Identification in these incidents was mainly based on molecular genetic investigations. The tsunami in Thailand in 2004 resulted in ~230 000 victims. Only a very small part of the fatalities was handled according to the Interpol identification recommendations (predominantly European victims in the main tourist areas). The major part of the German victims was identified by forensic odontology. Due to severe decomposition of the dead bodies, molecular genetic methods played a secondary role. A completely different situation was encountered in the World Trade Center terrorist attack in which identification without DNA-based methods was inconceivable.

 http://dx.doi.org/10.1016/B978-0-12-802718-9.00016-8

Figure 1 Obsequies, Kaprun disaster, Austria, 2000. Source: Prof. Dr E. Tutsch-Bauer, Department of Forensic Medicine, Salzburg University.

Figure 2 Scene before recovery of the bodies, Kaprun disaster, Austria, 2000. Source: Prof. Dr E. Tutsch-Bauer, Department of Forensic Medicine, Salzburg University.

The airplane accident involving the Air France Flight in 2007 resulted in a majority of missing person cases with a few victims recovered some years later. As for the Haiti earthquake in 2010, which claimed ~200 000 lives, no information regarding a global identification effort has been reported. The identification procedures after the Fukushima disaster have also been awkward to handle, published forensic measures being very scarce.

In general, it can be ascertained that a high number of victims in a less developed area results in a lack of forensic identification activities. Contrary to this, a smaller number of victims (particularly from regions of Europe, the United States, or other highly industrialized states), in an incident happening in a highly developed area, leads to intensive forensic procedures executed over a long period of time (more than 1 year in the Thailand tsunami tourist areas).

Usually, forensic pathologists are embedded in disaster victim identification (DVI) teams, which have been constituted in most countries and which consist of a number of different specialists to cover the majority of possible situations. DVI teams have been established following different problems in the identification of mass disaster victims in the 1960s and 1970s. They are mostly governmental organizations, but recently, private companies have also begun to offer comparable services.

After an incident, the initial chaotic phase is usually followed by organization and acquisition of responsibilities by the different departments in charge. Typically, the police is the leading authority, but cooperation with experts in different fields is also sought. Incident management implies the establishment of command structures, emergency systems, law enforcement operations, family assistance, media operations, and logistical support, as well as financial and administrative support. A competent incident management requires experience and authority. The forensic pathologist can be a valuable advisor concerning all decisions belonging to the recovery and identification procedures.

The complexity and character of the operations depend necessarily on the specific situation and mainly on the estimated number of fatalities and the condition of the bodies. Also, responder protection has to be considered according to the individual nature of the incident.

Airplane crashes and major fires may be the most frequently emerging problems for the forensic pathologist in this field. In crashes involving light planes, the number of bodies is small and the identification procedure is much easier. In crashes involving big commercial aircraft, especially in collisions occurring at a flying altitude of more than 30 000 ft (9000 m), grave problems can be expected with the number of dismembered bodies being high. DNA analysis will play a decisive role in these incidents.

In all kinds of accidents involving means of transportation, investigations concerning evidence of explosion should be carried out.

In all kinds of accidents involving means of transportation, investigations concerning evidence for explosion should be made. Nowadays, the first question is the differentiation between an accident due to a technical defect (possibly resulting in civil action) and a terrorist attack. Also, the examination of victims, for example, for evidence of fire or bomb residue, may prove helpful.

Besides the passengers, there is a special interest concerning the pilot or the train driver. These victims require a full autopsy with a screening for diseases, signs of intoxication, or trauma not related to death. In all cases in which a mistake on the part of the personnel is discussed as the reason for the crash, a detailed knowledge of any impairment of health is essential for all further deliberations. Prior to this, of course, is the identification of the pilot or other staff members by individual criteria and personal effects (dress code). Also, special injury patterns such as severe hand or foot injuries in a pilot may be helpful for identification; however, with the modern technology in airplanes, this kind of injury-dependent identification has lost its relevance in most cases.

Very important are all kinds of natural diseases which may lead to an individual mistake as a reason for an accident: predominantly all kinds of heart disease, especially typical coronary atherosclerosis leading to a myocardial infarction, but also neuromuscular diseases or hypoxic changes in the central nervous system leading possibly to unconsciousness are to be discussed. Another aspect may be an impairment of a pilot due to alcohol consumption or intoxication with legal or illegal drugs. For this reason, analysis of blood alcohol concentration and toxicology are indispensable in all investigations concerning airplane crashes or train accidents.

Eisenmenger reported one case of a helicopter crash, in which the pilot was found intoxicated with a BAC of 2.3 g%. There were also reports on airplane crashes where the pilot was found with a lethal gunshot injury or a stab wound, obviously having been killed by one of the passengers in the context of a homicide–suicide event.

Tasks

The main tasks involving activities of forensic pathologists are as follows:

- Recovery of the bodies at the scene.
- DVI.
- Assistance in criminalistic incident reconstruction.

Recovery of the Bodies at the Scene

In the scenario of a mass disaster, the forensic pathologist is an auxiliary force working in line with the police after rescue operations have been terminated. The recovery of the bodies from the scene is an important step in successful catastrophe management. Not only in airplane crashes but also in other mass incidents, severe fragmentation of the bodies may be one of the main problems. The forensic pathologist's opinion and an anatomic classification of the body fragments, as well as the differentiation between human and animal tissue, will be helpful. This applies accordingly to exhumations from mass graves and major fire accidents. A compartmentation of the scene in different sectors with the establishment of a grid system is usually a police exercise. The same applies to the documentation of bodies or body fragments, concerning the place of discovery and personal effects.

Tags should be applied to the bodies or body bags to ensure a continuous flow of the case file. The body has to pass through the different morgue stations (see below), and all the data have to be documented and compared afterward. Up to the final report and the release of the body, the case file and the identification form have to be signed by the leading police officer, the authorized forensic pathologist, and dentist.

Disaster Victim Identification

Identification is the main task of the forensic pathologist. The necessity for identification results from different laws. Inheritance is usually connected to the declaration of the death of a person. The same applies to the payment from insurance companies, the possibility of a new marriage, and so on. Identification may result from different measures and is basically a duty of the police. One possibility often used in countries without a functioning forensic system is the ascertainment of identity by inspection of human remains by means of visual recognition. Also, a comparison with photographs is sometimes used for identification. In case of severe traumatic damage or decomposition of the bodies, this procedure is not feasible. However, even in cases with well-preserved bodies, erroneous identification has been reported, obviously very difficult situation for relatives or loved ones. For this reason, other methods for identification must be preferred.

Personal effects

Personal effects are items from victims' pockets, clothing, ID papers, or jewelry. The value for identification usually is limited to single items with an outstanding individual character. A certain commingling of all these items may happen due to events prior to the incident as well as by the catastrophe itself. However, important is also the return of personal items to the relatives and families of the victims. Special police units usually arrange for recovery, processing, and handing over of personal effects. In the Kosovo Identification Project, individual clothing was the main factor for identification success.

Fingerprinting

Fingerprints as well as footprints and palm prints are very useful for identification. Fingerprint experts are a part of all DVI teams. If possible, all opportunities should be taken to get fingerprint evidence processed by forensic experts belonging to the police. Fingerprints, footprints, and palm prints can be compared with prints taken from the assumed person's home or objects, which can be assigned to a certain person. Sometimes, this can be difficult and the assignment may be doubtful. Comparative prints may be missing (e.g., in a plane crash in a residential zone; after destruction of the accommodations, it will usually be impossible to get comparative prints). In some countries, fingerprints are part of identity papers. Sometimes, a police investigation with a documentation of fingerprints years ago can be helpful. Nowadays, computer processing of all the evidence is possible, facilitating comparison in cases with large numbers of individuals.

Fingerprinting becomes less important in cases of severe fragmentation of the bodies or in major fires. Amputation of fingers or hands for fingerprint or palm processing at the fingerprint station as was commonly recommended in the past should be avoided because commingling is possible afterward and control of the results becomes more difficult.

Autopsy

Depending on the local factors, setting up of a temporary morgue might be necessary. Additional resources (staff, equipment, etc.) are required to perform autopsies, but the results of a forensic autopsy are essential for a complete identification procedure and for control purposes. The autopsy may be limited to special findings or procedures according to the condition of the bodies and the needs of identification. The main findings are related to gender, age estimation, body height, race, and individual markers such as tattoos, scars, results of a medical treatment, or malformations.

Gender determination: Depending on the condition of the bodies, primary or secondary sex organs may be detectable; body hair may be shaved; for skeletonized individuals, molecular genetics is an option.

Age estimation: Will cover only the biological age, not the actual age. Degenerative changes in the ankles or the spine may help as well as the degree of atherosclerosis and abrasion in the dental state.

Race: Determination of race will be impossible in cases of major fire and severe decomposition. Comparable effects occurred after the Southeast-Asian tsunami in 2004 within a few days so that a differentiation between the local Asian people and Caucasian tourists was not possible any more.

Body constitution and nutritional status: May be difficult or impossible to assess in cases of fragmentation, burn injuries, and decomposition.

The report of the forensic pathologist should (if possible or determinable) contain the following information: body number, condition of the body, weight of the remains, missing body parts, gender, height, diagram of injury pattern, foreign bodies, hair color and length, eye color, scars and tattoos, birthmarks or other typical features, operations and prostheses, diseases, and X-rays or photographs of important findings.

The value of the findings depends on the condition of the body and the number of individuals who have to be identified. In cases of severe decomposition, fragmentation, or mass grave exhumation, the help of a forensic anthropologist is necessary. If the victims are skeletonized, special discriminant analyses can be helpful even if they are very complex. Prostheses and implants, as well as artificial lenses, radiological seeds, or pacemakers, may offer a quick, simple, and safe means of identification of single individuals. An autopsy also offer the possibility to retain body fluids for further investigations (DNA, alcohol, toxicology) and to find some evidence concerning the cause of death or items such as projectiles, bomb fragments, or other injuries important for the criminalistic reconstruction. Interpol forms for antemortem and postmortem investigations offer the possibility for an exhaustive data collection and a safe comparison of the findings even when supported by computer technology. However, in a number of mass disasters, only a small portion of the individual parameters can be collected as the Interpol forms require. In this situation, priorities have to be established.

In general, two possible ways can be discussed for identification: The first one is a systematic comparison between all antemortem and postmortem data, after all the operations have been completed. The second one is the selective identification and release of single bodies in a good condition with characteristic identity features and a stepwise narrowing of the group of unknown individuals. Additionally, a classification in different age groups or concerning gender may be helpful. In incidents involving a large number of victims and cooperation between DVI teams of different nationalities, a systematic approach is mandatory.

Forensic odontology

DVI requires support of experienced forensic odontologists. Teeth are very resistant concerning physical influences. Especially in cases of severe decomposition or major fire, pathological findings or DNA may be difficult, even impossible, to assess. In a large number of cases, therefore, forensic odontology has been a key to successful identification. However, with the growing number of individuals and the higher quality of dental care, the number of individuals without individualizing findings will become the norm. Finally, in children, a precise estimation of age will be possible by evaluating the dental state; however, typical findings for individualization such as inlays with different materials, crowns, or prostheses will be missing. Perhaps signs of an orthodontic treatment may be helpful.

In conclusion, DVI without forensic odontology is hardly conceivable. Odontology exclusively in scenarios with a high number of victims will miss a lot of important findings.

Radiology

The comparison of radiographs can be a very effective means for identification. However, a quick acquisition of X-rays for comparisons may be difficult. Also, the path of rays has to be comparable in standard pictures. The application of CT technology is possible, but very expensive and not feasible in all conditions. Interpretation of CT scans for identification purposes needs an expert in forensic radiology and cannot be done by a clinical radiologist. X-rays of the skull (frontal sinus), fractures, prostheses, and malformations are applicable for comparisons. Dental radiography has a special relevance in identification.

Molecular genetics

DNA fingerprinting is the most important development in forensics during the last 25 years. The benefit of DNA technologies for identification is very high since international agreements made investigations and results comparable. The advantages are the high degree of safety concerning distinction between different individuals and the possibility to identify even very small portions of tissue in cases of extreme fragmentation of bodies. Familial relationships can be detected and complete families can be assigned (however, sometimes, a familial relationship is assumed before the incident, which is, in actual fact, missing—problem of the cuckoo's child). The disadvantages are that it is a time-consuming process compared with forensic odontology and involves the impairment of the results by decomposition and degradation of DNA. Also, for DNA analysis, the comparison samples may be difficult to attend. Furthermore, the DNA analysis usually cannot be carried out in the morgue and needs special laboratories. The success of DNA techniques is not always guaranteed as experiences in the Thailand tsunami disaster demonstrated. However, in DNA results, computer technology makes the data processing easier.

Forensic anthropology

The forensic anthropologist can be helpful in distinguishing human remains from nonhuman remains and in determining the gender, age, race of the victims, and the number of fatalities. Also an experienced anthropologist can easily detect certain abnormalities or individualizing traits in skeletal elements. Sometimes, additional X-ray investigations may be necessary.

Criminalistic Reconstruction

Cause and time of death investigation may belong to the data needed for a criminalistic reconstruction as well as miscellaneous findings such as foreign bodies with evidence for an explosion or toxicological analysis. Carbon monoxide may give evidence that a person was alive when a fire was set. In incidents resulting from toxic agents, special analysis may be helpful in the therapy of survivors if there is a lack of information concerning the toxic substance or even a specific misinformation, as was reported in the Bhopal incident in 1984: an estimation of between 4000 and 25 000 victims.

In conclusion, victim identification after a mass disaster is an important task for police forces and all forensic, medical, and biological experts and normally a basic necessity from the ethical point of view. Cooperation between the different forensic disciplines with varying focuses concerning the type of incident will be the base of success.

See also: **Biology/DNA:** Disaster Victim Identification; **Forensic Medicine/Clinical:** Traffic Injuries and Deaths.

Further Reading

Blau, S., Robertson, S., Johnstone, M., 2008. Disaster victim identification: new applications for postmortem computed tomography. Journal of Forensic Sciences 53 (4), 956–961.

Brkic, H., Strinov, D., Kubat, M., Petrovecki, V., 2000. Odontological identification of human remains from mass graves in Croatia. International Journal of Legal Medicine 114, 19–22.

Budimlijy, Z.M., Prinz, M.K., Zelson-Mundorff, A., et al., 2003. World Trade Center human identification project: experiences with individual body identification cases. Croatian Medical Journal 44, 259–263.

Budowle, B., Bieber, F.R., Eisenberg, A.J., 2005. Forensic aspects of mass disasters: strategic considerations for DNA-based human identification. Legal Medicine 7, 230–243.

Dedouit, F., Temon, N., Costagiola, R., et al., 2007. New identification possibilities with postmortem multislice computed tomography. International Journal of Legal Medicine 121, 507–510.

DiMaio, V.J., DiMajo, D., 2001. Forensic Pathology, second ed. CRC Press, Boca Raton/London/New York/Washington.

Eisenmenger, W., 2004. Massenkatastrophen. In: Brinkmann, B., Madea, B. (Eds.), Handbuch Rechtsmedizin. Springer- Verlag, Berlin/Heidelberg/New York.

Fullerton, C.S., Ursano, R.J., Wang, L., 2004. Acute stress disorder, posttraumatic stress disorder and depression in disaster or rescue workers. American Journal of Psychiatry 161 (8), 1370–1376.

Jensen, R.A., 2000. Mass Fatality and Casualty Incidents: A Field Guide. CRC Press, Boca Raton, FL.

Knight, B., 1996. Forensic Pathology, second ed. Arnold, London/Sydney/Auckland.

Leclair, B., Shaler, R., Carmody, G.R., et al., 2007. Bioinformatics and human identification in mass fatality incidents: the world trade center disaster. Journal of Forensic Sciences 52, 806–819.

Meyer, H.J., 2003. The Kaprun cable car fire disaster – aspects of forensic organisation following a mass fatality with 155 victims. Forensic Science International 138 (1), 1–7.

Meyer, H.J., Chansue, N., Monticelli, F., 2006. Implantation of radio frequency identification device (RFID) microchip in disaster victim identification (DVI). Forensic Science International 157, 168–171.

Moody, G.H., Busuttil, A., 1994. Identification in the Lockerbie air disaster. American Journal of Forensic Medicine and Pathology 15 (1), 63–69.

Peschel, O., Eisenmenger, W., 2010. Rechtsmedizinische Aspekte bei Großschadensereignissen. In: Schutzkommission beim Bundesministerium des Inneren. Katastrophenmedizin, Bonn.

Peschel, O., Lessig, R., Grundmann, C., Peter, J., Tsokos, M., 2005. Tsunami 2004. Rechtsmedizinische Erfahrungen aus dem Einsatz der Identifizierungskommission in den ersten Tagen in Thailand. Rechtsmedizin 15, 430–437.

Peschel, O., Müller, E., 2003. Beweissicherung und Identifizierung. Bericht über den Einsatz von Rechtsmedizinern im Kosovo 1999. Rechtsmedizin 13, 194–200.

Prinz, M., Carracedo, A., Mayr, W.R., et al., 2007. DNA Commission if the International Society for Forensic Genetics (ISFG): recommendations regarding the role of forensic genetics for disaster victim identification (DVI). Forensic Science International: Genetics 1, 3–12.

Rutty, G., Robinson, C., Morgan, B., Black, S., Adams, C., Webster, P., 2009. Fimag: the United Kingdom disaster victim/forensic identification imaging system. Journal of Forensic Sciences 54 (6), 1438–1442.

Schuller-Götzburg, P., Suchanek, J., 2007. Forensic odontologists successfully identify tsunami victims in Phuket, Thailand. Forensic Science International 171, 204–207.

Soomer, H., Ranta, H., Penttila, A., 2001. Identification of victims from the M/S Estonia. International Journal of Legal Medicine 114, 259–262.

Stevens, P.J., 1970. Fatal Civil Aircraft Accidents. Wright, Bristol.

Tsokos, M., Lessig, R., Grundmann, C., Benthaus, S., Peschel, O., 2006. Experiences in tsunami victim identification. International Journal of Legal Medicine 120, 185–187.

Key Terms

Accident, Accident investigation, Accident reconstruction, Adversary truth, Air bag, Air France 447, Airplane crash, Audio recording, Black box data, Black boxes, Blunt force, Car, Causation, Collision, Criminalistic reconstruction, CVR, Death, Derailment, Digital versus analog, Disaster victim identification, EgyptAir 990, Electrically erasable programmable read-only memory (EEPROM), Electrocution, Employee injury, Event data recorders, Event recorders, Evidence, FAA, FDR, Forensic odontology, Forensic pathology, GPS, Head, Human factors, Impact, Injury, Level crossing, Mass disaster, Nonvolatile memory (NVM), Nonvolatile random-access memory (NVRAM), NTSB, Pedestrian, Platform gap, Probable versus proximate cause, Rail, Railroad, Recorded data disclosure, Recorder color, Recorder hostility, Recorder survivability, Recovery of bodies, Rolling stock, Route tracing, Scuff-type marks, Signals, Skull fractures, Specific injury patterns, Speed, Subdural hematomas, Tachograph chart falsification, Tachographs, Technical truth, Thorax, Track, Traffic accident investigation, Video recording, Whiplash.

Review Questions

1. What kinds of accidents can a railroad accident investigator examine?
2. What, by US law, does every leading locomotive have installed in it?
3. What do locomotive event recorders *not* record?
4. What other sources of information might a railroad accident investigator use? Name at least three.
5. What is the NTSB?
6. What is the difference between approximate and proximate causes?
7. What is an aircraft flight data recorder ("black box")?
8. What is a ULB?
9. Were automotive airbags a success at first?
10. What are the three phases of airbag deployment?

11. What kinds of injuries can be suffered from airbags?
12. In a black box, what is the ECU?
13. What is a digital tachograph? What does it do?
14. What is a speed trace?
15. What is wheel slip and how can it effect accident reconstructions?
16. Why are low-speed incidents difficult to interpret with a tachograph?
17. What constitutes the single leading cause of forensic relevant deaths in the world?
18. What is decollement?
19. What types of injuries result from seat/shoulder restraints?
20. What is a WAD?

Discussion Questions

1. What human factors can contribute to a railroad accident? What mechanical factors can? How are they related?
2. How would you correlate black box data with any eyewitness accounts? Which you give greater weight to?
3. What additional considerations would there be for a large-scale scene, such as an airplane crash or a train derailment, over a smaller scene? What kind of planning do you think would be necessary or prudent?
4. Why is black box data encrypted? What is the point? What would be the consequences of it not being encrypted?
5. Consider a car–train collision. What factors would you consider to determine if it was an accident or a suicide? Could you rule out homicide? What evidence would you want to collect and examine to rule out your multiple hypotheses?

Additional Readings

Brady, T., 2014. The aircraft accident investigation that never was. Journal of Aviation/Aerospace Education & Research 23 (2), 1.

LeHardy, P.K., Moore, C., September 2014. Deep ocean search for Malaysia airlines flight 370. In: 2014 Oceans-St. John's. IEEE, pp. 1–4.

Müller, R., Drax, C., 2014. Aircraft accident investigations. In: Aviation Risk and Safety Management. Springer International Publishing, pp. 139–148.

San Kim, D., Yoon, W.C., 2013. An accident causation model for the railway industry: application of the model to 80 rail accident investigation reports from the UK. Safety Science 60, 57–68.

Underwood, P., Waterson, P., 2014. Systems thinking, the Swiss Cheese Model and accident analysis: a comparative systemic analysis of the Grayrigg train derailment using the ATSB, AcciMap and STAMP models. Accident Analysis & Prevention 68, 75–94.

Section 4. Fire Investigations

Crime scenes are complicated enough but a fire scene is far more so because of the destruction of the fire and the additional damage cause by fighting it. In the United States, over 16 000 fires are intentionally set each year, resulting in hundreds of deaths and over half a billion dollars in damage[1]. Who processes a fire scene (arson is a legal designation) varies by jurisdiction, politically complicating an already difficult situation. Fire departments, fire marshals, forensic personnel, and police all may have some authority in the aftermath of a blaze. More work remains to ensure that proper validated science is used to investigate fire scenes and analyze the evidence from them.

Chemistry of Fire

E Stauffer, Commissariat d'Identification Judiciaire, Police Cantonale Fribourg, Fribourg, Switzerland
N NicDaéid, University of Strathclyde, Glasgow, UK

Glossary

Autoignition Ignition in the absence of an external source of ignition, also known as nonpiloted ignition. This can result from the ambient temperature or from a self-heating process of the fuel (known as spontaneous ignition) bringing the fuel to its autoignition temperature.

Autoignition temperature The minimum temperature to which a substance must be heated in air to ignite independently of the heating source.

Combustion An oxidation reaction that occurs at a rate sufficient to produce heat and light.

Conduction Process of transferring heat through a material or between materials by direct physical contact from a region of higher temperature to a region of lower temperature.

Convection Process of transferring heat by movement of a fluid.

Evaporation The physical change of state of a liquid into a vapor at a temperature below the boiling point of a liquid.

Exothermic reaction A chemical reaction that releases heat.

Fire point The minimum temperature, generally a few degrees above the flash point, at which a liquid produces a sufficient quantity of vapors to sustain continuous flame when ignited.

Fire triangle A symbol used to describe the necessary requirements for fire to occur in which the three sides represent fuel, oxidizer, and a source of ignition. The connection of the sides to form a triangle symbolizes the chain of reaction.

Flash point The lowest temperature at which a substance will give off a vapor that will flash or burn momentarily when ignited.

Glowing combustion The rapid oxidation of a solid fuel directly with atmospheric oxygen creating light and heat in the absence of a flame.

Ignition The initiation of combustion.

Pyrolysis A process by which a solid (or a liquid) undergoes chemical decomposition into smaller molecules due to heat, without interaction with oxygen or any other oxidizer.

Spontaneous heating An exothermic chemical or biological process that may generate enough heat to ignite the reacting material (spontaneous ignition).

Vapor The gaseous phase of a substance simultaneously present with its liquid and/or solid state at a temperature below its boiling point.

[1] www.usfa.fema.gov.

 http://dx.doi.org/10.1016/B978-0-12-802718-9.00017-X

Introduction

Fire has always fascinated people. It can be beautiful, such as the lighting of the Olympic flame at the opening ceremony. It has definitely been useful since the first Cro-Magnon man discovered it and used it to cook food. It can occur as part of a natural phenomenon, as with a volcano eruption or a lightning strike. Nevertheless, it can be dangerous, destructive, and even deadly, as the Great Fire of London in 1666, for example.

Because fires are often the result of negligence or criminal activity, and because they lead to the destruction of property and the loss of lives, many fires are investigated by law enforcement. The goal of such an investigation is usually twofold: determining (1) the origin and (2) the cause of the fire. This is performed by observing the fire scene, tracing the fire back to its point or points of origin through the study of fire patterns, and, finally, identifying the first fuel ignited and the ignition source. In order to do so, the underlying theories developed in fire chemistry and physics are used. The transfer of heat, the ignition of a combustible material, the way fire propagates, and its interaction with fuels are all examples of events that are regulated through fire chemistry and physics. This phenomenon is actually much more complicated than it may first appear.

Fire investigators and fire debris analysts must fully understand the underlying chemistry and physics of fire. Without that knowledge, it is impossible to conduct a sound fire investigation, or to perform a sound fire debris analysis. The full understanding of the phenomenon regulating the birth of a fire and its subsequent propagation is fundamental. This chapter provides an introduction to fire chemistry. It starts with the basic conditions necessary for a fire to exist and develops the interactions between these conditions.

Conditions for a Fire

Fire is the result of combustion that occurs at a rate high enough to generate heat and light. There are other combustion reactions, such as rusting, for example, that are not considered as "fire" because they do not provide heat and light. Combustion is an exothermic chemical reaction between a combustible substance (called fuel) and an oxidizer. In order for a fire to occur, three conditions must be present and interacting together:

- Fuel (combustible material)
- Oxidizer
- Source of heat (activation energy)

These conditions are usually represented in a graphic called the fire triangle (see **Figure 1**). Each side of the triangle represents one condition and the fact that the triangle is assembled symbolizes the interaction between the three conditions and, thus, the birth of a fire.

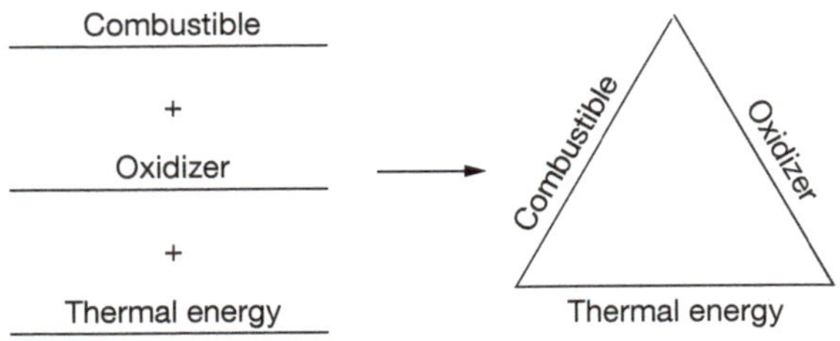

Figure 1 Fire triangle. When the three sides interact together, fire occurs. Reproduced from Stauffer, E., Dolan, J.A., Newman, R., 2008. Fire Debris Analysis. Academic Press, Burlington, MA, p. 86. © Elsevier.

Other graphical representations exist, such as the fire tetrahedron (where the fourth side is the chemical chain reaction); however, they are not as widespread as the fire triangle. If the fire triangle symbolizes the birth of a fire, it is also used to explain fire extinguishment. As a matter of fact, extinguishing a fire simply consists of suppressing one of the sides, or breaking the interaction between two sides. By using water on a fire, one removes the heat, and thus the fire dies. By using a carbon dioxide fire extinguisher, one suppresses the oxidizer, and the fire extinguishes.

Fuel comprises any material that can be oxidized when in the presence of a suitable ignition source as well as sufficient oxidizer. Fuels are almost everywhere, such as in wood, plastics, paper, gasoline, and propane. However, it is important to realize that many substances that are not thought of as fuels at first are actually very good fuels. For example, some metals or some other inorganic materials can greatly contribute to a fire, once ignited. Combustibles are found in all three phases (gaseous, liquid, and solid), which can seriously complicate a fire investigation.

The oxidizer is readily available through the oxygen contained in the air (21%). However, oxygen is not the only oxidizer that can be used to feed a fire. For example, chlorine is also an excellent oxidizer, and it is possible to start a fire without any oxygen by using chlorine and methane. However, in almost all fires that a fire investigator will be called upon to investigate, oxygen will be the oxidizer involved. The few exceptions that exist are typically found in chemical plants, where particular chemical substances are used. As a result, the presence or absence of an oxidizer is often not a decisive factor in the determination of the cause of a fire. However, it highly influences the fire's development.

The source of heat is the activation energy that raises the temperature of the fuel high enough so that it will ignite under the appropriate circumstances. While this statement appears simple, the issue of the source of energy is quite complex and is often the key element in a fire investigation. As a matter of fact, combustibles are in constant contact with oxygen in everyday life, and things do not catch fire spontaneously: The missing element is the source of energy on, which the fire investigator often will focus. It is also important to understand that not all sources of energy are suitable for all fuels. Even if they contain sufficient energy, some are not capable of transferring their energy to the fuel in question. As an example, although it is possible to light a newspaper with a lit cigarette, it is impossible to ignite gasoline with a lit cigarette. Conversely, an electrical spark can readily ignite a proper gasoline/air mixture, but it will be relatively inefficient in igniting a book.

Fire as a Chemical Reaction

As previously indicated, combustion is a chemical reaction between a combustible material and an oxidizer. It is actually a redox reaction during which the oxidizer is reduced and the combustible material is oxidized. The formula, as follows, provides an example of such a redox reaction, with the combustion of ethane:

$$2C_2H_6 + 7O_2 \rightarrow 4CO_2 + 6H_2O$$

In this example, the oxidation number of carbon goes from −III (in ethane) to +IV (in carbon dioxide) and in oxygen goes from 0 (in molecular oxygen) to −II (in carbon dioxide and water). Thus, carbon is oxidized and oxygen is reduced.

The above formula shows the stoichiometric reaction between ethane and oxygen and is referred to as a complete combustion. In practice, few fires lead to a proper stoichiometric ratio. As a result, the chemical reaction does not go to completion. Two types of reactions, complete and incomplete, are thus possible.

In a complete reaction, the stoichiometric ratios are reached and, thus, enough oxidizer is available to fully oxidize the fuel into the simplest products: carbon dioxide and water in this example. In real life, it is almost impossible for a fire in an uncontrolled environment to reach the stoichiometric ratio. Complete combustion is the result of a controlled burn, such as with an oxyacetylene torch, where it is possible to provide sufficient oxygen to guarantee the stoichiometric ratio of the reaction. This is shown as an example in the following equation:

$$2C_2H_2 + 5O_2 \rightarrow 4CO_2 + 2H_2O$$

In such a case, the flame will be blue, as no carbon particles are generated. This is shown in **Figure 2**.

If the amount of oxygen available is lower, an incomplete reaction will occur, such as

$$2C_2H_2 + 3O_2 \rightarrow 4CO + 2H_2O$$

In this example, carbon monoxide is produced instead of carbon dioxide. Carbon monoxide is a colorless and odorless toxic gas. It is extremely dangerous to human beings and is often the cause of asphyxiation in fires. Additionally, carbon monoxide is a combustible gas, since it can burn with oxygen to produce carbon dioxide. As such, carbon monoxide has resulted in explosions.

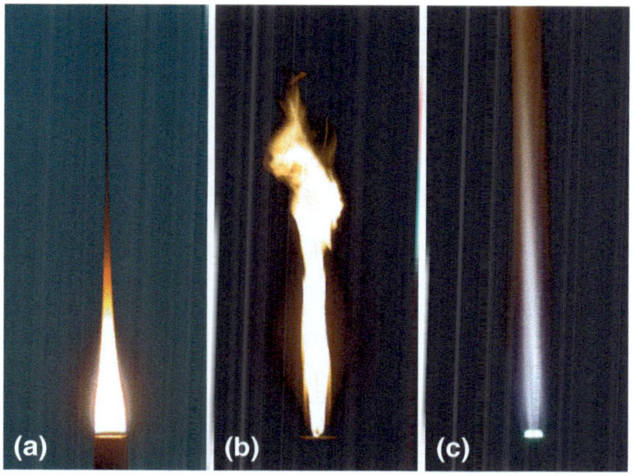

Figure 2 Example of incomplete, partial, and complete combustion with an oxyacetylene torch where the availability of oxygen is increased from left to right. (a) Notice the generation of heavy soot, indicative of an incomplete combustion. (b) The bright yellow flame also indicates an incomplete combustion, however, without the release of soot. (c) A complete combustion is shown as a blue flame. Reproduced from Stauffer, E., Dolan, J.A., Newman, R., 2008. Fire Debris Analysis. Academic Press, Burlington, MA, p. 92. © Elsevier.

If even less oxygen is available, carbon will be produced:

$$2C_2H_2 + O_2 \rightarrow 4C + 2H_2O$$

In this situation, a heavy release of soot will be seen, because there is not enough oxygen to oxidize carbon.

The well-known orange color of the flame is a typical indicator of incomplete combustion. This indicates clearly that incomplete combustion is much more common than complete combustion, since the orange color is almost always associated with flaming combustion. A complete combustion usually leads to a blue flame (depending on the fuel), such as the one shown in **Figure 2**.

If oxygen is completely removed from the equation (and given that there are no other oxidizers), then fire will extinguish. This situation is called an oxygen-controlled fire as opposed to a fuel-controlled fire, where the fire is controlled according to the available supply of fuel. The availability of oxidizer is an important factor in the development of a fire. When the fire is in its initial growing phase, it consumes more and more oxygen. When the concentration of oxygen is of least 10%, it is possible to have a flaming fire (i.e. a fire with visible flames). If the concentration drops below 10%, it is no longer possible to have flames, and the fire smolders. In a smoldering fire, the reaction between the combustible material and the oxidizer is a solid-to-gas reaction, meaning that the combustible material is a solid and the oxidizer is a gas in its surroundings. One can thus deduct that a smoldering fire can only occur with solids. A typical example of a smoldering fire is glowing charcoal on a grill. In a flaming fire, a gas-to-gas chemical reaction occurs. This means that only gases can burn in flames; thus, solids and liquids must first transform into gas.

If a flaming fire is in a confined environment, it will run out of oxygen relatively quickly and the amount of soot produced will greatly increase. Eventually, the level of oxygen will drop below 10%, at which point the flames will die and the fire will smolder. At that point, the fire is oxygen-controlled, because it is a lack of oxygen that prevents it from spreading further. If one creates an opening and a sudden supply of oxygen is provided, the fire will consequently burst into flames. This phenomenon is known as backdraft and is extremely dangerous to fire fighters.

In practice, a fire is an extremely complex chemical reaction with many different reactants and many levels of incomplete combustion, which generates many different radicals, as shown in **Figure 3**.

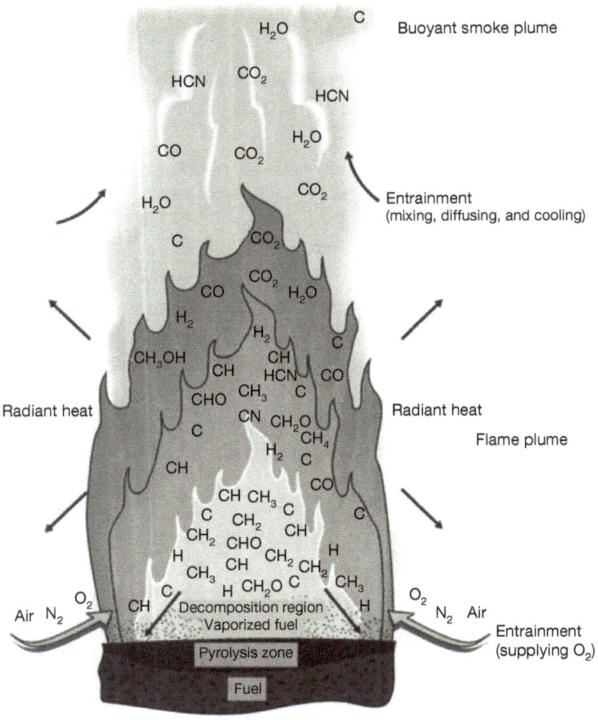

Figure 3 Representation of a fire plume with the presence of many different molecules and radicals. Reproduced from DeHaan, J.D., Icove, D.A., 2011. Kirk's Fire Investigation, seventh ed. Pearson Education, Upper Saddle River, NJ. © Prentice Hall.

Phase Change and Pyrolysis

As stated previously, a fuel must be in a gaseous form in order to burn in a flaming fire. This means that because solids (except for smoldering fire) and liquids do not burn, they must be transformed into a gaseous phase. This is an important point in fire investigation, because one must determine how the fuel got in its proper form before it was ignited.

In order to achieve a gaseous state, liquids and solids can undergo different physical transformations as shown in **Figure 4**.

Besides physical phase changes, a solid (or a liquid, less often) can also undergo a chemical change, called pyrolysis, to become a gas. Pyrolysis is the decomposition of a larger molecule into smaller molecules as a consequence of exposure to heat, without undergoing any oxidation.

Pyrolysis is a very important phenomenon in fire investigation and, more importantly, in fire debris analysis. As a matter of fact, pyrolysis is the main way by which a solid gives off ignitable gases. As such, pyrolysis products can be found in fire debris samples, thus rendering their interpretation much more complex. There are three main pyrolysis mechanisms: random scission, side-group scission, and monomer reversion. Often, a substance is subjected to more than one mechanism. This is dictated by the strengths of the bonds where the bond with the lowest bond dissociation energy will often break apart first.

In random scission, the carbon-to-carbon bonds are randomly broken, thus creating smaller molecules. With the example of polyethylene, random scission will create a series of

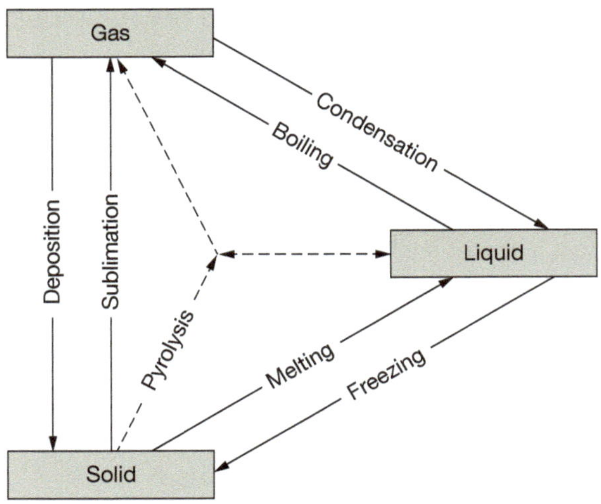

Figure 4 Passage of a solid or liquid to the gaseous phase by chemical and physical changes. Reproduced from Stauffer, E., Dolan, J.A., Newman, R., 2008. Fire Debris Analysis. Academic Press, Burlington, MA, p. 96. © Elsevier.

saturated and unsaturated hydrocarbons (alkanes, alkenes, and alkadienes). In the side-group scission mechanism, the carbon-to-side element bond is broken, thus leaving an unsaturated carbon chain that rearranges into a series of aromatic compounds, such as benzene, toluene, and styrene. Polyvinylchloride is a perfect example of a polymer undergoing mainly side-group scission. In monomer reversion, the polymer undergoes a reverse phenomenon, thus decomposing into its monomeric form. Polymethylmetacrylate is a typical example of a polymer undergoing monomer reversion. This pyrolysis mechanism usually results in one molecule; thus, it does not produce a significant interference in fire debris analysis.

Heat Source and Transfer

Fundamentally, only five heat sources exist:

- Electrical, as with the heat produced by an electrical arc between two conductors
- Mechanical, as with the heat generated by the friction of a stone on metal, such as in a lighter
- Chemical, as with the heat generated when sodium is mixed with water
- Biological, as with the biological activity in hay
- Nuclear, as with the reaction of fissile material

Thus, all heat sources found on earth are derived from one of these five fundamental sources. Once ignition has occurred, an open flame is available.

Heat can be transferred through three different modes: convection, conduction, and radiation. Convection is the mode of transfer occurring with fluids (liquids and gases). Energy transfers from one place to another through the motion of fluids. Hot gases released by a fire are a good example of convection. Because a fluid's density usually decreases with the increasing temperature, a fluid warmer than its environment moves upward. In the case of a typical room fire, convection is the phenomenon that creates the hot layer of gas located at the ceiling of the compartment.

In conduction, heat is transferred within a solid, such as through a copper wire through the interaction of vibrating atoms from particle to particle. By heating one end of the wire, the other end experienced an increase in temperature. If a solid is a good conductor, conduction will occur with significant intensity. If the solid is a poor conductor, the phenomenon of conduction will barely take place.

In radiation, heat is transferred through electromagnetic waves. The main difference in this heat transfer mode is that it does not require matter in order to take place. With convection, heat can only rise upward. With conduction, heat can only go through the conductor in contact. With radiation, heat can be transferred in all directions. In a typical room fire, it is the radiant heat that increases the temperature of the objects on the floor. The body of hot gases at the ceiling (created through convection) reaches a temperature high enough to radiate all the way down to the floor and bring the nearby objects to their autoignition temperature. As the fire develops, radiative heat becomes the dominant heat transfer mechanism.

Flammability Limits, Flash Point, and Fire Point

While a combustible material is represented by one side of the fire triangle, its mere presence is not a sufficient condition for fire to exist. The combustible material must be present in a sufficient concentration so that it can be ignited. This is called the lower flammability limit (LFL). Conversely, one cannot have too much fuel; otherwise, it will not ignite either. This is called the upper flammability limit (UFL). **Figure 5** shows the concept of LFL and UFL with the examples of gasoline and acetylene. Each fuel exhibits its own LFL and UFL.

When a liquid is at equilibrium, a given vapor concentration is found at its surface, known as the vapor pressure. This vapor pressure is highly dependent on the temperature of the liquid. When the vapor pressure equals the LFL, ignition can occur. At that point, the temperature of the liquid is defined as the flash point. In other words, it is the temperature at which a liquid gives off a sufficient vapor concentration so that it can be ignited using a piloted ignition source. At the flash point, when ignition occurs, the air/vapor mixture above the substance will burn very quickly (flash flame); however, fire will not sustain.

Flash points vary greatly from one substance to another and are very relevant in terms of both fire safety and fire investigation. The flash point of a material is the lowest temperature at which an ignition can occur if a suitable ignition source is brought within the ignitable vapor/oxidizer mixture. As a practical example, gasoline, with a flash point of approximately −40 °C, will readily ignite when a lighter is swept above its surface. However, diesel fuel, with a flash point of approximately 40 °C, will not ignite when sweeping a lighter above its surface.

For a fire to sustain once initial ignition has occurred, one would have to reach the fire point of the substance. The fire point can thus be defined as the lowest temperature at which a material will generate a sufficient concentration of vapors so as to form an ignitable mixture that will sustain combustion, once ignited. In general, the fire point is a few degrees above the flash point.

Ignition

Similarly to the presence of a combustible material, the mere presence of a heat source in the fire triangle does not imply that

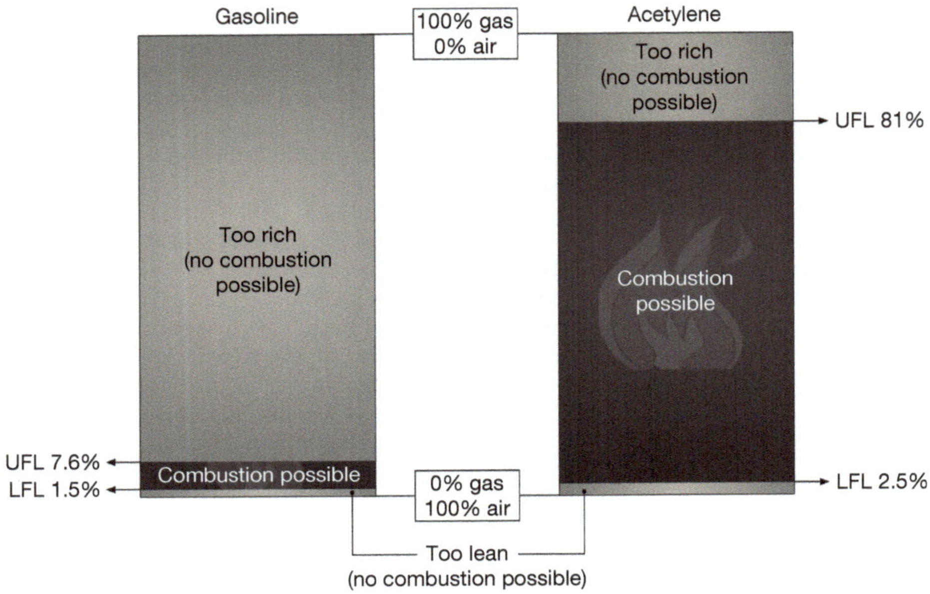

Figure 5 Illustration of the concepts of lower flammability limit (LFL) and upper flammability limit (IFL). Reproduced from Stauffer, E., Dolan, J.A., Newman, R., 2008. Fire Debris Analysis. Academic Press, Burlington, MA, p. 117. © Elsevier.

ignition will occur. The heat source must be suitable to the combustible material and have sufficient energy for fire to occur.

Ignition of a material can be reached through different modes. The material can be heated to the point at which it ignites, which is called autoignition. Alternatively, the material can be brought to its flash point or fire point, and its vapors then ignited (piloted ignition). The flash point, fire point, and the autoignition temperature are three parameters that characterize a material and with which a fire investigator must be familiar.

When a material is heated, in an oven, for example, at some point, the material will ignite by itself. This is called autoignition, or nonpiloted ignition, which is the ignition of a material without the addition of a piloted source, such as a flame. As an example, with an autoignition temperature of approximately 100 °C, carbon disulfide will ignite when in contact with boiling water.

Autoignition occurs quite often in fires. For example, the phenomenon of flashover, where a compartment suddenly bursts into flames, is the result of the autoignition of materials. The small original fire gives off hot gases that accumulate at the ceiling. They produce radiant heat that increases the temperature of the material on the floor. At some point, the material on the floor reaches its autoignition temperature, at which point it ignites. Because the autoignition temperatures of most common goods found in a household are in the same range, the entirety of the room bursts into flames.

Another important phenomenon is spontaneous ignition, which is the autoignition of a material through the heat generated by itself. This means that it does not receive energy from an external source, but it produces the increase of heat by itself. A typical example of spontaneous ignition is the ignition of a rag due to the exothermic reaction of drying linseed oil present on it. Spontaneous ignition is a specific type of autoignition, for which the production of heat is particular; however, it does not change the fact that the material must reach its autoignition temperature before it ignites.

Finally, piloted ignition is the ignition of a material through a direct external source (piloted source), such as a flame or a spark. Practically, the application of this ignition source to the material brings a portion of the material to its autoignition temperature; however, the rest of the material may be of much lower temperature.

Conclusion

The chemistry of the fire phenomena is a complex interaction of many physical and chemical characteristics of the materials involved. We have attempted to discuss the most important of these factors in this brief chapter. It is essential that those involved in the investigation of fires fully understand these properties to ensure that correct interpretation of postfire indicators can be made.

See also: **Chemistry/Trace/Fire Investigation:** Analysis of Fire Debris; Interpretation of Fire Debris Analysis; Physics/Thermodynamics; Thermal Degradation.

Further Reading

Babrauskas, V., 2003. Ignition Handbook. Fire Science Publishers, Issaquah, WA.

Bond, J., 1991. Sources of Ignition: Flammability Characteristics of Chemicals and Products. Butterworth-Heinemann, Jordan Hill, England.

DeHaan, J.D., Icove, D.J., 2011. Kirk's Fire Investigation, seventh ed. Pearson Education, Upper Saddle River, NJ.

Drysdale, D., 1985. An Introduction to Fire Dynamics. Wiley, Chichester, England.

Ettling, B.V., Adams, M.F., 1971. Spontaneous combustion of linseed oil in sawdust. Fire Technology 7 (3), 225–236.

Faraday, M., 1993. The Chemical History of a Candle. Cherokee Publishing Co, Atlanta, GA.

Haessler, W.M., 1974. The Extinguishment of Fire. National Fire Protection Association, Quincy, MA.

Holleyhead, R., 1996. Ignition of flammable gases and liquids by cigarettes: a review. Science and Justice 36 (4), 257–266.

Lentini, J.J., Waters, L.V., 1991. The behavior of flammable and combustible liquids. Fire and Arson Investigator 42 (1), 39–45.

National Fire Protection Association, 2011. NFPA 921 Guide for Fire and Explosion Investigation, seventh ed. National Fire Protection Association. Quincy, MA.

Nic Daéid, N., 2004. Fire Investigation. CRC Press, Boca Raton, FL.

Quintiere, J.G., 1997. Principles of Fire Behavior. Delmar Publishers, Albany, NY.

Simmons, R.F., 1995. Fire chemistry. In: Cox, G. (Ed.), Combustion Fundamentals of Fire. Academic Press, San Diego, CA, pp. 403–473.

Stauffer, E., 2003. Basic concept of pyrolysis for fire debris analysts. Science and Justice 43 (1), 29–40

Stauffer, E., Dolan, J.A., Newman, R., 2008. Fire Debris Analysis. Academic Press, Burlington, MA.

Physics/Thermodynamics

JHC Martin and RS Pepler, Institut de Police Scientifique et de Criminologie, Université de Lausanne, Lausanne, Switzerland

This chapter is reproduced from the previous edition, volume 2, pp. 928–933, © 2000, Elsevier Ltd

Introduction and Overview

Within the various domains of criminalistics, the investigator compares traces found at the scene of a crime with those recovered from a person, tool, or place. By contrast, the work of the fire investigator is not purely comparative in nature; extensive practical experience of the examination of as many fire scenes as possible combined with a sound scientific knowledge, especially in the areas of physics and thermodynamics, is paramount to the success of the investigation. These two criteria enable him or her to identify the origin of the fire and determine its cause despite having to work on sites that have been destroyed by the fire and further disrupted during the extinguishing process.

This chapter explores the relevant background areas of physical thermodynamics; the role of thermodynamics in fire investigation; fire ignition and propagation; thermodynamic classification of ignition sources; and the phenomena of smoldering and flaming combustion.

Physical Thermodynamics: The Relevant Background

Before considering the role of thermodynamics in fire investigation, it is essential to remember that the fundamental subject under scrutiny is concerned with the transformation of a macroscopic system which is dependent on one of the basic elements of physics: temperature. It is also important to understand that an intensive thermodynamic variable is one which is dependent on the amount of a chemical substance in the system, whereas an extensive variable does not.

Physical Systems

A physical system is defined as that part of the universe under the influence of thermodynamics which has been identified for study. It can be:

- isolated: no exchange of heat, work, or matter with the surroundings is possible. The system is said to be ideal and can be assumed in various situations;
- closed: exchange of heat and work, but not matter, is possible with the surroundings;
- open: exchange of heat, work, and matter is possible with the surroundings.

Thermodynamic Principles

Two of the fundamental laws of thermodynamics can be described in the following manner:

The first law of thermodynamics

All systems possess an internal energy U, which is a state variable, signifying that U is independent of the history of the system. When a system undergoes a change from state 1 to state 2, the difference in thermodynamic energy, ΔU is:

$$\Delta U = U_2 - U_1$$

While respecting the principle of the conservation of energy, the ΔU of a closed system can be expressed as:

$$\Delta U = \Delta Q + \Delta W$$

where ΔQ and ΔW are, respectively, the applied heat and work.

For an isolated system, U is constant; in keeping with the conservation of energy, U_1 is therefore equal to U_2. There is no energy change between the system and its surroundings therefore:

$$\Delta Q = \Delta W = 0$$

The second law of thermodynamics

This law concerns change and entropy. Entropy, S, is an extensive state variable which denotes the degree of disorder of the system. During reversible changes, the system and surroundings are in constant equilibrium. When considering such a change from a state 1 to a state 2, the entropy of a system is:

$$S = \int dQ/T$$

irrespective of the pathway taken.

However, energy dissipated in the form of heat is not an ideal process; it is spontaneous and irreversible which consequently means that the entropy of such a system increases.

The Role of Thermodynamics in Fire Investigation

An understanding of fundamental thermodynamics is important in fire investigation in considering possible ignition processes. The most important factors are identified below.

 http://dx.doi.org/10.1016/B978-0-12-802718-9.00018-1

Temperature

Temperature is an intensive state variable which can be measured for any given system by the use of a thermometer. The Celsius, or centigrade, scale of temperature is governed by the triple point and the boiling point of water. At 0 °C, under standard pressure, water exists simultaneously in solid, liquid, and gaseous states; at 100 °C only the existence of the liquid and gaseous states is possible. There are two other temperature scales which are extensively used: Kelvin (K) and Fahrenheit (F). The conversion formulae between the three scales are as follows:

$$0°C = 273.15K = 32°F$$

$$T(K) = T(°C) + 273.15$$

$$T(°F) = 9/5T(°C) + 32$$

At any given atmospheric pressure there are certain temperature values, which are specific to a particular substance, that indicate whether the ignition of this substance is possible.

- Flash point: the lowest temperature at which a sufficient quantity of vapor is produced above a solid or liquid which can be ignited by a flame, spark, or localized heat source. At this temperature combustion does not endure;
- Point of inflammation: the lowest temperature at which the production of flammable vapors above a solid or liquid is such that a flame, spark, or other localized heat source will cause their ignition; combustion, and its accompanying flames are sustained;
- Point of autoignition: the lowest temperature at which a solid or liquid will spontaneously ignite without the presence of an additional heat source (flame, spark, or other).

Heat

Heat is a form of energy which has a number of effects on a system. It can be generated by the transformation of energy, be it mechanical, electrical, chemical, or other, into calorific energy. A temperature gradient exists between two systems at different temperatures, along which calorific energy (heat) is transferred in order to reach a state of thermal equilibrium.

The fundamental unit of calorific energy is the Joule (J); however, a more widely recognized terminology is the calorie (cal) or kilocalorie (kcal). These units are defined as the quantity of heat required to raise 1 g, or 1 kg respectively, of water from 14.5 °C to 15.5 °C.

$$1 \quad \text{kcal} \equiv 10^3 \text{J} \quad 1 \quad \text{cal} \equiv 4.185 \ \text{J}$$

Physical Properties

Heat capacity, C, is the heat transferred to a body in order to raise its temperature by one degree (K):

$$C = \frac{dQ}{dT} \quad \text{units:} \quad \text{J K}^{-1}$$

Specific heat capacity, c, is the heat capacity of 1 kg of a substance which, in general, is dependent on temperature:

$$c = \frac{1}{M}\frac{dQ}{dT} \quad \text{units:} \quad \text{J kg}^{-1}\ \text{K}^{-1}$$

One of the fundamental laws of physical chemistry is the ideal gas law:

$$pV = nRT$$

where p = pressure, in Pa (1 atm = 101325 Pa) V = volume, in m^3; n = moles; T = temperature, in K; R = gas constant = 8.315 J K^{-1} mol^{-1}.

When an ideal gas makes the transition from state 1 to state 2, the following combined gas law applies:

$$\frac{p_1V_1}{T_1} = \frac{p_2V_2}{T_2}$$

This expression can be used for real gases on the understanding that the data obtained will only be an indication of and not the exact values desired.

Fire: Ignition and Propagation

Fire is an oxidation–reduction (redox) reaction accompanied by the production of heat and light. The two chemical requirements are an oxidizing agent (or combustive agent)—usually the oxygen in air—and a reducing agent (or fuel), which must be in a gaseous state in order to mix with the air. An activation energy, always in the form of heat, is required to ignite the gaseous mixture and combustion is maintained by the calorific energy produced during the exothermic reaction.

Activation Energy

The limit of flammability of a gas (or gaseous mixture) is expressed as a percentage which indicates the proportion of gas present in the air. The ignition of a homogeneous mixture of air and a gas is possible when the lower limit of flammability has been attained; however, ignition is impossible above the upper limit of flammability as the air is too saturated. Understanding and establishing the activation energy that is required for ignition of a gas/vapor (other than autocombustion) is fundamental to fire and explosion investigations. The scene examination is often complex as several calorific energy producing systems can be present, of which some could be weak heat producers and others situated some distance from the combustible material. In any situation, the means of ignition has to originate from one of these sources. In order to eliminate each ignition possibility until left with the most probable, a second factor must be taken into consideration: the thermal or calorific power.

Thermal Power and Reaction Temperature

The thermal or calorific power of a system is the calorific energy produced over a period of time. The investigator can only estimate the thermal power which is at the origin of and aids the development of a fire. The relationship between the thermal power of the heat source, W_T, and that dissipated in the surrounding environment, w_T, constitute the thermodynamic foundation of the whole investigation. This can be seen from the following equation derived from the Arrhenius expression:

$$W_T = \Delta H_c n A \exp(-E_a/RT)$$

where: ΔH_c = enthalpy of combustion; A = preexponential factor, E_a = activation energy and:

$$w_T = hA\Delta T$$

where: h = convective heat transfer coefficient; A = surface area.

The thermal power of a heat source is exponentially dependent on temperature, and the only circumstances when a body will not emit heat is at absolute zero temperature (0 K); $W = 0$ if, and only if, $T = 0$.

This observation is important in that any system not at $T = 0$ K will produce calorific energy. If the energy dispersion into the surrounding environment is restricted, the system's temperature will inevitably increase. Ignition frequently occurs where the temperature of a combustible material has been augmented to the point of autocombustion.

The possibility of a given heat source (spark, flame, fermentation, etc.) igniting a specific material (wood, insulation, paper, etc.) in a certain environment (quasi-isolated, closed or open) must always be considered before the source can be eliminated from the investigation. However, calculations of W_T and w_T for real systems are impractical and inexact, except for certain types of heating; instead, the essential factor that should be taken into consideration when examining possible ignition sources is the influence of W_T and w_T on the system. Therefore, the first step of a fire/explosion investigation is to locate the place where the activation energy was produced; in other words, to determine the foyer, or seat, of the fire/explosion. Once a potential heat source has been identified, the implications of the thermal power produced and the heat dissipated in its immediate vicinity should be estimated. This enables the investigator to verify whether or not the proposed source is a plausible source of ignition of the material situated at the (presumed) foyer.

Combustion Speed

The phenomena of smoldering fires, flaming fires, and explosions differ only in their rate of combustion.

Smoldering is a flameless combustion involving the slow oxidation of a porous combustible material, which occurs in an environment where oxygen can easily penetrate, yet is sufficiently isolated so that heat dissipation to the surroundings is minimal.

The speed of combustion of a flaming fire can vary enormously. The chemical reaction occurs without a perceptible pressure increase within the immediate vicinity.

An explosion is an extremely rapid combustion. When occurring in a confined environment, it is accompanied by a considerable augmentation of pressure. The reactants are present within their limits of flammability and are usually in a gaseous state; however, very fine droplets of liquid, such as those produced by an aerosol, or fine powders/dust can also combust explosively. After ignition, the entirety of the explosive material undergoes the exothermic reaction of combustion almost instantaneously; the ambient air subsequently expands very quickly resulting in a pressure increase in the immediate environment.

Thermodynamic Classification of Ignition Sources

Fires can be classified according to their ignition source whether due to a spark, naked flame, or localized heat source.

Heating

Heating is a transfer of calorific energy to a physicochemical system without the presence of a spark or flame. If the system generates heat by itself to augment its temperature to the point of inflammation and subsequently to autoignition, this is known as spontaneous combustion. With the exception of this latter form of heating, heat is transferred to the material via the methods outlined below.

Conduction

This is the mode of heat transfer without displacement of matter, due to electronic movement within a solid. Where two faces of a solid are at different temperatures, T_1 and T_2, the heat flux, q', through the material is:

$$q' = \frac{k}{x}(T_1 - T_2) \quad T_1 > T_2$$

where: x = thickness of material; k = thermal conductivity.

If a steady-state heat flux traverses two (or more) layers of material:

$$q'_1 = q'_2 = q'_{1,2}$$

$$q'_{1.2} = \frac{T_1 - T_2}{x_1/k_1 + x_2/k_2}$$

Convection

Convection occurs in liquids and gases and is a transfer of heat via mechanical movement due to changes in density. This process can, however, transfer heat to a solid. For example, if a solid surface of area A is in contact with a liquid/gas at temperature T, the heat flux, q'', transferred to the surface is:

$$q'' = hADT \quad \text{units:} \quad \text{Watts}$$

where, h = convective heat transfer coefficient; A = surface area.

Radiation

Electromagnetic radiation in the visible and infrared regions is radiated from all bodies with a temperature above absolute zero. When a system is in thermal equilibrium, its components emit and absorb radiated heat at the same rate. Although this radiation is rarely quantified it is important to note that:

- radiation is the prevalent mode of heat transfer between flames and combustible material; it is therefore the principal cause of vaporization and consequently the main means of the ignition of fresh material;
- before suspecting human intervention after a large fire where several foyers have been identified, it is essential to verify that the separate ignition sites are not a consequence of thermal radiation. Fire may have spread in this manner from one combustible material to another despite the lack of an apparent physical link.

When a body has two identical surface areas, A, respectively at temperatures T_1 and T_2, a heat flux, f, flows from surface 1 to surface 2 if $T_1 > T_2$:

$$f = a_1 a_2 \sigma A\left(T_1{}^4 - T_2{}^4\right)$$

where, a = heat absorption or reflection factor; σ = Boltzmann constant.

Heating appliances

When a heating appliance is identified as being the origin of a fire, any of the three modes of heat transport can constitute the transfer of calorific energy to the combustible material. As the correct use of these devices is for the generation of heat, the cause of the fire has to be a result of an inappropriate usage or a malfunction of the apparatus.

Electrical sources

Fires of electrical origin are often complex to investigate and difficult to prove. Heat is produced when an electrical current, I, passes through a conductor with resistance, R, over a time period, t, according to Joule's law:

$$\Delta E_{\text{electric}} = I^2 Rt \quad \text{units:} \quad A^2 \cdot \Omega \cdot \text{sec} = \text{J}$$

The electrical power is therefore:

$$W = I^2 R \quad \text{units:} \quad \Omega A^2 = W$$

Friction

The generation of heat by the friction between two objects is proportional to the speed at which the two bodies move against each other and their respective weights. The nature of the surfaces in contact, which is indicated by the friction coefficient, will also have a bearing on the quantity of calorific energy produced; those with a high friction coefficient generating more heat.

Sparks

Material sparks

Material sparks are produced when solid particles with a mass of milligram magnitude are emitted into an open system existing at high temperature. These tiny fragments of incandescent or burning matter can constitute the origin of a fire if:

- when smoldering, the distance between the source and the combustible material is very short; due to their miniscule mass the particles lose heat rapidly;
- the spark enters into contact with a highly flammable material or one with a small heat capacity in which a smoldering fire can manifest.

A spark of mass m, and heat capacity c, emitted with a temperature T_2, into an environment at temperature T_1, produces heat, Q, according to:

For example, when grinding a metal, the metallic fragments produced are of milligram mass, which generate an approximate heat of $Q \approx 1.3$ J; this is of sufficient magnitude to ignite a flammable gas/vapor. Another example is the fusion of a metal/metal oxide when welding; the resulting droplets have a diameter of 2–5 mm and generate heat between the values of: $37 < Q < 578$ J. The first value is more than sufficient for the ignition of a gaseous mixture, whereas the second can initiate the combustion in a nearby combustible material such as plastic isolation or wood, provided that it is within the boundaries of the quasi-isolated system.

Electrical sparks and arcs

In the space separating two charged objects, there is an electrical field. If an avalanche of electrons occurs due to this field, it is visible (to the naked eye) as a spark. Because of its short lifetime, an electric spark produces a relatively small amount of heat; nevertheless, the millijoules of energy generated can ignite a homogeneous air/flammable gas mixture resulting in an explosion.

The difference between an electrical arc and spark is the duration of the electrical discharge: that of an arc being markedly longer than a spark, which can be said to be instantaneous. Logically, it follows that the energy produced by an arc is of greater magnitude and can subsequently inflame a combustible material rather than uniquely gaseous mixtures as for a spark.

The above phenomena will occur under the following conditions.

- A defective contact (of a switch, relay unit, etc.) in an electric circuit, with or without a coil, will result in an increase in resistance and will thus generate heat. This default can also occur with the use of adapters and extension cables.
- In the presence of an electrolyte (water, humid air, etc.), a defective insulation of a cable will enable the passage of an electric current between two conducting wires at different electric potentials. An augmentation in resistance will subsequently generate calorific energy at a precise point which can be sufficient to ignite the insulation or a combustible material in close proximity.

Smoldering

In a smoldering solid, the heat conduction is difficult to calculate, as are the speed and propagation of combustion. This type of fire obviously still requires an ignition source; however, the thermal power of the source can vary enormously. For example:

- after most fires, smoldering is the natural progression of flaming combustion. This latter form of combustion can be reinitiated when a great quantity of smoldering material is remnant after a large fire. This eventuality only applies to one aspect of fire investigation: that where flames are rekindled from smoldering remains, even after extinction by the fire brigade;
- when the heat source is weak, smoldering combustion commences at the surface of the material. In this case the only conceivable progression of combustion is to the interior of the material, otherwise self-extinction will occur.

Smoldering combustion is self-maintained if an equilibrium is established between the thermal power of the exothermic (combustion) reaction and that dissipated into the surroundings:

If the difference, ΔW, between the former thermal power values is small, the fire will neither self-extinguish nor reach an open flame state; the smoldering combustion will simply continue to burn fresh material with a random progression that is due to the inhomogeneous environment. In order to maintain a constant value of ΔW, or one that varies within a very narrow range, the oxygen supply must be adequate; nevertheless, it must be noted that even in an oxygen-deficient environment (proportion of O_2 in the air $<16\%$), it is possible for this type of combustion to be sustained.

Smoldering material will burst into flames if there is sufficient heat buildup and enough oxygen available. For example, smoldering combustion will progress to the flaming stage if the burning material gets into contact with air. A pertinent example of this is when the exterior surface temperature of a material is increased by heat transferred from internal smoldering combustion; when the surface, which is in contact with the ambient air, reaches the point of inflammation, flaming combustion will result and will spread at a rapid rate.

Flames

If a combustible material is exposed to prolonged contact with a flame its temperature will eventually reach the point of inflammation and ignition will occur. The resulting flames will subsequently raise the temperature of the surrounding material which will also attain its inflammation point and start to burn. The propagation of fire by direct flame contact is a natural progression which does not require further explanation.

Hot gases produced by burning material are propelled upward. Due to their inferior density the buoyancy forces within the smoke plume cause the hot gases to rise with a speed that is dependent on the temperature of the fire. At high temperatures, for example, directly above the flames, the gases are dispersed vertically very rapidly and have little sideways diffusion into the ambient air. This effect diminishes as the ambient temperature decreases, thus the gases have greater horizontal convection the further they are from the flames. They can therefore be seen to rise in the form of a three-dimensional "V" (an inverted cone). This plume of hot gases transport soot particles which are deposited on vertical nonflammable supports; the characteristic triangular pattern of these deposits can often be used to locate the seat of a fire.

The smoke resulting from a fire contains combustion products and small unburned residues which are, by definition, nonoxidized vapors that are transported by the diffusion of the other hot gases. These vapors can accumulate in enclosed spaces and spontaneously ignite, under the right conditions, even at considerable distances from the original foyer.

Fire spread can often be deduced with relative ease if the initial heat source has been identified and there is evidence of the presence of combustible material throughout the entire burned area. However, when justifying a source of ignition, problems can arise when the heat source is found to be separated from the combustible material by nonflammable material such as walls, floors and ceilings.

With an open fire, the vertical flame movement and the air supply are practically limitless. At the exterior of a stone or brick building, a fire situated at ground level can extend to roof height. If the roof is constructed from combustible material such as wood or thatch, it will eventually catch fire (due to the prolonged contact with the flames and hot gases) even if separated from the seat of the fire by several floors of nonflammable material.

If the fire is restricted by a wall or other incombustible barrier, the supply of air may well be limited. The flames will consequently spread toward the direction of the air source. When vertically extending flames are hindered by the ceiling of a room, the plume of hot gases is redirected along its surface. The horizontal spread can be of considerable distance as the area directly below the ceiling is at a highly elevated temperature, thus sustaining the capacity of the gases to ignite fresh combustible material. Therefore:

- if the entrainment of air and the diffusion of flammable gases/vapors in the vertical part of the flame is sufficient, there will be limited flame spread: combustion will occur at ceiling level;
- with a low ceiling or a vapor-rich smoke plume, the hot gases will spread out beneath the ceiling and the flames can spread to material situated some distance from the original fire seat.
- It must be remembered that a corridor, ventilation shaft, or other form of passageway can aid the propagation of fire between rooms, apartments, or even separate buildings.

Conclusion

This chapter does not attempt to list each and every possible heat source capable of initiating a fire. Instead it indicates the most significant sources and analyzes them thermodynamically, consequently highlighting the role of thermodynamics and that of fundamental physics in verifying the conclusions of the fire investigator. Nevertheless, it must be noted that numerical values are rarely calculated when considering the various exchanges of heat or the laws of thermodynamics, as the fires under investigation are not performed in the laboratory under controlled conditions. The equations and mathematics are simply used to give an indication of the magnitude of the process in question. The work of the fire investigator is therefore to identify the origin of a fire and to report its cause by exploiting an extensive practical experience and intricate knowledge of the laws of physics and chemistry.

See also: **Chemistry/Trace/Fire Investigation:** Chemistry of Fire; **Investigations:** Fire Patterns and Their Interpretation; Major Incident Scene Management; Types of Fires; **Management/Quality in Forensic Science:** Risk Management.

Further Reading

Atkins, P.W., 1990. Physical Chemistry, sixth ed. Oxford University Press, Oxford.
Bush, L.S., McLaughlin, J., 1996. Introduction to Fire Science. Fire Engineering. Penn Well, Tulsa, USA.
Chaussin, C., Hilly, G., Barralis, J., 1969. Chaleur et Thermodynamique. Dunod, Paris.
Clavin, P., Linau, A., 1985. Theory of Gaseous Combustion. In: NATO ASI Series B, vol. 116, pp. 291–338.
DeHaan, J., 1991. Kirk's Fire Investigation, third ed. Prentice Hall, Brady.
Drysdale, D., 1986. An Introduction to Fire Dynamics, second ed. John Wiley & Sons, Chichester.
Noon, R., 1992. Introduction to Forensic Engineering. CRC Press Ltd, Boca Raton.
Phillipps, C.C., McFadden, D.A., 1986. Investigating the Fireground. Fire Engineering. Penn Well, Tulsa, USA.
Schultz, N., 1985. Fire and Flammability Handbook. Van Nostrand, New York.
Yallop, H.J., Kind, S.S., 1980. Explosion Investigation. The Forensic Science Society and Scottish Academic Press, Harrogate.
Yereance, R.A., 1989. Electrical Fire Analysis. Charles C. Thomas, Springfield, IL.

Thermal Degradation

K Grimwood, University of Technology Sydney, Sydney, NSW, Australia

Introduction

Thermal degradation describes the chemical decomposition of materials via the application of heat. In fire investigation, the terms thermal degradation, thermal decomposition, and pyrolysis are often used interchangeably; however, this is not wholly correct. Generally, pyrolysis is the thermal degradation of solid fuels and describes the stage without which flaming combustion cannot occur, while evaporation is the thermal degradation of liquid fuels. There are exceptions, however, with some ignitable organic liquid fuels undergoing pyrolysis, rather than evaporation.

In fire investigation, thermal degradation occurs in an oxygenated or reduced oxygen environment, although thermal degradation is not a phenomenon specific to fire investigation. In chemistry, materials are pyrolyzed, or undergo thermal degradation, in air as well as in inert atmospheres, such as nitrogen. The mechanism of thermal degradation of materials can be monitored with the use of thermogravimetric instrumentation, and pyrolysis products can be identified using analytical techniques such as gas chromatography coupled with mass spectrometry.

There is extensive representation of materials in structures or vehicles that will undergo thermal degradation at elevated temperatures. A broad range of examples have been provided in this section that may be commonly encountered in fire investigation. It is also pertinent to understand that not all materials commonly found in structures or vehicles undergo thermal degradation according to the description provided here. Glass, for example, undergoes state transition at elevated temperatures and so has not been discussed in this section.

Thermal Degradation Effects

Wood

Wood is a composite material consisting of cellulose, lignin, and hemicelluloses. Wood can undergo thermal degradation and combustion; however, the extent of thermal degradation is dependent on a number of conditions. The species of wood, its pH, and water content, in addition to the degree of thermal assault the wood is exposed to, will dictate the extent of degradation. A flaming fire will occur when combustible pyrolysis products are ignited and can independently sustain flame.

Experiments on the thermal degradation of chemically untreated wood have demonstrated the following general trends:

1. From 65 to 100 °C, the strength of wood can be significantly and irreversibly affected. The reduction in strength is most likely due to depolymerization in the form of carbohydrate loss. However, this is dependent on the species, duration of exposure to heat, and thickness of the wood.
2. Between approximately 100 and 200 °C, pyrolysis yields water vapor (H_2O), carbon dioxide (CO_2), and formic and acetic acid as well as additional noncombustible liquids and gases. At this temperature, significant char can be produced if the exposure period is sufficient.
3. The progression of events occurs from 200 to 300 °C that are initiated, in part, at lower temperatures but continue throughout this temperature range and potentially into the next. Dehydration early in this temperature gradient leads to the significant pyrolysis of hemicelluloses and lignin, producing further noncombustible liquids and gases. Most notable is the significant generation of carbon monoxide (CO) and tar. Toward the end of this temperature gradient, the depolymerization of cellulose occurs more readily and results in the formation of volatile pyrolysis products that are combustible in addition to significant char.
4. Temperatures of 300–450 °C result in the generation of the majority of the combustible volatiles. These are produced from the depolymerization of cellulose.
5. At temperatures in excess of 450 °C, the phenomenon referred to as afterglow occurs, where further thermal degradation of the wood, now in the form of char, yields CO_2, CO, and H_2O.

Wood that has been chemically treated to inhibit ignition, heat release, and flame spread has, for this reason, an alternative thermal degradation mechanism compared to untreated wood. There are different chemical compounds in use to aid flame retardation; however, the fundamental goal of each is the same. The period during thermal degradation when noncombustible gases and H_2O are the major pyrolysis products is maintained for a longer time. The result of flame retardation is an increase in smoke production and a higher char yield.

Wood char is a product of the thermal degradation, pyrolysis, of chemical components in wood. Historically, deep charring on wood was widely accepted as an indicator of rapid and intense fire. Considerable research regarding wood char

 http://dx.doi.org/10.1016/B978-0-12-802718-9.00019-3

has shown that the level of char is dependent on several factors including, but not limited to, the extent of thermal exposure, ventilation, density of the wood, its specific chemical composition, grain size, and moisture content. Recognizing that char yields differ depending on a number of variables is essential in fire investigation as the extent of char is often employed as a tool to compare fire spread within a scene.

Polymeric Materials

The term polymeric material refers to the repetition of many monomeric units linked together to form a chain. Polymeric materials are abundant in nature and examples include silk, latex, cellulose, and wood. Polymers can also be synthesized to produce different materials that display a vast range of physical and chemical properties. Synthesized polymeric materials include polyurethane, polyvinyl chloride, and polystyrene. The diversity in the physical properties of polymeric materials ensures that their use in building and vehicle construction is prolific. In addition to construction, polymeric materials are widely used for the internal coverings and furnishings of buildings and vehicles. Furthermore, polymeric materials play an important role in the textiles industry.

Pyrolytic thermal degradation of polymeric materials generally occurs in elastomeric and thermoplastic materials to produce noxious gaseous products. This is of particular interest during a fire investigation as inhalation of the toxicants can impact on the occupants' ability to self-egress or ultimately could lead to death. Considerable amounts of noxious gases are produced by the pyrolysis (or combustion) of polymeric materials that include, but are not limited to, CO_2, CO, hydrogen cyanide (HCN), oxides of nitrogen, hydrocarbons, and nitrogen- or oxygen-containing organic compounds. There has been significant research regarding the production and toxic effects of CO as a toxicant in fires. More recently, since the advent and subsequent prolific use of nitrogen-containing polymeric materials, there has been an increase in the investigation of the production and toxic effects of HCN.

In addition to the toxic effects of the thermal degradation of polymeric materials, the evolution of pyrolysis gases in a compartment can contribute to the onset of flameover, a phenomenon that can lead to flashover. Factors that should also be considered during a flameover or flashover include ventilation, fuel load, and compartment size.

Human Remains

The discovery of human remains at a fire scene can occur during the course of suppression or investigation. Recognizing that remains will have been impacted thermally, and the potential for thermal degradation to have occurred, is essential to how the remains are treated and interpreted. Most agencies will have specialized personnel to process this type of evidentiary material; however, having an understanding of what information can be gathered and how its destruction can be avoided can aid the progress of the investigation. The effects of thermal degradation on human remains can be viewed at a macro- and microscopic level, although at the scene a macroscopic interpretation will be more achievable.

At a fire scene, human remains can range from appearing unaffected by heat and flame to being so completely desecrated that recognition and subsequent recovery become the domain of specialized personnel.

Visual indicators that may occur as a result of thermal degradation can mimic those caused by other means. Appreciating that different causes can produce similar results is vital to avoid inaccurate assumptions pertaining to the state of the remains prior to the fire.

Soft tissue

The soft tissue of humans includes, but is not limited to, the skin (epidermal and dermal layer), connective tissue, adipose tissue, muscles, veins, nerves, and hair follicles. Postmortem blistering can occur on the surface of the skin due to thermal degradation of soft tissue. This type of blister is gas filled, unlike liquid blisters formed antemortem from heat or chemical burns or postmortem from heating. Their presence, in the absence of other blisters, can potentially help determine the victim's state at the time of the fire when considered contextually.

Thermal degradation in the form of dehydration contributes to the contraction of tendons and muscles. Uneven relative contraction between the flexor and extensor muscles results in characteristic posturing widely known as the pugilistic pose. Formation of this pose can cause the victim to move during the fire from their original position prior to the fire. This is a fact that needs to be taken into consideration by the investigator when documenting and assessing the position of the remains.

Bone

Bone is a composite material consisting of both organic and inorganic components. The structural stability of bone is reliant on the interaction of these phases, and as the thermal degradation of the bone progresses, the structural stability decreases. It is important for those investigating the fire scene to recognize the fragility of the remains in order to ensure they are suitably protected prior to removal. The analysis of bone that has undergone thermal degradation can benefit fire investigation as the morphological change associated with the degree of thermal degradation can potentially present a temperature guide to the level of exposure.

The extent of thermal degradation of bone is dependent on a number of variables, including the health, age, and type of bone in addition to the degree of protection from thermal assault, temperature, and length of exposure of the bone to thermal interference. Further, the composite nature of bone ensures that thermal degradation is a nonuniform process meaning different degrees of degradation may be expressed on a single piece of bone.

During the progression of a fire, bones undergo a change in color and stability due in part to the thermal degradation of the bones organic and inorganic components.

Experiments involving the thermal degradation of bones have demonstrated the following general trends:

1. From 50 to 220 °C, the dehydration of bones occurs marking the initial stages of reduced stability.
2. At temperatures up to 600 °C, organic components of the bone are undergoing thermal degradation. The result is further loss of stability, and the characteristic coloration of the bone is reported as gray to blue-gray. It is worth noting that in an oxygen-deficient environment, the pyrolysis of organic components in the bone leads to carbonization of the bone, which is recognizable from its black appearance.
3. Between 600 and 900 °C, the thermal degradation of the remaining organic components comes to an end. In addition, the calcium hydroxyapatite component of the bone degrades and produces CO_2. The resultant calcinated material is white in appearance and has lost significant stability.

Although it is not within the scope of this chapter, it is worth noting that other color changes to the bone have been recorded such as pink, red, green, and yellow. These color changes are most likely associated with the presence of metals and alloys in the surrounding environment at the time of the fire.

In addition to the alteration in color and stability from thermal degradation of bone, other morphological aspects undergo modification as well. The length and width of bone are reduced as a result of thermal degradation. Bone shrinkage and deformation are dependent, like other items present in a fire, on a number of variables. In this instance, the variables associated with bone include, but are not limited to, the age, sex, type and health of the bone, the mineral content, and the degree of protection in addition to the temperature and length of thermal exposure.

The multiple variants associated with shrinkage in bone mean that the literature data lack standardization. The reduction reported is between 1% and 18% for the length measurements and up to 27% for diameters. There is, however, general agreement with regard to bone that has been subjected to temperatures in excess of 800 °C that suggests that the shrinkage is significant (reported in places as greater than 30%) and hinders the visual estimation of sex and age.

When the visual analysis has been exhausted at the scene and during the postscene investigation, complementary forms of bone analysis can be used to yield useful information. In addition to the existing methods, significant work using Fourier transform infrared spectroscopy has been undertaken to estimate the temperature that individual bones have been subjected to based on the thermal degradation of water and collagen present in the bone. X-ray diffraction has also been used in recent research to investigate the possibility of differentiating between species of bone that has been burned based on the level of thermal degradation of the mineral component of bone, calcium hydroxyapatite.

While some of the techniques discussed in this section require specialist knowledge and training that may not be available to those in the position of physically investigating a fire, the knowledge that they exist and how the techniques can be used to attain information is crucial. It is the responsibility of those in fire investigation to be aware of the tools available to them, which could aid in determining cause and origin

Blood

Accepted methods exist to test for trace amounts of blood in an evidentiary capacity. Blood is a nonhomogeneous substance and testing methods target different components of blood. The tests, the majority of which are colorimetric, focus on the protein and enzymatic portions of blood. Although the testing methods have been developed for use at ambient temperatures, some may be applied post suppression to aid in the detection of blood at a fire scene.

At temperatures below 50 °C, thermal degradation effects are insignificant and will not hinder the use of accepted testing mediums on recently created stains. Colorimetric tests that target the enzymatic portion of blood are generally still effective when the stain has not been subjected to temperatures in excess of 100 °C, as the degradation of the enzyme has not been extensive enough. Prior to 200 °C, protein-specific dyes, such as phenolphthalein, can remain effective as the proteins' response has not been stopped by degradation due to heat.

It is generally accepted that if a stain is subjected to temperatures in excess of 200 °C, thermal degradation of the target components will have been significant enough for conventional mediums used to test for blood to yield positive results. However, like many aspects of evidentiary material, and especially material exposed to elevated temperatures, variables exist, such as length of exposure, that contribute to the degree of thermal degradation.

It should be considered that protein in bodily fluids such as semen and deoxyribonucleic acid (DNA) will follow a similar thermal degradation pathway as blood.

Trace Evidence

It is thought, often by the criminal element, that fire will destroy all evidence of a crime having been committed. This is not the case, and the recovery of trace evidence including, but not limited to, fingermarks and ignitable liquid residue is possible.

Fingermarks

Fingermarks are secretions that include a combination of water, proteins, and sebaceous and eccrine elements, which are deposited when a finger touches a surface. Latent fingermarks (those that require enhancement to view) and patent fingermarks (those that are visible in white light) are susceptible to thermal degradation, but can still be present, located, and enhanced post suppression of a fire.

When a fingermark is subjected to elevated temperatures, thermal degradation of the proteins in the mark occurs to yield amino acids. Existing chemical enhancement techniques for the detection of latent fingermarks, such as ninhydrin, work by reacting with the protein and amino acid components in the fingermark. Therefore, this method of detection may still yield useful information if the mark has not undergone extensive thermal degradation. The remaining components of the fingermark, sebaceous and eccrine secretions, will maintain their integrity to higher temperatures than proteins. Consequently, accepted methods to detect this portion of latent fingermarks can still be used. Accepted techniques for the detection of latent fingermarks can generally be used at a fire scene when the fingermark has not been subjected to temperatures in excess of 200 °C. In scenarios where the temperature reaches between 200 and 600 °C, the number of methods available to detect the mark declines in response to the increase in thermal degradation of the components in the fingermark. The detection of latent fingermarks at a fire scene is dependent on a number of variables including the substrate the mark is on, the level of thermal assault, and thermal degradation the mark has been subjected to, in addition to the water solubility of the remaining components of the fingermark.

Research has been conducted, and is gaining greater acceptance, regarding the development of latent fingermarks on specific substrates via the application of heat. Latent fingermarks can undergo development to produce patent fingermarks on substrates such as paper exposed to heat. The mechanism of development is thought to be due to the thermal degradation of the substrate itself, which is accelerated in areas where fingermark residue is present, producing contrast between the fingermark and the substrate. Some researchers claim that the thermal degradation of the fingermark itself also contributes to the development.

Patent fingermarks may be observed as a depression in another substance, such as dust or paint, a deposit made from a combination of finger secretions and other substances, such as blood or paint, or a deposit of another substance, such as soot deposits, onto a latent fingermark so that it becomes patent. In cases such as this, the potential for thermal degradation is dependent not only on the nature of the fingermark secretions, but also on the other substance(s) present. The development or enhancement of these fingermarks using existing methodology is subject to the degree of thermal degradation the mark, adhering substance, and other substance(s) present have undergone.

The most significant challenge hindering the recovery of fingermarks at a fire scene, apart from heat and flame, is the excessive use of water during the suppression effort, residual condensation post suppression, and physical destruction of the mark during the overhaul and investigation process. Preserving the areas of the least amount of thermal degradation of the fingermark is paramount to aid the collection and further development of this type of trace evidence. Areas of the scene that have been protected from direct flame and have least exposure are the most likely to yield exploitable fingermarks. It is worth noting that similar observations and considerations are applicable to alternative marks such as palm, lip, ear, and footmarks.

Ignitable liquids

The term ignitable liquid includes organic and inorganic liquids that are flammable or combustible as well as substances that can be liquefied and ignited. In some cases of arson, ignitable liquids are used as accelerants to aid the initiation of flame and the propagation of fire. For this reason, it is beneficial for the fire investigator to be aware of the effect of thermal degradation on ignitable liquids to aid the collection and possible identification of the unknown liquid or residue.

The thermal degradation of ignitable liquids involves pyrolysis, evaporation, or a combination of both. The portion of the ignitable liquid that remains after thermal degradation is significant for the fire investigation when trying to determine the likelihood of the residual components detected being present at that specific location. If it is suspected that ignitable liquids are present at a fire scene and the reason for their presence cannot be satisfactorily explained, then further investigation is warranted. The amount of residue remaining will depend on the level of thermal degradation the liquid has undergone. The residue will mimic a sample that has been weathered, partially evaporated, and will be compared with known weathered samples accordingly.

Ignitable liquid residue may be found in fire debris located in areas that have been protected from thermal impact and potentially at the seat or seats of fire. It should also be noted that ignitable liquids can absorb into substrates including, but not limited to, carpet, fabric, and soil. Collecting substrates in and around areas that are suspected to have had ignitable liquids poured can help establish, via analytical methods, if ignitable liquid residues are present. It is possible that the degradation of the liquid will be sufficiently extensive that the detection of ignitable residues, if detectable at all, will be below the detection limits of accepted analytical methods. It is worthwhile noting that there has been significant research, and subsequent discussion, regarding the use and ability of accelerant detection canines to identify thermally degraded ignitable liquids at fire scenes.

Summary

The purpose of this section was to provide an overview of materials commonly encountered at fire scenes that may have been affected by thermal degradation. While the provision of an exhaustive list of thermally degradable materials would potentially be beneficial as an additional tool, without discussion, the list alone could lead to misinterpretation of the material being investigated. It should be noted that the majority of the aforementioned temperature ranges have been experimentally derived in a controlled environment. Consequently, when materials are thermally impacted in a fire scenario, the extent of thermal degradation will depend heavily on a number of factors.

See also: **Chemistry/Trace/Fire Investigation:** Analysis of Fire Debris; Chemistry of Fire; Interpretation of Fire Debris Analysis; Physics/Thermodynamics; **Investigations:** Evidence Collection at Fire Scenes; Fire Patterns and Their Interpretation.

Further Reading

Bradshaw, G., Bleay, S.M., Deans, J., Nic Daeid, N., 2008. Recovery of fingerprints from arson scenes: Part 1 – latent fingerprints. Journal of Forensic Identification 58, 54–82.

Brown, A.G., Sommerville, D., Reedy, B.J., Shimmon, R.G., Tahtouh, M., 2009. Revisiting the thermal development of latent fingerprints on porous surfaces: new aspects and refinements. Journal of Forensic Sciences 54, 114–121.

DeHaan, J.D., 2007. Kirk's Fire Investigation, sixth ed. Prentice-Hall, Upper Saddle River, NJ.

Dominick, A.J., Nic Daeid, N., Bleay, S.M., Sears, V., 2009a. The recovery of fingerprints on paper exposed to elevated temperatures – Part 1: comparison of enhancement techniques. Journal of Forensic Identification 59, 325–339.

Dominick, A.J., Nic Daeid, N., Bleay, S.M., Sears, V., 2009b. The recovery of fingerprints on paper exposed to elevated temperatures – Part 21: natural fluorescence. Journal of Forensic Identification 59, 340–355.

Mark, P., Sandercock, L., 2007. Fire investigation and ignitable liquid residue analysis – a review: 2001–2007. Forensic Science International 176, 93–110.

Mayne Correia, P.M., 1997. Fire modification of bone: a review of the literature. In: Haglund, W.D., Sorg, M.H. (Eds.), Forensic Taphonomy: The Postmortem Fate of Human Remains. CRC Press, Boca Raton, FL, pp. 275–293.

Moore, J., Bleay, S.M., Deans, J., Nic Daeid, N., 2008. Recovery of fingerprints from arson scenes: Part 2 – fingerprints in blood. Journal of Forensic Identification 58, 83–108.

National Fire Protection Association, 2011. NFPA 921 – Guide for Fire and Explosion Investigations, 2011 ed. National Fire Protection Association, Quincy, MA.

Raja, S., Thomas, P., Stuart, B., Guerbois, J., O'Brien, C., 2009. The estimation of pig bone age for forensic application using thermogravimetric analysis. Journal of Thermal Analysis and Calorimetry 98, 173–176.

Song, D.F., Sommerville, D., Brown, A., Shimmon, R., Reedy, B., Tahtouh, M., et al., 2011. Thermal development of latent fingermarks on porous surfaces – further observations and refinements. Forensic Science International 204, 97–110.

Stauffer, E., 2003. Concept of pyrolysis for fire debris analysts. Science & Justice 43, 29–40.

Types of Fires

JB Crippin, Western Forensic Law Enforcement Training Center, Pueblo, CO, USA

Glossary

Backdraft An explosion resulting from rapid/sudden influx of oxygen into an oxygen deficient area containing superheated products of an incomplete combustion reaction.

Combustible liquids Combustible liquids have flash points that are greater than 100 °F.

Combustion products Heat, gases, solid particulates, and liquid aerosols produced by burning.

Conduction Heat transfer to within an object or to another object by direct contact.

Convection Heat transfer by circulation within a medium such as a liquid or gas.

Fairies Light mass combustible material that ignites while floating on air currents prior to ignition of heavier mass/larger materials. In certain instances, if large quantities of this type of material are present, it can lead to an explosion.

Fire point Refers to the temperature at which a material gives off enough vapors to sustain combustion. This is usually slightly above the flash point temperature.

Flammable liquids Flammable liquids have flash points that are less than 100 °F.

Flammable/explosive limits This refers to a concentration range where too little fuel will not support combustion and too much fuel will displace oxygen that will prevent the reaction from taking place. There are exceptions that can occur when a fuel supplies its own oxygen.

Flammable/explosive range This refers to the range in which materials can then ignite and either combust or explode. This range will vary depending on the flammable material in question.

Flash point This refers to the lowest temperature at which a liquid (or in some cases a solid) will give off vapors that can ignite in combination with oxygen.

Fuel A material that yields heat (i.e., energy) as a result of combustion.

Ignition temperature The ignition temperature of a substance (solid, liquid, or gaseous) is the minimum temperature to which the substance exposed to air must be heated in order to start combustion.

LEL/UEL (LFL/UFL) Lower explosive (flammability) limit/upper explosive (flammability) limit. These are the upper and lower limits of the flammable range.

Pyrolysis The changing of material into other compounds by heat. In many instances, this process can precede actual combustion.

Vapor density Vapor density is a term used to describe the weight of vapors. The weight is always referenced against air which has density of 1. Lower densities rise, while greater densities sink.

Theory of Fire

The chemical reaction of fire, or combustion, is actually an oxidation reaction. Oxidation is the chemical combination of oxygen with any substance (rust is an example of oxidized iron). Very rapid oxidation of a substance is called combustion or fire. Complete combustion occurs when materials burn in oxygen and they can produce by-products that are limited by the fuels involved in the process. If a hydrocarbon combusts, carbon dioxide and water are the products produced. Oxides are the most common by-products when certain types of elements combust. Examples would be carbon to carbon dioxide, nitrogen to nitrogen dioxide, iron to iron oxide, and sulfur to sulfur dioxide, to name a few.

The most visible part of a fire is the flame. It is produced as the fuel burns/combusts. The color of the flame is sometimes dependent on the temperature of the fire; however, there are other factors that can cause different colors to appear in the flames. Certain materials will impart specific colors to the flame. Because of this, flame color cannot be considered as a definitive indicator of temperature (**Tables 1 and 2**).

Another visual product of fire is smoke. Smoke is an unwanted by-product of fire. It is a mixture of solid/liquid particulates and gases that can be released when a material undergoes combustion or pyrolysis. The items mix with air and

 http://dx.doi.org/10.1016/B978-0-12-802718-9.00020-X

Table 1 Flame color during combustion versus temperature

Faint red	900 °F
Blood red	1050 °F
Dark cherry red	1175 °F
Cherry red	1365 °F
Salmon red	1650 °F
Orange	1725 °F
Light yellow	1975 °F
White	2200 °F
Blue white	2550 °F

Table 2 Flame color from chemical compounds

Crimson	Strontium
Yellow	Sodium
White	Magnesium
Green	Boron
Blue	Copper
Purple	Potassium
Red	Rubidium

become airborne. Different materials can produce different colors of smoke during combustion (**Table 3**).

As the temperatures rise above the ambient temperature of a material, pyrolysis takes place. Pyrolysis is the chemical decomposition of matter through the action of heat. It can take place in the absence of oxygen. Pyrolysis is derived from the Greek phrase "pyr" meaning fire and "lysis" meaning separating and literally means to separate by fire.

Fires, as previously stated, are a chemical reaction. This type of reaction is exothermic (produces heat) and has several distinct characteristics:

Decomposition—Fire causes a decomposition of the material "burning" into other products. This decomposition of combustible material slowly gives off gases including water vapor (these combustible gases are not ignitable during the early stages of combustion process). The gas evolution (i.e., production) continues with some of the gas becoming ignitable (as the temperature increases the gas production also increases). As the temperatures and gas production increase, the reaction speeds up and eventually becomes self-sustaining as long as suitable combustible materials are present.

Table 3 Smoke color from combustion

Vegetation	White
Wood	Gray to brown
Paper	Gray to brown
Cloth	Gray to brown
Gunpowder	Yellow to brownish yellow
Petroleum products	Black

Heat—As stated earlier, fire is an exothermic reaction. Early man made use of this to keep himself warm and cook food. Heat as it relates to fire is how fire is propagated. Once enough heat is applied to an object, it will generally start to burn. However, in order for this to occur, the heat energy must be transferred to the "target" material so that combustion can occur.

Transfer of heat—There are three ways to transfer heat energy: conduction, convection, and radiation.

Conduction is the transfer of heat through molecular activity within a material or medium. Usually it is in reference to a solid. Generally, items must be in direct contact with another for conduction to take place.

Convection is different from conduction in that it is a transfer of heat energy by a circulating medium, such as a gas or liquid. This transfer of heat energy is more widespaced than in conduction and objects need not be in direct contact. This also includes boiling liquid evaporation explosions (BLEVES).

Radiation transfers heat energy by way of energy wavelengths. These wavelengths of energy move in the same way a light does. These "rays" travel in straight lines and when strike an object, are absorbed or reflected by the surface of the object.

Physical States of Fuel

The fuel necessary for a fire can be found in three different states of matter:

Solid—Solids have a definite shape and volume. They can easily be measured. Their combustibility is in direct proportion to their surface area available for combustion. For example, a solid block of wood is much harder to burn than if it is reduced to sawdust. In this instance, the sawdust can burn (i.e., combust) at such a rapid rate that it can actually cause an explosion. It should be noted, however, that in all actuality, solids themselves do not burn. They must be converted to combustible gases first.

Liquid—Liquids will assume the shape of their container. They exhibit a definite volume. If left out and uncovered, they can evaporate (turn to a gas). Liquids also exhibit two properties referred to as flash point and fire point. The term "flash point" means lowest temperature at which a liquid (or in some cases a solid) will give off vapors that can ignite in combination with oxygen and "fire point" means temperature at which a material gives off enough vapors to sustain combustion.

Gas—Gases are those materials whose molecules are in constant rapid and random movement. These materials have no definite shape or volume. They assume the constraints of their current container. Gases have the ability to be compressed or expanded, dependent on their surroundings. Their volume is directly dependent on pressure and temperature. An increase or decrease in temperature will cause a proportional increase or decrease in volume if the pressure remains constant. An increase or decrease in pressure will cause a proportional increase or decrease in volume, if the temperature remains constant. Gases are also able to diffuse. This term refers to their ability to uniformly distribute their molecules through another substance. This ability is measured by their vapor density. Air has a VD of 1. Gases also are permeable. This is the ability of other materials to pass through or permeate a gas.

The Fire Triangle

The old theory of fire utilized what was called the "fire triangle." The triangle was originally composed of three sides: Fuel, oxygen, and heat. All three must be present in order for "fire" to occur (see diagram below). Remove any one "side" and fire goes out. Pour cold water on a camp fire and you are removing heat and sometimes oxygen. Spray carbon dioxide on a fire and you are removing the oxygen (**Figure 1**).

The Fire Tetrahedron

The new theories of fire have built on the fire triangle and have replaced it with the fire tetrahedron. It consists of four components: fuel, oxygen, heat, and chemical reaction (**Figure 2**).

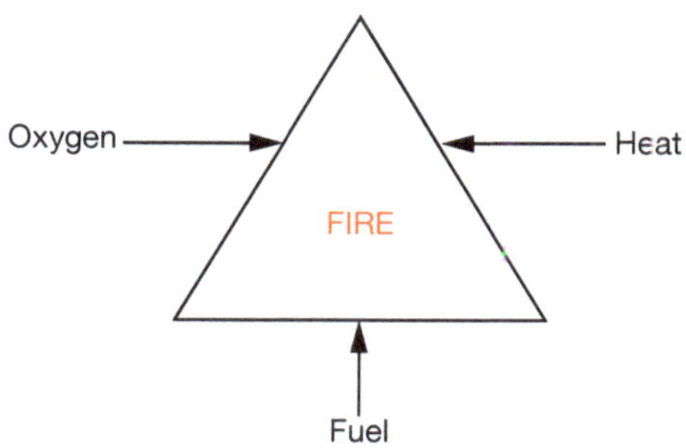

Figure 1 The fire triangle.

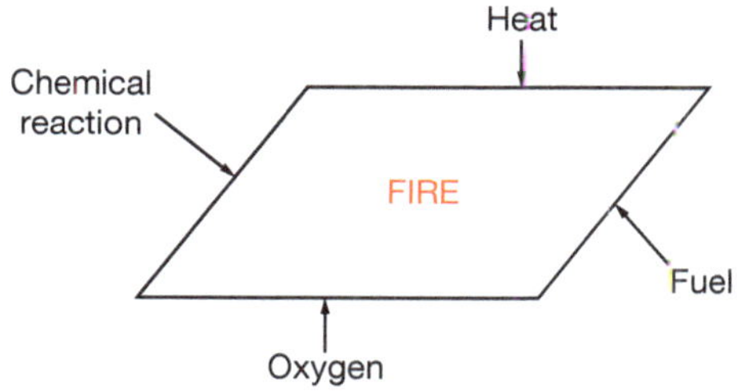

Figure 2 The fire tetrahedron.

The new theory of fire addresses the actual chemical reactions that take place producing combustion or fire. This theory was developed with the advent of chemical extinguishers that disrupt the actual chemical combustion process

Classes of Fire

Fire/combustion can be broken down into several distinct classes or types. They are as follows: Class A, Class B Class C, and Class D.

Class A fires are fires in "ordinary" materials (organic materials) that can burn to ash. The combustible materials that fall under this type of fire include wood, trash, nonmelting plastic, rubber, paper, and cloth. These fires are generally dealt with by cooling the fire below its ignition temperature. Class A fires can be extinguished with a water fire extinguisher, which contains air-pressurized water that will successfully snuff out flames as it cools down the combustible materials.

Class B fires occur in flammable substances that do not leave embers or ashes when they combust. These materials include gasoline, jet fuels, propane, paint thinner, kitchen grease kerosene, oils, paint, turpentine, grease, tar, and so on. Class B fires are extinguished by providing a barrier between the burning substance and air or oxygen. Class B fires require the use of a dry chemical or a carbon dioxide fire extinguisher to put it out. Dry chemical extinguishers contain foam or powder, while carbon dioxide extinguishers contain highly pressurized carbon dioxide. Utilizing a Class A extinguisher would merely spread a Class B fire.

Class C fires occur in live electrical materials. These types of fires involve electrical equipment. Class C combustible materials include electrical outlets, appliances, circuit breakers, and wiring. As with Class B fires, the best way to suppress a Class C fire is to remove the oxygen. Some type of nonconducting extinguishing agent is necessary to put out these types of fires If any of the other types of extinguishing agents are used, there is the possibility of an electrocution hazard being produced if a Class A extinguisher is used. Use dry chemical or powder fire extinguishers when confronting Class C fires instead of a carbon dioxide fire or Halon extinguisher on a Class C fire involving electronics, such as TVs and computers, because a powder extinguisher can leave a damaging residue and these types of extinguishers do not.

Class D fires are those that involve combustible (i.e., reactive) metals, such as titanium, magnesium, aluminum, zirconium, sodium, potassium, lithium, and so on The greatest hazard exists when these metals are in the molten state or in finely divided forms of dust, trimmings, or shavings. These types of fires are best extinguished by covering the combusting material with a special dry powdered or granular materials that exclude oxygen and do not react with the combusting metal. Class D fire extinguishers often contain sodium chloride, graphite-bearing compounds, or sodium carbonate. A powdered copper metal fire extinguisher on a Class D fire involves lithium and lithium alloys. Class D fires do not react well to water and in many cases the water may increase the intensity of the fire. It can be very dangerous to attempt to put out a Class D fire with an extinguisher that is not labeled as a Class D.

A fifth class, Class K fires, has been noted by some. These types of fires involve some type of cooking oil or fat. This class is common in the UK and referred to as Class F fires. However, in the US, these materials are listed as fuels in Class B fires.

Phases of Fire

Fire can be in any of several phases or forms. The *Incipient* phase can last from mere seconds to days. This phase is dependent on the type of fuel and ignition source. The *Emergent* phase may show slightly elevated levels of combustion products. There are no real open flames at this point. *Free* or *open burning* will show open flames and also show a rapid rise in temperature. At this point, "flashover" can occur throughout the room. Shortly before this happens, "fairies" will occur indicating flashover is ready. All fires are controlled by oxygen and fuel. Low oxygen levels can cause the fire to become semidormant until more oxygen is introduced, while shortages in fuel can actually extinguish the reaction.

See also: **Investigations:** Evidence Collection at Fire Scenes; Fire Patterns and Their Interpretation; Fire Scene Inspection Methodology.

Further Reading

De, H., John, D., Kirk, P.L., 1996. Kirks Fire Investigation. Prentice Hall, Upper Saddle River, NJ, p. 446.
Keyworth, R.J., 2010. Fires... Accidental or Arson?: Fire Investigations. Outskirts Press, Denver, p. 166.
Lentini, J.J., 2006. Scientific Protocols for Fire Investigation. CRC Press, Boca Raton, FL, p. 608.
NFPA, 2011. NFPA 921: Guide for Fire and Explosion Investigations. NFPA, p. 347.
Saferstein, R. (Ed.), 2001. Forensic Science Handbook, vol. 1. Prentice Hall, Upper Saddle River, NJ, p. 767.
Shakelford, R., 2008. Fire Behavior and Combustion Processes. Delmar Cengage Learning, New York, p. 320.
TWGFEX/NCFS, 2000. Fire and Arson Scene Evidence: A Guide for Public Safety Personnel. DOJ, p. 78.

Evidence Collection at Fire Scenes

JJ Lentini, Scientific Fire Analysis, LLC, Big Pine Key, FL, USA

Glossary

Accelerant Any substance, usually a flammable or combustible liquid, that is used to initiate or spread a fire.

Flash point The temperature at which a still pool of liquid will produce sufficient vapors to ignite when exposed to a competent ignition source.

Ignitable liquid A liquid that is capable of burning, often a petroleum-based product. Ignitable liquids may be flammable or combustible. Flammable liquids are those having a flash point below 37 °C. Combustible liquids are those having a flash point above 37 °C.

Introduction and Overview

Evidence collected from fire scenes can take many forms. The most common type of evidence collected at a fire scene is material suspected of containing ignitable liquid residues (ILRs). Other evidence that may be collected for ILR analysis includes comparison substrate samples, liquid samples for comparison, or suspect's clothing and shoes.

Suspected ignition sources, such as incendiary devices or appliances suspected of malfunctioning, are also frequently collected in an attempt to test the field investigator's hypothesis about the origin and cause of the fire.

The identification of an ILR in a sample of fire debris can support a field investigator's hypothesis regarding the incendiary nature of a fire. The identification of such a liquid residue in a fire scene, however, does not necessarily lead to the conclusion that a fire was incendiary in nature. Further investigation may reveal a legitimate reason for the presence of flammable or combustible liquids.

Due to the volatility of ignitable liquids and the variations in sampling techniques, the absence of detectable quantities of flammable or combustible liquid residues does not necessarily lead to the conclusion that flammable or combustible liquids were not used to accelerate the fire.

In order to have any utility, samples collected for ILR analysis must be properly packaged and labeled prior to submission to the forensic science laboratory.

Sample Selection and Documentation

In arson cases, it is common for the arsonist to accelerate the fire by pouring ignitable liquids throughout the structure. Typically, the liquids are poured on the floor, so flooring, if it survives, is what fire investigators usually sample. Selecting a sample that is likely to test positive for ILR (assuming there is any present) requires care, practice, and luck.

Where ignitable liquid has been used, it is often present in overabundance and it may be easily detectable by odor, once the floor has been cleared of fallen debris. It is also possible to see where the ignitable liquid was poured by observing the damage patterns on the floor.

On carpeted floors, there is often a distinct line between burned and unburned areas, although such lines are common in cases where no ignitable liquid was present. Further, because of the way air is entrained into the fire plume of a piece of burning carpet, there tends to be a "doughnut" pattern produced. This results in carpet at the center of the area saturated with ignitable liquid being less damaged than carpet at the edge of the area. **Figure 1** shows a typical doughnut pattern. Obviously, the "doughnut hole" presents the best sample

Figure 1 "Doughnut" pattern produced by gasoline burning on carpet.

 http://dx.doi.org/10.1016/B978-0-12-802718-9.00021-1

When no "doughnut hole" exists, the sample should include both burned and unburned carpet from the edge of the pattern to its middle.

On hardwood floors, ignitable liquids tend to run in the cracks, as shown in **Figure 2**. Vinyl floors also may exhibit extra burning where the ignitable liquid has penetrated into seams or lower parts of a textured floor, as shown in **Figure 3**.

Sample size is usually dictated by the size of the available sample containers. A sample container of 1–5 l will usually be adequate. It should be noted that because the laboratory analysis will usually involve sampling the headspace in a container, filling the container completely is not advisable. If the container is filled completely, it will be necessary for the laboratory to repackage the debris into a larger container for analysis.

When collecting samples, it is imperative that the investigator takes steps to avoid cross contamination or even the appearance of cross contamination. If the tools used for sampling are the same tools used for debris removal, those tools should be decontaminated prior to sampling using a detergent solution. Tools should be cleaned off in between samples to avoid carrying liquid residue from one place to another. If the investigator has any sense of where in the scene there is a higher concentration of ILR, that area should be sampled after the less concentrated areas are sampled.

Figure 2 Burning in the cracks of an oak parquet floor.

Figure 3 Burning in the cracks of a vinyl floor.

If arson is suspected, gasoline-powered tools should not be brought into the fire scene. If power is required, a generator should be used for electric tools and the generator should be kept outside. Fire extinguishment crews should be asked whether gasoline-powered fans were used on the fire, and whether it was necessary to refuel the fan at any time. Such data should be documented.

The most important attribute of any fire debris sample is its location. When the investigator makes the decision to sample a particular area, that area should be thoroughly photographed, and the location of each sample should be documented on the investigator's fire scene sketch. Once the sample has been placed in the container, another photograph should be taken showing both the area from which the sample was removed and the sample inside the container.

Disposable latex or nitrile gloves should be used to collect the samples and a new pair of gloves should be used for each sample. The use of gloves should be documented, preferably with a photograph of the discarded gloves at each sample location. Gloves should not be placed in the sample container.

Random sampling is not advisable. There should be a logical reason for the collection of samples. At times, this may mean nothing more than finding the location of doorways and hallways in a structure that has burned to completion.

Plastics and synthetic fibers, because of the way they respond to heat, tend to trap ILR in the melted polymer matrix. Wood flooring also tends to hold onto ILR by adsorbing the residues on the charcoal that is produced as wood burns.

Soils can retain some ILRs, particularly heavy petroleum distillates, almost indefinitely, but some residues in soils are subject to microbial degradation. Soil samples should be refrigerated if there is any significant delay (more than a few days) between collection and submission to the laboratory, or between submission and analysis. Concrete has been known to retain detectable quantities of ILR, but the overlying floor covering (if any) is more easily sampled.

In addition to his or her own olfactory senses, the investigator may employ a hydrocarbon sniffer, a mechanical device that senses hydrocarbons in the air. These devices are not very selective, however, and should not be used in place of laboratory analysis. False positives are common.

An "accelerant-detecting canine" is probably the most effective aid to collecting samples that have a higher probability of testing positive in a laboratory. As with any tool, canines are subject to limitations. The canine responds to scents that it has previously been rewarded for alerting to. If, on a particular scene, a canine responds to pyrolysis products, rather than foreign ignitable liquids, it is likely to respond inappropriately for the rest of the day because it has been rewarded for doing so.

Canines are incapable of identifying different ILRs. This can only be determined by laboratory analysis. There are some investigators who insist that because their canine has demonstrated extreme sensitivity, a negative laboratory test of the sample selected by the canine can be ignored. This is not true. It is the considered opinion of the scientific community (including responsible canine handlers) that canine alerts unconfirmed by laboratory analysis do not constitute valid evidence of the presence of an ILR. Some canines have confirmation rates over 90%, but others fall below 50% on a regular basis. The canine is only a tool.

One useful attribute of this tool is that the animal is willing to spend all day sniffing out a large area, something that most fire investigators are unwilling to attempt. If the scene has been properly cleared prior to the canine's arrival and there is any ILR to be found, the canine will likely find it.

Comparison Samples

Beginning in the 1980s, forensic science laboratories acquired the ability to detect smaller and smaller quantities of ILRs. The lower limit of detection dropped by at least two orders of magnitude. The result of this improved technology has been that the laboratory can now detect "background levels" of ILR that are normal to the area being sampled. **Table 1** provides a list of common background materials that may be the source of ILRs. For example, the solvents used in hardwood floor-finishing products persist indefinitely in the thin polyurethane or other polymer film. For this reason, comparison samples are often necessary.

Because the fire investigator usually gets only one chance to sample the scene, comparison samples should be collected as a matter of routine. The samples should consist of the same substrates, such as carpet, flooring, and concrete, as the suspect samples. Obviously, the comparison samples should be collected from areas where, in the fire investigator's judgment, no ignitable liquid would be expected. Except in cases where there is a large excess of ILR present on the sample, the absence of a comparison sample may render a positive finding from the laboratory invalid.

Packaging Options

Fire debris samples are typically packaged in metal cans, glass jars, or plastic evidence bags specifically designed for fire debris. Each type of packaging has advantages and disadvantages.

Metal cans are usually impervious to hydrocarbon residues, are commonly available, and resist breaking. Laboratory analysts sometimes prefer cans because the cans can be heated directly in the oven. One problem with cans is that they tend to corrode over time. Fire debris tends to contain water and acids that can cause perforation of an uncoated metal can in a matter of weeks. For this reason, lined cans, the type used for water-based paints, are recommended over unlined cans. Each time an investigator purchases a new batch of cans, an empty can should be submitted to the laboratory to verify that the can lining is not contributing significantly to the extract.

Glass jars, such as Mason jars, do not corrode, but they require careful packaging to avoid breakage and the cross contamination that may result. The two-piece lids on Mason jars were not designed for hydrocarbons, and the gasket can deteriorate if exposed to high concentrations of ILR. It is recommended that the gasket be turned away from the sample.

Two-layer nylon/polyacrylonitrile-co-methacrylate-cobutadiene bags, specifically designed for fire debris analysis, are available and have been shown to make an excellent container for fire debris, assuming they are properly sealed. Research has shown these bags to be superior to nylon or polyester bags that have been used for evidence collection. The evidence bags can withstand extraction oven temperatures of up to 70 °C. The bags have the disadvantage of being subject to punctures, and fire debris often contains materials capable of causing a puncture.

However it is packaged, the fire debris should be labeled in accordance with industry practices. The label should include a unique identifier that allows a chain of custody to be established. A label on a fire debris sample should also include an identification of the person who collected it, the date it was

Table 1 Sources of "background" ignitable liquids

Substance	*Potential source*
Toluene, xylenes, C_3- and C_4-alkylbenzenes, indenes	Adhesives, polymer decomposition, shoes, clothing
Medium petroleum distillates and heavy petroleum distillates	Adhesives, insecticides, polishes lubricants, magazines, newsprint, shoes, clothing, asphalt smoke, plasticizers, floor finishes
Isoparaffins	Moisture barriers, adhesives, plasticizers
Normal paraffins	Vinyl flooring, carbonless forms, copier and laser printer toner
Alcohols, acetone	Combustion products

collected, a brief description of the substrate, and a brief description of the location of the sample. More elaborate descriptions of the sample can be provided on an evidence transmittal, such as the one shown in **Figure 4**.

Clothing and Shoes

Occasionally, law enforcement officers encounter a suspect shortly after a fire has been set and collect the suspect's clothing

APPLIED TECHNICAL SERVICES, INCORPORATED

1190 Atlanta Industrial Drive Marietta, GA 30066 (770) 423-1400 Fax (770) 424-6415 e-mail fire@atslab.com

EVIDENCE TRANSMITTAL

Date:

Verbal Report To:

Phone #:

Submitted By:

Written Report To:

File #:

Invoice To:

Insured:

Claim Number:

Date of Loss:

ATS Reference #

Description of Evidence: Container, Size Type of Material, Condition of Material (burned or unburned)

Location Collected

1.
2.
3.
4.
5.
6.

Special Instructions:

Chain of Evidence
(Signature Required)

From: To: Date: Time:
From: To: Date: Time:
From: To: Date: Time:
From: To: Date: Time:

ATS 800 (10/99)

Professional Engineers
Design • Consulting • Testing and Inspection
Members in AAFS, ACS, ASM, ASME, ASNT, ASQC, ASTM, AWS, FSCT, IAAI, NACE, NCSL, NFPA, SAFS
GEORGIA SOCIETY OF PROFESSIONAL ENGINEERS, NATIONAL SOCIETY OF PROFESSIONAL ENGINEERS

Figure 4 A typical chain of custody form for submission of fire debris samples to a private laboratory.

and shoes for ILR analysis. This can be a very useful means of linking a suspect to a fire scene, assuming it is done properly.

Clothing and shoes should be packaged separately to avoid contamination of the clothing with the background compounds on the shoes.

The left and right shoes should likewise be packaged separately. Shoes often contain significant quantities of ILRs because of the dyes and adhesives used in their manufacture. If an accelerant is splashed on an arsonist's shoes, it is unlikely that each shoe will be contaminated with the same amount. Thus, one shoe could act as a comparison sample for the other. It may be necessary to try to find a new pair of shoes of the same manufacture as the suspect's shoes, in order to determine what should and should not be present on the shoes.

Liquids for Comparison

Fire investigators will sometimes find containers of ignitable liquids at fire scenes that they wish to submit for comparison with residues extracted from test samples. This can be accomplished, but it is necessary to take affirmative steps to avoid contaminating the test samples with the comparison liquids. The liquid should be packaged and shipped separately from the test samples. Suitable containers for liquids include flint glass jars or small evidence cans containing surgical gauze pads onto which a liquid is absorbed. Note that "nonstick" gauze pads may contain petroleum products and are not suitable for absorbing liquids.

There are times when liquid absorbed onto a gauze pad is not suitable. Sometimes it is necessary to conduct a flash point test to determine whether a liquid meets statutory requirements. For example, some jurisdictions define a Molotov cocktail as a breakable container filled with a flammable liquid. In such a case, a sample of at least 5 ml of liquid is required and preferably 100 ml.

The laboratory procedures necessary to "match" an ignitable liquid to a particular source are beyond the scope of this chapter, but readers should be aware that, except in the most unusual cases, the laboratory will only be capable of excluding a particular source, rather than positively identifying the comparison liquid as the source of the residue found on the test samples.

Evidence Collection for Other Types of Testing

Spontaneous Heating

The forensic science laboratory is capable of performing other chemical and physical analyses in addition to testing for ILR.

Oily rags, suspected of having undergone spontaneous ignition, are collected and packaged in the same way as fire debris submitted for ILR analysis. The investigator needs to tell the laboratory that spontaneous heating is suspected because the analytical procedure for finding the kinds of products that undergo spontaneous heating (usually vegetable oils) is different from the procedure used for detecting ILR. It is possible to conduct the testing for vegetable oils on the same sample used to test for ILR. ILR can be isolated using a nondestructive headspace concentration technique, and then the vegetable oils can be extracted with a solvent.

Occupancies that are subject to fires caused by spontaneous heating include commercial laundries, restaurants, and health spas, as well as buildings under construction or renovation where oil-based paints and stains are used. Spontaneous heating fires tend to occur in clothes dryers when the laundry has not been thoroughly cleansed of vegetable oils. These fires happen after the dryer has stopped, and sometimes after the laundry has been removed and stacked. Spontaneous fires associated with painting typically occur as a result of improper disposal of rags or filters contaminated with products that contain linseed or tung oil.

Laboratory analysis for vegetable oils can only determine whether the oils are present. The laboratory analysis will not allow a determination that the oils actually underwent spontaneous heating or combustion.

Electrical Malfunctions

Most public-sector forensic science laboratories do not have in-house capabilities for examining appliances or other electrical devices suspected of causing a fire. An engineering laboratory will usually be required for such analyses.

If a product malfunction is suspected, the fire investigator is advised to ensure that any entity that might be a party to litigation has the opportunity to participate in the inspection. When possible, this should include making any suspect device available both on the fire scene and in the engineering laboratory.

The appliance or device should be fully documented on the scene and then labeled to provide a chain of custody. If there are exemplars of appliances at the fire scene, they should be collected as well. Additionally, appliances in the general area of origin, which the investigator does not believe started the fire, should be collected for purposes of elimination. Appliances and devices damaged in fires are typically delicate. Conductors and other parts may have melted or become brittle. Care should be taken to avoid further damage in the packaging and transportation of damaged appliances to the laboratory.

Laboratory inspection will usually involve the use of a magnifying glass or a stereomicroscope. X-ray examination is desirable prior to performing any destructive disassembly of the appliance or device. Real-time X-ray is usually preferable to examination with industrial X-rays and film because it can be accomplished in a more reasonable time frame.

An experienced fire investigator or electrical engineer can usually recognize evidence of electrical activity. The hallmarks of electrical activity are that the damage to the conductor is highly localized and complementary, that is, one can identify both the high-voltage and low-voltage (or zero-voltage) side of the arcing event. On small-diameter stranded wires, such as low-amperage appliance power cords, it is possible to become confused between environmental melting and electrical melting. Scanning electron microscopy is often useful, especially when different metals were involved in the arcing event.

Except in cases where a loose connection results in "series arcing," electrical arcs are almost always a result of an energized conductor being exposed to the fire. The insulation on the conductor carbonizes and becomes conductive, resulting in "arcing through char." Such arcing may help to pinpoint the origin of the fire, but it is always a result of the fire.

There is currently no valid laboratory analysis that is capable of identifying an arc that caused a fire. Scanning electron microscopy and auger electron microscopy have both been tried, but neither is accepted as valid for making the identification of an electrical arc that started a fire.

Heat-producing devices are typically equipped with safety features that, surprisingly, may survive a fire in testable form. This is because the devices are designed to withstand a hot environment. One-time fuses can be tested to see if they are electrically open. Resettable bimetallic thermal cutoff devices can be examined to see if they are open or closed, and they can be heated with a hair dryer to determine whether they still respond to heat.

Some metals respond to heating by undergoing microscopic changes in their grain structure. These changes are capable of being identified by a metallurgist. The testing required to visualize the grain structure involves cutting and polishing a section of the metal, so appropriate steps regarding destructive testing need to be taken.

Other Forensic Analyses

Because of the destructive nature of fires, investigators often overlook traditional forensic evidence such as fingerprints, blood, toolmarks, and other kinds of trace evidence. Much of this evidence can actually survive the fire and, particularly in incendiary fires, is capable of associating a suspect with the scene.

In fatal fires, the body constitutes significant evidence. A toxicological evaluation of the victim's blood can determine whether the victim was alive during the fire. A victim with no carbon monoxide in his or her blood was probably dead before the fire began. Recent research has indicated that there is a relationship between carbon monoxide levels and the victim's location relative to the origin of the fire. Victims who die from thermal injuries because they were close to the origin typically have lower carbon monoxide levels than victims found in rooms some distance from the origin, particularly when the room of origin progressed beyond flashover.

See also: **Chemistry/Trace/Fire Investigation:** Analysis of Fire Debris; Chemistry of Fire; Interpretation of Fire Debris Analysis; Physics/Thermodynamics; Thermal Degradation; **Investigations:** Fire Patterns and Their Interpretation; Fire Scene Inspection Methodology; Types of Fires.

Further Reading

ASTM E1188, 2005. Standard Practice for Collection and Preservation of Information and Physical Items by a Technical Investigator. ASTM International, West Conshohocken, PA.

ASTM E1459, 2005. Standard Practice for Physical Evidence Labeling and Related Documentation. ASTM International, West Conshohocken, PA.

ASTM E1618, 2010. Standard Test Method for Ignitable Liquid Residues in Extracts from Fire Debris Samples by Gas Chromatography-Mass Spectrometry. ASTM International, West Conshohocken, PA.

Lentini, J.J., 2006. Scientific Protocols for Fire Investigation. Taylor & Francis, Boca Raton, FL.

NFPA, 2011. NFPA 921: Guide for Fire and Explosion Investigations. National Fire Protection Association, Quincy, MA.

Olenick, S.M., Klassen, M.S., Roby, R.J., Ma, T., Torero, J.L., 2010. The behavior of liquid fuel on carpet (porous media): a case for the inclusion of science in fire investigation. Fire Technology 46 (4), 843–852.

Putorti, A.D., 2001. Flammable and Combustible Liquid Spill/Burn Patterns. NIJ Report No. 604-00. National Institute of Justice, Washington, DC.

Stauffer, E., Dolan, J., Newman, R., 2008. Fire Debris Analysis. Academic Press, Burlington, MA.

Relevant Website

CFITrainer.net—IS the eponymous name of the website.

Fire Scene Inspection Methodology

JJ Lentini, Scientific Fire Analysis, LLC, Big Pine Key, FL, USA

Glossary

Flashover A transition phase in the development of a fire in a structure, in which all combustible surfaces exposed to thermal radiation ignite more or less simultaneously and fire spreads rapidly throughout the space, resulting in full room involvement. The transition is from "a fire in a room" to "a room on fire."

Fuel-controlled fire A fire in which the heat release rate and growth rate are controlled by the characteristics of the fuel, such as quantity and geometry, and in which adequate air for combustion is available.

Full room involvement A compartment fire in which the entire volume is involved in fire, but flaming will only occur when there is sufficient oxygen available.

Spoliation Loss, destruction, or material alteration of an object or document that is evidence or potential evidence in a legal proceeding by one who has the responsibility for its preservation.

Ventilation-controlled fire A fire in a postflashover compartment, where all fuels are involved, and the heat release rate of the fire is controlled by the availability of oxygen.

Introduction and Overview

Because each fire is different, specifying an exact methodology for a fire scene inspection is difficult, but there exist sufficient similarities among fires that make it necessary for the investigator to develop a standard approach. Not every step described in this chapter will be necessary for every fire, and many of the routine steps described can only be carried out by the "first responder."

Procedures will vary according to the problem a fire investigator is asked to address. Usually, the problem is to determine the origin and cause of the fire, but often there are additional questions that need to be addressed, such as the reason for a victim's failure to escape, the reason for a fire detection or suppression system's failure to function, or the reason for an unusually rapid spread of the fire.

In many cases, the origin and cause of the fire are obvious, and it is only the secondary questions that require any significant skills to address.

Although many fire investigators began their careers as firefighters, there are significant differences in the skill sets required to extinguish a fire and those required to investigate a fire. Extinguishment officers are 100% effective at putting out fires. Fire investigators, on the other hand, must sometimes accept the reality that the determination of the origin and cause is an impossible task, due to the complexity of fire behavior.

First Assumptions

More than 80% of fires are accidents, and because the consequences of declaring an accidental fire to have been intentionally set are so serious, there is a lively debate in the fire investigation community about what should be presumed. In some states, it is the law that all fires are presumed accidental until proven otherwise, but the fire investigation community has taken the position that fire investigations should be approached without any presumptions.

Planning the Investigation

It is necessary to gather sufficient information to formulate a workable fire scene inspection plan. When the assignment for an investigation is first received, the investigator needs to know what resources will be necessary to process the scene. Additionally, it is necessary to identify "interested parties," and to learn the use (occupancy) of a structure. Certainly, public sector investigators have different responsibilities from private sector investigators. Often, the public sector investigator's work is done once the determination has been made as to whether the fire was the result of a criminal or an accidental cause.

It is often necessary to arrange for additional personnel such as fire protection engineers or structural and electrical engineers. In commercial or industrial occupancies, heavy equipment may be required to process the scene.

 http://dx.doi.org/10.1016/B978-0-12-802718-9.00022-3

It is a good idea to learn as early as possible whether there were any eyewitnesses to the fire and whether any of them photographed or videotaped the fire in progress. If a security system exists in the building, records of alarms or even videotapes of the fire in progress may be obtainable from the alarm-monitoring company.

It is also important to ensure that the investigator has the authority and the right to enter the property to conduct the inspection. Such authority usually comes in the form of permission from the owner to enter the building, but a search warrant may be necessary in some cases.

Initial Evaluation: Can This Inspection Be Conducted Safely?

Fire scenes are dangerous places and must be approached with caution. Because the mission of the investigator is not urgent, there is no need to compromise safety. The emergency is over. The fire has been extinguished and there is nobody to rescue. If possible, fire scene inspections should not be conducted at night. The scene should be secured until there is adequate light to conduct the fire inspection safely and to make the necessary observations.

Obvious fire scene hazards include structural instability, energized electrical circuits, and leaking fuel gases, but there exist less obvious hazards that the investigator must consider. These include fire gases such as carbon monoxide and hydrogen cyanide and hazards such as asbestos and biohazards.

The initial survey usually begins with the exterior of the building and then moves to the interior. It is best to move from the area of least damage toward the area of greatest damage to avoid developing a case of "tunnel vision" and to make certain that all of the building has been examined by someone. It is during this initial walkthrough that the investigator can plan for the documentation and reconstruction of the fire scene. At this point, however, the investigator should avoid forming hypotheses about the fire if at all possible.

Documentation

Documentation is the most important task in a fire investigation. At some point, a jury may be asked to review that documentation and agree with the fire investigator's interpretation, but documentation is important for its own sake. Regardless of when the investigator arrives at the scene, be it minutes, hours, or months after the fire, the investigator should document the entire scene prior to making any changes.

Photographs represent the most important form of documentation, and now that images are practically free, there is no reason to limit the number of images collected. Digital cameras collect data about the order in which photographs are taken and the amount of time passed between the first and last photograph. A logical, well-organized investigation is reflected in the photographic record.

Close-up photographs are necessary in many cases, but need to be placed into context. This is accomplished by taking an overall view of the item of interest, prior to zooming in on it.

Other forms of documentation are also necessary for a proper investigation. Written notes allow the documentation of those data not suited to photography, and video images frequently help to orient the photographs.

A critical piece of documentation that is required on all but the simplest fire scenes is the sketch. In most instances, the minimum sketch should be a floor plan. Ready-made sketches may be found in the form of escape plans posted in commercial or business occupancies. A sketch is absolutely essential in order for readers of an investigator's report to understand what they are looking at. Graph paper, ruled in bold lines in 2.5 cm^2 and fine lines in 2.5 mm^2, makes the task of sketching much easier.

The level of detail required in the sketch will be dictated by the needs of the fire investigation. In many cases, a plan view showing the rooms and doorways is all that is necessary. If ventilation is an issue, and it usually is, windows are not difficult to add. If computer modeling is contemplated, however, it is not only necessary to draw the doors and windows, but to record the height of each opening; the height of all ceilings, sills, and overhangs; and the composition of the interior finish throughout. Additionally, the location of each fuel package in the room must be recorded.

Sketches should be oriented to the earth using a north arrow. An approximate scale is also essential to any sketch.

During the documentation of the scene prior to any disturbance, the investigator can make the observations necessary to help him or her decide how to reconstruct the scene. It is in this preliminary documentation that the utilities should be observed. Interior and exterior finishes can also be documented at this point. The use of a checklist of common building features is recommended so that both the important and unimportant data can be recorded. It may be that the natural gas water heater had nothing to do with the fire, but if the investigator is unable to correctly answer a question about the water heater, his credibility is likely to be impaired.

Reconstruction

Reconstruction is carried out in order to recreate, as nearly as possible, the position of various structures and fuel packages prior to the fire. Fire patterns on items that have been moved can be much more easily understood if these items are put back into place.

Although it is sometimes necessary to use heavy equipment to remove fallen structural materials, in most cases, the best tool is the flat-headed shovel. It may take 10 such shovels and several wheelbarrows to remove the debris and expose the original interior surfaces. Such work is dirty and time-consuming, but it is during this reconstruction process that valuable evidence is uncovered.

The most valuable evidence is that which shows sequential data. A fire scene records everything that happened during the course of the fire from beginning to end, but most patterns do not come with time stamps. Often, the best that can be hoped for is to determine which object fell down first or what position the object was in when it was sprayed with water. Sections of ceiling material that have fallen down can be examined to determine whether there was more damage on the top or the bottom. Protection patterns caused by falling debris can also provide clues as to how the fire progressed.

Once debris has been moved, items should be placed back in their original locations, if possible. This can be particularly difficult with square or round-bottomed objects, as it is possible to reposition them exactly backward from their original location.

Once reconstruction has been completed, any portion of the scene that has been changed during the process must be documented again.

Once reconstruction is completed and documented, the fire investigator likely has all of the physical evidence there is to see. The work that has been done should allow for the development of a credible hypothesis as to where the fire started and a second credible hypothesis as to what caused it. But there are other steps that intervene.

Inventory

It is often necessary to make an inventory of the items that were present in a structure at the time of the fire. Such an inventory may serve to document the fuel load. Also, the investigator can determine whether the inventory is consistent with the reported occupancy.

Inventory may also reveal the prior removal of contents or even the substitution of contents. The unexplained absence of valuable items or the substitution of valuable items with replacements of lesser value raises the possibility that someone had prior knowledge of the fire.

The interpretation of the inventory requires considerable caution. Fires destroy combustible items, and many of them may be completely consumed or unrecognizable after a fire. This does not necessarily mean those items were not present. It is far easier to prove the presence of an item than to prove its absence.

Avoiding Spoliation

Fire scenes are evidence, and evidence needs to be protected. This is particularly true with respect to civil cases where one party may seek to hold another party liable for monetary damages caused by a defective product or service. When an investigator reaches the point where it appears that he or she can identify a responsible party, fairness and the law require that he stop and allow the potentially responsible party an opportunity to review the evidence in place. This may require delaying further investigation for several days or even weeks, particularly if multiple parties are suspected of playing a role in causing or spreading the fire.

The fire investigation community has carefully considered the concept of spoliation and has concluded that certain activities, such as debris removal, or opening a suspected appliance in order to identify the manufacturer should not be considered spoliation. Additionally, it may be necessary to move potentially causative agents from the scene in order to protect them from further damage.

Spoliation is an issue that is generally not applicable in criminal fire investigations, although there have been some cases where public agencies have been held responsible for allowing evidence to be destroyed and depriving a potential defendant of his right of confrontation under the sixth amendment.

Origin Determination

The determination of the point of origin is the fire investigator's most important and most difficult task. It is axiomatic that if the origin of the fire is not determined correctly, the cause will also be incorrectly determined. Generally, there are three categories of evidence that must be considered by the investigator in helping to locate where the fire started. These include the following:

1. Witness observations
2. The analysis of fire patterns including patterns created on the electrical system
3. The consideration of fire dynamics

Witness observations may be crucial for determining where the fire started. Particularly in the case of full room involvement, a witness statement may be the only evidence that allows the origin to be pinpointed. Witnesses may also help the investigator understand the arrangement of furniture and other items in the structure prior to the fire.

Investigators should be cautious when their interpretation based on fire patterns varies significantly from what witnesses have observed. Some experiments on the validity of origin determination using fire patterns have produced disturbingly high error rates. Other data that can be characterized as witness statements include video of the fire in progress and alarm system functions.

A word about the proper use of witness statements is in order. When determining the origin and cause of the fire, the investigator should consider only those statements that have a bearing on the fire science. The power of science in proving the truth rests on the independence of the evidence. This independence can be compromised by statements having nothing to do with the physical evidence, such as inculpatory statements made by a suspect to a third party either before or after the fire. The consideration of such statements results in contextual bias and can lead to errors in interpretation of the artifacts at the fire scene.

The mechanisms by which fire patterns are created and how they might be interpreted to lead an investigator to the origin are discussed elsewhere in this book. Confinement patterns, movement patterns, and intensity patterns can all be useful indicators of how a fire moved. It is useful to place arrows on the sketch indicating the movement of a fire.

It is very important to consider which patterns may be the result of postflashover ventilation-limited burning, which may well have absolutely no meaning with respect to the origin.

The following is a checklist of activities to be conducted when determining the origin of the fire. This checklist follows the basic format of the scientific method.

1. Data collection
   Basic site data
   Determination of prefire conditions
   Fire department information
   Alarm, detector, and security data
   Documentation of postfire conditions
   Witness statements and observations
   Excavation, examination, and reconstruction of the scene
   Documentation of postreconstruction conditions
2. Data analysis
   Damage assessment (Which rooms became fully involved?)
   Fire pattern analysis
   Heat and flame vector analysis
   Depth of char and calcination surveys
   Electrical arc mapping
   Event sequencing
   Fire dynamics considerations
   Building construction and occupancy considerations
3. Hypothesis development
   Initial origin hypothesis
   Working origin hypotheses
   Alternate hypotheses

Once an origin hypothesis has been developed, it must be tested against the evidence. There are four questions that an investigator should ask once he or she has formed a hypothesis as to the origin of the fire. These include the following:

1. Is there a competent ignition source at the origin?
2. Does the origin explain the data?
3. Are contradictions resolved?
4. Does an alternate origin explain the data equally well?

If there is no competent ignition source found at the origin, it is likely that the proposed origin is incorrect. The same is true for the absence of a competent fuel source at the hypothesized origin. There is a temptation on the part of some investigators to conclude that the ignition source must have been an open flame that was carried away by the person who set the fire, and the fuel source was an ignitable liquid poured on the floor, which was completely consumed by the fire. It is this type of determination that results in serious errors. In cases of full room involvement, it is often not possible to define the origin as an area smaller than the entire room.

The next question asks whether an origin in the hypothesized location explains how the fire spread from there and damaged the rest of the structure.

It is an unusual fire scene that does not contain some contradictory information. The investigator needs to attempt to resolve conflicting information; for example, if the electrical system shows signs of failure far away from where the origin is hypothesized and there is no evidence of electrical activity in the hypothetical origin. Regardless of the fire's cause, energized electrical conductors in the area of origin are likely to exhibit evidence of melted conductors.

Finally, the investigator needs to consider alternate origins that might explain the data equally well.

Cause Determination

A fire cause is defined as the circumstances, conditions, or agencies that bring together a fuel, ignition source, and oxidizer (usually air) resulting in combustion. The determination of a fire cause requires the identification of those factors that were necessary for the fire to have occurred. Those factors include the presence of a competent ignition source, the type and form of the first fuel ignited, and the circumstances that allowed the factors to come together and start the fire.

As with origin determination, cause determination must follow the scientific method. The investigator should carry out the following activities.

1. Data collection
   Document the fuels in the area of origin
   Document all potential ignition sources
   Identify the oxidizing agent (usually this is atmospheric oxygen)
   Identify the circumstances that brought the fuel, ignition source, and oxidizer together
2. Data analysis
   Analyze the fuel (How much is present? How easily is it ignited?)
   Analyze ignition source (Does it have sufficient energy to cause ignition? Did it exist for a sufficient period of time to cause ignition?)
   Analyze the oxidizer, especially if other than air
   Analyze potential ignition sequences

3. Hypothesis development
   Propose a separate hypothesis for each potential ignition source
   Consider absent ignition sources
   Propose a first fuel for each ignition source
   Consider alternate hypotheses

Once a cause hypothesis has been developed, it must be tested against the evidence. These are the questions that an investigator should ask once he or she has formed a hypothesis as to the cause of the fire.

1. Is (or was) the hypothesized ignition source located at the origin?
2. Can the hypothesized ignition source ignite the first fuel?
3. Did the hypothesized ignition source have sufficient time?
4. Is the hypothesized cause consistent with all known facts?
5. Are contradictions resolved?
6. Does another cause hypothesis explain the data equally well?

The determination of the fire's cause should be based on evidence, rather than an absence of evidence. There has been a tendency on the part of some investigators to declare that because they could find no accidental cause for the fire, it must have been intentionally set, even when there is no evidence of incendiary activity. Such determinations are referred to as "negative corpus" determinations. Negative corpus determinations are a violation of the scientific method. The proper classification of the cause of a fire for which there is no evidence to support a classification of either accidental or incendiary cause is undetermined.

See also: **Chemistry/Trace/Fire Investigation:** Analysis of Fire Debris; Chemistry of Fire; Interpretation of Fire Debris Analysis; Physics/Thermodynamics; Thermal Degradation; **Investigations:** Evidence Collection at Fire Scenes; Fire Patterns and Their Interpretation; Types of Fires.

Further Reading

ASTM E1188, 2005. Standard Practice for Collection and Preservation of Information and Physical Items by a Technical Investigator. ASTM International, West Conshohocken, PA.

ASTM E2332, 2004. Standard Practice for Investigation and Analysis of Physical Component Failures. ASTM International, West Conshohocken, PA.

ASTM E678, 2007. Standard Practice for Evaluation of Scientific or Technical Data. ASTM International, West Conshohocken, PA.

ASTM E860, 2007. Standard Practice for Examining and Preparing Items that Are or May Become Involved in Criminal or Civil Litigation. ASTM International, West Conshohocken, PA.

Carman, S.W., 2008. Improving the Understanding of Post-flashover Fire Behavior. Proceedings of the International Symposium of Fire Investigation NAFI, Sarasota, FL.

Lentini, J.J., 2006. Scientific Protocols for Fire Investigation. Taylor & Francis, Boca Raton, FL.

NFPA, 2011. NFPA 921: Guide for Fire and Explosion Investigations. National Fire Protection Association, Quincy, MA.

US Department of Justice, 2000. Fire and Arson Scene Evidence: A Guide for Public Safety Personnel. US Department of Justice NCJ 181584, Washington, DC.

Relevant Websites

http://www.Fire.NIST.gov—Building and Fire Publications, National Institute of Standards and Technology.

http://www.CFITrainer.net—Certified Fire Investigator Training, sponsored by the IAAI.

Fire Patterns and Their Interpretation

JJ Lentini, Scientific Fire Analysis, LLC, Big Pine Key, FL, USA

Glossary

Flashover A transition phase in the development of a fire in a structure, in which all combustible surfaces exposed to thermal radiation ignite more or less simultaneously and fire spreads rapidly throughout the space, resulting in full room involvement. The transition is from "a fire in a room" to "a room on fire."

Fuel-controlled fire A fire in which the heat release rate and growth rate are controlled by the characteristics of the fuel, such as quantity and geometry, and in which adequate air for combustion is available.

Full room involvement A compartment fire in which the entire volume is involved in the fire, but flaming will only occur when there is sufficient oxygen available.

Ventilation-controlled fire A fire in a postflashover compartment, where all fuels are involved, and the heat release rate of the fire is controlled by the availability of oxygen.

Introduction and Overview

A fire pattern, also called a burn pattern, is the visible or measurable physical changes or identifiable shapes formed by a fire effect or group of fire effects. Fire effects are the changes in or on a material as a result of exposure to the fire.

The term "fire effects" describes the artifacts left behind by many different processes, including dehydration, melting, color changes, oxidation, charring, loss of material, alloying, spalling, expansion and deformation, annealing, soot and smoke deposition, and clean burn. Even damage to an electrical system can constitute a fire pattern. Fire investigators are thus presented with a vast array of different patterns, all of which require some level of understanding of how they are created and what they mean.

Accurate interpretation of fire patterns can lead the fire investigator to the correct origin of the fire, and hopefully, to the correct identification of the fire's cause. However, accurate interpretation is often an elusive goal.

For many years, fire investigators followed the simplistic maxim that heat rises, and fire burns up and out. While it is true that heat rises, it does so only until it reaches the ceiling. At this point, the behavior of the fire changes.

When the rising smoke and heat reach the ceiling, the upward movement is stopped and a ceiling jet is created, sending the products of combustion in a horizontal direction, which then become confined by the walls of the compartment.

A hot gas layer develops at the ceiling, and gradually, this layer becomes thicker and more energetic. It radiates heat in all directions, including downward. When the temperature of the hot gas layer reaches 500–600 °C, all exposed combustible surfaces in the room ignite more or less simultaneously. This transition from "a fire in a room" to "a room on fire" is known as flashover. It is the fire investigator's goal to determine the first fuel package to ignite. That ignition necessarily occurred some minutes before flashover. Postflashover burning, however, is usually far more energetic than preflashover burning, and it has the capability to obscure or even obliterate patterns that existed prior to flashover.

By visualizing the interaction of a conical plume with a vertical surface, one can understand the generation of some fire patterns. Such patterns are common fire artifacts, and often help fire investigators determine the origin of small fires. As a fire progresses to full room involvement, however, the "rules" of fire pattern interpretation change.

Unfortunately, many fire investigators fail to understand underventilated fires and may misinterpret the most heavily damaged area of the room as necessarily being the origin of the fire. This is explained further when ventilation-generated patterns are discussed.

Plume-Generated Patterns

Fire patterns can result from several different kinds of interactions between the fire and its surroundings. The best-understood interactions are those that occur between the fire plume and a nearby vertical surface. Such patterns are often called "truncated cone patterns." (Fire investigators refer to "cones" and "inverted cones," but their definition is the exact opposite of the classic geometric definition of a cone. In fire investigation, the apex of the cone is at the bottom.)

 http://dx.doi.org/10.1016/B978-0-12-802718-9.00023-5

These patterns are typically recorded on vertical surfaces, and fire investigators look for what they call "V" patterns.

Fire patterns evolve over the life of a fire. When the fire is new, the pattern on an adjacent vertical surface will assume the shape of the flame, and will exhibit what fire investigators call an inverted cone pattern. This triangular pattern is characteristic of young fires. **Figure 1** is a schematic representation of how such a pattern is created, and **Figure 2** shows a triangle-shaped pattern produced by a test fire. As the fire continues to grow, the pattern becomes columnar, and the sides of the pattern are roughly perpendicular to the floor. Columnar pattern production is shown schematically in **Figure 3** and an actual columnar pattern is shown in **Figure 4**. Columnar patterns have a very short life span, and change as soon as the fire begins to interact with the ceiling.

Once the fire plume encounters the ceiling, or an intervening horizontal surface, a "V" pattern is the result. A semicircular pattern is created on the ceiling. **Figure 5** shows how the pattern is produced when the burning fuel package is close to the wall, and **Figure 6** shows an actual V pattern from a fire scene. As can be seen, the schematics are quite a bit cleaner than the actual patterns, and sometimes fire investigators will disagree about whether a pattern is actually a V (or half of a V).

When the fuel package burns farther away from the wall, the interaction of the fire plume with the wall occurs at a higher level, and the bottom of the intersection of the cone and the wall is not as sharp. This results in a U-shaped pattern on the wall, and a circular pattern on the ceiling. **Figure 7** shows

Figure 2 Triangle-shaped pattern, also called an inverted cone pattern, produced in a test fire. Copyright 2012 from Scientific Protocols for Fire Investigation, second ed. by John J. Lentini. Reproduced by permission of Taylor and Francis Group, LLC, a division of Informa plc.

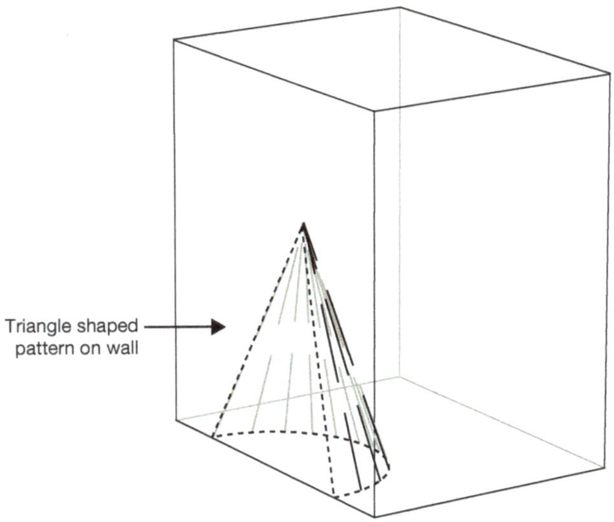

Figure 1 Intersection of an incipient fire with an adjacent wall. Copyright 2012 from Scientific Protocols for Fire Investigation, second ed. by John J. Lentini. Reproduced by permission of Taylor and Francis Group, LLC, a division of Informa plc.

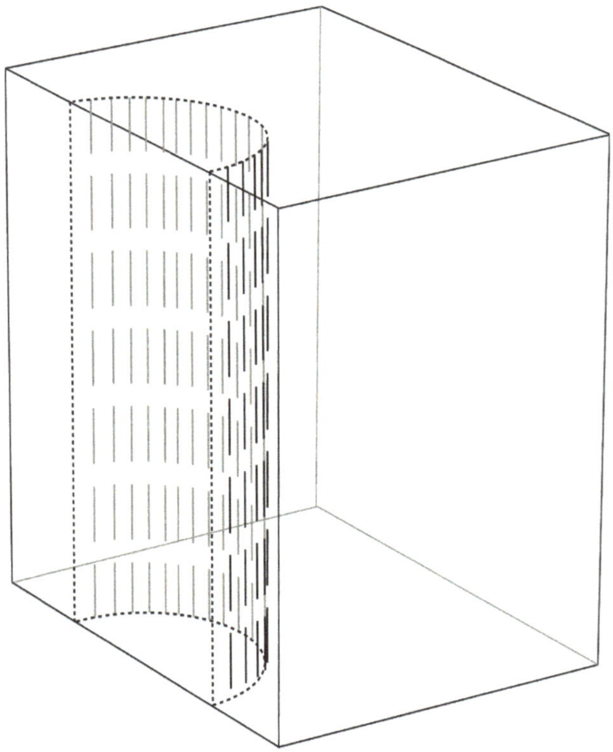

Figure 3 Intersection of a columnar fire with an adjacent wall. Copyright 2012 from Scientific Protocols for Fire Investigation, second ed. by John J. Lentini. Reproduced by permission of Taylor and Francis Group, LLC, a division of Informa plc.

Figure 4 A columnar fire pattern produced by a fire that was extinguished prior to the formation of a V-shaped pattern. Copyright 2012 from Scientific Protocols for Fire Investigation, second ed. by John J. Lentini. Reproduced by permission of Taylor and Francis Group, LLC, a division of Informa plc.

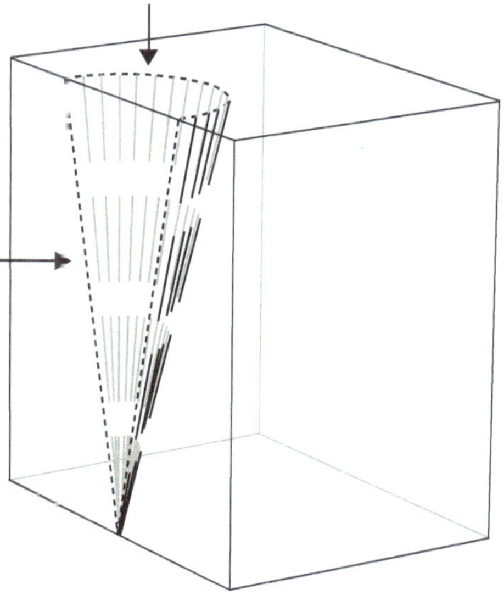

Figure 5 Intersection of an established fire plume with an adjacent wall. Copyright 2012 from Scientific Protocols for Fire Investigation, second ed. by John J. Lentini. Reproduced by permission of Taylor and Francis Group, LLC, a division of Informa plc.

schematically how a U-shaped and circular patterns are created, and **Figure 8** shows a classic U-shaped pattern from a fire scene. Patterns on ceilings are often more difficult to recognize, because ceilings are likely to collapse sooner than walls.

Figure 9 shows a semicircular pattern above the origin of a fire that originated inside a wall.

As a fire progresses, more fuel packages become involved, and may result in new "V" patterns or "U" patterns being produced. In the past, some fire investigators have interpreted multiple "V" patterns as evidence of multiple origins, but this is an interpretation that can rarely be justified once a fire becomes fully involved. Such interpretations are not justified in cases of full room involvement or in any fire scene where all of the fire damage is contiguous. It is only when there is no communication between the fire patterns generated by separate fire plumes that one can state with any validity that there were two (or more) origins.

Plume pattern generation is simple to understand and explain; the difficulty lies in locating the first plume pattern generated. Further, once a compartment progresses beyond flashover, interaction with the fire plume may not be the dominant means of pattern generation on building surfaces.

Confinement Patterns

Like plume-generated patterns, confinement patterns are easily explained. Confinement patterns are caused by the hot gas layer being trapped beneath a ceiling, and interacting with the walls. The result can be either a smoke horizon or a heat horizon, depending on how well developed the fire becomes in the compartment where the pattern exists. **Figure 10** shows a smoke horizon and **Figure 11** shows a heat horizon.

Confinement patterns can provide a fire investigator with sequential data because if there is a smoke horizon or a heat horizon on a wall, then in all likelihood, that pattern was generated before the failure of the ceiling above it. This can be very useful for determining the origin in a multilevel structure, and in some cases, the confinement pattern may be the only pattern remaining. Even the absence of a confinement pattern can help to eliminate a particular level of the structure as the location of the origin.

Consider a structure where everything has collapsed into the basement. If there is a smoke or heat horizon on the basement wall, it is unlikely that the fire entered the basement late in its development. Conversely, the absence of a smoke or heat horizon in the basement tends to eliminate the basement as the origin.

A different kind of confinement pattern is sometimes produced when the heat of the fire was confined in a wall space between two vertical studs. While the pattern on the side of the

Figure 6 Typical V pattern in a residential fire. The fire originated in a plastic trash can located against the wall in the background. Copyright 2012 from Scientific Protocols for Fire Investigation, second ed. by John J. Lentini. Reproduced by permission of Taylor and Francis Group, LLC, a division of Informa plc.

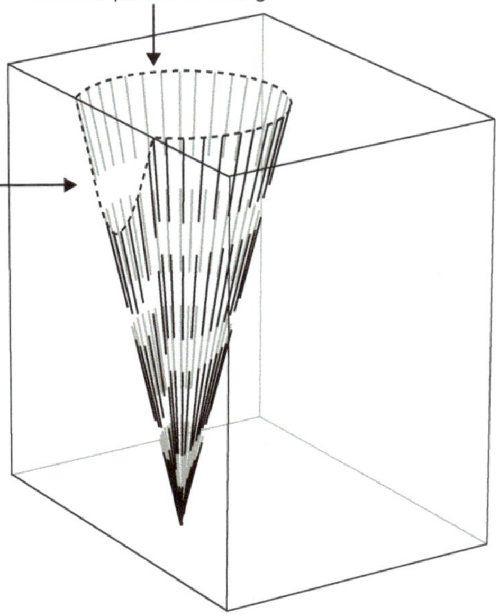

Figure 7 Intersection of a fire plume located some distance from the wall with the wall and ceiling. Copyright 2012 from Scientific Protocols for Fire Investigation, second ed. by John J. Lentini. Reproduced by permission of Taylor and Francis Group, LLC, a division of Informa plc.

wall where the fire originated may be unclear, the pattern on the opposite side of the wall may show a clear delineation of the location of the studs. **Figures 12 and 13** show both sides of such a confinement pattern. Similarly, fires burning upward through a floor will be influenced by the structural members under the floor and patterns with parallel edges may be the result. A fire penetrating downward through the floor, on the other hand, is unlikely to be significantly influenced by structural members below the floor.

Movement Patterns

When the fire moves from one room to another, it frequently records movement patterns at or near the doorways. An example of such a pattern is shown in **Figure 14**. Movement patterns are useful for tracing a fire back to its origin and for helping to eliminate the room into which the fire moved as the compartment of origin. Movement patterns typically appear as diagonal patterns moving upward from a doorway.

Irregular Patterns

Irregular patterns are typically found on floors in compartments where fires have gone to flashover and burned for sometime thereafter. For decades, fire investigators thought that they could identify irregular patterns on floors as having been caused by the presence of ignitable liquids burning there. This misconception resulted from training exercises designed to teach participants to "recognize arson." These training exercises almost invariably involved the extinguishment of the fire before it became fully involved. In such a case, it is indeed possible to recognize a pattern caused by a flammable liquid. **Figure 15** shows such a pattern, which can be characterized as "an obvious pour pattern."

Figure 16 also shows an irregular fire pattern, but there was no accelerant involved in the test fire that generated this pattern. This pattern was produced when the carpeted floor was exposed to radiant heat from above. The carpet failed by randomly tearing open and shrinking back in response to the heat. Some parts of the floor were protected by the carpet and other parts were exposed. The lines of demarcation between burned and unburned areas were sharp, continuous, and irregular.

It is these alternating exposure and protection patterns that occur in fully involved compartments that are capable of confusing fire investigators, particularly those who have not watched test fires burn beyond flashover. The patterns shown in

Figure 8 Typical U pattern in a residential fire. The fire originated behind the door at the left side of the photo. Copyright 2012 from Scientific Protocols for Fire Investigation, second ed. by John J. Lentini. Reproduced by permission of Taylor and Francis Group, LLC, a division of Informa plc.

Figure 9 Semicircular pattern on the underside of the floor above the origin. The fire originated at an overheated switch inside the wall cavity. Copyright 2012 from Scientific Protocols for Fire Investigation, second ed. by John J. Lentini. Reproduced by permission of Taylor and Francis Group, LLC, a division of Informa plc.

Figure 10 An example of a typical smoke horizon. Photo courtesy of Mick Gardiner, Gardiner Associates, with permission.

Figure 11 An example of a typical heat horizon. Photo courtesy of Mick Gardiner, Gardiner Associates, with permission. Copyright 2012 from Scientific Protocols for Fire Investigation, second ed. by John J. Lentini. Reproduced by permission of Taylor and Francis Group, LLC, a division of Informa plc.

Figure 12 An example of a horizontal confinement pattern. Hot embers from above fell onto the horizontal member, where they caused damage to the drywall. Copyright 2012 from Scientific Protocols for Fire Investigation, second ed. by John J. Lentini. Reproduced by permission of Taylor and Francis Group, LLC, a division of Informa plc.

Figure 13 Opposite side of the wall shown in **Figure 12**. The outline of the stud space can be clearly seen. Copyright 2012 from Scientific Protocols for Fire Investigation, second ed. by John J. Lentini. Reproduced by permission of Taylor and Francis Group, LLC, a division of Informa plc.

Figures 17 and 18 were both interpreted by fire investigators as indicating that flammable liquids burned on the surface of the floor. Even though the laboratory samples tested negative for the presence of ignitable liquid residues, arson and homicide charges were brought against survivors of both fires. Convictions resulted in both cases.

A peer-reviewed study of test fires set with gasoline and kerosene revealed that patterns produced by ignitable liquids on floors often appear different than what fire investigators might expect. The pattern shown in **Figure 19** is similar to one that an investigator characterized as "puddle-shaped," and therefore caused by flammable liquid burning on the floor. An actual gasoline fire on a similar parquet floor that resulted in a pattern with an entirely different appearance is shown in **Figure 20**.

The attribution of holes in the floor to the burning of an ignitable liquid on top of the floor is almost always in error. Ignitable liquids are incapable of existing at temperatures above their boiling point, so as long as liquid remains, the temperature on the floor will be less than the boiling point of the flammable liquid. Once the liquid has completely vaporized, there will be no more heat from that source. Arsonists use ignitable liquids to cause a fire to spread more rapidly than it would otherwise. The ignitable liquids themselves tend to burn out very quickly, and in the greater scheme of the overall fire, provide relatively little fuel or energy.

It has been the misidentification of patterns similar to those shown in **Figure 16** that have caused numerous false convictions for arson.

Laboratory analysis is required to accurately identify all but the most obvious "pour patterns," and even in the case of obvious patterns, a laboratory analysis certainly does no harm.

Spalling

Spalling is the chipping or pitting of concrete or masonry surfaces in response to heat. The spalled area can range in size from a few square centimeters to a few square meters. Because

Figure 14 Typical movement pattern through a doorway. Photo courtesy of Mick Gardiner, Gardiner Associates, with permission. Copyright 2012 from Scientific Protocols for Fire Investigation, second ed. by John J. Lentini. Reproduced by permission of Taylor and Francis Group, LLC, a division of Informa plc.

the result of spalling is a depression in a concrete surface, spalling is often "puddle shaped." For years, fire investigators attributed spalling of concrete to the intense heat generated by the burning of an ignitable liquid. This is incorrect on a number of levels. First, spalling can be caused by the differential expansion of the aggregate and the cement matrix, by the presence of steel reinforcement bars, or by the vaporization of the waters of hydration that are a part of the concrete.

More importantly, a flammable or combustible liquid burns no more intensely than other carbonaceous fuels such as wood. Burning ignitable liquids are no more likely to cause spalling than they are to burn holes in wooden floors, installing is frequently seen on surfaces where no flammable liquid could exist, such as a concrete ceiling.

Numerous experiments have been conducted to try to cause spalling using ignitable liquids, and most of those experiments have failed. On the other hand, setting a well-ventilated wood fire above a concrete floor will almost invariably result in spalling.

Electrical Damage

When a fire damages the insulation of an energized electrical conductor, the first thing that happens is the formation of carbonized insulation, which can conduct electricity from one conductor to another. The resulting current flow may or may not be sufficient to activate the overcurrent protection, that is, trip a circuit breaker or melt a fuse.

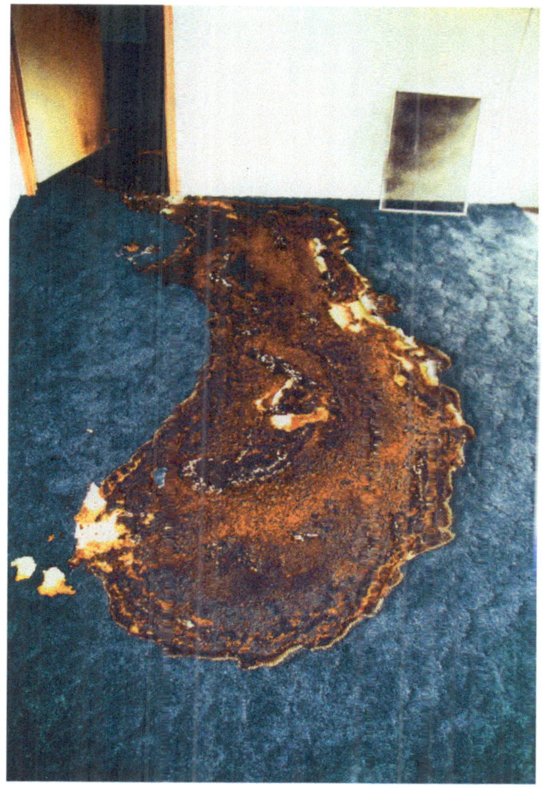

Figure 15 Irregular pattern caused by the burning of a medium petroleum distillate on a carpet. This is a rare instance where the cause of the pattern can be interpreted as having been caused by ignitable liquids on the basis of appearance alone. Copyright 2012 from Scientific Protocols for Fire Investigation, second ed. by John J. Lentini Reproduced by permission of Taylor and Francis Group, LLC, a division of Informa plc.

A fire investigator may observe multiple areas where this "arcing through char" has occurred during the course of a fire. Arcing through char will leave behind artifacts in the form of globules melted copper. The first conductors that are likely to be compromised by a fire are power cords to electrical appliances. Branch circuit conductors behind walls and ceilings can also be damaged.

By noting the location of arcing artifacts on various circuits, the investigator may be able to infer the sequence of the arcing events. Certainly, all the arcing events will have taken place before the disconnection of power to the structure, and the events on a particular circuit will have occurred before the activation of any overcurrent device protecting that circuit. If a particular conductor is severed, then any arcing event that is electrically downstream of the point of severance can be logically inferred to have occurred before the event that caused the conductor to be severed.

Figure 16 A more typical irregular pattern caused by the random failure of the carpet, resulting in alternate areas of exposed and protected subflooring. Copyright 2012 from Scientific Protocols for Fire Investigation, second ed. by John J. Lentini. Reproduced by permission of Taylor and Francis Group, LLC, a division of Informa plc.

This mapping of arcs can result in a pattern that allows an investigator to determine which parts of the electrical system were compromised first, and provide clues to the origin of the fire.

Clean Burn

Clean burn is a fire pattern on surfaces where the soot has either burned away, or was never deposited because the surface was too hot. It stands to reason that if the surface had soot deposited on it and then the soot was burned away, chances are the fire originated elsewhere. The problem of fire patterns is that they do not come with time stamps on them. If the fire begins close to a wall, it may heat the wall to the point where soot is never deposited on it, yet there is a white area that is indistinguishable from a clean burn that occurred later in the fire.

Figure 21 shows a clean burn located directly above the origin of a test fire, on the left wall between the bed and the chair. It is unlikely that there was ever any significant amount of soot deposited in the white area. Video of the test showed that the area indicated by the arrow was an area of early flame contact with the wall. The white areas to the right of the nightstand and to the right of the bed on the rear wall were actually not burned. The wall above the nightstand caught fire before the nightstand. This test fire was extinguished after only 10 s of postflashover burning. **Figure 22** shows another ventilation-generated pattern from a similar test fire. It is interesting to note that the pattern assumes a V-shape, but the pattern was caused by a cloud of fuel-rich gases burning from the top downward.

Figure 17 This fire pattern was erroneously attributed to the burning of ignitable liquids. The unburned areas were protected by objects on the floor. Photograph courtesy of David M. Smith, Associated Fire Consultants, with permission.

Figure 18 This fire pattern was erroneously attributed to the burning of ignitable liquids. The unburned areas were protected by objects on the floor.

Figure 19 A "puddle-shaped" pattern, similar to one that was attributed to the burning of a puddle of ignitable liquid. Actually, this pattern was caused by the burning of a cardboard box. Copyright 2012 from Scientific Protocols for Fire Investigation, second ed. by John J Lentini. Reproduced by permission of Taylor and Francis Group, LLC, a division of Informa plc.

Figure 20 A fire pattern on a parquet floor that was caused by the burning of 1 l of gasoline.

Figure 21 Clean burns resulting from a short-lived test fire. The wall between the bed and the chair never accumulated any soot. The wall and ceiling above the nightstand were damaged by the burning of a fuel-rich cloud of smoke produced by the chair. Photo courtesy of ATF Fire Research Laboratory and Steven W. Carman, Carman Fire Investigations, with permission.

Figure 22 Clean burn resulting from a fire that burned for approximately 2 min beyond flashover. The V-shape developed from the top downward. The fire originated between the bed in the chair. Photo courtesy of ATF Fire Research Laboratory and Steven W. Carman, Carman Fire Investigations, with permission.

Intensity Patterns

Fires produce patterns as a result of the response of materials to heat exposure. The various heat effects on materials can produce lines of demarcation. These lines of demarcation may be helpful in determining the characteristics and quantities of fuel materials, as well as the direction of fire spread.

Intensity patterns may be very confusing, because of the difficulty in determining whether the pattern was caused by a fire of longer duration or of higher intensity. Attributing more intense damage to longer duration frequently will result in determining an incorrect area of origin. The test fire shown in **Figure 23** was set in the kitchen, which is at the left side of the photo. Because of more fuel and better ventilation, however,

Figure 23 Although this fire started in the kitchen at the left, the living room at the right exhibited far more damage. Copyright 2012 from Scientific Protocols for Fire Investigation, second ed. by John J. Lentini. Reproduced by permission of Taylor and Francis Group LLC, a division of Informa plc.

there was much more severe damage in the living room, shown at the right side of the photo.

Often, fire patterns are a combination of movement, intensity, ventilation, and plume-generated patterns.

Ventilation-Generated Fire Patterns

In the early stages of a fire, the rate of heat release increases as more fuels become involved. The plumes from these burning fuel packages pump smoke and energy into a hot gas layer trapped beneath the ceiling. As the fire grows, the hot gas layer becomes thicker and hotter. It radiates heat in all directions, including downward. When the temperature of the hot gas layer reaches 500–600 °C, it radiates sufficient heat downward (on the order of $20\,kW\,m^{-2}$) to ignite every exposed combustible surface in the compartment. This begins the transition phase known as flashover.

The burning of all of this exposed fuel dramatically increases the heat release rate within the compartment, but even more importantly, it consumes most of the oxygen. At this point, the fire can burn only in those places that have sufficient oxygen. This fire is known as a "ventilation-controlled" or "ventilation-limited" fire.

The compartment in this case is full of fuel in the form of combustible vapors, aerosols, and gases, as well as particles of unburned fuel. The entire atmosphere is flammable. In an atmosphere of air, one can create a flame by introducing and igniting a flammable gas such as methane or acetylene. If the atmosphere is methane, however, as occurs on some extraterrestrial bodies, one could create a flame by introducing a stream of oxygen. This is exactly what happens in ventilation-limited fires.

There is likely to be more intense burning around ventilation openings such as doors and windows. What has been learned lately, however, goes far beyond that and has important ramifications for the validity of fire origin determination. Patterns are likely to be created not only around the ventilation openings, but, because of the way a ventilation-limited fire behaves, they may also be created on walls and other surfaces opposite to the ventilation openings as a result of a jet of air being drawn in by the fire.

Ventilation-generated patterns can create significant confusion. Fire investigators very often will use the level and intensity of the fire patterns to determine the area of origin. The "lowest and deepest char" is often singled out as the point where the fire must have burned the longest, but this is not true.

In 2005, agents of the Bureau of Alcohol, Tobacco and Firearms (ATF) conducted an experiment at a fire investigation seminar in Las Vegas, Nevada, to replicate similar experiments they had conducted at their training facility in Glynnco, Georgia. They built two identical rooms furnished as bedrooms, and then set them on fire and allowed them to burn for 7 min.

They then asked 53 participants in the seminar to observe the fire patterns and determine which quadrant of the room contained the origin. Three of the participants identified the correct quadrant in room number one and three different participants identified the correct quadrant in room number two. Although it was argued that some of the participants were not qualified fire investigators, and that time and resources were limited and investigators were not allowed to touch the evidence, the low success rate of the participants was a cause for concern.

The ATF agents conducting the experiment formed a hypothesis that the amount of postflashover burning would influence the investigator's ability to correctly identify the quadrant of origin. The experiment was repeated with some modifications.

In 2007, three burn cells were constructed in Oklahoma City, and 70 fire investigators were asked to identify the quadrant of origin. In the first burn cell, the fire was allowed to burn 30 s beyond flashover, and 64 of 70 investigators correctly identified the quadrant of origin.

In the second burn cell, the fire was allowed to burn for 70 s beyond flashover. In this case, 6 investigators declined to identify the quadrant of origin but of the 64 who did attempt to identify where the fire started, only 44 correctly identified the quadrant of origin.

In the third burn cell, the fire burned for 3 min beyond flashover. Seventeen investigators declined to make a determination. Of the 53 investigators who responded, only 13 correctly identified the quadrant of origin. This amounts to a 75% error rate, which is no better than if 53 untrained individuals had picked the quadrant of origin at random. Clearly, these snapshots of origin determination accuracy indicate that there are serious problems with the training of fire investigators.

In an effort to understand what was causing the confusion, the experimenters reconstructed three burn cells at the ATF laboratory Fire Research Laboratory in Ammendale, Maryland. The burn cells were equipped with thermocouples and radiometers and the burning was carefully monitored. The results indicated that what was happening was that the fire was burning most intensely at a point directly across from the single opening, a doorway, as a result of the vitiation (lack of oxygen) of most of the compartment and a jet of clean air coming in at the bottom of the door.

The test fire was modeled using FDS (fire dynamics simulator) and the output was translated using smokeview, a program that allows visualization of the numerical output of the model. **Figure 24** shows that the radiant heat flux opposite the doorway was $150\,kW\,m^{-2}$ on the wall. **Figure 25** shows a "slice" of the smokeview output, where red indicates oxygen at a concentration of 21% (air) and blue indicates a concentration close to zero.

These results help us to understand what is happening in the production of fire patterns after flashover, but far more

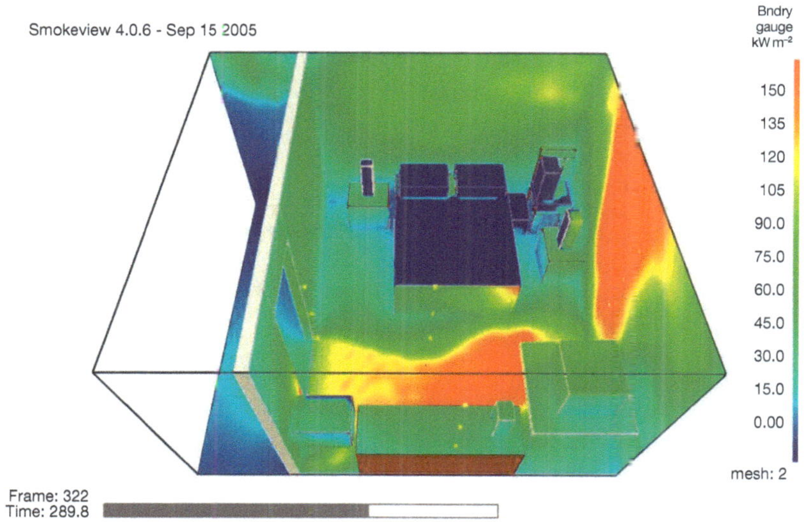

Figure 24 Smokeview output of a fire dynamics simulator model, showing the effects of ventilation. The fire originated to the right of the headboard, but air flowing in through the bottom of the door resulted in a radiant heat flux of more than 150 kW m^{-2} on the wall opposite the door. Photo courtesy of ATF Fire Research Laboratory and Steven W. Carman, Carman Fire Investigations with permission.

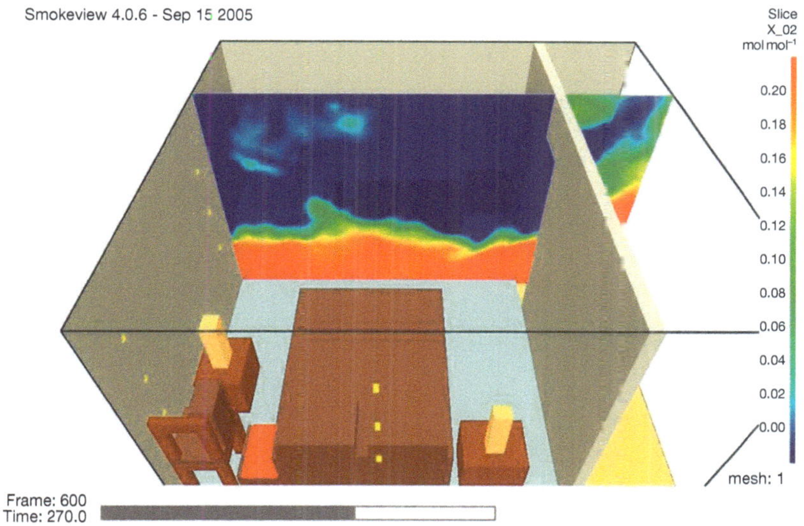

Figure 25 Smokeview slice showing the oxygen concentration in the plane of the door. Photo courtesy of ATF Fire Research Laboratory and Steven W. Carman, Carman Fire Investigations, with permission.

research is needed. This was a simple test with a single opening and a single compartment. Simple tests similar to this are necessary to understand what is happening, but the conditions of this test seldom obtain in a typical structure fire. Many times, the initial ventilation source (before the windows break) is from inside the structure. Many times, there are multiple ventilation sources, especially after the windows break.

The consumption of all the oxygen in the room can cause the fire at the origin to burn much more slowly there than elsewhere. If the origin is located in a place where a ventilation-generated pattern is overlaid, then information about the origin may be lost.

Fire investigators' confusion about how to correctly locate the origin of the fire is not a trivial matter. It is generally accepted that if one does not correctly identify the origin, one is likely to incorrectly identify the cause. Even more concerning is the fact that when a fire investigator identifies an area of origin and then finds no competent ignition source or even a competent fuel source in that area, he is likely to conclude that the fire was set using an open flame that was removed from the scene, and that the initial fuel was an ignitable liquid or some other highly flammable fuel that was introduced and completely consumed. A false determination of arson is likely.

This is new knowledge for many fire investigators, and upon being confronted with it, they are likely to reject the scientific basis of the experiments and insist on their ability to determine an origin even in a fully involved compartment.

What should happen is that once a compartment is identified as having been fully involved, every potential ignition source in that room should be examined, unless there is some compelling evidence that allows the origin to be narrowed further than the entire confines of the compartment. An eyewitness to the beginning of the fire or perhaps a videotape of the fire might meet these criteria. Certainly, the simple traditional interpretation of fire patterns in such a situation is likely to lead to error.

See also: **Chemistry/Trace/Fire Investigation:** Analysis of Fire Debris; Chemistry of Fire; Interpretation of Fire Debris Analysis; Physics/Thermodynamics; Thermal Degradation; **Investigations:** Evidence Collection at Fire Scenes; Fire Scene Inspection Methodology; Types of Fires.

Further Reading

Carman, S.W., 2008. Improving the Understanding of Post-flashover Fire Behavior. Proceedings of the International Symposium of Fire Investigation. NAFI, Sarasota, FL.
Lentini, J.J., 2006. Scientific Protocols for Fire Investigation. Taylor & Francis, Boca Raton, FL.
NFPA, 2011. NFPA 921: Guide for Fire and Explosion Investigations. National Fire Protection Association, Quincy, MA.

Relevant Websites

Fire.NIST.gov—The National Institute of standards and Technology Bureau of Fire Research Website.
CFITrainer.net—Sponsored by the International Association of Arson Investigators.

Analysis of Fire Debris

N NicDaéid, University of Strathclyde, Glasgow, UK
E Stauffer, Commissariat d'Identification Judiciaire, Police Cantonale Fribourg, Fribourg Switzerland

Introduction

When a fire investigator suspects the presence of an ignitable liquid at a fire scene, he/she will proceed to the collection of fire debris in order to obtain a confirmation through a laboratory analysis. The analysis tests for the presence of ignitable liquid residues (ILR) within the sample; however, the presence of such ILR may not necessarily be correlated to the use of an ignitable liquid. In addition, despite the enormous variety of ignitable liquids available on the market, the analysis performed focuses mostly on petroleum-based ignitable liquids. Thus, it may not detect other, less commonly used ignitable liquids unless the analyst is advised to look for them. This emphasizes the importance of a clear communication between the fire debris analyst and the fire investigator at the scene.

Laboratory-based analyses require the appropriate skill and knowledge of relevant scientific instrumentation, proper laboratory practice in dealing with crime scene evidence, and an understanding of the nature of flammable materials and their pyrolysis and combustion products as well as an ability to interpret the results of their analysis. Knowledge of the nature of fire and the mechanisms of combustion is also essential.

ILR cannot be analyzed in situ from the fire debris sample. They must first be extracted from the sample and prepared into a suitable form for injection and analysis by gas chromatograph–mass spectrometry (GC-MS). **Figure 1** indicates the different steps taken during the examination of a fire debris sample.

The examination steps are preceded by the work of the fire investigator at the fire scene, in particular by the collection of the debris.

Evidence Collection—Sampling Containers

The correct collection of evidence for laboratory examination (specifically for the identification of any ignitable liquids that may have been used to accelerate the fire) from a fire scene is of great importance. However, not every scene will require the collection of fire debris for such analysis. Specifying how to gather fire debris from every scene is difficult because each scene will differ in its nature. However, in general, samples must be placed in sealable, gas-tight containers of which there are various types in common use around the world.

Unused tin cans (similar to paint cans) predominate in the United States and Australia/New Zealand. The cans come in a variety of sizes and types and can be unlined or lined with an epoxy coating (designed for the storage of water-based paints). When a lining is present, it can interfere with the chromatographic response of some ignitable liquids.

The sample is placed in the can and the lid firmly shut in place. It is then sealed and labeled. Cans have the advantage of not being permeable to any vapor and are generally very durable and resistant to puncture. However, sample size is restricted by the size of the can, which may become problematic with large samples. Given certain samples or conditions, corrosion and rusting can also be a problem, which may occur relatively quickly.

In many countries around Europe, glass jars as well as tin cans are used. Glass jars have the obvious disadvantages of fragility as well as constraining sample size but otherwise prove to be effective containers. They have the advantage of being able to see the sample without opening the container.

Also in Europe and elsewhere in the world, a popular method of sample storage involves the use of solvent-free nylon bags. Samples are placed within the bag, which is sealed by a variety of means including knotting the neck tightly, tying with a cable tie, or heat sealing. Adhesive tape is not generally used to seal such containers because of poor sealing efficiency. Bags have the obvious disadvantage of being easily punctured although this is minimized by "double bagging." These containers have, however, been shown to be permeable to alcohols. Research has demonstrated that from time to time hydrocarbon residues are detected within nylon bags, which may lead to erroneous results.

Some European countries, particularly Germany and Scandinavia, use different types of bags for fire debris. These include Kapak bags (polyethylene/polyolefin composite material, manufactured in the US), laminated bags (polyethylene–aluminum/polyamide), duo bags (polypropylene/polyethylene), and CLS bags (polypropylene/polyamide, manufactured by CLS, Israel). Duo bags are preferred in Sweden although they have been shown to leak decane and branched alkanes after about a week. Similar to nylon bag, the seal also has an effect on leakage rates with aluminum tie seals showing leakage within 6 h, cable tie seals allowing leakage after 24 h, and heat seals showing leakage only after about 6 weeks.

 http://dx.doi.org/10.1016/B978-0-12-802718-9.00024-7

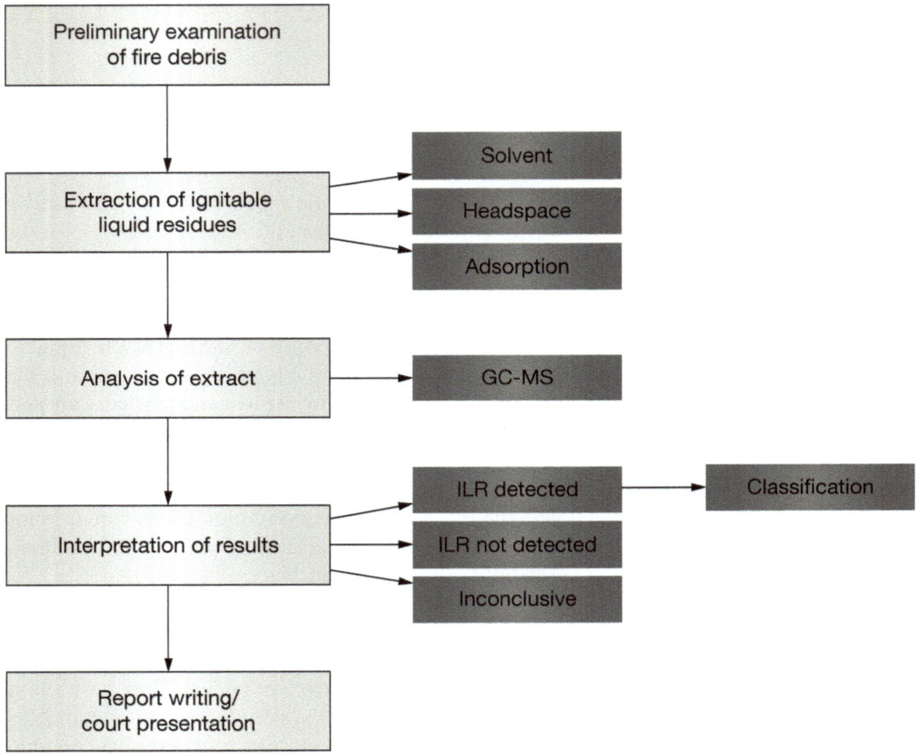

Figure 1 The different steps in the examination of fire debris samples. *GC-MS*, gas chromatograph–mass spectrometry; *ILR*, ignitable liquid residues. Adapted from Stauffer, E., Dolan, J.A., Newman, R., 2008. Fire Debris Analysis. Academic Press, Burlington, MA, p. 10; © Elsevier.

Whatever container is chosen, the procedure for sample gathering is the same. In a best-case situation, three samples should be collected for laboratory analysis:

- A container control sample—this shows any volatiles that may be within the container itself (sometimes known as a negative control sample). Often, a few containers by batch are analyzed, thus suppressing the need to send an empty container with every sample.
- A comparison sample—this is the same material as that of the sample but taken from an area away from the one suspected to contain ignitable liquid. Such sample is used in the interpretation of the data to discriminate between the interfering products and the possible ILR from the original sample. A comparison sample is always difficult to choose because one never knows the history of the substrate and, in many cases, the investigator may not be able to find a suitable comparison sample.
- The fire debris sample.

Each sample should be clearly labeled with the case name, location, date of recovery, and person responsible for recovery.

Occasionally, positive control samples are prepared for a particular case. These contain a sample spiked with a volatile flammable liquid packaged and treated in the same manner as the other samples pertaining to the case. These samples allow the fire debris analyst to evaluate the influence of the matrix against the content of the ILR. The positive control is generally analyzed at the end of the analytical sequence to reduce the risk of contamination.

Preliminary Examination of Fire Debris Samples

Items sent for analysis should be recorded and signed into the laboratory. It is essential that correct record keeping is carried out to ensure that the continuity of evidence is complete.

Before conducting an extraction, the analyst must take a first look at the sample. The goal of this step is to gather information about the nature of the sample, which is very important for the interpretation of the data. Likewise, observing the sample allows the analyst to choose the best extraction technique. Finally, fire debris samples may contain other forensic evidence, which may need to be separated prior to extraction.

Contrary to some popular beliefs, one will not lose all the ILR of a fire debris sample by opening the container, spreading some of its content on a clean sheet of paper, and examining it.

Alternatively, it may not be necessary to remove the content from the container if one can see through. However, in any case, it is crucial that the fire debris analyst has a clear understanding of the sample's content.

In some laboratories, a preliminary sample of the headspace gases associated with the fire debris sample is taken. This gives a first indication of the presence of ILR and their relative quantity. Instruments such as portable flame ionization detectors (FIDs) are sometimes used for this purpose.

Extraction and Sampling Techniques

Among the many different techniques used to extract ILR from fire debris samples, three are most commonly used: direct headspace, dynamic headspace concentration, and passive headspace concentration. These are also the only techniques that still have valid standards through ASTM International. Distillation was widely used in the past; however, because of the efficiency and, most importantly, nondestructive nature of the adsorption techniques, it has been abandoned and is not described here. Solvent extraction entails soaking the debris in a given solvent, so that the analytes of interest will preferentially migrate into the solvent rather than remain adsorbed on the matrix. ASTM standard E1386 describes this technique. Solvent extraction is destructive and can coextract matrix compounds. As such, solvent extraction is not routinely used, but rather when a particular application is required.

Direct Headspace Analysis

Direct headspace analysis is commonly used especially for samples thought to be rich in volatile components. This technique requires that the sample be contained within a vessel such as a nylon bag and heated to some degree to release volatile components from the sample into the headspace above the sample (see **Figure 2**). A sample of this headspace is directly analyzed using GC-MS. Heating of the sample is essential as higher molecular weight compounds such as long-chain hydrocarbons have lower vapor pressures and are less likely to be detected without heating the samples first.

The general procedure, described in ASTM standard E1388, for direct headspace analysis requires the sample in its packaging container to be placed in an oven heated to 60–90 °C for about 30 min. A sample of the headspace is then extracted using a gas syringe (previously flushed with air and heated). The gas sample is then injected directly into the analytical instrument. This method is easily susceptible to contamination and should be monitored by the injection of air blanks between each sample. Because of the sampling procedure, low concentrations of volatile materials may not be detected and a technique allowing for sample concentration may be required. Fire debris may also (and often do) contain water,

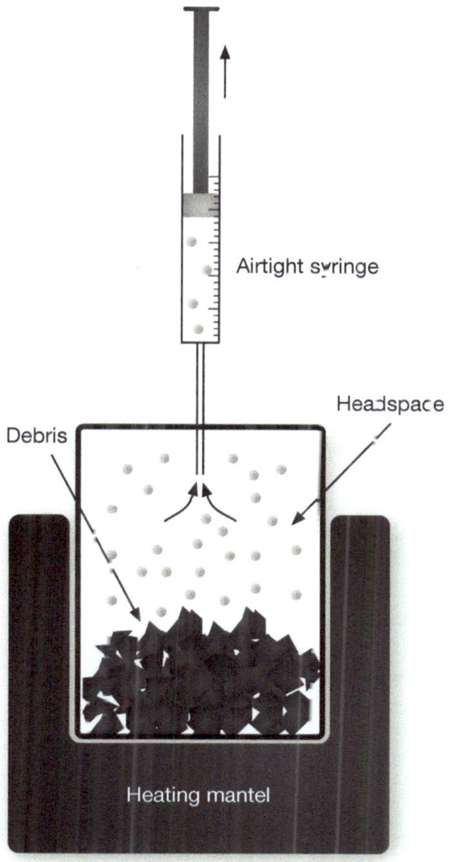

Figure 2 Principle of direct headspace extraction. Source: Stauffer, E., Dolan, J.A., Newman, R., 2008. Fire Debris Analysis. Academic Press, Burlington, MA, p. 393; © Elsevier.

which influences the vapor pressure of other compounds and results in lower detection levels of ILR.

Passive/Dynamic Headspace Analysis

These methods of sampling are among the most commonly used. They involve the extraction and concentration of vapors onto a solid phase by either an active or passive adsorption method involving heating of the sample to facilitate release of volatile components. Once extracted on the adsorbing medium, this one is desorbed using a solvent or heat before injection into the GC-MS.

Two main adsorption media are commonly used in fire debris analysis Tenax and activated charcoal. Tenax is an organic polymer (2,6-diphenyl-*p*-phenylene oxide), which has a high affinity for hydrocarbons and low affinity for water. Tenax must be thermally desorbed because it resists very little to solvents. Activated charcoal is most commonly desorbed using pentane, carbon disulfide (although this is very toxic), or diethyl ether. Alternatively, it can also be thermally desorbed.

Passive headspace analysis, which is described in ASTM standard E1412, involves placing the adsorbent within the sample container, which is heated to 80–100 °C for a number of hours. Extraction time varies depending on laboratory practice but is usually set between 12 and 16 h. Following this, the sample is removed and the adsorption media desorbed with the solvent. The sample's extract is then injected into the GC-MS with appropriate solvent controls and blanks.

Dynamic headspace analysis, which is described in ASTM standard E1413, involves drawing the heated headspace vapors through the adsorbent by use of a vacuum over a set period of time (see **Figure 3**). Desorption is accomplished as described before.

Solid-phase microextraction (SPME) techniques have also been increasingly used. In this case, volatiles are adsorbed onto the surface of a fiber and thermally desorbed within the injection port of the instrument. Unfortunately, in addition to SPME fibers being extremely fragile, SPME exhibits a serious problem of displacement, thus rendering the extraction of ILR extremely inefficient. As such, SPME has not yet become a routine technique in fire debris analysis and cannot be used as a standalone extraction technique according to ASTM standard E2154.

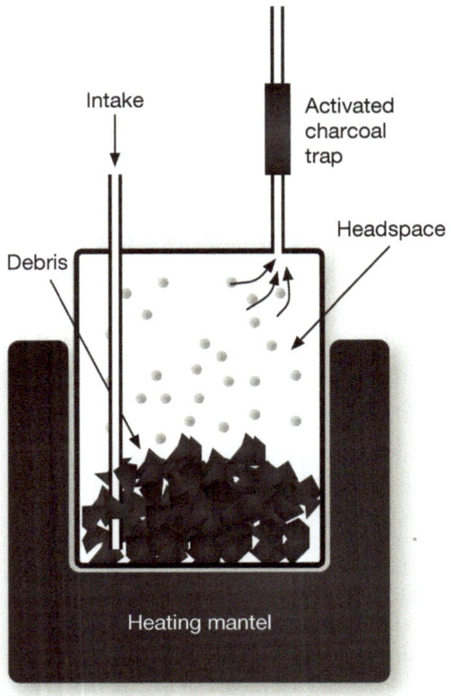

Figure 3 Principle of passive headspace concentration. Source: Stauffer, E., Dolan, J.A., Newman, R., 2008. Fire Debris Analysis. Academic Press, Burlington, MA, p. 404; © Elsevier.

Effectiveness of Extraction/Concentration Methods

Each of the methods described has advantages and disadvantages. Overexposure of the adsorbent materials to debris containing volatile hydrocarbons will usually favor the adsorption of heavier components through the displacement of lower molecular weight components. Conversely, underexposure can result in the nonappearance of the heavier end of any volatile materials that may be present in the debris. This is why the preliminary examination of the fire debris must be thoroughly conducted, so that the analyst can determine which technique and parameters will result in the most efficient extraction.

Analysis

Instrumental Analysis

The analysis of fire debris samples is carried out by GC-MS. Because an ignitable liquid is usually composed of tens or hundreds of different compounds, it is necessary to first separate them before they can be identified. Because ILR are volatile compounds, gas chromatography is the most suitable separation technique available. Mass spectrometry, which is a detector commonly hyphenated to GC, provides the necessary structural information to identify most compounds encountered in fire debris analysis. Thus, the hyphenation of GC and MS results in a powerful analytical technique that can separate, characterize, and identify volatile components within complex mixtures, such as the ones encountered in fire debris analysis.

The proper functioning of the instrument may be checked using a standard mixture of compounds before and after analysis of samples when necessary. It is essential to ensure (and be able to prove) that the instrument is clean and that the observed peaks are due to the sample being analyzed, and not carried over from previous injections. A record of the order in which samples are injected, with the negative container control being injected first, should be kept.

While GC-FID was widely accepted in the past for fire debris analysis, the evolution of technology and the wide availability of inexpensive GC-MS led to the demise of GC-FID. GC-MS is now promoted as the gold standard for fire debris analysis. As such, ASTM has withdrawn its standard on fire debris analysis by GC--ID and only analysis by GC-MS has a valid ASTM standard (E1618).

Gas Chromatography

The sample is introduced into the gas chromatographic system through the injector port, is then swept through the column by the carrier gas, and ends up in the detector. **Figure 4** shows the general GC setup.

Fire debris analysts often use helium as a carrier gas. In order to guarantee an effective separation, it is important to choose

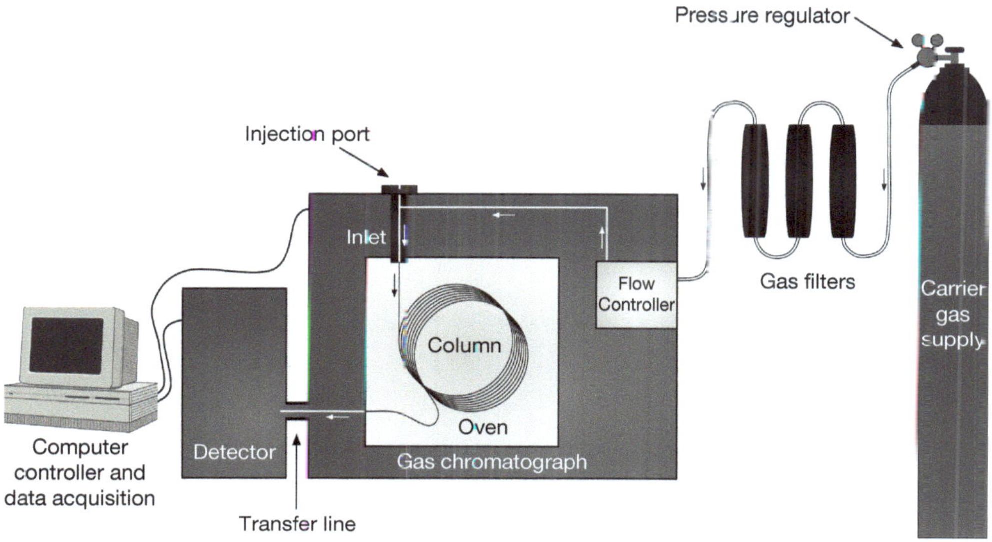

Figure 4 A chromatograph and its main components. Source: Stauffer, E., Dolan, J.A., Newman, R. 2008. Fire Debris Analysis. Academic Press, Burlington, MA, p. 246; © Elsevier.

the right column and the correct temperature settings. A bonded, nonpolar capillary column (100% dimethylpolysiloxane or 5% phenylpolysiloxane/95% dimethylpolysiloxane) or equivalent is required for ignitable liquid analysis as recommended in ASTM E1618. Automated temperature programming allows the oven to gradually increase its temperatures within a specified time so that the analytes separate at different rates.

Mass Spectrometer

The mass spectrometer is comprised of three major elements, which are the ionizer for ionizing and fragmenting the sample, the mass analyzer for separating the ionized particles, and the electron multiplier, which detects the ionized particles.

Electron impact (EI) is the ionization technique of choice in fire debris analysis. EI produces a charged molecule or radical cation known as the molecular ion or parent ion and denoted as $M^{+\bullet}$.

$$M(g) + e^- \rightarrow M^{+\bullet} + 2e^-$$

Continuous bombardment causes the molecular ion to break down further producing more fragment ions or daughter ions.

$$M^{+\bullet} \rightarrow m_1^+ + m_2^\bullet \ldots \text{ or}$$
$$M^{+\bullet} \rightarrow m_1^+ + m_2^\bullet$$

All of these ions are then separated according to their mass to charge ratio (*m*/*z* ratio) by the mass analyzer and detected in proportion to their abundance. Mass spectral data are usually collected in scan acquisition mode, which provides a wide range of ion detection, commonly set between 10 and 400 amu in fire debris analysis. This results in a total ion chromatogram (TIC). Alternatively, it is possible to use selected ion monitoring (SIM), which only scans for preselected ions to detect compounds of interest. SIM mode increases the detector sensitivity.

GC-MS Data

GC-MS typically generates data under the form of a TIC. The characterization and identification of ILR are based on visual pattern recognition. The TIC pattern and relative peak ratio obtained from the sample are visually compared to the TIC from an ignitable liquid reference collection or database to identify the class of ignitable liquid present.

In order to properly interpret the chromatograms, the fire debris analyst must be very familiar with the types of compounds found in ILR, as shown in **Table 1**.

Most ignitable liquids are composed of various aliphatic components such as octane or trimethylpentane and aromatic components such as trimethylbenzenes, xylenes, and toluene.

The use of selective ion monitoring (SIM) and extracted ion chromatograms to distinguished ignitable liquid samples, ILR, and interfering products also brings undeniable advantages. By looking at the single ions rather than the TIC, it is possible to filter the pattern, which greatly facilitates the comparison.

Table 1 Compounds commonly found in ignitable liquid residues

Type of compounds	*Specific structure*	*Description*	*Examples*
Alkane	Normal alkanes (*n*-alkanes)	Straight chain hydrocarbons (C_nH_{2n+2})	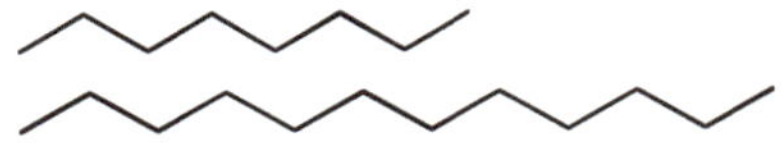
	Isoalkanes (isoparaffins)	Branch chained hydrocarbons (C_nH_{2n+2})	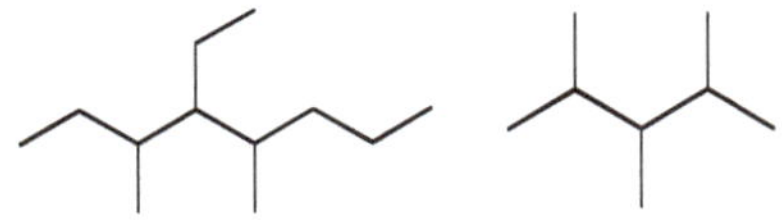
	Cycloalkanes (cycloparaffins, naphthenics)	Cyclic hydrocarbons (C_nH_{2n})	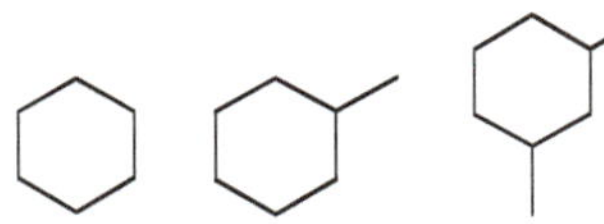
Aromatic	Simple	Alkyl-substituted benzenes	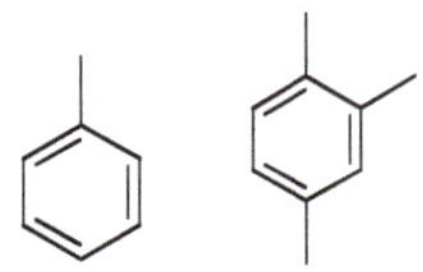
	Polynuclear (naphthalenes)	Multiple fused benzene ring compounds	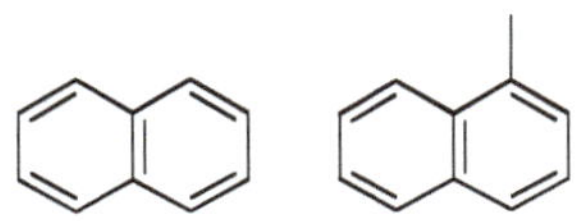
	Indane	Benzene ring fused to a cyclopentane	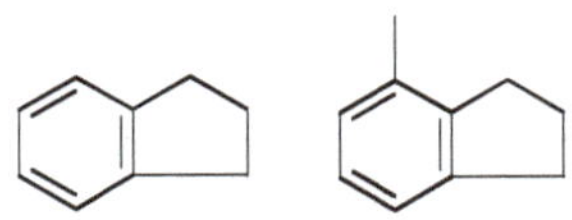

Table 2 Ions representative of ignitable liquid residues compounds

Hydrocarbon class	*Ions*				
Alkanes (iso and normal)	43	57	71	85	99
Cycloalkanes and alkenes	55	69	83	97	
Simple aromatics	91	105	119	133	
Polynuclear aromatics	128	142	156		
Indanes	117	118	131	132	

Table 2 shows the ions most representative of the different types of compounds present in ILR.

Once ILR analysis is performed, the interpretation of the results starts. This is the most difficult part of fire debris analysis and it is briefly described in other chapters of this book.

See also: **Chemistry/Trace/Fire Investigation:** Chemistry of Fire; Interpretation of Fire Debris Analysis; Physics/Thermodynamics.

Further Reading

ASTM International, 2010. ASTM E1618-10 Standard Test Method for Ignitable Liquid Residues in Extracts from Fire Debris Samples by Gas Chromatography-Mass Spectrometry. ASTM International, West Conshohocken, PA. Annual book of ASTM Standards 14.02.

DeHaan, J.D., Icove, D.J., 2011. Kirk's Fire Investigation, seventh ed. Pearson Education, Upper Saddle River, NJ.

Newman, R., Gilbert, M., Lothridge, K., 1997. GC-MS Guide to Ignitable Liquids. CRC Press, Boca Raton, FL.

Nic Daéid, N., 2004. Fire Investigation. CRC Press, Boca Raton, FL.

Stauffer, E., Dolan, J.A., Newman, R., 2008. Fire Debris Analysis. Academic Press, Burlington, MA.

Interpretation of Fire Debris Analysis

E Stauffer, Commissariat d'Identification Judiciaire, Police Cantonale Fribourg, Fribourg, Switzerland
N NicDaéid, University of Strathclyde, Glasgow, UK

Abbreviations

GC	Gas chromatography or gas chromatograph
IL	Ignitable liquid
ILR	Ignitable liquid residues
MS	Mass spectrometry or mass spectrometer

Glossary

Aliphatic An organic compound which is not aromatic; organic compounds that are alkanes, alkenes, and alkynes and their derivatives.

Alkane A saturated hydrocarbon compound having the general formula C_nH_{2n+2}.

Aromatics A class of unsaturated organic compounds that have a benzene ring or that have chemical properties similar to benzene as part of their structure.

Combustion products The set of products that are released during the combustion reaction of materials. These products are the result of both complete and incomplete combustion but not of the pyrolysis process.

Crude oil Naturally occurring oil consisting primarily of hydrocarbons with some other elements such as sulfur, oxygen, and nitrogen. Source material of nearly all petroleum products.

Fire debris A generic term commonly used to describe material collected at a fire scene and submitted to the laboratory for ignitable liquid residue analysis.

Gasoline A mixture of several hundreds of volatile hydrocarbons ranging from C_4 to C_{12} used in an internal combustion engine.

Ignitable liquid A liquid fuel that is either flammable or combustible.

Ignitable liquid residues The remaining portion of an ignitable liquid on a substrate after undergoing physical and/or chemical changes.

Interfering products The set of chemicals found in a sample that interferes with the proper identification of ignitable liquid residues.

Isoparaffinic products An ASTM class of petroleum distillate almost exclusively composed of branched alkanes.

Microbial degradation The decomposition of petroleum products by bacterial action that can diminish some components relative to others resulting in an altered chromatographic pattern that may not allow for a definitive characterization.

Petroleum distillates An ASTM class of products obtained primarily from the fractionation of crude oil.

Pyrolysis products The set of products generated by the process of pyrolysis only.

Substrate The sample material from which a substance of interest (analyte) is removed for analysis.

Introduction

Once the chromatogram has been obtained, it is time to conduct the most difficult part of fire debris analysis: the interpretation of the results. It is necessary to distinguish the interpretation of chromatograms obtained from neat liquids from the ones obtained from fire debris samples. In the first case, the neat liquid is simply diluted and injected. As such there are almost no influences to take into account in the interpretation. In the second case, the debris is first subjected to an extraction (passive headspace, solvent, etc.) and then analyzed. In addition, interfering products are coextracted with

 http://dx.doi.org/10.1016/B978-0-12-802718-9.00025-9

ignitable liquid residues (ILRs). Thus, the interpretation of the chromatogram is much more complicated.

The goal of the interpretation of the results is to determine whether or not ILRs are present in the fire debris sample. In order to achieve this, one will have to study the chromatogram for patterns exhibited by known ignitable liquids (ILs).

Because thousands of different ILs with different components exist, a system of organizing them into groups and finding common patterns exhibited within each group had to be devised. This led to a classification system, now described in the ASTM standard test method for ILR in extracts from fire debris samples by gas chromatography–mass spectrometry (GC-MS) E1618. As a result, these patterns are well known and the process of interpretation is clearly described.

Classification

While one may think that hundreds of thousands of different ILs potentially used at fire scenes may exhibit as many different patterns, this is not the case. First of all, most ILs are petroleum based, that is, they are derived from crude oil. As such, most of them are composed exclusively of aliphatic and/or aromatic compounds. Second, because the processes of transforming crude oil into refined products are not very diverse, patterns exhibited by petroleum-based ILs can be placed into six different classes of ILs, each with a characteristic set of patterns. Finally, when dealing with nonpetroleum-based ILs, even though all possibilities are open, these liquids usually consist of only a few different components. Thus, the resulting chromatograms are not as complicated as those from petroleum-based ILs, which can exhibit several hundreds of components.

ASTM standard E1618 proposes a classification system based on eight different classes as shown in **Table 1**.

The examples for each class are not exhaustive; they are just the most commonly encountered products found in each particular class. In his/her findings, the fire debris analyst does not report a finished product itself, but rather the ASTM class. Then, products found in that class can be cited as examples, to guide the fire investigator.

Because the difference between the classes relies on the chemical composition of the IL, the ASTM system introduces a second dimension of classification to refine the different categories: the boiling point range. By using the boiling point range, one can refine the different classes and create more pertinent categories. In practice, the light subclass ranges approximately from 0 °C to 150 °C, the medium from 120 °C to 240 °C, and the heavy from 120 °C to more than 350 °C. The only exception to that subclassification is gasoline, whose

Table 1 ASTM E 1618–10 ignitable liquid classification scheme

Class	*Light (C_4–C_9)*	*Medium (C_8–C_{13})*	*Heavy (C_8–C_{20+})*
Gasoline	Fresh gasoline is typically in the range of C_4–C_{12}		
Petroleum distillates (including dearomatized)	Petroleum ether Some cigarette lighter fluids Some camping fuels	Some charcoal starters Some paint thinners Some dry cleaning solvents	Kerosene Diesel fuel Some jet fuels Some charcoal starters
Isoparaffinic products	Aviation gas Some specialty solvents	Some charcoal starters Some paint thinners Some copier toners	Some commercial specialty solvents
Naphthenic paraffinic products	Cyclohexane-based solvents/products	Some charcoal starters Some insecticide vehicles Some lamp oils	Some insecticide vehicles Some lamp oils Industrial solvents
Aromatic products	Some paint and varnish removers Some automotive parts cleaners Xylene-based products Toluene-based products	Some automotive part cleaners Specialty cleaning solvents Some insecticide vehicles Fuel additives	Some insecticide vehicles Industrial cleaning solvents
Normal alkane products	Solvents: pentane, hexane, heptane	Some candle oils Some copier toners	Some candle oils Carbonless forms Some copier toners
Oxygenated solvents	Alcohols Ketones Some lacquer thinners Fuel additives Surface preparation solvents	Some lacquer thinners Some industrial solvents Metal cleaners/gloss removers	
Others/miscellaneous	Single-component products Some blended products Some enamel reducers	Turpentine products Some blended products Some specialty products	Some blended products Some specialty products

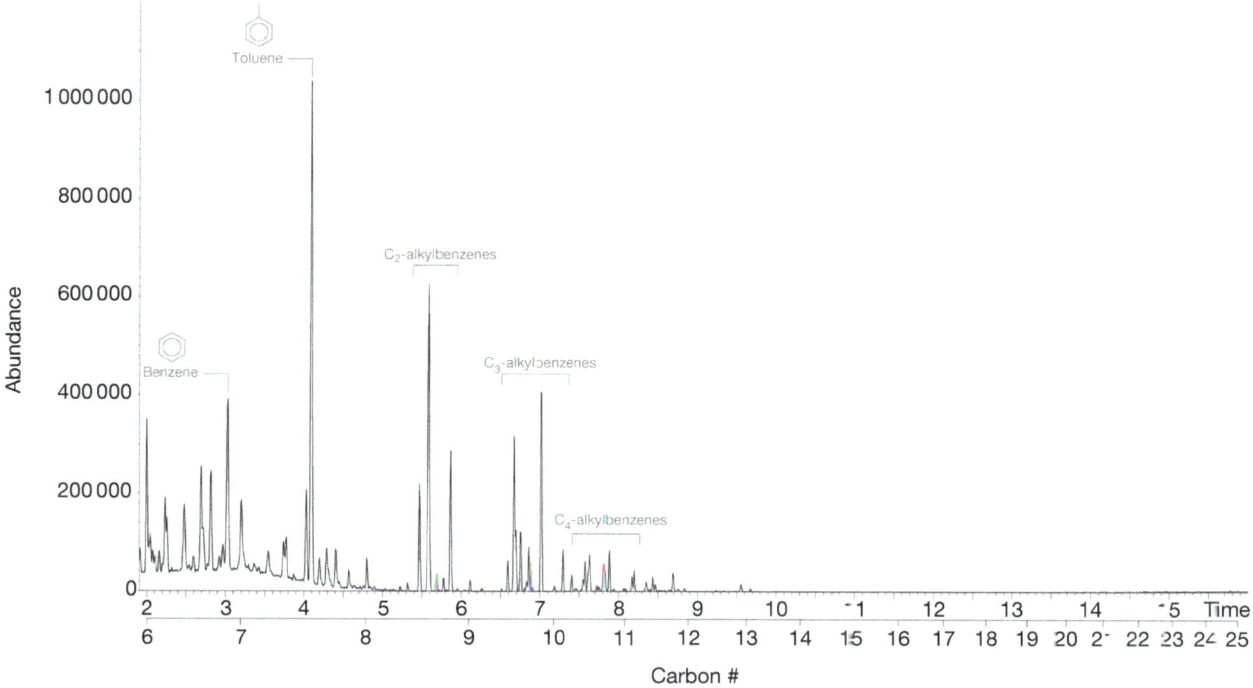

Figure 1 Chromatogram of a neat gasoline (unweathered). Time in minutes. Reproduced from Stauffer, E., Dolan, J.A., Newman, R., 2008. Fire Debris Analysis. Academic Press, Burlington, MA, p. 323. © Elsevier.

boiling point range does not vary greatly. This classification system works perfectly well with the separation and analysis obtained by GC-MS as this instrument separates the compounds based on their boiling points and the MS provides identification of their chemical nature.

Interpretation of Neat Liquids

With petroleum-based products, the basic principle of interpretation is to evaluate the chromatogram for the presence, distribution, boiling point range, and relative abundance of all saturated aliphatics and all aromatics. With nonpetroleum-based products, the analyst looks at all compounds present in the chromatogram and evaluates whether or not they could originate from an IL.

While an advanced knowledge of crude oil-refining processes is necessary to fully understand the reasons behind the chemical compositions of the different ASTM classes, this goes beyond the scope of this chapter. Nevertheless, it is possible to provide a rapid overview of the different classes and their compositions.

Gasoline is mostly composed of aromatic compounds ranging from C4 (4-carbon'long chain) to C12 (12-carbon'long chain). It also contains some alkanes; however, they are normally not abundant. **Figure 1** shows an example of a chromatogram of gasoline.

Petroleum distillates are the closest products to crude oil, as they have undergone a minimum of refinement. They contain both aliphatics and aromatics in a normal (Gaussian) distribution with spiking *n*-alkanes. Some petroleum distillates said to be dearomatized have no aromatic content. **Figure 2** shows examples of light, medium, and heavy petroleum distillates.

Isoparaffinic products are exclusively constituted of isoalkanes as shown in **Figure 3**. They have no aromatics, *n*-alkanes or cycloalkanes.

Naphthenic paraffinic products are comprised of cycloalkanes and isoalkanes (see **Figure 4**). Basically, a naphthenic paraffinic product is a petroleum distillate in which the *n*-alkanes and the aromatic content have been removed.

Aromatic products are composed exclusively of aromatic compounds. As a matter of fact, such a product is constituted of the aromatic fraction that was isolated from crude oil. In general, they exhibit a narrow boiling point range as shown in **Figure 5**.

n-Alkane products represent the simplest pattern: a narrow fraction of *n*-alkanes (usually not spanning more than four or five carbons). **Figure 6** shows an example of light, medium, and heavy *n*-alkane products.

Oxygenated solvents include all ILs containing at least one oxygenated compound in large excess of the rest of the components (at least one order of magnitude in the chromatogram). Oxygenated solvents may also contain other ILs

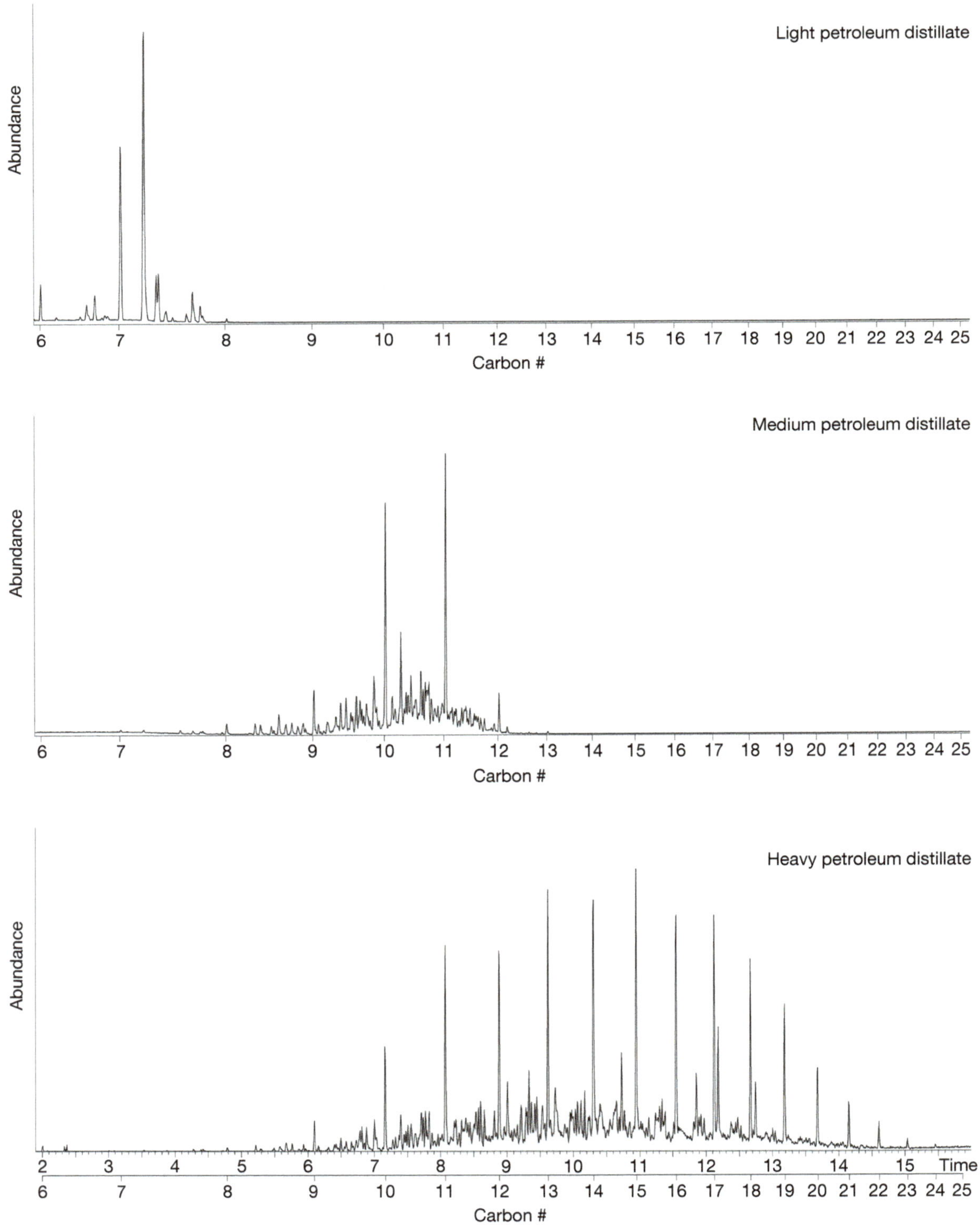

Figure 2 Chromatograms of light, medium, and heavy petroleum distillates. Time in minutes. Reproduced from Stauffer, E., Dolan, J.A., Newman, R., 2008. Fire Debris Analysis. Academic Press, Burlington, MA, p. 328. © Elsevier.

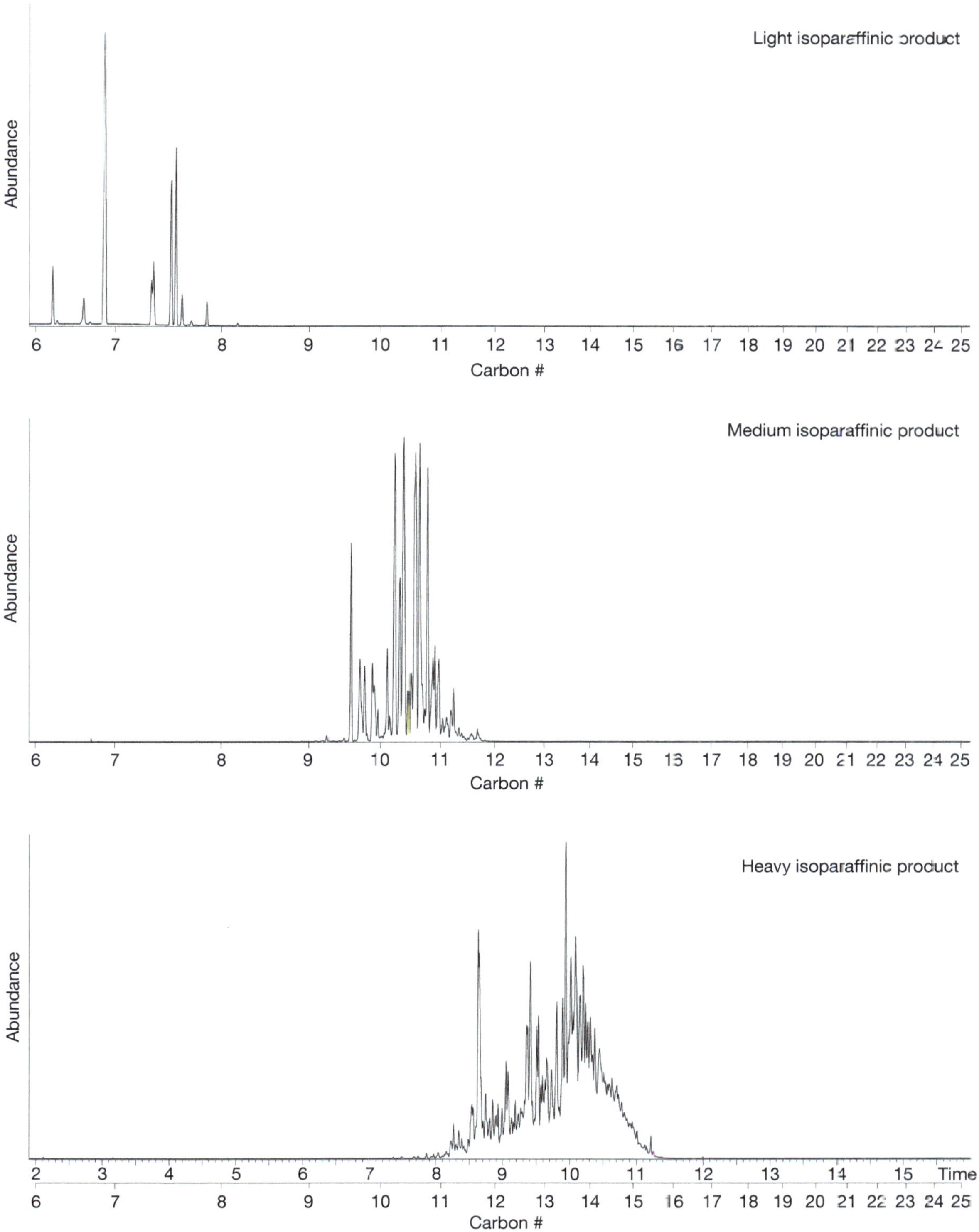

Figure 3 Chromatograms of light, medium, and heavy isoparaffinic products. Time in minutes. Reproduced from Stauffer, E., Dolan, J.A., Newman, R., 2008. Fire Debris Analysis. Academic Press, Burlington, MA, p. 332.© Elsevier.

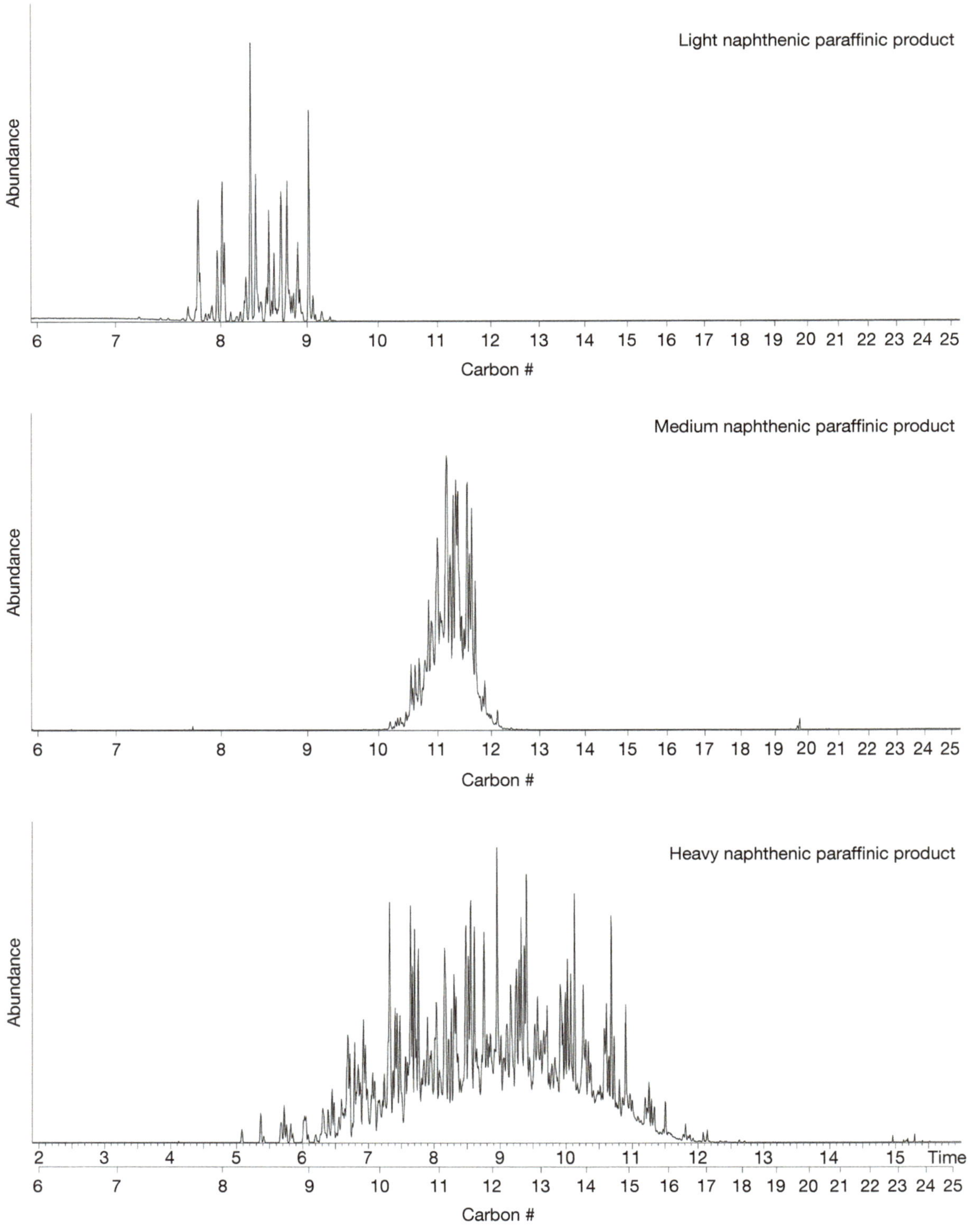

Figure 4 Chromatograms of light, medium, and heavy naphthenic paraffinic products. Time in minutes. Reproduced from Stauffer, E., Dolan, J.A., Newman, R., 2008. Fire Debris Analysis. Academic Press, Burlington, MA, p. 335. © Elsevier.

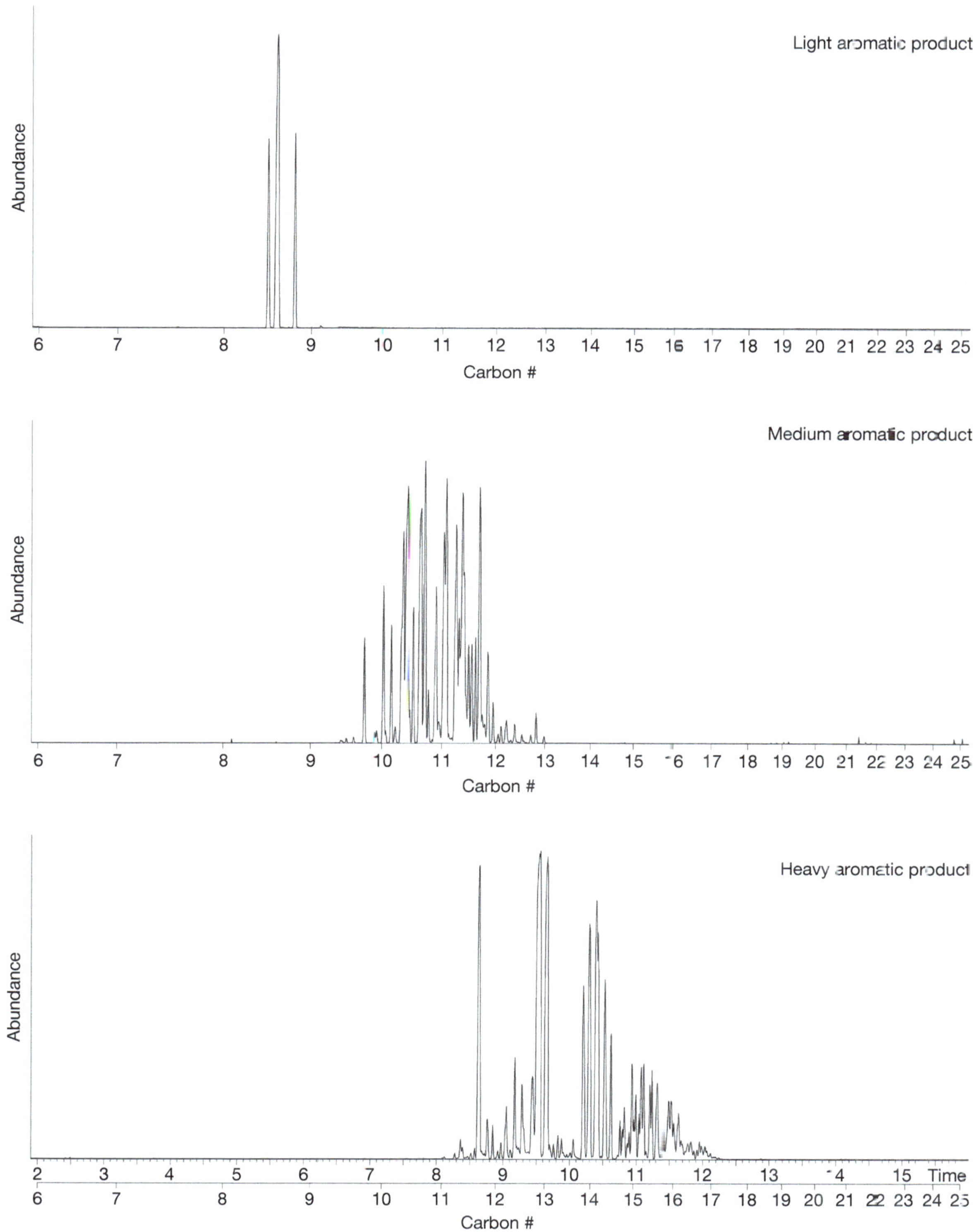

Figure 5 Chromatograms of light, medium, and heavy aromatic products. Time in minutes. Reproduced from Stauffer, E., Dolan, J.A., Newman, R., 2008. Fire Debris Analysis. Academic Press, Burlington, MA, p. 339. © Elsevier.

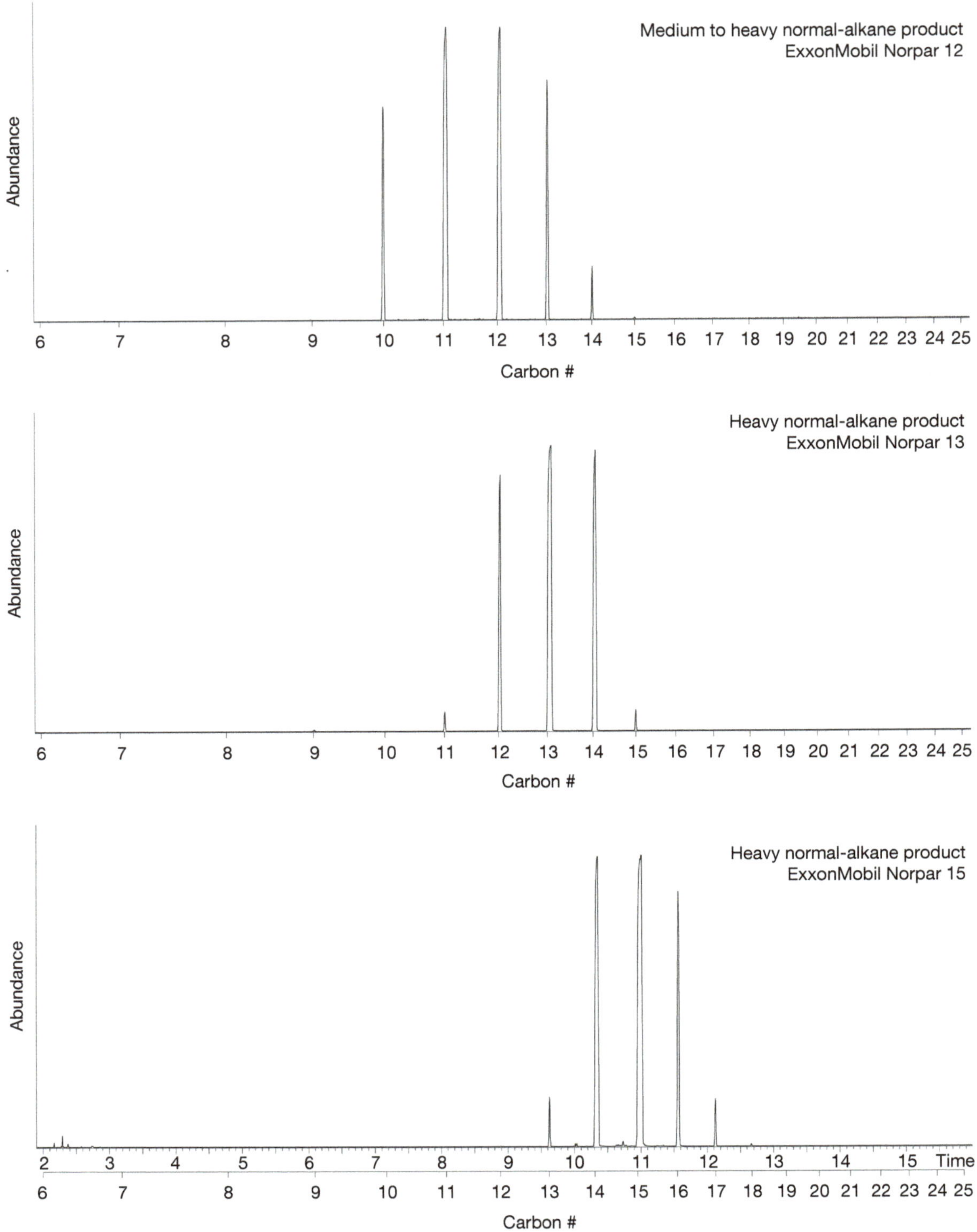

Figure 6 Chromatograms of light, medium, and heavy *n*-alkanes. Time in minutes. Reproduced from Stauffer, E., Dolan, J.A., Newman, R., 2008. Fire Debris Analysis. Academic Press, Burlington, MA, p. 337. © Elsevier.

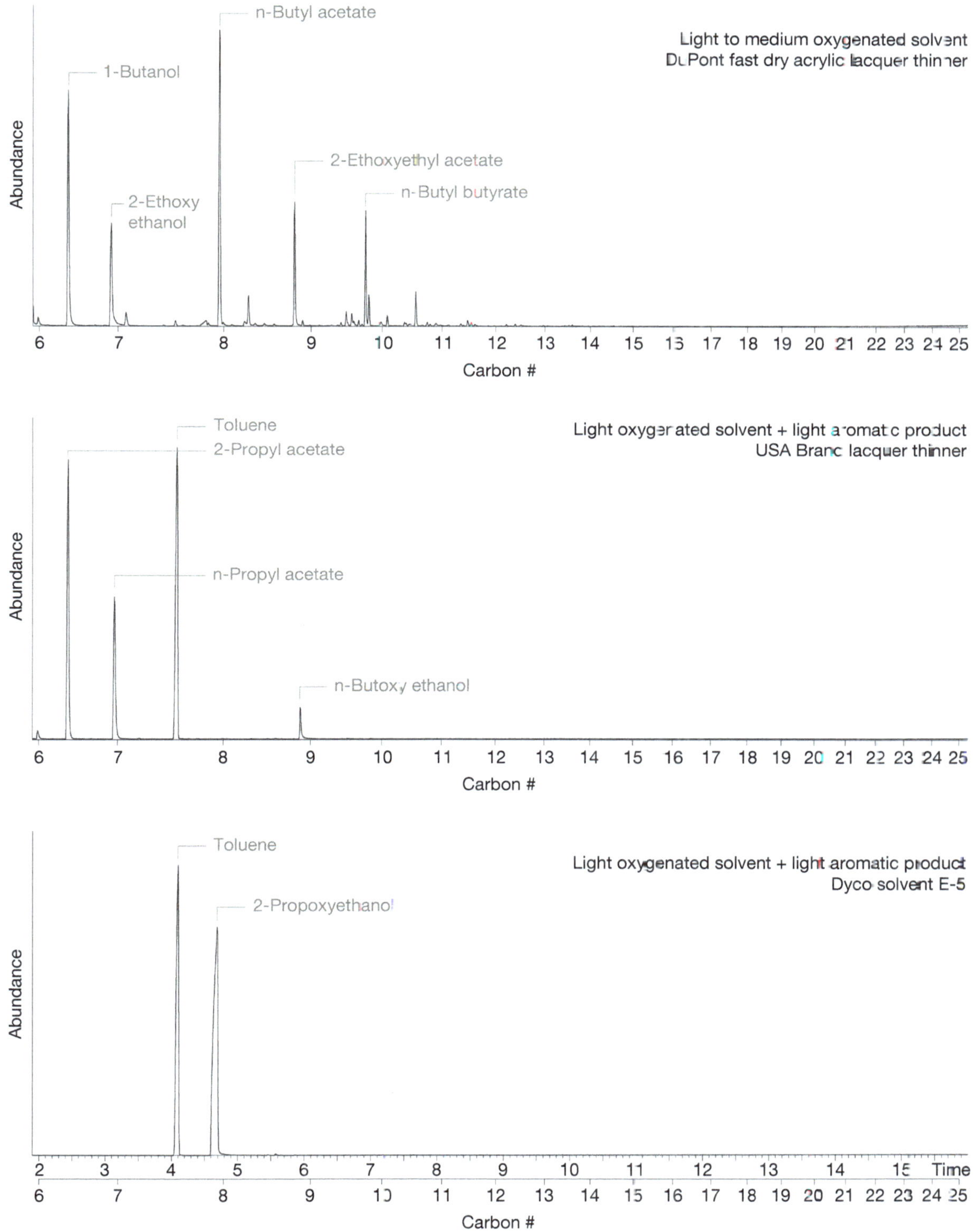

Figure 7 Chromatograms of light, medium, and heavy oxygenated solvents. Time in minutes. Reproduced from Stauffer, E., Dolan, J.A., Newman, R., 2008. Fire Debris Analysis. Academic Press, Burlington, MA, p. 342. © Elsevier.

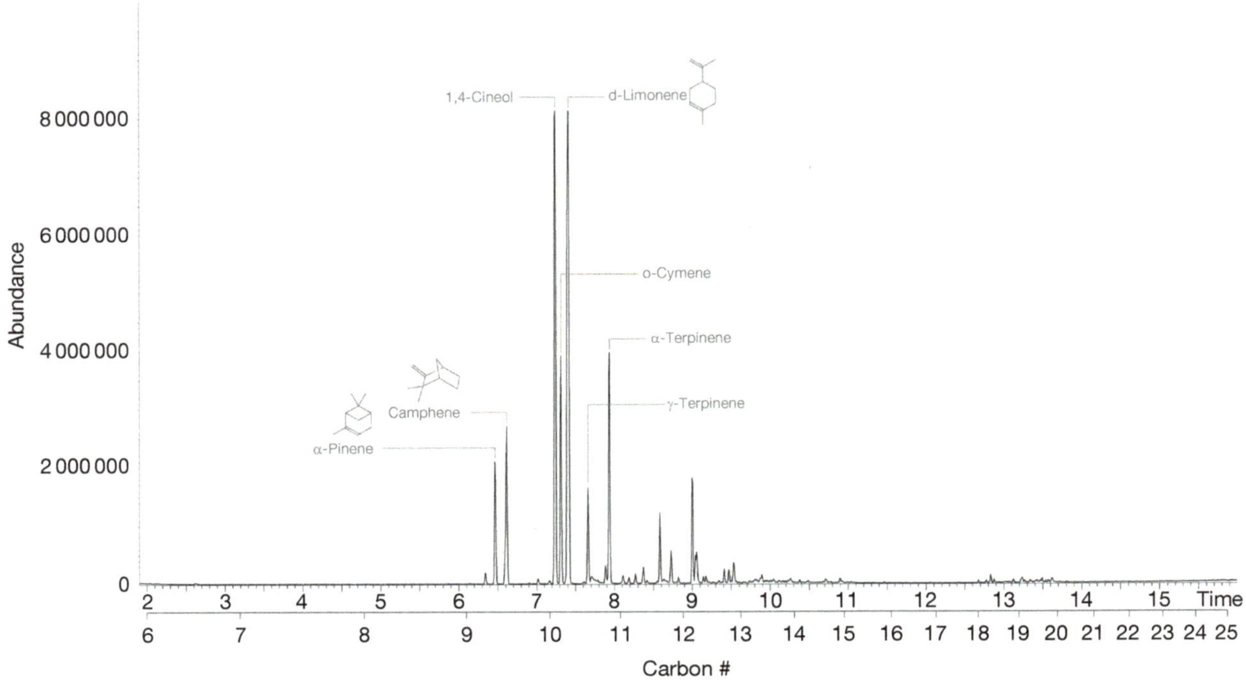

Figure 8 Chromatogram of turpentine product, classified as miscellaneous. Time in minutes. Reproduced from Stauffer, E., Dolan, J.A., Newman, R., 2008. Fire Debris Analysis. Academic Press, Burlington, MA, p. 344. © Elsevier.

such as medium petroleum distillates. **Figure 7** shows three examples of oxygenated solvents.

Miscellaneous products include ILs that do not fit in any of the categories previously described. **Figure 8** shows an example of a turpentine, which is classified as miscellaneous.

In summary, **Table 2** shows the different components found in each ASTM class and subclass.

While the interpretation and classification of chromatograms may appear relatively complex at first, it is in fact quite easy with neat liquids. **Figure 9** provides a guide to the proper

Table 2 The different components found in each ASTM class and subclass

Class	*Alkanes*	*Cycloalkanes*	*Aromatics (including indanes)*	*Polynuclear aromatics*
Gasoline	Present, less abundant than aromatics;	Present, less abundant than aromatics	Abundant	Present
Petroleum distillates	Abundant, normal (Gaussian) distribution	Present, less abundant than alkanes	Present, less abundant than alkanes (absent in dearomatized distillates)	Present (depending on boiling point range), less abundant than alkanes (absent in dearomatized distillates)
Isoparaffinic products	Branched alkanes abundant, *n*-alkanes absent or strongly diminished	Absent	Absent	Absent
Naphthenic paraffinic products	Branched alkanes abundant, *n*-alkanes absent or strongly diminished	Abundant	Absent	Absent
Aromatic products	Absent	Absent	Abundant	Abundant (depending on the boiling point range)
Normal alkanes products	Abundant	Absent	Absent	Absent
Oxygenated solvents	Composition may vary, presence of oxygenated organic compounds			

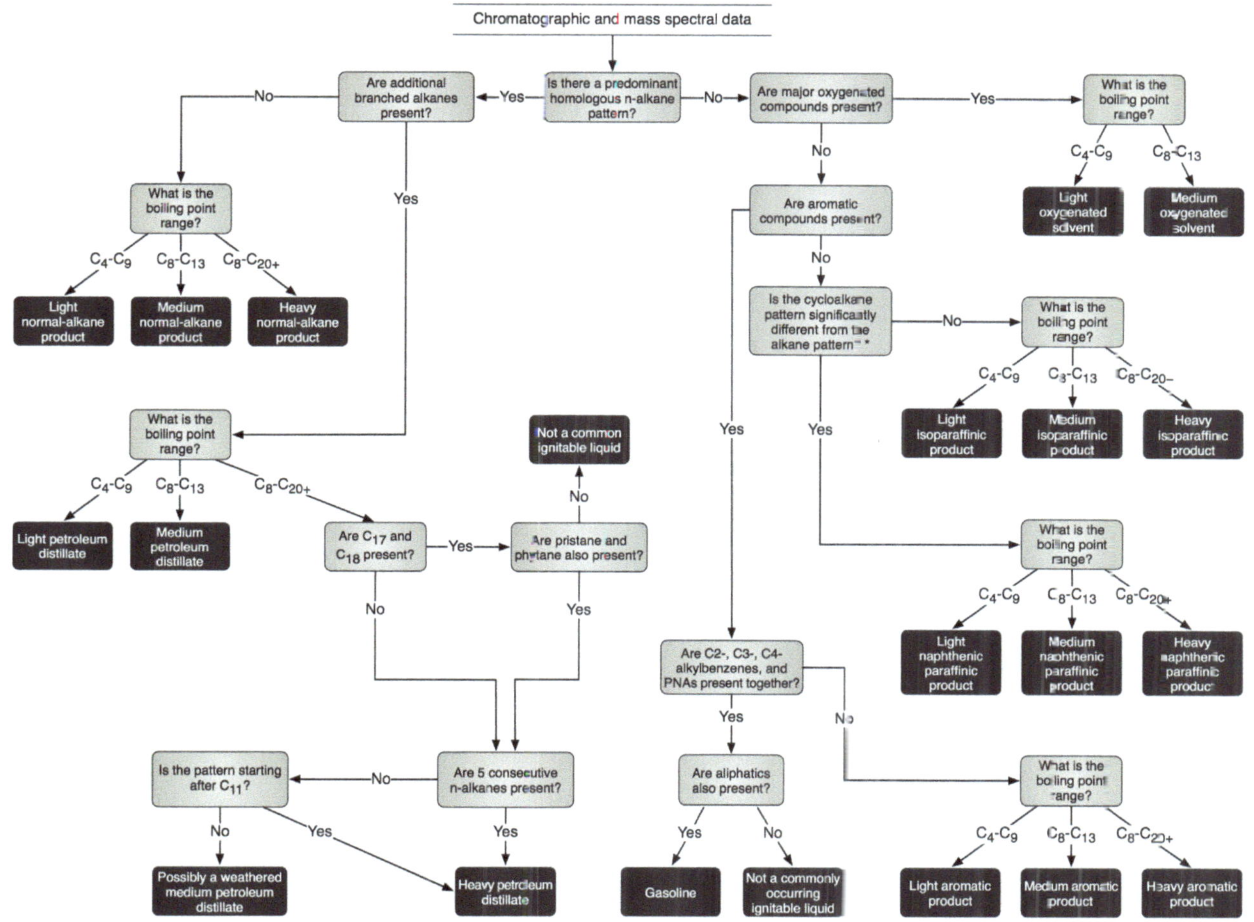

Figure 9 Petroleum-based ignitable liquid flow chart. Every question related to the presence of specific compounds implies that these compounds must be present in the proper pattern (as compared to a pattern of these compounds from a reference liquid analyzed on the same system). * Or "Are cycloalkanes distinctively present in the extracted ion chromatograms"? Reproduced from Stauffer, E., Dolan, J.A., Newman, R., 2003. Fire Debris Analysis. Academic Press, Burlington, MA, p. 345. © Elsevier.

interpretation of chromatograms. If the analyst follows this guide, there should be no problem in correctly identifying neat liquids. Unfortunately, it is a whole other story with ILR from fire debris samples.

Interpretation of ILRs

When an IL is poured onto a substrate, and then set on fire, extinguished, collected, and finally extracted, one can easily imagine that it no longer exhibits the same chromatographic pattern as when it was neat. This is the reason why the interpretation of ILR is much more complicated than that of mere IL, in addition to the fact that the analyst does not know at first whether or not ILR are present in the debris.

There are several parameters influencing the composition of the ILR extract from a fire debris as shown in **Figure 10**.

First, the substrate itself may already contain some IL or at least some compounds that are found in IL. These are called precursory products and they may be due to the raw material constituting the substrate, to its manufacturing process, to the setup in its final position/use, and to some natural or accidental contaminations. For example, some woods contain terpenes, compounds typically found in thinners. Most outsoles are glued to the shoe and this glue usually contains an aromatic compound, such as toluene. A carpet is also often glued to the floor; however, this glue usually contains a medium petroleum distillate. When one sprays insecticides around a baseboard, these become contaminated with a naphthenic paraffinic or an aromatic product. These are all examples of precursory products

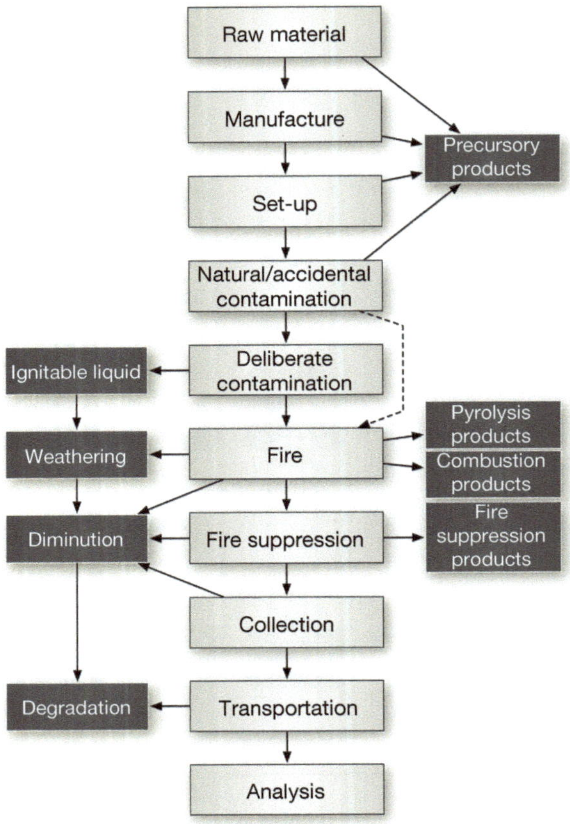

Figure 10 The different steps (in the middle) to which the fire debris sample is subjected from its creation to its analysis, along with the influences (on the left) on the potential ignitable liquid present in the debris and the different interfering products created (on the right). Reproduced from Stauffer, E., Dolan, J.A., Newman, R., 2008. Fire Debris Analysis. Academic Press, Burlington, MA, p. 443. © Elsevier.

that may be found on a substrate before it is even deliberately contaminated with an IL used as an accelerant.

Second, once an IL is poured onto a substrate, it will undergo three effects, which will influence its composition: weathering, diminution, and degradation. Weathering is the effect of the evaporation of an IL. Because an IL is commonly made of many different compounds of different boiling points, not all compounds will evaporate at the same rate. As such, the chromatographic pattern of a neat IL (unweathered) is different from that of its 50% evaporated version. As a matter of fact, as the weathering increases, the chromatogram moves to the right, meaning that the light compounds disappear and the heavy compounds become more and more dominant. Diminution represents the uniform loss of the different compounds of an IL. In practice, it occurs simultaneously with weathering, but it may also be due to poor evidence collection or to fire suppression activities. Finally, degradation occurs when the substrate, mostly soil, contains proteobacteria, which are capable of degrading petroleum-based IL. These bacteria, depending on their type, will selectively degrade aliphatics or aromatics. As a result, the composition of an IL may drastically change, not based on its boiling point range, but rather on its chemical characteristics.

Third, when a substrate burns, pyrolysis and combustion products are created. Pyrolysis products consist of compounds that are often the same as the ones found in petroleum products. As a result, they strongly interfere with the chemical composition of an IL, making it impossible in some instances to properly identify an ILR. Most commonly encountered pyrolysis products are toluene, styrene, naphthalene, benzaldehyde, ethylbenzene, indene, phenylethyne, *m,p*-xylenes, 1- and 2-methylnaphthalene, acetophenone, and the series of alkane–alkene–alkadiene ranging from C10 to C16. **Figure 11** shows an example of pyrolysis products created with burned polyethylene.

Combustion products usually do not interfere as much as pyrolysis products with ILR because they are oxidized products, which are not often found as IL components. Because they are very light compounds, they tend not to be trapped in substrates.

Finally, fire suppression agents may also be used by the intervening fire department. Some of these agents may also contribute to the fire debris extract, such as some foams that use D-limonene or some alcohol-based compounds.

Interfering products, then, are a set of products comprising precursory, pyrolysis, combustion, and fire suppression products. When interpreting chromatograms of extracts from fire debris samples, one must account for the presence of these interfering products. In addition, one must not forget the different effects occurring directly on the IL. This is why a systematic approach was developed.

Systematic Approach

Because the interpretation of chromatograms for ILR identification is quite complicated, it is important to follow a systematic approach, which is constituted of the six following steps:

1. Identify the sample and its substrate.
2. Estimate the typical contribution from that substrate.
3. Determine to which influences the substrate was subjected.
4. Estimate the effect of these influences.
5. Study the chromatogram from start to finish, including peak identification.
6. Study extracted ions in the regions of interest, including peak identification.

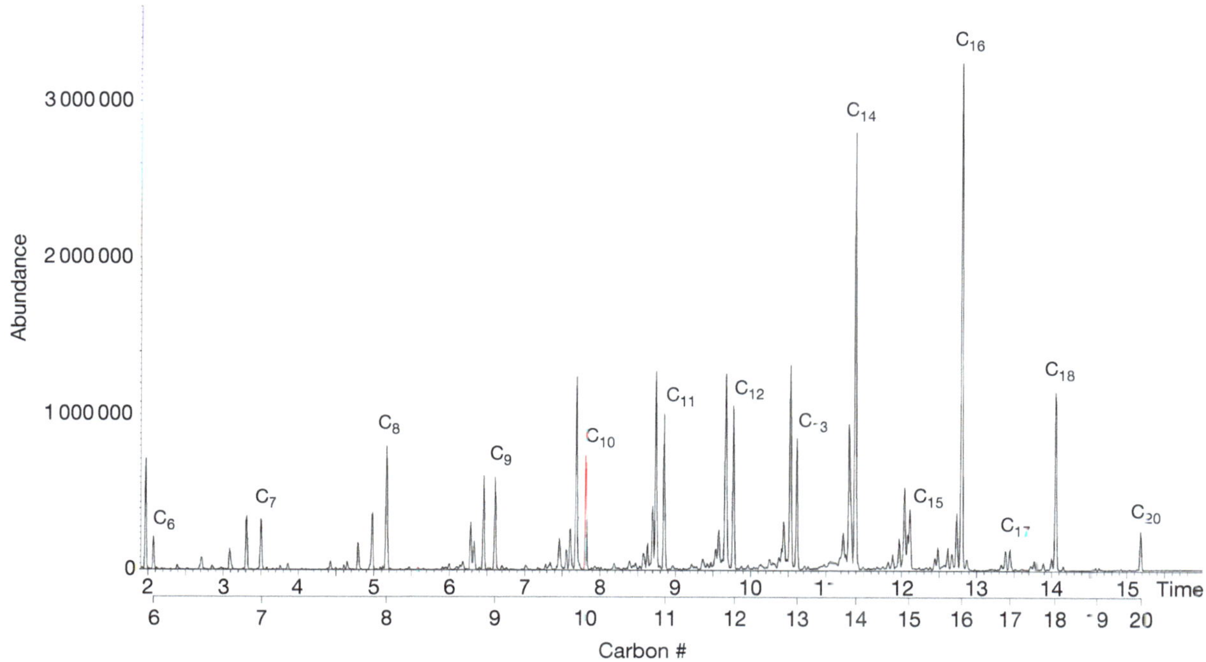

Figure 11 Chromatogram of polyethylene pyrolysis products. Time in minutes. Reproduced from Stauffer, E., Dolan, J.A., Newman, R., 2008. Fire Debris Analysis. Academic Press, Burlington, MA, p. 459. © Elsevier.

Even if the fire debris analyst did not work on the fire scene, he/she must have some clear basic knowledge about the sample in question, particularly in regard to its composition, its environment at the time of the fire, and the different steps it underwent. Knowledge of the sample's history is crucial, too. The preliminary examination of fire debris samples is, thus, a very crucial step that should never be undermined.

Significance of Findings

Fire debris analysis is an extremely complex science and the reason is twofold. First, the interpretation of chromatograms is rendered very difficult due to the numerous components of many different IL and the presence of interfering products. The second reason is that the simple presence of ILR in a debris does not imply at all that it was used as an accelerant in the fire. This last determination requires the experience of both the fire debris analyst and the fire investigator, as well as a very good knowledge of the fire scene and the circumstances surrounding the fire.

See also: **Chemistry/Trace/Fire Investigation:** Analysis of Fire Debris; Chemistry of Fire; Thermal Degradation; **Methods:** Gas Chromatography; Gas Chromatography–Mass Spectrometry; Mass Spectrometry.

Further Reading

ASTM International, 2010. In: ASTM E1618-10 Standard Test Method for Ignitable Liquid Residues in Extracts from Fire Debris Samples by Gas Chromatography-Mass Spectrometry, Annual Book of ASTM Standards 14.02. ASTM International, West Conshohocken, PA.

Byron, D.E., 2002. The effects of surfactants and microbes on the identification of ignitable liquids in fire debris analysis. Fire and Arson Investigator 53 (1), 50ss.

DeHaan, J.D., Brien, D.J., Large, R., 2004. Volatile organic compounds from the combustion of human and animal tissue. Science and Justice 44 (4), 223–236.

DeHaan, J.D., Icove, D.J. 2011. Kirk's Fire Investigation, seventh ed. Pearson Education, Upper Saddle River, NJ.

Gilbert, M.W., 1998. The use of individual extracted ion profiles versus summed extracted ion profiles in fire debris analysis. Journal of Forensic Sciences 43 (4), 871–876.

Leffle, W.L., 2000. Petroleum Refining in Nontechnical Language, third ed. PennWell Corporation, Tulsa, OK.

Lentini, J.J., Dolan, J.A., Cherry, C., 2000. The petroleum-laced background. Journal of Forensic Sciences 45 (5), 968–989.

Mann, D.C., Gresham, W.R., 1990. Microbial degradation of gasoline in soil. Journal of Forensic Sciences 35 (4), 913–923.

McGee, E., Lang, T.L., 2002. A study of the effects of a micelle encapsulator fire suppression agent on dynamic headspace analysis of fire debris samples. Journal of Forensic Sciences 47 (2), 267–274.

Newman, R., Gilbert, M., Lothridge, K., 1997. GC-MS Guide to Ignitable Liquids. CRC Press, Boca Raton, FL.

Nic Daéid, N., 2004. Fire Investigation. CRC Press, Boca Raton, FL.
Spreight, J.C., 1999. The Chemistry and Technology of Petroleum, third ed. Marcel Dekker, New York, NY.
Stauffer, E., 2003. Basic concept of pyrolysis for fire debris analysts. Science and Justice 43 (1), 29–40.
Stauffer, E., Dolan, J.A., Newman, R., 2008. Fire Debris Analysis. Academic Press, Burlington, MA.
Trimpe, M.A., 1991. Turpentine in arson analysis. Journal of Forensic Sciences 36 (4), 1059–1073.

Suspicious Deaths

J Horswell, Approved Forensics Sendirian Berhad, Selangor, Malaysia

Cause of Death

To understand suspicious death, the crime scene investigator should understand the various causes of death and the manner in which death may occur. The cause of death refers to the disease or injury responsible for initiating the chain of events that result in death. Some examples of causes of death include cancer, heart disease, pulmonary embolism, gunshot wound(s), stab wound(s), and asphyxiation by strangulation or hanging. They may be acute or chronic. The cause of death should reflect the first event that sets into motion a number of sequential events, such as broken bones from a blunt injury leading to immobilization resulting in pulmonary embolism, in which, the blunt trauma is the original cause, whereas the pulmonary embolism is the final cause of death. One would expect the cause of death to read "Cause of death = Pulmonary embolism due to blunt trauma injury to the skeletal system [here nominate the bones broken]."

Manner of Death

The manner of death relates to the circumstances in which the death occurred and is generally divided into five categories:

- "Natural" death is due to a natural process or disease.
- "Accidental" death is due to the unintentional or inadvertent actions of the deceased or another person.
- "Suicidal" death is due to an intentional act or acts of the deceased who anticipates his/her resultant death.
- "Homicidal" death is due to the direct action of another, which results in the death of a person.
- "Undetermined or unknown" death is where the surrounding circumstances cannot be determined with reasonable certainty.

It is important for the crime scene investigator to be knowledgeable about the death scene and for the forensic pathologist to attend the scene; however, this is not always possible, particularly, in rural or remote communities.

A heavy burden is therefore placed on the crime scene investigator who must have the knowledge and skills to understand fully the various causes of death and the manner in which the decadent met death. This can only be achieved by visiting as many death scenes as possible. Scenes provide important information that may indicate cause of, and means of, death. The environment at the scene may help explain the condition of the body when it was discovered. If the deceased died while lying close to a heater, the body will decompose faster than usual. If the deceased collapsed onto an irregular or rough surface, this may explain minor injuries that are found on the body. Postmortem injuries should not be confused with antemortem injuries. Cause of death may be obvious, for example, a gunshot wound to the head; however, to determine the manner of death, further information will be required, such as whether this was an accident, suicide, or homicide.

Death scenes should be treated as homicide scenes and dealt with as such until the contrary is proved. This can only be achieved through detailed examination and investigation.

There are now two distinct tasks ahead of the crime scene investigator. The first is the technical recording and retrieval of potential physical evidence from the scene. Just as important is the second, the reconstruction in the mind of the crime scene investigator of the events and circumstances surrounding the death. This chapter will discuss the technical issues first and the reconstruction issues second.

Death Scene

When violent or suspicious death is discovered the death scene should be secured immediately so that no one will have an opportunity to change it. Indoor scenes may be easy to secure and protect. Outdoor scenes can be difficult. The more urban the scene, the more difficult it is to secure, as there will be a need for several scene guards and the setting up of cordons. The more remote the scene, the easier it is to secure. The weather and movement of predatory animals through the scene common approach paths, and the maintenance of a scene log to record visits to the scene, add a dimension to the processing of a crime scene and the condition in which the deceased may be found. Outdoor scenes may be cooler than indoor scenes, and the weather can have an impact as can the movement through the scene by animals. Any disturbances must be recorded.

The crime scene investigator should take photographs and video recording of the scene immediately before anything is moved. "Stepping plates" may be used to secure a route to the body preserving any evidence that may be under each plate for later recovery as each plate is removed. Notes and a sketch should be made at this time. The deceased's location relative to other objects and structures within the scene is very important. The actual position of the deceased is plotted: the head and crutch are good points on the body to use for plotting its position. Accurate measurements of the deceased relative to other potential physical evidence and the scene fixtures must be made to place items appropriately within the sketch made at the scene. Great care should also be taken to preserve any footwear marks or other significant evidence types around the body.

The deceased is the most valuable piece of potential evidence at any death scene. Hence, the systematic and thorough examination of the body should be undertaken at the scene. Blood, weather conditions, location, and poor lighting may mask faint injuries and trace evidence on the body; therefore, the crime scene investigator should document in writing, by sketch, and by photographic recording all information about the body that can be gathered at the scene. The remainder will have to wait until the postmortem examination. The attendance of other experts at the scene may also be considered, for example, biologist, entomologist.

The environment in which the body was found will affect the rate of body cooling. The wind conditions, the temperature, and the presence of rain should be noted. The crime scene investigator will need to develop a general description of the deceased, including, where possible, the gender, race, age, height, and weight.

One of the first most important questions that will require an answer is, did death occur at this locus? The position in which the deceased was discovered is of particular importance, as it will provide an indication as to whether or not the body was moved before being discovered. The presence or absence of rigor mortis or stiffness of the body, whether it is absent, minimal, moderate, advanced, or complete, will help the crime scene investigator and the pathologist determine if the person had died at that locus, in the position, as found. The crime scene investigator should inspect the body for postmortem lividity, hypostasis, or livor mortis. A pink-purple discoloration is usually present at the lowest point of the body. This is due to the settling of the blood by gravitation. Its location and state of fixation should be noted and photographed. For example, unfixed livor mortis blanches white when moderate pressure is applied as opposed to fixed livor mortis, which remains the same color when pressure is applied. If livor mortis is noted on the deceased in areas not consistent with it, forming in the lowest parts of the body, the crime scene investigator should consider the possibility that the deceased was moved after death.

The blood flow patterns should match the position of the body. If the scene is one of apparent violence, the blood flow or splash patterns can indicate the use of a weapon and in some cases indicate the type of weapon.

Is the trace evidence at the scene consistent with the death having occurred at this locus? Does the body contain any trace evidence that is unusual for this locus, for example, mud on soles of shoes and/or grass and/or seed material found on the clothing when the body has been located inside a house? Is the death one that can be attributed to natural causes? Is there any external sign of violence? Is there anything about the scene that arouses the crime scene investigator's suspicions?

The crime scene investigator should consider several hypotheses and then set out to prove or disprove each one. The physical evidence present, together with the known facts, should be sufficient to enable the crime scene investigator to develop a reasonable hypothesis as to what has happened at the scene; however, this is not always possible. Suspicions may not be roused until the postmortem examination reveals something that was not apparent at the scene. The pathologist may, however, provide the investigating officer with a definitive indication as to the cause of death, giving investigators lead(s) to start their lines of inquiry.

Before the deceased is moved from the scene, the crime scene investigator should begin the examination of the body systematically, locating potential physical evidence that may be lost during transportation, noting and photographing trauma before removing any identified potential physical evidence, and packaging each item separately. The taking of body swabs and tape lifts may be considered at this stage. The deceased's hands, feet, and head should be bagged using paper bags, as the use of plastic and any subsequent refrigeration will cause plastic bags to sweat. The bags should be large enough to be taped securely around the wrist, ankle, or neck and should allow "ballooning" over the area to be protected. The body should then be placed into a new, clean body bag. If face down, it should not be turned over during this process so as to avoid "flooding" wounds and thus damaging evidence. A thorough examination of the area that was previously obscured by the body may now be conducted. The deceased should then be transported to the mortuary for a full postmortem examination. This is then the realm of the forensic pathologist.

Postmortem Examination

The postmortem examination is normally carried out by an experienced forensic pathologist. The crime scene investigator should be present, as should the investigating officer or his/her delegate.

The deceased should be taken from the body bag and all packaging items should be retained for a subsequent search for trace evidence in the forensic science laboratory. The crime scene investigator should then record photographically the clothing and the condition of the body and take a facial photograph for use in identification processes. The forensic pathologist should be taking notes and making diagrams of the overall body and any item of evidence or wound(s) or external markings located.

The clothing should be inspected and any obvious trace evidence removed and then the item should be packaged. Each item of clothing should be removed and packaged separately. It is at this time that wallets and property should be removed from the deceased's pockets and the contents noted and the items packaged separately. It may well be that these items are of no forensic value, as such, and thus may be released into the custody of the investigating officer. All items of clothing should remain in the custody of the crime scene investigator.

An external examination should be conducted next. Questions may already exist that need answers. Are there any marks of violence, such as stab or gunshot wounds, ligature or restraint marks, injection sites, or blunt injury wounds? Are these injuries incompatible with life, or could the person have lived and moved for a time after sustaining the injuries, thus explaining some of the evidence located at the scene? Could the deceased have inflicted these injuries? Were there other persons involved? Do the injuries appear to have been made antemortem, at the time of death, or postmortem? Bleeding into the tissues (bruising) usually means that the deceased was alive at the time of the injury. Are there any injuries that could be regarded as "defense" wounds? These would indicate that the deceased was alive, alert, and aware that she/he was being assaulted and aware of the life-threatening nature of the assault.

The crime scene investigator should search the body using oblique lighting. This is a method highlighting minute particulate matter and/or bruising that, under normal lighting, is difficult to see.

At this stage of proceedings, the forensic pathologist should take fingernail scrapings, with each digit's scrapings being placed into a separate and labeled container. The next sample to be obtained is the hair samples, which should be recovered from a site of interest and individually packaged in labeled containers. These samples should remain in the custody of the crime scene investigator. A complete series of radiographs should then be taken. These may reveal broken bones, projectiles, foreign bodies, prostheses that have been inserted into the person while alive and major deformities.

Identification may be an issue. Identification is important in both civil and criminal law. It provides proof of the person's death. It allows for

- notification of next of kin,
- completion of the death certificate,
- eventual disposition of the remains,
- settlement of the deceased's estate, and
- identification in criminal proceedings.

Visual identification, although the most frequently used method of identification, is subjective and can be unreliable, as many people bear a resemblance to each other. The person making the identification can also be so shocked by seeing the deceased that she/he can make a false identification owing to being in a state of denial. Forensic identification techniques should also be employed. These include the following:

- A general description of the deceased: color of hair and eyes, height, the presence of scars (accidents or operations), marks, and tattoos.
- A general description of the clothing: type, color, and fabric of each item.
- A forensic odontologist should be requested to X-ray, chart, and make any casts necessary of the deceased's teeth prior to the postmortem examination begins. These may be later used for comparison purposes.
- DNA samples (see below).
- Contents of the deceased's wallet.
- A medical records search for any major operation, broken bones, or the addition of any prosthesis fitted during life (if a name can be suggested for the deceased).
- The last issue to be dealt with before the deceased is released is that she/he be fingerprinted. A fingerprint technician should attend the mortuary and take inked impressions from each digit and area where there are ridge patterns on the deceased's hands. These fingerprints can later be used to search through criminal and civil fingerprint databases such as the National Automated Fingerprint Identification System (NAFIS), and the digits can be used to identify the deceased.

The postmortem examination continues with items of interest being recorded by the pathologist in his/her notes and diagrams and photographically recorded by the crime scene investigator, who should use a scale rule when taking any close-up photographs. It is also at this stage of the postmortem examination that samples should be taken by the pathologist: blood for DNA profiling, alcohol determination, and drug and poison screens. Samples should also be taken for histopathology and microbiology if the forensic pathologist thinks that these are necessary.

Violent crimes typically result in a great deal of physical evidence being deposited on the deceased and the suspect and left at the scene. A thorough search should be made, as well, of the living suspect and his/her environment, for any evidence, which may be expected to be extremely transient and easily destroyed or contaminated. Depending on the type of crime, for example, shooting, stabbing, rape/murder, or blunt injury beating, different forms of evidence may be present. Additionally, the specific area on the body in which evidence is found is of extreme importance; for example, dried semen in the pubic hair of a deceased female or firearm discharge residue on and around the entry point to the chest is indicative of a contact wound. Consideration may also be given to other experts attending the postmortem examination (e.g., firearms expert).

Summary

After all the information is to hand from the crime scene and the postmortem examination, those involved should be able to work out the cause and manner of death from the facts available. In the interpretation of the cause of death, those involved must identify the initial factor(s) that may have led to a subsequent cause of death. For example, pulmonary edema can and will cause death; however, it may not be the root cause of death. Another set of circumstances, such as receiving severe burns in an explosion, may have occurred in the first instance, which led to the onset of pulmonary edema and the deceased's ultimate death.

Although modern forensic investigation is advanced, there will be times when the crime scene does not provide information and the postmortem examination does not reveal a definitive cause of death. These are the difficult cases.

See also: **Forensic Medicine/Causes of Death:** Sudden Natural Death; **Forensic Medicine/Pathology:** Early and Late Postmortem Changes; **Foundations:** Overview and Meaning of Identification/Individualization.

Further Reading

Bevel, T., Gardner, R.M., 2008. Bloodstain Pattern Analysis, with an Introduction to Crime Scene Reconstruction, third ed. CRC Press, Boca Raton, FL.

Brogdon, B.G., 1998. Forensic Radiology. CRC Press, Boca Raton, FL.

Brogdon, B.G., Vogel, H., McDowell, J.D., 2003. A Radiological Atlas of Abuse, Torture, Terrorism, and Inflicted Trauma. CRC Press, Boca Raton, FL.

Di Maio, V.J., Di Maio, D., 2001. Forensic Pathology, second ed. CRC Press, Boca Raton, FL.

Dix, J., 2000. Colour Atlas of Forensic Pathology. CRC Press, Boca Raton, FL.

Dix, J., Calaluce, R., 1998. Guide to Forensic Pathology. CRC Press, Boca Raton, FL.

Dix, J., Ernst, M.F., 1999. Handbook for Death Scene Investigations. CRC Press, Boca Raton, FL.

Dix, J., Graham, M., 2000. Time of Death, Decomposition and Identification: An Atlas. CRC Press, Boca Raton, FL.

Fisher, B.A.J., 2004. Techniques of Crime Scene Investigation, seventh ed. CRC Press, Boca Raton, FL.

Gebeth, V., 1983. Practical Homicide Investigation: Tactics, Procedures, and Forensic Techniques. Elsevier, Amsterdam.

Hanzlick, R., 2007. Death Investigation Systems and Procedures. CRC Press, Boca Raton, FL.

Horswell, J. (Ed.), 2004. The Practice of Crime Scene Investigation. CRC Press, Boca Raton, FL.

Ludwig, J., 2002. Handbook of Autopsy Practice, third ed. Humana Press, Totowa, NJ.

National Medicolegal Review Panel, 1999. Death Investigation: A Guide for the Scene Investigation – Every Scene Every Time. US National Institute of Justice, Washington, DC.

Pomara, C., Kaarch, S.B., Fineschi, V. (Eds.), 2010. Forensic Autopsy: A Handbook and Atlas. CRC Press, Boca Raton, FL.

Thali, M.J., Dirnhofer, R., Vock, P., 2009. The Virtopsy Approach, 3D Optical and Radiological Scanning ND Reconstruction in Forensic Medicine. CRC Press, Boca Raton, FL.

Wagner, S.A., 2004. Colour Atlas of the Autopsy. CRC Press, Boca Raton, FL.

Wagner, S., 2009. Death Scene Investigation, a Field Guide. CRC Press, Boca Raton, FL.

White, P. (Ed.), 2010. Crime Scene to Court – The Essentials of Forensic Science. Royal Society of Chemistry, Cambridge.

Youngner, S.J., Arnold, R.M., Schapiro, R., 1999. The Definition of Death: Contemporary Controversies. The Johns Hopkins University Press, Baltimore.

Key Terms

Accelerant, Accelerant detection, Activation energy, Air quality standards, Aliphatics, Analytical methods, Aromatics, Arson, ASTM, Burning, Canines, Carbon dioxide, Chemical analysis, Chemical Engineering, Chemicals, Combustible, Combustion, Compartment fires, Contamination, Corrosion, Death scene examination, Evidence collection, Exothermic, Explosions, Fire, Fire debris, Fire dynamics, Fire investigation, Fire modeling, Fire patterns, Fire tetrahedron, Fire triangle, Flames, Flammable, Flashover, Fuel, Gasoline, GC-MS, Hazards, Heat, Heat release, Human remains, Ignitable liquid, Ignitable liquid residues, Ignition, Incendiary, Interfering products, Isoparaffinic, Labeling, Manner and cause of death, Microbial degradation, Naphthenic paraffinic, Normal alkanes, Origin determination, Oxidation, Oxidizer, Packaging, Petroleum distillates, Plume, Pollution, Polymeric material, Postmortem examinations, Pyrolysis, Pyrolysis products, Safety, Smoldering, Sniffers, Spalling, Spark, Spontaneous combustion, Substrate, Suspicious deaths, Thermal, Thermal degradation, Thermodynamics, Toxic substance, Trace evidence, Ventilation, Venting, Wood.

Review Questions

1. What is fire? What kind of a phenomenon is it? Why is it orange?
2. What is required for fire? What is the "fire triangle"? How does it differ from a "fire tetrahedron"?
3. What is backdraft? What is flash point? What is flashover? How do these relate to the analysis of fire debris?
4. How does ignition occur?
5. Define thermal degradation. Give three examples of materials and how they degrade thermally.
6. List the classes of fires and their definitions.
7. Why is random sampling at fire scenes not advisable?

8. List the main types of packaging for fire debris samples at a scene, explaining the advantages and disadvantages of each.
9. What is the most important task in a fire investigation? Why?
10. What is the determination of cause in a fire investigation and what is required to determine it?
11. What is "spalling"? What is a "pour pattern"? What is a "clean burn"?
12. List the ways of sampling for fire debris for ignitable liquid residues. What are the strengths and weaknesses of each?
13. What are some of the artifacts that can interfere with the interpretation of ignitable liquid residue results?
14. Why don't the thousands of ignitable liquids available produce thousands of analytical results? Why are only a few "patterns" seen? What are those classes and how are they defined?
15. How should neat samples be handled differently from debris samples?
16. What is the difference between a fire and an explosion?
17. List the types of ignition sources and define them.
18. What is the difference between fire and pyrolysis?
19. What is the difference between cause and manner of death? Whose responsibility is it to determine cause and manner?
20. What is the most valuable piece of potential evidence at a death scene?

Discussion Questions

1. Why does the environment determine the body's rate of cooling? How is this affected by a fire? How would you determine time since death in a fire-related scene, assuming the body was burned?
2. Describe several myths of fire investigation. How have they been shown to be wrong? What are the implications of these myths in the criminal justice system?
3. Why is the scientific method applicable to fire scene investigations? How should this change the approach of the investigation?
4. How do the scene and laboratory approaches vary between criminal investigation and engineering investigation (the article by Perlmutter in this section)? Why would they differ?
5. A fire scene involves a utility shed and its contents, including typical gardening and lawn care materials and gas-powered equipment (lawn mower). The investigator is called out at 10 p.m. to conduct the scene investigation. What should the investigator's strategy and planning be? What resources are required?

Additional Readings

St. Pierre, K., Desiderio, V., Hall, A., 2014. Recovery of oxygenated ignitable liquids by zeolites, part I: novel extraction methodology in fire debris analysis. Forensic Science International 240.

Waddell, E., Song, E., Rinke, C., Williams, M., Sigman, M., 2013. Progress toward the determination of correct classification rates in fire debris analysis. Journal of Forensic Sciences 58 (4), 887–896.

Section 5. Other Methods

> Successful engineering is all about understanding how things break or fail. Henry Petroski.

The profession of engineering is very broad and covers a wide range of specialized fields, each with a particular emphasis on a physical phenomenon, technology, or process. From the Roman aqueducts to the Phoenix Mars Lander, engineering encompasses a staggering breadth of knowledge and application. Some of this makes its way into forensic investigations. Audio recordings, slip-and-fall incidents, structural collapses, and industrial incidents may not seem to be connected but they could all be the subject of a forensic engineering investigation.

Audio Enhancement and Authentication

C Grigoras and JM Smith, University of Colorado Denver, Denver, CO, USA

Glossary

Artifact A visual/aural aberration in an image, video, or audio recording resulting from a technical or operational limitation. For example, speckles in a scanned picture, "blocking" in JPEG compressed images, unnatural "birdy" noises as a result of MP3 audio compression.

Audio enhancement Processing of recordings for the purpose of increased intelligibility, attenuation of noise, improvement of understanding the recorded material, and/or improvement of quality or ease of hearing.

Dynamic range The ratio of the strongest nondistorted signal to that of the weakest discernible signal in a unit or system as expressed in decibels (dB); a way of stating the maximum signal-to-noise ratio.

Forensic audio A subdiscipline of digital and multimedia evidence, which involves the scientific examination, analysis, comparison, and/or evaluation of audio.

Media Physical objects on which data can be stored.

Media authentication The process of substantiating that the data in an image, video, or audio recording is an accurate representation of what it purports to be.

Multimedia evidence Analog or digital media, including, but not limited to, film, tape, magnetic and optical media, and/or the information contained therein.

Introduction

Audio forensics is a discipline within the digital and multimedia evidence (DME) field of forensics. DME includes digital evidence such as computers and mobile phones as well as recorded evidence in the form of audio, video, or still images. This confluence comes as a result of the ubiquity of digital technology and the commonality of media used to store digital data. For example, an external hard drive that is collected at a crime scene may contain audio recordings relevant to a case while the digital video recorder-based security system in the next room recorded audio and video to a hard disk drive during the events being investigated. Or, a cell phone found on a suspect may have been used to record voice notes and videos

 http://dx.doi.org/10.1016/B978-0-12-802718-9.00027-2

all related to a case under investigation. Because these items of evidence coexist in a digital realm that is not tangible as most other forensic sciences are known to be, the necessities of handling digital evidence must be respected throughout the DME discipline.

Audio evidence is commonly found during the course of investigation when circumstances such as these arise: a victim records audio for their protection, perpetrators use recordings to carry out their crimes, or phone call logging systems such as emergency assistance record a violent crime that occurred during a call. Similarly, investigators employ audio recording technology in covert operations or with a cooperative subject during an interview. The possible scenarios are too numerous to list here, but it is obvious that recorded audio evidence can be an important part of both criminal and civil proceedings where recordings of speech or other acoustic events can be very useful.

Because most forensic audio recordings are made in nonideal environments and the circumstances under which they are made are hard to control, quality, and intelligibility often suffer. The details that have been recorded are often hard to understand or not clear enough to make confident decisions. This is when it may be necessary to enhance or clarify the recorded audio by removing interfering noises and distortions or adjusting the amplitude of low-level information. Likewise, the authenticity of audio recordings must be established if they are to be given any weight in criminal prosecution or in the settlement of a civil dispute. This is necessary in order for the Trier of Fact to be assured that the recorded information accurately represents the events under investigation. A brief background on the topics of forensic audio enhancement and authenticity is provided in the sections below.

Technology

There are many elements to consider when presented with audio evidence before processing or conducting an analysis. Chiefly, is the recorded audio analog or digital? If dealing with analog recorded sound, it is imperative to understand analog media in order to choose the proper playback equipment. Once identified, all equipment used for the playback, analysis, and digitization of analog material should be properly maintained and calibrated according to the manufacturer's specifications. If working with recorded audio that is digital, it is no less imperative to understand the equipment and media used to record the audio digitally. It is also crucial to understand the variety of digital audio formats that are available and the difference between those that are native uncompressed and those that are compressed using either lossy or lossless compression algorithms. While an in-depth description of these principles will not be presented here, the following outline will give one a basic understanding.

Sound can be described as the compression and rarefaction of a medium (in most cases air) as a result of an acoustic impulse or vibration. In a relevant example, one can imagine the human mechanism that produces speech where air is forced through the larynx by the lungs causing the vocal folds to vibrate. The periodicity of this vibration produces sound with a perceived pitch while the modulation of this sound by the tongue and teeth establish language communication. Further modulation to the sound is caused by the acoustic space in which the sound is made by either absorbing or reflecting certain frequencies lending greater or lesser energy to the speech impulses. The acoustic space can obviously vary; it could be an office, auditorium, stairwell, or in an extreme example an anechoic chamber which absorbs all frequencies of sound resulting in reverberant energy close to zero. Acoustic impulses such as speech are fleeting and unless recorded, last no longer than a fraction of a second.

The ability to record and playback sound has been around since Thomas Edison perfected the phonograph in 1878. The principles of sound have obviously not changed since this time but methods to record sound have undergone many technical advances and every day we see innovations in sound recording technology. In the broadest and most basic sense, recording technology can be summarized as a microphone that acts as a transducer converting the acoustic energy into an electrical signal, which may be recorded on some sort of medium. Once an acoustic sound is represented in this way, it may be referred to as audio and recorded in one of two ways: as an analog or a digital signal. Although becoming more and more obsolete, the media used to record analog audio is most commonly in the form of a cassette or microcassette tape. On tape-recorded media such as this, the analog waveform is recorded as varying fluctuations of magnetic energy representing the original compression and rarefaction of sound. When played back, variation of the stored magnetic flux is converted back to an electrical signal and, in a process opposite that of a microphone transducer, the speaker driver acts as a transducer converting electricity back into a series of acoustic impulses. While digital audio may be played back in a similar fashion, it is recorded onto media as a representation of the acoustic waveform as a finite set of binary values, or samples. In this case, sound is quantized at a specific sample rate and bit depth. The sample rate refers to the number of samples per second used to quantize the waveform and the bit depth is the number of bits per sample available to represent the amplitude of the waveform. Once sampled and stored in some sort of file format, digital audio is no different than any other file on a computer and may be saved for later use and playback. Digital audio may be saved as a common file format such as .WAV or .AIFF or bit compressed in some manner to reduce the size of the recorded file. Compression algorithms used to achieve smaller file sizes or bit rates vary but will either be lossless like the Free Lossless Audio Codec or lossy such as MP3, AAC, WMA, and many more. With lossy encoding algorithms, the digital audio is typically limited in its frequency range and/or perceptually encoded resulting in the permanent degradation of the original, uncompressed digital waveform. It is crucial when working with digital audio to understand the principles behind audio sampling and encoding.

Competency and Proficiency

In order to work in the field of audio forensics, as with all forensic sciences, a level of education and training must be met. This is crucial because of the role the forensic expert plays in the courtroom and in the litigation process. It is common for an examiner to prove competency with a relevant educational background and up-to-date training and/or scientific research in order to conduct examinations. Once deemed competent and approved by either a court or by their agency, the analyst may work with evidence and provide scientific and unbiased opinion in matters related to their field.

The foundation of knowledge one should possess regarding the science of sound, audio technology, and laboratory practices is broad and may be garnered through various avenues including but not limited to training in audio production and recording or electrical engineering and signal processing. For an inclusive list of the minimum areas one should know and demonstrate competency and proficiency in, refer to documents from working groups such as the Scientific Working Group on Digital Evidence, European Network of Forensic Science Institutes—Forensic Speech and Audio Analysis Working Group, etc. A summarized list of these areas would include: audio recording technology, audio evidence collection and handling, audio laboratory configuration to include analog and digital signals, the repair and recovery of analog and digital audio media, and a working knowledge of the ethical and legal issues surrounding the processing, analysis, and interpretation of audio evidence.

Evidence Handling

As stated earlier, the handling of audio evidence is crucial. Thorough documentation at each stage of analysis is important to establish and maintain the provenance of recorded media upon seizure and throughout the term in which it is in possession. Some important considerations follow. A chain of custody or audit trail shall be initiated and maintained for all items of evidence. Digital evidence media shall be bit-stream imaged to ensure provenance and integrity of data. Digital analyses and processing shall be carried out on a working copy of recordings so that the original unprocessed files may remain unchanged. All actions taken to enhance or process evidence shall be logged so that a similarly trained examiner may reproduce results. The audio as it was originally recorded must be provided to the examiner for analysis. In cases where this is not possible, the examiner should be aware of and make others aware of limitations including those related to enhancement and authentication.

Forensic Audio Enhancement

Forensic audio recordings are typically made in nonideal environments in circumstances that are hard to control leading to poor fidelity and intelligibility. The details that have been recorded are often hard to understand or not clear enough to make confident decisions. This is when it may be necessary to enhance or clarify the recorded audio by removing interfering noises and distortions or adjusting the amplitude of low-level information. This can either be done in real time while monitoring the audio of events or in a postprocessing phase after the recorded events have taken place. Either way, the application of signal processing in the digital domain is standard practice. There is often a trade-off where one can increase intelligibility by increasing the signal-to-noise ratio only to a certain point before the artifacts created by signal processing algorithms themselves interfere. Therefore, care should be taken to not over process material and abide by a general "less-is-more" rule of thumb.

A simplistic approach to enhancing forensic audio can be followed as such: file preparation, critical listening and analysis, audio processing, and preparation of output file for distribution.

File Preparation

It is necessary to prepare audio for analysis and enhancement prior to examination. This means taking steps necessary to digitize, transfer, or convert recordings into an uncompressed format (most commonly WAV) and at a relevant sample rate, bit depth, and channel configuration (mono or stereo). Before describing this process, a brief introduction will be given on digital sampling with regard to sample rate and bit depth.

Digital audio represents an acoustic sound wave as a finite set of binary values or samples. When an analog signal is received by an analog to digital converter, the acoustic waveform is quantized at a specific sample rate and bit depth. The sample rate refers to the number of samples per second used to quantize the waveform and the bit depth is the number of bits per sample available to represent the waveform's amplitude at that moment in time. The resolution of the recording, then, is a function of the rate at which the waveform is sampled according to Nyquist's theorem that defines the upper bound of a sampling system. As two discrete sample points are required to represent the waveform of a specific frequency, the sampling rate must be twice that frequency. In other words, an 8 kHz audio recording only represents frequencies up to 4 kHz. Thus, the higher the sample rate, the more accurately the acoustic waveform is represented. See **Figure 1** below demonstrating one period of a 1 kHz sine wave sampled at two different sample rates.

For analog media like cassette and microcassette tapes, it is necessary to digitize the audio so that is may be processed,

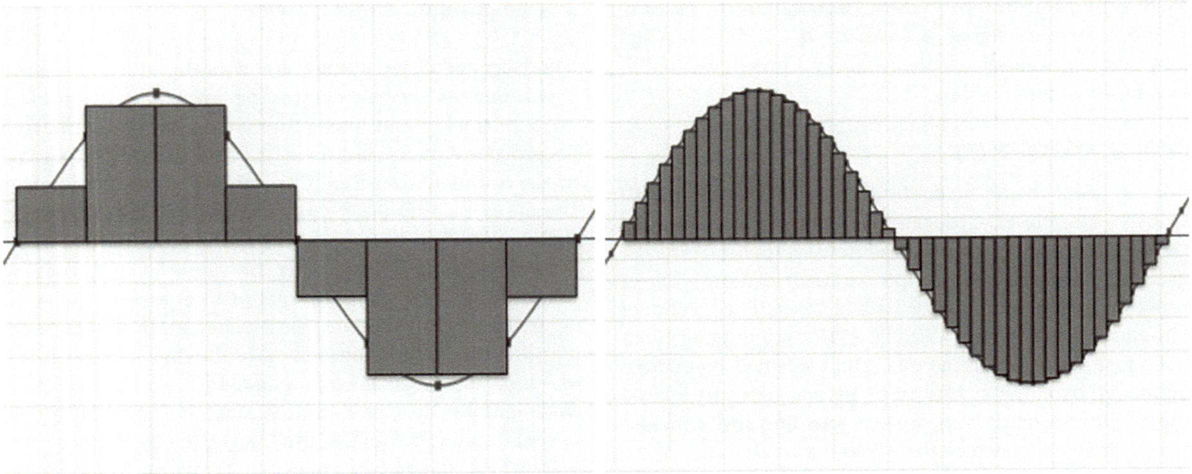

Figure 1 1 kHz sine wave sampled at 8 kHz (left) and at 44.1 kHz (right).

distributed, and played back more conveniently. For this process, all analog playback equipment shall be maintained and calibrated regularly according to the manufacturer's specifications and the analog to digital converter used should be external from the host computer and connected via Firewire, USB, or similar. The sample rate and bit depth used to digitize analog tape should be a minimum of 44.1 kHz, 24 bits (or 16 bits in the unlikely event that 24 are unavailable).

It is also possible to receive recordings on digital media that will require a transfer. These types of legacy media include digital audio tapes, Minidisc, and others. As the recordings that reside on these devices are native digital, it is best to digitally transfer the audio to the host computer using digital outputs on the media player. This maintains the originally quantized waveform and avoids redigitizing a quantized waveform if one were to use analog outputs on the digital media player. When performing a digital transfer, the original sample rate and bit depth should be used rather than resampling during the process in order to preserve the originality of the recordings.

The third general type of recording submitted would be one that was recorded as a digital audio file. There are many devices capable of recording digital audio files with the most common being hand-held recorders and cell phones. The digital audio files may be recorded in an uncompressed format such as WAV or AIFF. More commonly, however, recorded files are bit compressed in some manner to reduce the size of the recorded file. This process employs compression algorithms in order to achieve a smaller file size or bit rate and will either compress the audio waveform losslessly, causing no detrimental effect to the sound when played back, or in a way that causes permanent degradation of the original digital waveform. The term "lossy" is used for these codecs (short for compression/decompression) where the digital audio is typically limited in its frequency range and/or perceptually encoded. Audio codecs include AMR, MP3, WMA, ADPCM, etc. If the digital audio being submitted is in a digital format that is compressed, it is necessary to convert it to an uncompressed format such as WAV in order to achieve the best possible results. In this case, the original sample rate should be maintained and bit depth maximized for processing. When it is submitted to a laboratory for enhancement or clarification and prepared in these ways, it can be easily processed to clarify the content as much as possible.

Critical Listening and Analysis

The second step to enhancing audio material is to listen to the recording while analyzing the time, frequency, and amplitude content in order to characterize the types of noises or distortions that are masking the target signal (most commonly speech). For this process, software should be used that is not only capable of audio playback and displaying the digital waveform (**Figure 2**) but also display of the fast Fourier transform (FFT) spectrum (**Figure 2**) and spectrogram (**Figure 2**). The FFT is the algorithm employed to convert the time-based audio signal to the frequency domain for frequency analysis. By listening to the material while conducting visual analyses, interfering noises and distortions masking the intended material can be observed and measured.

Generally, an interference masking the target signal can be characterized as either a noise or a distortion. These differ in that "noises" are acoustic or electromagnetic events captured while the recording was made. "Distortions" on the other hand are changes to the content of an audio signal during transmission or recording that disrupt the original waveform, much like a reflection on water distorted by ripples that would otherwise be accurate. In this same example, noise would be

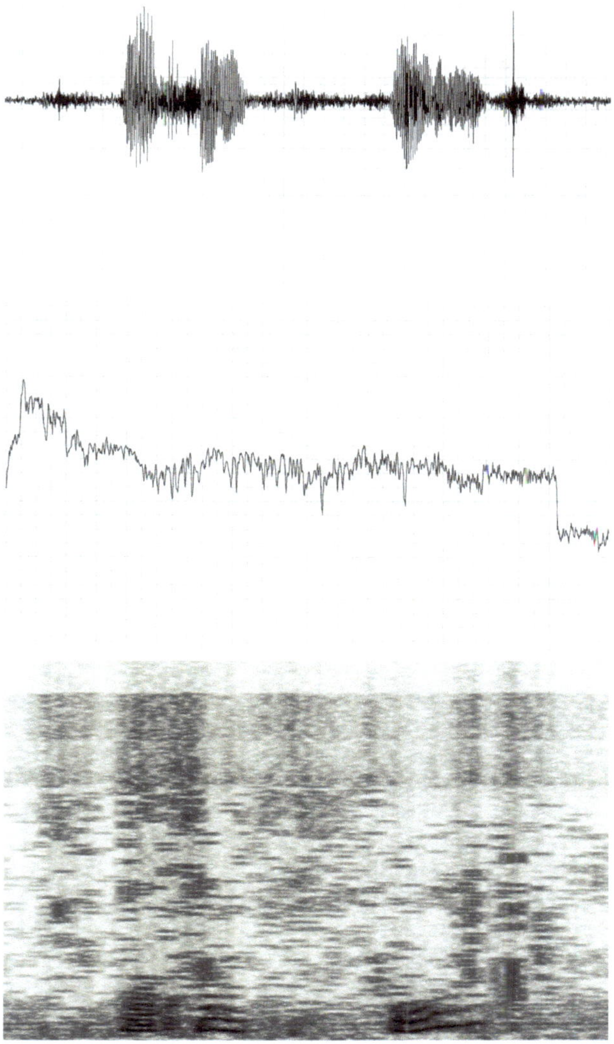

Figure 2 Waveform (top) with time on the *X*-axis and amplitude on the *Y*-axis. Fast Fourier transform spectrum (middle) with frequency on the *X*-axis and amplitude on the *Y*-axis. Spectrogram (bottom) with time on the *X*-axis, frequency on the *Y*-axis, and intensity of color (or gradient of grayscale) representing amplitude.

leaves floating on the surface of the water masking the reflection itself. Once noises and distortions are classified and measured, digital processing can be applied to remove them as much as it is possible.

Common types of noise in forensic audio recordings can be grouped into four categories: tonal, continuous broadband, variable broadband, and convolved. Tonal noise commonly

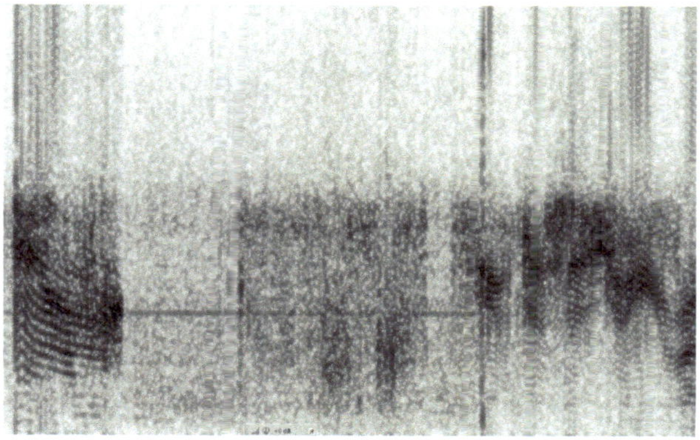

Figure 3 Spectrogram showing speech with tonal noise (the *vertical line* at approximately 1.6 kHz).

comes from mains power hum, when an audio signal is not properly shielded, or acoustic noises from machines that oscillate at a specific frequency or set of frequencies (**Figure 3**). Continuous broadband noise (**Figure 4**) occupies a broad range of frequencies with little spectral change in time. For instance: tape hiss, machine or channel noise, or building ventilation systems. Variable broadband noise is a term used to describe interferences that occupy a broad range of frequencies with variable spectral content. For instance wind, rain, and speech other than the target signal (**Figure 5**). Lastly, convolved noise is excessive reverberation interfering with the source signal (**Figure 6**).

Distortions disrupt a signal and prevent the acoustic waveform from being accurately recorded due to interference or faulty equipment. One common example includes mobile phone interference (GSM bursts shown in **Figure 7**). In another example, **Figure 8** shows artifacts due to the application of aggressive lossy compression.

Audio Processing

Once the various interfering noises and distortions are identified, specific filters and algorithms can be applied in an attempt to remove or attenuate them. At this stage, the recorded audio may be enhanced for increased intelligibility or ease of listening. Not all noises can be removed nor all distortions fixed. In cases where the ratio of target signal amplitude to noise amplitude is too small, content in the waveform representing the target signal is not adequately represented. Therefore, it must be noted that some audio is recorded so poorly and with such compromising conditions that speech is not recoverable. In these instances, the submitter of the recording must be made aware of the limitations when working with forensic audio. Specific filters and algorithms designed to remove noises and fix distortions are provided here. While this

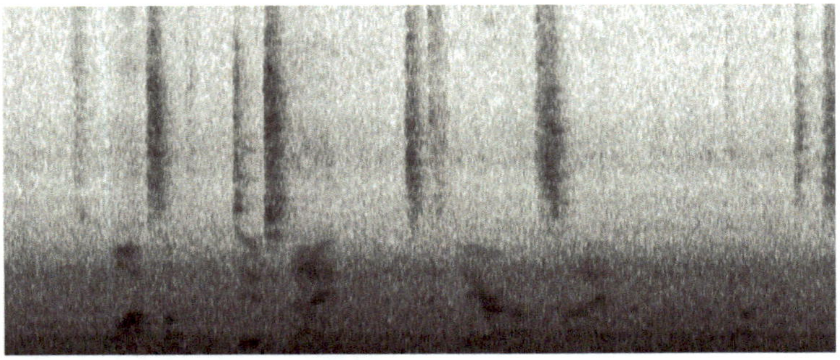

Figure 4 Spectrogram showing speech in the presence of continuous broadband noise.

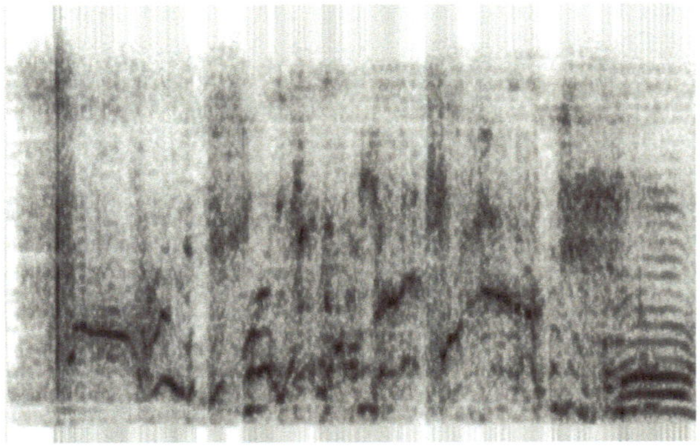

Figure 5 Spectrogram showing speech in the presence of variable broadband noise.

section is not entirely exhaustive, the elements provided represent commonly employed techniques. In most cases, distortions should be fixed before attempting to remove noise.

- Clipped audio may be repaired manually or by using a *De-clipping* algorithm that analyzes the distorted waveform with flattened or clipped peaks and interpolates the waveform that should otherwise be sinusoidal in nature.
- *Mobile phone interference* is induced into audio equipment electromagnetically as mobile devices transmit packets of data in very short bursts. These bursts are tonal in nature but use of notch or comb filters is not typically successful because of the interference has multiple harmonics of very high amplitude. It is possible to remove the bursts by determining the fundamental frequency of the distortion (based on the carrier network) and removing them with an adaptive filter. Modern filters built for this purpose will then interpolate the information obscured by the bursts.
- *Time domain distortions*, such as tape wow and flutter, can be compensated for by applying time compression or expansion on appropriate segments of recorded audio.
- *Lossy compression artifacts* can in some cases be removed and the missing information reconstructed using algorithms designed for musical recordings. However, with forensic audio where speech is typically the target signal and other noises are present, these solutions are not very useful.
- *Notch, comb, and band-pass filters* are digital filters that are designed to attenuate or remove a specific frequency or frequencies. They are therefore useful in the removal of tonal noises. While a notch filter can be used to attenuate

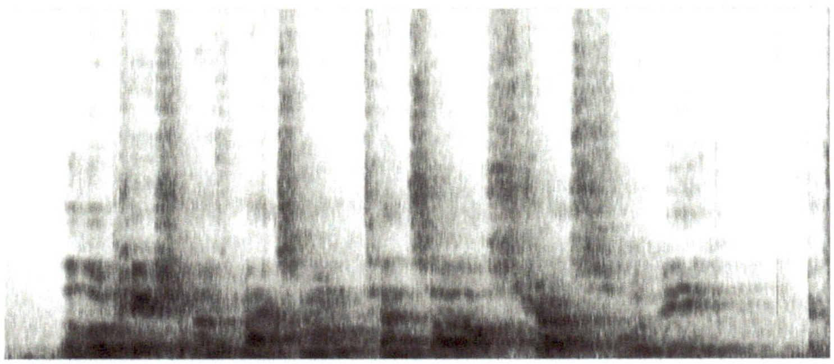

Figure 6 Spectrogram showing speech recorded in excessively reverberant conditions. The clarity of speech formants are smeared in time due to reflections from the environment.

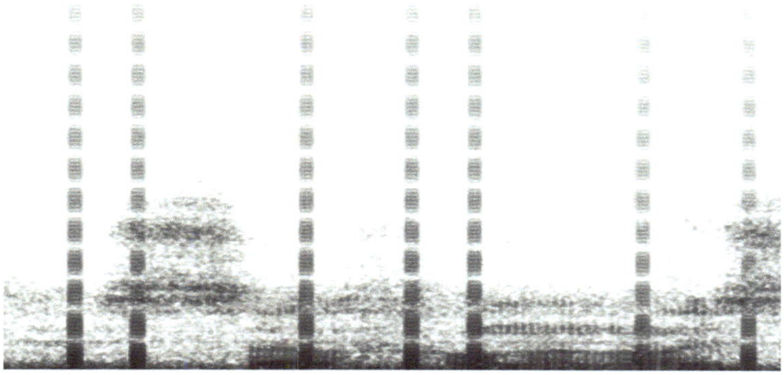

Figure 7 Spectrogram showing speech obscured by GSM burst interference.

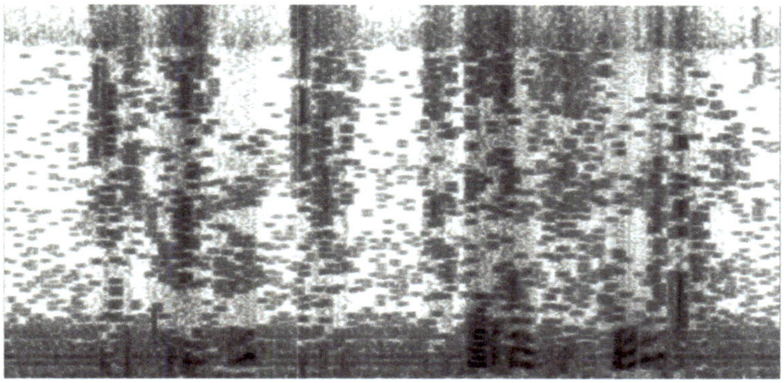

Figure 8 Spectrogram showing speech recorded with lossy WMA compression.

one specific frequency (such as 1.6 kHz in **Figure 3**) comb filters are designed to attenuate a fundamental frequency and its harmonics. This is especially useful in the case of induced mains power hum (at 60 or 50 Hz depending on the country) where harmonics at multiples of the fundamental are present. By selecting the fundamental frequency, harmonics are automatically calculated and attenuated. Band-pass filters on the other hand will attenuate a range of frequencies. For instance, a high-pass filter can easily be applied to attenuate frequencies from 0 to 150 Hz in order to reduce low-frequency rumble and make speech more intelligible.

- *Spectral subtraction* is very useful in removing continuous broadband noise where the spectral content of noise does not vary greatly over time. In a simple arithmetic procedure, this algorithm subtracts an input noise profile from recorded audio to effectively remove noise and enhance intelligibility.
- *Adaptive filters* can work in both the time and frequency domains and achieve spectral subtraction on short time windows that change as the noises in the environment change. These filters develop predictive coefficients to continually update the input noise profile to be subtracted and can therefore be useful in dealing with variable broadband noise like outdoor environments where background conversations, rain, etc. may be present.
- *Convolved noise*, or excessive reverberation, can be removed with some success using algorithms designed for this purpose. While many methods have been proposed, typical techniques that are commercially available use adaptive dereverberation algorithms where spectral subtraction of convolved elements is achieved based on predictive coefficients.
- *Stereo source separation* can be very powerful when available. In this situation, a stereo recording is provided that has speech in the presence of, most likely, variable broadband noise. If the target speech signal is present on only one channel, the second channel can be used to inform the algorithm of the noise profile to be removed. For example, for a stereo recording of a noisy bar where the background voices, music, and other acoustic impulses are captured in relatively equal amplitude on both channels and speech is on one. Alternatively, when dealing with mono recordings, it is possible to provide the algorithm with the second

reference channel containing identical noise information as that which is present on the source. For instance, when presented with a mono recording of speech obscured by commercially released music or a media broadcast, the music or broadcast can be obtained to build the reference channel for subtraction.

- In some cases, conversations are recorded in which one person is at a higher amplitude than another. This situation is referred to as *near-party/far-party* and can easily be compensated for by either manually adjusting the volume of the digital file during quiet portions or by applying amplitude compression/limiting to bring loud and quiet material to within the same range of volume. Some systems provide an automatic gain control feature that will expand and compress portions of audio to achieve a more balanced overall level.

A forensic audio recording that requires enhancement may be processed with one or many of the above techniques. Depending on the noises and distortions present, it is up to the examiner which ones will be appropriate and in which order. This is done by comparing and contrasting the audio before and after each step of processing. It is advised to save each iteration of processing as an independent audio file to maximize efficiency if any alterations to previous steps must be taken. For any processing that is applied, it is necessary to document filters and settings used so that identical results may be achieved at a later time by the same examiner or by another adequately trained examiner. This helps to establish the provenance of the original material and can demonstrate, if necessary, that the content or meaning of the audio was not changed during the course of enhancement.

Preparation of Output File

Once noises and distortions have been compensated for satisfactorily, further processing may be needed to make the recording appropriate for playback in a courtroom: adjusting the overall amplitude of the audio, for example. Or it may be desired to further shape the frequency response of the recording by applying equalization. Once complete, it is necessary in this last stage of the forensic audio enhancement procedure to prepare the audio in a format as deemed appropriate by the analyst and the submitter of evidence. For example, due to technology limitations of the contributor or the court, it may be requested to provide an audio CD that can be played back in a CD player requiring the output audio to be at 44.1 kHz 16 bits. If the file is to be converted in any way from the resolution that it was processed at, the unconverted output file should be also provided as it represents the most original form of the enhanced evidence. Finally, the original unprocessed audio should also be provided to the contributor especially in cases where the originally supplied recording is not in a format useful for the contributor or the courts.

Forensic Audio Authentication

Commonly, a review of audio authenticity is requested to determine whether the material presented accurately represents the events recorded with requests coming either through investigation or as a request from the court or its advocates. An audio authenticity examination, as defined by the Scientific Working Group on Digital Evidence, seeks to determine if a recording is original, unaltered or continuous, and/or consistent with the manner in which it is alleged to have been produced. Many elements from a recording may be analyzed and the methodology employed will depend on whether the

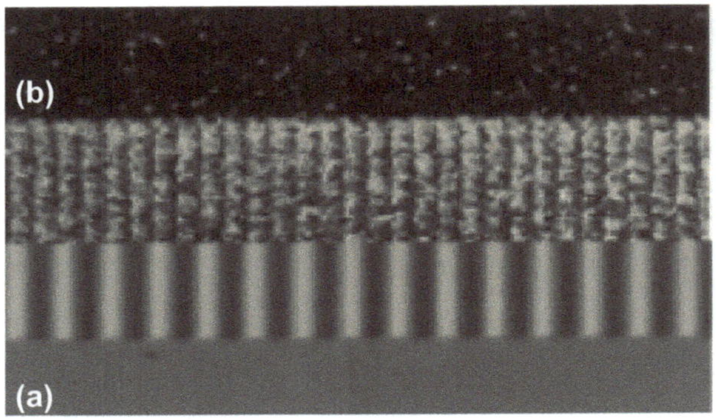

Figure 9 (a) 1 kHz sine wave imaged using a magneto-optical system. (b) 1 kHz sine wave imaged using Freon-based ferrofluid. Photo courtesy of Jonathan Broyles www.iasforensics.com.

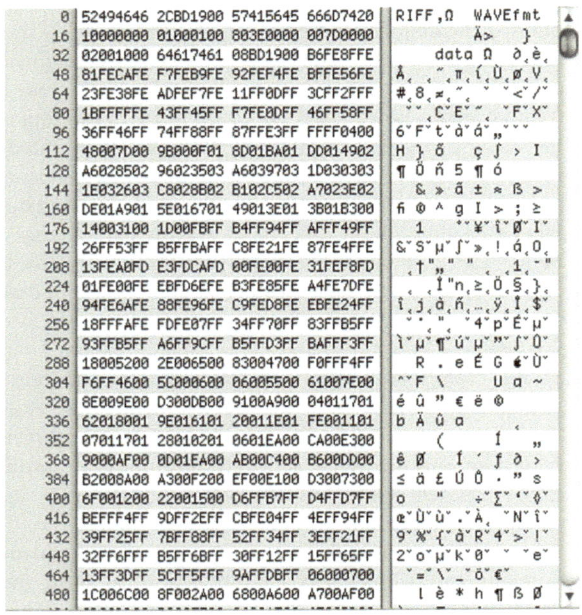

Figure 10 Hexadecimal display of audio data.

recording was made digitally or on analog media. In either case, the questioned original recording on its original media must be analyzed and, if available, the original recording device must be obtained. This is crucial in establishing the provenance of the recording; otherwise analyses and conclusions are compromised. If the original recorder is not available, a recorder of the same model and brand may be used to produce test recordings. Test recordings are necessary to produce known data from which comparisons to the questioned recording can be made.

Analog Tape Authentication

Analysis methods in tape authentication examinations first began with the Watergate Scandal of 1972. Since that time, the methodology employed to analyze analog tape recordings has not changed greatly nor have the tools for analysis. Once the questioned recording has been obtained, an analysis of the magnetic material on the tape is conducted along with digital analysis of the audio material. In essence, the examiner must analyze any inconsistencies in the questioned recording and make comparisons with test recordings in order to determine whether they are a result of normal operation or possible editing.

Bruce Koenig published a paper in 1990 describing a formal methodology for conducting analyses of analog recordings that employs Freon-based ferrofluid to conduct visual examinations of the magnetic waveform stored on the tape (**Figure 9**). An alternative method of visualizing magnetic flux that provides higher frequency resolution and less physical contact with the tape itself employs a magneto-optical imaging system (**Figure 9**).

Digital Audio Authentication

An analysis of digitally recorded audio does not differ in scope from an analog examination in that the examiner seeks to

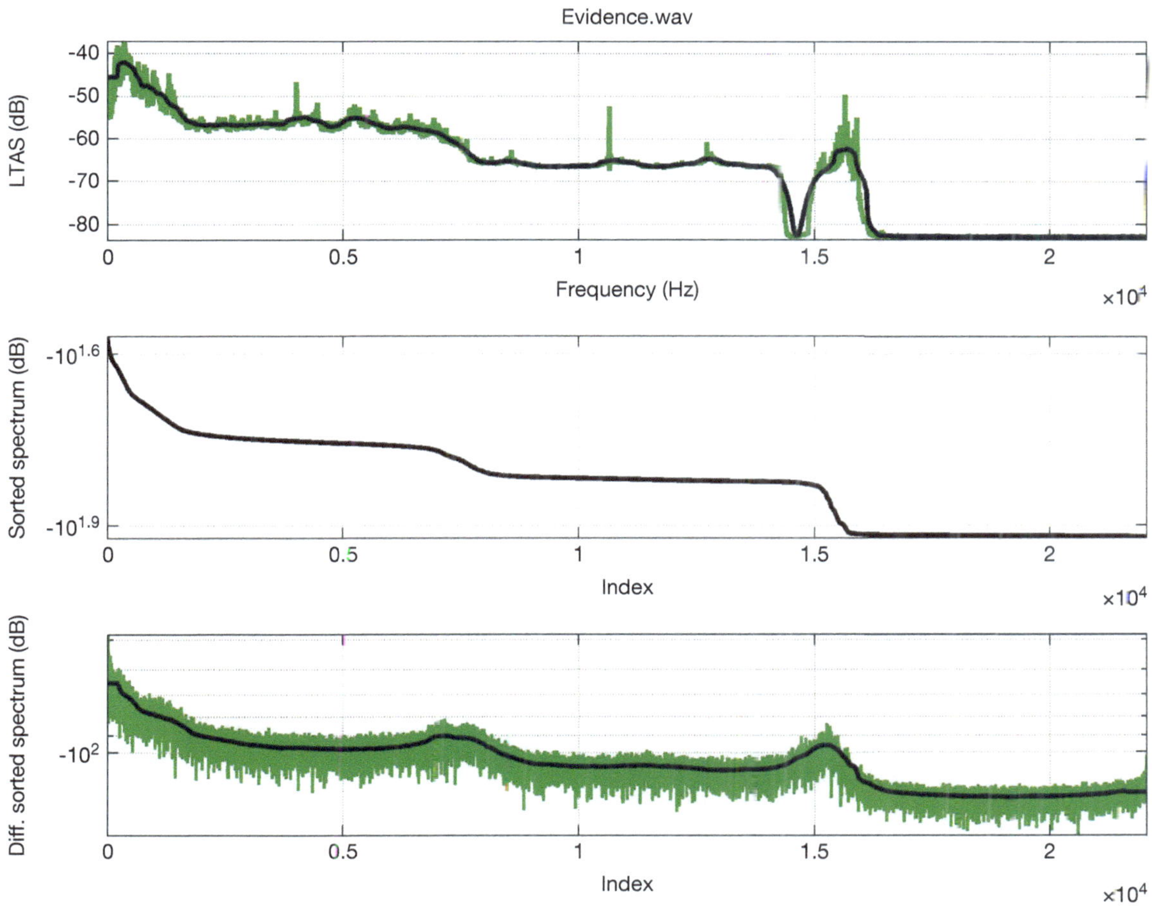

Figure 11 Long-term power spectrum, sorted spectrum, and differentiated sorted spectrum of long-term average spectrum.

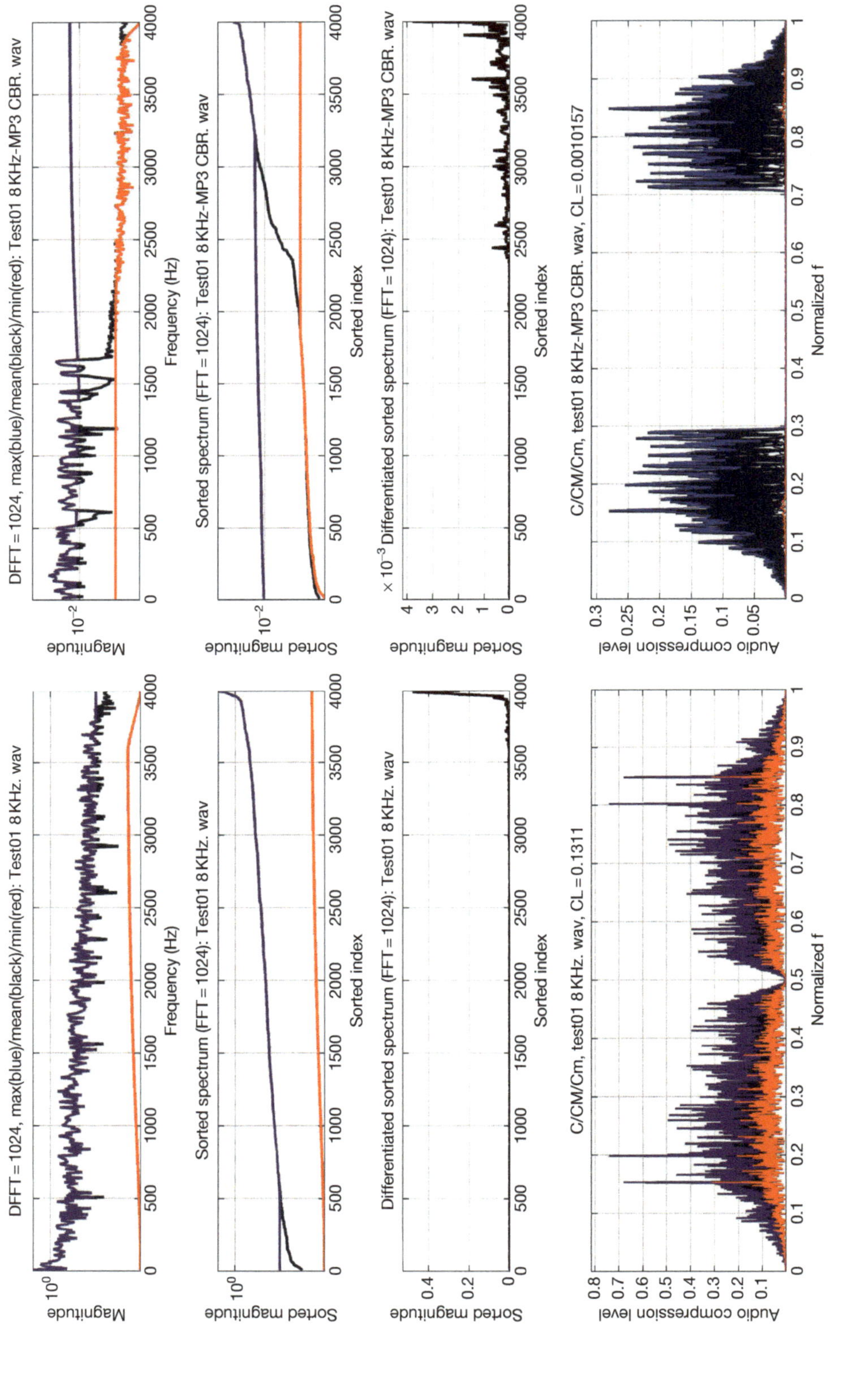

Figure 12 Compression level analysis original WAV PCM recording (left column), MP3 CBR (right column).

explain observed anomalies, through the analysis of test recordings, as a result of normal recorder operation or inconsistencies which may indicate alteration or editing. However, the techniques employed vary greatly. These should include but not be limited to analysis of: header and file structure, waveform, FFT spectrum, spectrogram, compression level, and electric network frequency (ENF). Koenig and Lacey provide a formal methodology for the authentication of digital audio recordings while information regarding analysis can be found in the references section as well. Different techniques have been developed for forensic audio authenticity analysis and this chapter presents some of them.

MAC, header, and file structure analysis can be carried out checking the MAC (Modify, Access, Creation time stamp of a file and/or digital media) and using software capable of viewing the individual bits and bytes that make up a file represented as hexadecimal, or hex, data (**Figure 10**). In viewing the hex data of a digital audio file, comparisons may be made between a questioned recording and test recordings to observe recorder and file write behavior. The internal arrangement of proprietary files can confirmed as well as the header structure. Header information may also include the date/time of the recording, recording duration, make/model/serial number of the recorder, file type, and compression scheme and rate. Any observed inconsistencies in this information may be a result of alteration or editing.

Power spectral analysis may be useful in identifying traces of file recompression or inconsistencies between the recording and the claimed recorder. If present, inconsistencies can be easily detected by comparing the long-term power spectrum, sorted spectrum, and differentiated sorted spectrum of unknown and test recordings as well as comparing histograms of their spectral power (**Figure 11**).

Compression level analysis may also be useful in identifying traces of file recompression or inconsistencies between the recording and the claimed recorder. This function proposed by Grigoras, applies the discrete fast Fourier transform on the second derivative of consecutive frames to

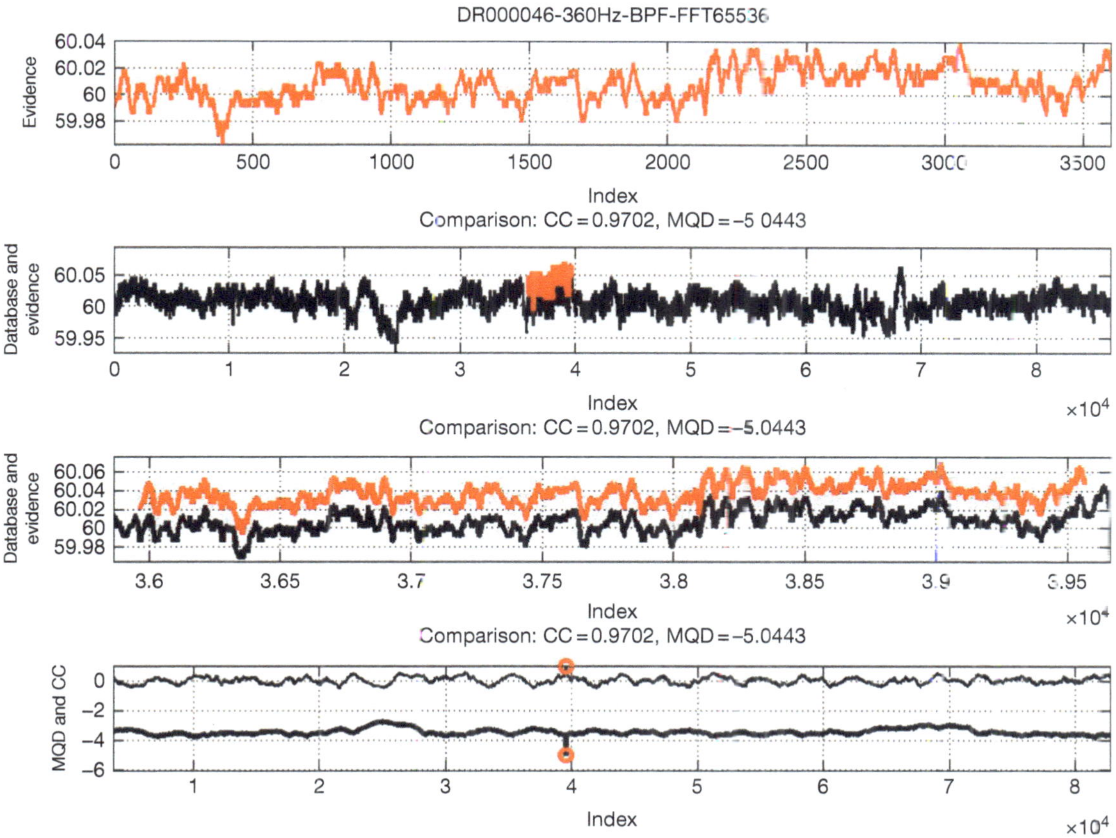

Figure 13 Electric network frequency analysis against a forensic database.

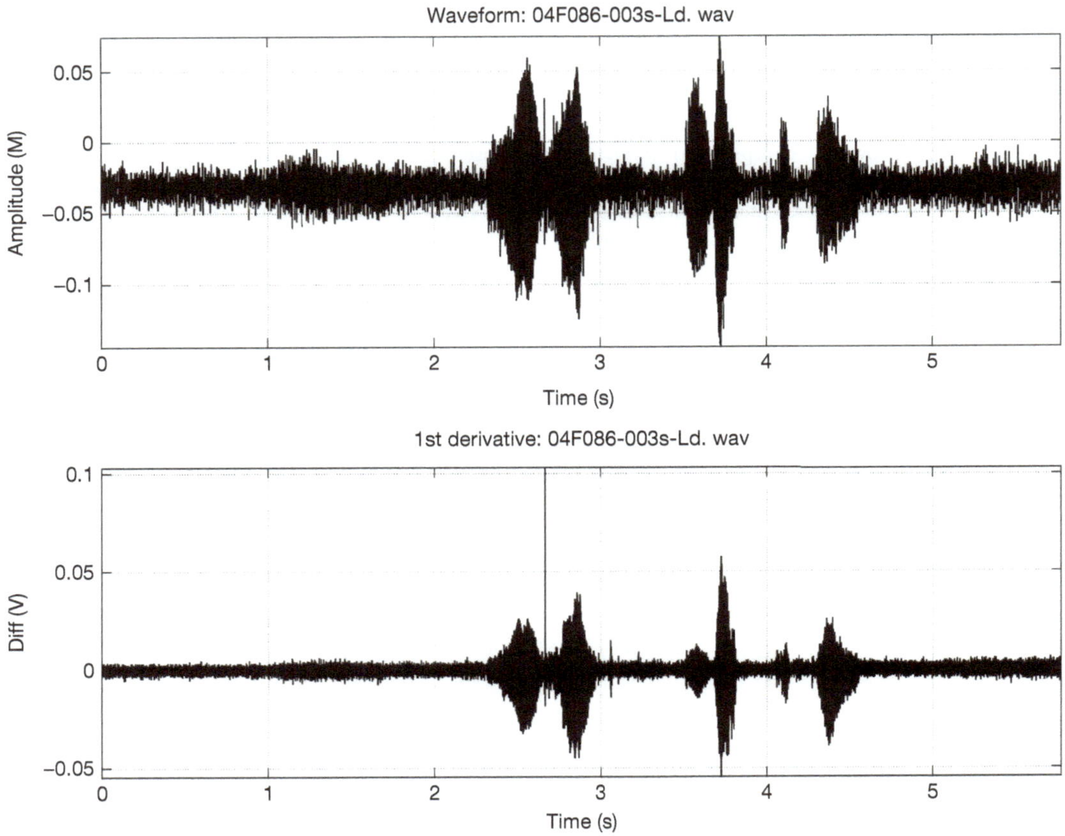

Figure 14 Butt-splice detected at 2.665 s.

produce a plot of the compression level. By comparing an unknown recording's compression level with test recordings of the claimed recorder, inconsistencies may be found if present (**Figure 12**).

ENF analysis involves the detection, extraction, and comparison of the ENF trace induced onto recording from mains power. If present, ENF may help an examiner do many things in addition to identifying the presence of edits, insertions, and deletions. With a reference database of ENF, it is possible to determine the date/time a recording was made, how the device was powered, and gross geographical location, all of which can further corroborate the authenticity of a recording (**Figure 13**).

Butt-splice detection can be useful in identifying the insertion or deletion of material without the application of cross-fades (**Figure 14**).

DC-level analysis may be used to determine the insertion of material or to compare an unknown recording to the recorder claimed to have made it (**Figure 15**).

In the examination of digital audio for authenticity, a sound methodology should be used that is based on validated methods and principles. While the specific analyses will depend on the situation, the examiner should not rely on one specific element or tool but seek to exhaust all possible techniques because an inconsistency observed through one analysis may not be present in another. Specific determinations from analyses will also vary depending on the situation but will generally conclude that the recording is consistent with an authentic recording, inconsistent with an authentic recording, or inconclusive.

See also: **Behavioral:** Forensic Linguistics; Forensic Phonetics; **Digital Evidence:** Digital Imaging: Enhancement and Authentication; **Digital Evidence/Photography and Digital Imaging:** Digital Imaging and Photography: An Overview; **Investigations:** Recording.

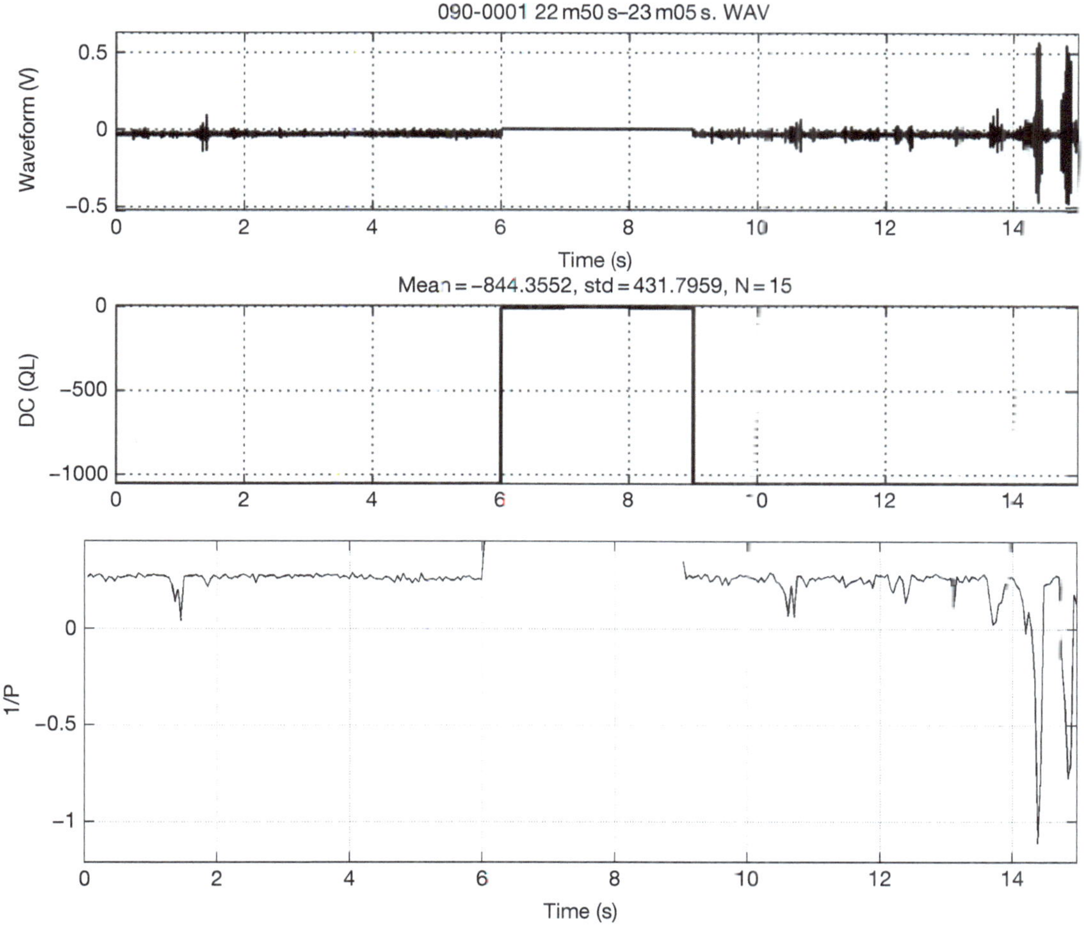

Figure 15 DC and power analysis showing a signal's discontinuity.

Further Reading

Bouten, J., van Rijsbergen, M., Donkers, S., 2007. Derivation of a transfer function for imaging polarimetry used in magneto-optical investigations of audio tapes in authenticity investigations. Journal of the Audio Engineering Society 22, 257–265.

Brixen, E.B., 2011. Audio Metering: Measurements, Standards, and Practice, second ed. Focal Press, Burlington, MA.

Cooper, A., 2010. Detecting butt-spliced edits in forensic digital audio recordings. In: Proceedings of the 39th International Conference of the Audio Engineering Society: Audio Forensics. Denmark, 2010.

Grigoras, C., 2009. Applications of ENF analysis in forensic authentication of digital audio and video recordings. Journal of the Audio Engineering Society 57, 643–661.

Grigoras, C., 2010. Statistical tools for multimedia forensics. In: Proceedings of the 39th International Conference of the Audio Engineering Society: Audio Forensics. Denmark, 2010.

Harris, C.M., 1998. Handbook of Acoustical Measurements and Noise Control, third ed. Acoustical Society of America, Woodbury, NY.

Koenig, B.E., 1990. Authentication of forensic audio recordings. Journal of the Audio Engineering Society 38, 3–33.

Koenig, B.E., Lacy, D., 2009. Forensic authentication of digital audio recordings. Journal of the Audio Engineering Society 57, 663–695.

Loizou, P.C., 2007. Speech Enhancement: Theory and Practice. Taylor and Francis Group, Boca Raton, FL.

Pohlmann, K.C., 2005. Principles of Digital Audio, fifth ed. McGraw-Hill, New York, NY.

Relevant Websites

www.enfsi.eu—Documents of the ENFSI Expert Working Group Forensic Speech and Audio Analysis.

www.swgde.org—Documents of the Scientific Working Group on Digital Evidence.

www.aes.org—The EOB Tape of June 20, 1972 by the Advisory Panel on White House Tapes.

Investigation and Analysis of Electrical Accidents

TP Shefchick, Shefchick Engineering, Sunnyvale, CA, USA

Glossary

Arcing A discharge of electricity through a gas.

Burn patterns The damage caused by fire.

Circuit breaker A device designed to open and close a circuit by nonautomatic means, and to open the circuit automatically on a predetermined overload of current.

Conductor A substance or body that allows a current of electricity to pass continuously along/through it.

Electrocution The destruction of life by means of electrical current.

Exemplar Something that serves as a model or example.

Fuse An electrical safety device consisting of or including a wire or strip of fusible metal that melts and interrupts the circuit when the current becomes too high.

Lock out To open an electrical circuit and place a lock on it so that it cannot be closed without removing the lock.

Overhead Aerial utility power supply equipment to buildings or structures.

Panel board A single panel or a group of panel units containing buses, switches, fuses, and/or circuit breakers for the control of light, heat, and power circuits.

Resistance The physical property of an element or device to rest/oppose the flow of electricity through it.

Service Electrical wires and/or equipment supplying electrical power to buildings or structures.

Specifications The design requirements.

Tripped The position of a circuit breaker's operating handle when it opens due to an overload current.

Underground Subterranean utility power supply wires and/or equipment to buildings or structures.

Investigation

The investigation of electrical phenomena began in Europe in the 17th century. The first electrical investigators were called electricians. One of the first electrical investigators was Benjamin Franklin. He made on-site inspections of churches which were struck by lightning. He wrote newspaper articles requesting information about lightning damage and read articles about lightning damage to churches to determine the effectiveness of church bell ringing, which was believed to disperse lightning. Franklin found that bell ringing during a thunderstorm is a very dangerous occupation. There was no great need for electrical investigations until the inventions of Alexander Graham Bell's telephone in 1876 and Thomas A. Edison's incandescent lamp in 1878. Now we are surrounded by electrical fields, wiring, and devices which can, and do, cause injury or property damage.

The on-scene electrical investigation should be conducted as soon as possible after an event, before evidence is moved, discarded, or altered by repairmen, rescue workers, or first responders. The electrical investigator utilizes the same scene investigation techniques as the crime scene investigator, except that electrical items are the focus of the investigation. As a result, it is imperative that the electrical investigator have substantial practical knowledge about electrical equipment, wiring, and devices. Practical electrical knowledge is obtained through education and experience in designing, installing, maintaining, and repairing electrical devices, appliances, and equipment.

Before visiting the accident scene, any available information about the accident or suspected cause should be obtained from eyewitnesses, private investigators, insurance adjusters, police officers, firemen, first responders, and/or attorneys. The information could be used to confirm or refute what is found at the scene. The suspected cause must be evaluated. Upon arrival at the scene, photographs should be taken of the area surrounding the accident. The electrical power service to the scene should be inspected and documented with photographs and/or diagrams. Is it a residential, commercial, or industrial overhead or underground electrical service? Is there any evidence of recent work and/or repair to the exterior electric service?

Next, the investigator must determine the origin of the electrical accident. Sometimes the origin is where the greatest amount of damage is found. However, combustible materials in the area could cause severe damage remote from the origin.

 http://dx.doi.org/10.1016/B978-0-12-802718-9.00028-4

This is where the information obtained from eyewitness accounts might be helpful.

If an area of origin cannot be determined, all the electric lines/wires passing through damaged areas should be inspected. The investigator should be looking for evidence of electrical arcing or short-circuiting on the wires. Electrical arcing or short-circuiting generates temperatures in the range of 3000–20 000 °C. It is the heat generated by the arcing or short-circuiting of electricity that causes most electrical damage and injury. The wires are normally made of copper or aluminum. Copper melts at 1083 °C and aluminum at 570–660 °C. Normally, fires do not generate sufficient heat to cause copper to melt. Therefore, the copper wiring will remain intact except for the insulation, which might have burned away. Evidence of melting on copper wiring indicates that electrical arcing or short-circuiting may have occurred at the melted location. Photographs should be taken to document the area surrounding the melted wiring and close-up photographs of the melted wires. Aluminum wiring is quite frequently found melted by a fire, as fires can, and do, generate sufficient heat to melt aluminum. Subsequently, melted aluminum wiring may not be evidence of electrical arcing or short-circuiting. Any melted aluminum wiring should be photographed in place. Wiring suspected of short-circuiting or arcing should be retained for further analysis. As much as possible the wiring should be retained, especially undamaged sections, which may contain manufacturer's labeling and specifications. All sources of electrical power to the accident area should be inspected and photographed. If damage is found upon power supply components, they should be retained for further analysis. All fuses should be tested with a continuity tester to determine if they have opened. The position of all circuit breakers should be documented as to whether they are "on/off" or in a middle "tripped" position. Evidence of overheating on power supply components, a blown fuse, or tripped circuit breaker might aid in the determination of the accident's origin. All electrical devices in the damaged area should be inspected for evidence of internal damage, which could indicate the origin of the problem. In addition, the installation and condition of undamaged electrical equipment in the vicinity of the accident should be inspected. The undamaged equipment might reveal an installation or component problem that caused the electrical accident.

If the area of origin can be determined, it should be photographed from various angles before anything is moved. Everything electrical should be retained from the area of origin and labeled to identify where it was found at the scene. Manufacturer's labels and specifications on electrical equipment should be documented. If a specific electrical device or product is suspected of causing the accident, the accident area should be searched for a similar device, product, or installation.

An exemplar product or installation could reveal what precipitated the accident. The source of electrical power to the origin should be inspected and traced back through the electrical distribution system to where it obtains electrical power. The fuse and/or circuit breaker protecting the distribution system should be inspected and photographed. Any damage to the fuse and/or circuit breaker would require that it/they be retained. They should also be retained if evidence of electrical arcing, short-circuiting, or conductor overheating was found in their vicinity. The distance from the fuse and/or circuit breaker to the area of origin should be documented/measured. If the distance is too long, the circuit breaker or fuse will not provide adequate protection. The resistance of a long circuit reduces the fault current through it and impedes the operation of a circuit breaker or fuse.

Electrical conductor overheating is another way electricity can cause damage by igniting combustible materials in contact with it.

Electrical conductor overheating is normally prevented by a fuse or circuit breaker that opens and removes electrical current from the conductor before it overheats. However, sometimes the fuse is too large (mismatched) for the conductor or the circuit breaker fails to properly operate.

The size of the fuses or circuit breakers should be documented, as well as whether any of the fuses opened or circuit breaker positions were "on/off" or "tripped."

The investigation should be conducted with no electrical power feeding into the accident area. The investigator should have a voltage detector to verify that there is no electrical power. A continuity tester would facilitate determining which fuse or circuit breaker fed electrical conductors in the area of origin or if a switch is open or closed. Small digital microscopes that operate with laptop computers provide excellent close-up photographs of electrical devices. A quality 3 charge coupled device progressive scan video camera can provide excellent video of the accident scene and show items from a different perspective than a still photographic camera. It will also bring your audience into the accident scene. Small hand tools are required to open up electrical devices such as panel boards, receptacles, switches, and fixtures. Wire cutters of various sizes are a necessity. A wire caliper would be useful in determining the size of electrical wiring. A very bright light source such as a high-power light-emitting diode flashlight will aid the investigation.

When photographing electrical wiring or components, a dark blue cloth used as a background will produce the best-quality photographs.

Electrical devices retained from the scene should be opened to inspect their internal components for evidence of electrical arcing, short-circuiting, or overheating. The exterior of the devices should be photographed before disassembly begins and during various phases of the disassembly. However, if a device cannot be disassembled without altering its appearance, it should be x-rayed to examine its internal components. X-rays will normally show evidence of electrical arcing or short-circuiting within the electrical devices. However, the X-rays will not show evidence of overheating. The X-rays will aid in the destructive examination/disassembly of the device.

If the electrical accident involves personal injury or death, the medical records should be obtained to determine the injuries sustained by each victim. Electrical entry and exit wounds, degree of burns, and their locations are normally reported. A low-voltage electrical shock (120 V alternating current or less) might not leave an entry or exit wound, especially if water is involved. The water cools the area and prevents burning of the flesh. The medical records may indicate significantly elevated serum creatine phosphokinase levels due to the muscle contractions caused by the electrical current if the victim remains in the circuit for seconds or minutes while alive. The clothing worn by the victim(s) at the time of the incident should be examined, including their footwear. Any protective clothing and gear should be retained for further examination/analysis. The type of flooring in the accident area should be documented along with the environmental conditions. Architectural plans, construction drawings, and permits of the accident area should be obtained.

An Internet search may reveal videos or photographs of the accident taken by eyewitnesses. Sometimes exemplars, manufacturer's operation manuals, and drawings can be obtained through the Internet. Complaints concerning the product or service might be posted online or at websites such as the Consumer Product Safety Commission.

Analysis

Electrical wiring can melt as a result of a eutectic reaction, which is the lowering of the melting temperature of a metal by a melted metal having a lower melting temperature. The eutectic melted wire can resemble an electrical short circuit or arcing damage. A metal such as aluminum, which melts due to fire exposure, can fall onto copper wiring and cause it to melt, as well. Elemental analysis of the melted areas by scanning electron microscopy/energy dispersive x-ray spectrometry can determine if there is foreign metal present in the melt. If electrical arcing or short circuiting is confirmed, "arc mapping" can be conducted to aid in the determination of the accident's origin. Arc mapping involves plotting the location of arcs or short circuits on a diagram of the accident area. The arc or short circuit that is found furthest downstream from the source of electrical power most probably is closest to the origin of the problem as it was the first location where the electrical system sustained damage. Upstream electrical short circuits would prevent the occurrence of downstream short circuits, if they occurred first. Consequently, electrical short-circuiting or arcing found at the point of the accident's origin downstream from the power source is most probably the initial location of the commencement of the accident. The reason for the arcing or short circuiting should be determined. Did the installation of the wiring meet the requirements of the local electrical codes? In most of the United States, the National Fire Prevention Association's (NFPA) National Electrical Code (NEC) applies to the installation of electrical wiring in residential and commercial buildings. The National Electrical Safety Code (NESC) applies to the installation and maintenance of electric utility wiring and distribution equipment. Many large cities and states have their own electrical codes which follow the requirements of the NEC and the NESC with more stringent requirements.

The x-rays of a suspected defective circuit breaker should be evaluated with design drawings or the x-rays of a properly functioning exemplar circuit breaker. If no discrepancies are found, the circuit breaker should be tested to determine whether it meets the requirements of its certifying laboratory (Underwriters Laboratory, Inc. for low voltage in the United States of America). The National Electrical Manufacturers Association, NEMA, and IEEE have standards which apply to low-voltage and high-voltage circuit breakers that could be of use in an investigation. The testing should be done prior to disassembly of the circuit breaker because it is very difficult to reassemble a circuit breaker exactly as it was prior to disassembly. After testing, the circuit breaker should be disassembled in order to inspect/examine the internal components for evidence of overheating and arcing, which might not be apparent in the X-rays.

Testing should be performed on protective components in electrical equipment and devices to verify that they function as designed. These safety components may be intended to protect against overcurrent, overvoltage, under voltage, temperature limiting, liquid level control, or ground fault protection.

Construction specifications, work invoices, and daily job site reports may be useful in the analysis. The building codes of NFPA, American Society for Testing and Materials, and The International Building Code may also assist in the analysis of the accident scene.

Casework/Examples

The first example involves an electrical accident or more appropriately an electrical incident which occurred at a sporting club. The incident originated at an outdoor racquetball court (**Photograph 1**). A 12 kV electrical power line had dropped down and made contact with the metal screen around the racquetball court (**Photograph 2**). The power line continued down and burned a hole in the metal floor of the racquetball court (**Photograph 3**). Luckily, no one was utilizing the racquetball court at the time of the incident. Severe damage was found underneath the racquetball court where oil heaters and fans were located to heat the court as it was located in a temperate climate location (**Photograph 4**). None of the fuel

Photograph 1 Electrical incident at an outdoor racquetball court.

Photograph 2 Electrical power line dropped down and made contact with the metal screen around the racquetball court.

Photograph 3 The power line burned a hole in the metal floor of the racquetball court.

Photograph 4 Severe damage was found underneath the racquetball court.

ignited even though there was evidence of electrical arcing surrounding the fuel storage tank (**Photograph 5**). An electrical panel board near the racquetball court was completely destroyed (**Photograph 6**). The metal siding on a club house located ~3 m from the court was damaged by electrical arcing (**Photograph 7**). The ground conductors of various appliances that were plugged in at the time of the incident were burned away. Ground conductors of wiring within the club house were damaged (**Photograph 8**) and evidence of electrical arcing was found on a fuel line (**Photograph 9**). Fortunately, the fuel in the line did not ignite.

Inspection of the area surrounding the sporting club found that the local utility was installing power line poles and power lines at the time of the incident (**Photograph 10**). There were no environmental factors that could have caused the line to fall as the weather was clear at the time of the incident.

Subsequently, the power line most probably fell down as a result of the work activity in the area by the utility company.

The damage was the result of installing an electric power line above the metal racquetball court in violation of electrical code requirements. The NESC requires a vertical clearance of 3 m above the metal racquetball court and a horizontal clearance of 2.4 m.

The next example involves a wildland fire. A wildland fire is a fire which originates in an uninhabited area of land. It might subsequently spread and damage/destroy inhabited buildings and areas of land. Inspection of the burned land areas indicated that the fire originated near a utility electric pole

Photograph 5 Evidence of electrical arcing surrounding the fuel storage tank.

Photograph 7 An electrical panel board near the racquetball court completely destroyed.

Photograph 6 The metal siding on a club house damaged by electrical arcing.

Photograph 8 Ground conductors of wiring within the club house were damaged.

(**Photographs 11 and 12**). The utility pole was made of wood (**Photograph 13**). Two levels of power lines were on the utility pole with 69 kV lines at the top (**Photograph 14**). Inspection of the power lines found splices in the power lines (**Photograph 15**). The power lines crossed over a gully where burning was still occurring underneath a 12 cm layer of ash at the time of the investigation (**Photograph 16**). The power lines were traced to the utility substation which controlled them. The wildland fire had destroyed vegetation and buildings in its path as it spread (**Photographs 17 and 18**). Electrical ground conductors were found severely heat-damaged in buildings that were not ignited by the fire as the result of an electrical power surge (**Photographs 19 and 20**).

The weather in the area was extremely dry with very little rainfall and at sunset the wind would pick up, sometimes exceeding 100 km h^{-1}.

Information was obtained from the utility company concerning protective devices in the utility substation meant to protect the power lines located in the area of the fire's origin. The lines were protected by an electrical recloser. A recloser is a device that senses an electrical problem on a power line such as contact with a tree branch. The recloser will momentarily remove electrical power from the line and then attempt to reclose it in order to restore electrical power if the problem has gone away. The number of attempts the recloser will make is determined by its settings. After a set number of attempts have

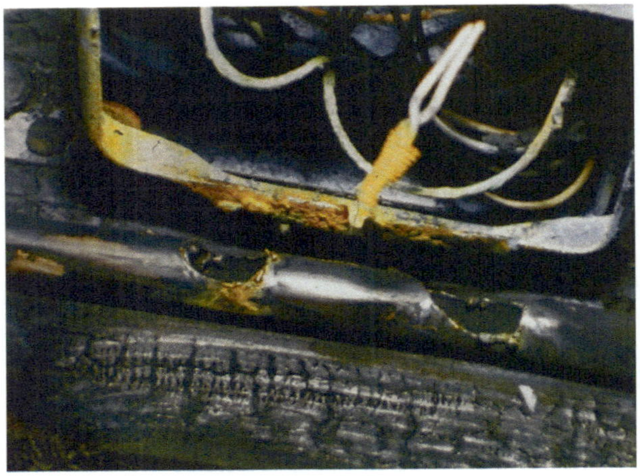

Photograph 9 Evidence of electrical arcing was found on a fuel line.

Photograph 10 Area surrounding the sporting club found that the local utility was installing power line poles and power lines at the time of the incident.

Photograph 11 Inspection of the burned land areas indicated that the fire originated near a utility electric pole.

Photograph 12 Closer perspective of burn patterns near utility poles.

Photograph 13 The utility pole made of wood.

Photograph 14 Two levels of power lines were on the utility pole with 69 kV lines at the top.

Photograph 15 Inspection of the power lines found splices in the power lines.

Photograph 16 The power lines crossed over a gully where burning was still occurring underneath a 12-cm layer of ash at the time of the investigation.

Photograph 17 The wildland fire had destroyed vegetation and buildings in its path as it spread.

Photograph 18 The contents of some buildings were completely destroyed as the fire spread.

Photograph 19 Electrical ground conductors were found severely heat-damaged in buildings that were not ignited by the fire as the result of an electrical power surge. Arrow in the photograph indicates the ground conductors.

been tried without success, the recloser will lock out and prevent further power from flowing through the power lines that it protects. To reenergize the power lines, a utility worker must manually reclose the recloser at the utility power substation. The records for the subject recloser indicated that it had locked out and been manually reclosed twice. Depositions of the utility workers revealed that on the night of the fire, one individual was at the utility power substation manually closing the recloser while another worker was on a hill looking to see where the power lines sparked or arced in the darkness. Both utility workers ran from the area after realizing they had ignited a wildland fire.

This fire was the result of the inappropriate action by the utility workers. The utility workers should have walked along the power lines inspecting them until they located the problem with the power lines.

Photograph 20 Heat-damaged ground conductors are indicated by the arrow in the photograph.

Manually closing the recloser after it had locked out without inspecting the power lines resulted in this fire.

The power line came down because the wind in the area after sunset quite frequently exceeded the capacity of the components on the wooden utility pole in the area of fire origin. The wooden poles were replaced with metal poles that have a higher wind capacity and they look like wooden poles from a distance.

The third case study involves a serious electrical injury to a maintenance worker. The maintenance worker had been assigned the task of cleaning the roof gutters on a large three-story building utilizing a motorized manlift (**Photograph 21**). The manlift was rented from a local company and they brought the manlift to the building location for the maintenance worker to use. The maintenance worker proceeded to clean the gutters of the roof and while lowering the manlift at one point, his head made contact with a 12 000 V power line. The 12 000 V power line was in close proximity horizontally to the building and ~2 m below the level of the roof gutters. The ground sloped away from the building, which required the manlift to have something to stabilize it when the platform is raised. The manlift had warning labels on it (**Photograph 22**) which read: "Danger electrocution hazard." "Do not operate this machine unless you have been trained in the safe operation of this machine." "Training includes complete knowledge of safety and operating instructions contained in the manufacturer's manuals." "This machine is not insulated." "You must maintain a clearance of at least 10 feet between any part of the machine or load and any electrical line or apparatus charged up to 50 000 V."

Photograph 21 The maintenance worker cleaning the roof gutters on a large three-story building utilizing a motorized manlift.

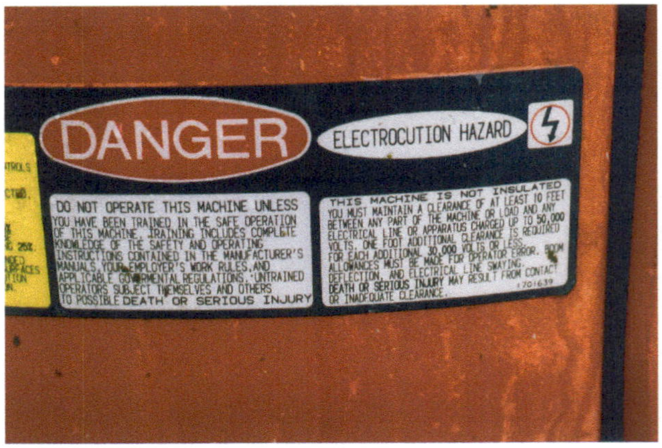

Photograph 22 Warning labels on the manlift.

The tires and base of the manlift were in contact with wet vegetation near the perimeter of the building, which provided a high-resistance current path from the lift's base to the ground (**Photograph 23**). This manlift did not have outriggers on it. Outriggers are devices which extend from the base of the lift in order to provide stability, particularly when the lift platform is raised. Outriggers are normally made of metal, which would have provided a low-resistance current path from the lift to the ground. The operator's platform could be entered through a metal gate on the platform (**Photograph 24**). Inspection of the interior side of the entrance gate revealed evidence of burning and clothing material adhering to it (**Photograph 25**).

Photograph 23 The tires and base of the manlift were in contact with wet vegetation near the base of the building.

Photograph 24 The operator's platform could be entered through a metal gate on the platform.

Photograph 25 Inspection of the interior side of the entrance gate revealed evidence of burning and clothing material adhering to it.

Photograph 26 Evidence of burning was also found on the frame of the platform near the operator's controls.

Evidence of burning was also found on the frame of the platform near the operator's controls (**Photograph 26**). The operator's controls were covered with spackling materials (**Photograph 27**). An unlabeled black box was found on the

Photograph 27 The operator's controls were covered with spackling materials.

Photograph 28 An unlabeled black box was found on the platform which was also covered with spackling materials.

platform which was also covered with spackling materials (**Photograph 28**). Cleaning the roof gutters did not involve the use of spackling materials. The spackling material covered the latches of the box and verified that it had not been opened by the maintenance worker. The unlabeled black box contained the manlift manufacturer's instruction book/user's manual for the safe operation of the lift.

Evidence of burning found on the interior side of the entrance gate and on the frame near the operator controls indicates that electricity exited the maintenance worker's back and left hand. Twelve thousand volts contacting the head of a person and exiting out their back and hand would normally electrocute them. However, the tires of the lift provided some isolation (high resistance) to the flow of electricity through the manlift to the ground. A manlift with outriggers should have been supplied by the rental company as the land around the perimeter of the building sloped away from it causing the manlift to be unstable when the platform was raised. However, metal outriggers would have provided a low-resistance path through the manlift to the ground and would have resulted in his electrocution. Instead, the worker was seriously burned and unconscious for more than a month. This incident was the result of the use of inappropriate equipment in a hazardous area by maintenance workers who were not aware of the hazards or not appropriately trained to perform the work.

The next case study is severe burn injuries that a building maintenance worker received from electrical arcing while attempting to change a door on a motor control center.

An electrical contractor had installed capacitors on a motor control center to improve the power factor of the electricity coming into the building (**Photograph 29**). Electric utilities quite frequently charge their customers additional fees for bad electrical power factors. Power factor is the ratio of total watts to the total root-mean-square of volt amperes. A disconnect switch for the capacitors was installed at the bottom of the motor

Photograph 29 An electrical contractor had installed capacitors on a motor control center to improve the power factor of the electricity coming into the building.

control center (**Photograph 30**). The electrical contractor did not install the correct door to operate the capacitor disconnect switch in the bottom unit. Subsequently, the contractor asked a maintenance worker of the building to replace the door with the correct unit. The motor control center had warning labels on it. One warning label stated: "DANGER: hazard of electrical

Photograph 30 A disconnect switch for the capacitors was installed in the bottom of the motor control center.

Photograph 31 Warning labels on the motor control center.

shock or burn. TURN OFF POWER supplying this equipment before working inside" (**Photograph 31**). Another warning label stated, "TURN OFF POWER AT UTILITY before working behind this cover. The line side bus of any main breaker or switch remains energized when the main is in the off position" (**Photograph 32**). A manufacturer's label on the motor control center indicated that the voltage within it was 277/480 V, three phase, and it had a current rating of 2000 A (**Photograph 33**).

Photograph 32 Another warning label.

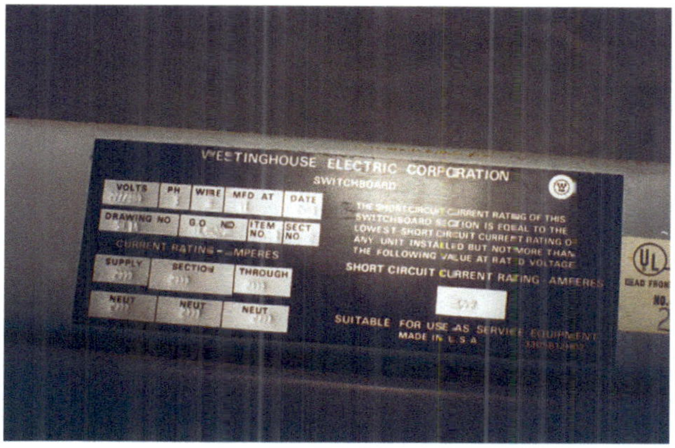

Photograph 33 Manufacturer's label on the motor control center indicating that the voltage within it was 277/480 V, three phase, and it had a current rating of 2000 A.

The cardboard covering the bottom opening of the motor control center was removed and evidence of burning and arcing was found within it (**Photograph 34**). The disconnect switch that was previously located in the burned opening had been retained (**Photograph 35**). Electrical arcing burned a hole through the steel at the top of the unit and through its back (**Photograph 36**).

The building maintenance worker was not an electrician. He had no knowledge of the hazards involved in working within this motor control center. He was not wearing any personal protective equipment such as eye goggles, face shield, insulating gloves, insulating sleeves, or insulated safety hat. When he was trying to change the door on the capacitor control compartment, his screwdriver slipped and made contact

Photograph 34 The cardboard covering the bottom opening of the motor control center was removed and evidence of burning and arcing was found within it.

Photograph 35 The disconnect switch that was previously located in the burned opening had been retained.

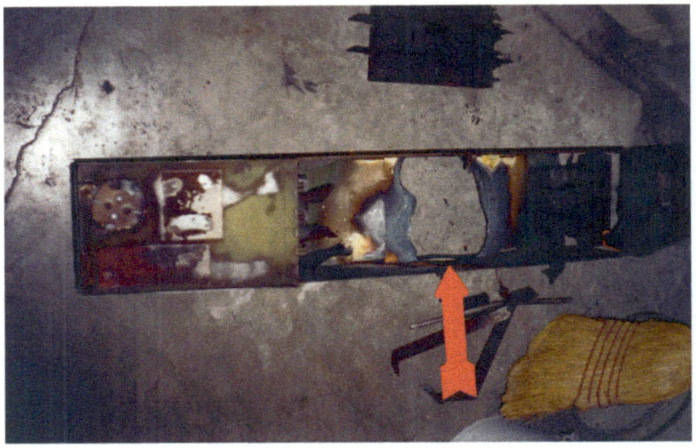

Photograph 36 Electrical arcing burned a hole through the steel at the top of the unit and through its back.

Photograph 37 The hotel had four ways of traveling from its registration desk to its guest rooms: walking, golf cart, boat, or electric train.

Photograph 38 A maintenance manager at the hotel had found a loose cover over the electric train track which passed through a pedestrian walkway.

between a charged 277 V component and the grounded enclosure. This caused an arc blast within the unit that vaporized copper and steel components of the compartment. As he was bending over in front of the compartment at the time, he was severely burned from the top of his head to his waist.

To safely change the door on the motor control center, electrical power would have to be turned off at the electrical utility's supply point and locked out. The capacitors should be discharged and the power supply conductors grounded to drain off any residual electrical charge within them, prior to attempting maintenance work.

The fifth example/case study is an electrocution which occurred at a luxury hotel. The hotel had four ways of traveling from its registration desk to its guest rooms; walking, golf cart, boat, or electric train (**Photograph 37**). A maintenance manager at the hotel had found a loose cover over the electric train track which passed through a pedestrian walkway (**Photograph 38**). The walkway was wet as the canal for the boat system was located in close proximity 2 m away. The maintenance manager was concerned that a guest might slip and fall and hurt themselves. Therefore, he attempted to tighten bolts which held the covers in place (**Photograph 39**). While reaching underneath the train track cover, he was electrocuted.

Examination of the underside of the train revealed that it received electrical power from electrical bus bars energized with 480 V of alternating current beneath it (**Photograph 40**). Electrical contact shoes (**Photograph 41**) would slide in

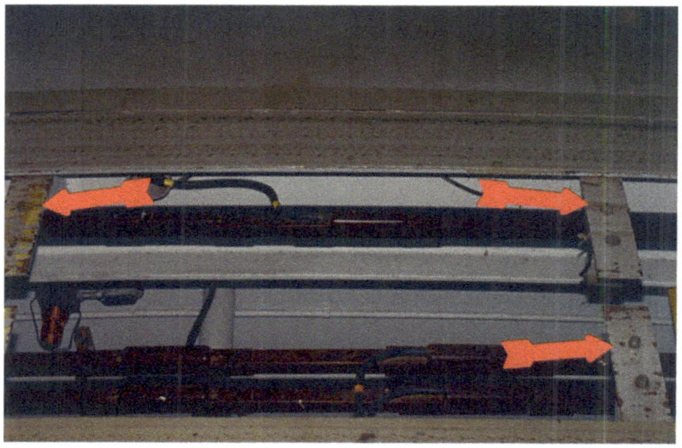

Photograph 39 The maintenance manager attempted to tighten bolts which held the covers in place.

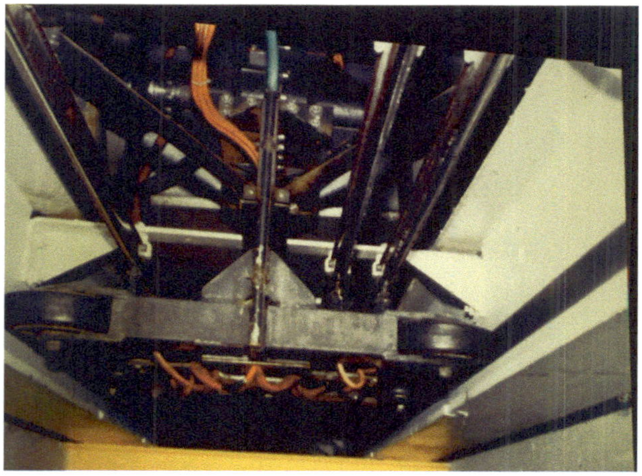

Photograph 40 Examination of the underside of the train revealed that it received electrical power from electrical bus bars.

Photograph 41 Electrical contact shoes.

Photograph 42 Electrical contact shoes and energized bus bars

a groove of the bus bars as the train moved along its track to obtain power for its traction motors (**Photograph 42**). An electrical joint compound was utilized to reduce the electrical resistance between the bus bars and electrical shoes (**Photograph 43**). No evidence was found that the manager made contact with the bus bars. However, an electrical joint compound which conducts electricity was found between the bus bars and bolts that held the train track cover in place. Subsequently, the man was electrocuted when he touched the electrical joint compound while kneeling on the wet concrete walkway which grounded him. Electrical current entered his body through his hand, passed through his heart, and exited out his knees, killing him.

The sixth and final case study is that of a church fire (**Photograph 44**). This church was utilized by two different denominations, with differing beliefs, and an arsonist was known to be active in the area. A clockwise inspection of the church exterior found smoke residue around an opening along its base near the northeast corner (**Photograph 45**). Examination of the burn patterns within the church indicated that the fire originated in the northeast corner below the first pew (**Photographs 46 and 47**). Electrical copper and aluminum wiring was found passing through the area of fire origin (**Photograph 48**). Fused disconnect switches located in the basement of the church supplied electrical power to the wiring which passed through the area of origin (**Photograph 49**). One fused disconnect switch was missing its fuses and the aluminum wiring previously attached to it had been cut off (**Photograph 50**). Heat damage was found on the cut-off wires

Photograph 43 An electrical joint compound was utilized to reduce the electrical resistance between the bus bars and electrical shoes.

Photograph 44 The sixth and final case study is that of a church fire.

(**Photograph 51**). The manufacturer's label on the inside of the fused disconnect switch was clearly heat-damaged (**Photograph 52**). The aluminum wiring was traced to a second-floor gymnasium attached to the church that was utilized to play basketball. The aluminum wiring was spliced within a junction box in the gymnasium that fed power to an electrical receptacle.

Photograph 45 A clockwise inspection of the church exterior found smoke residue around an opening along its base near the northeast corner.

Photograph 46 Examination of the burn patterns within the church indicated that the fire originated in the northeast corner below the first pew.

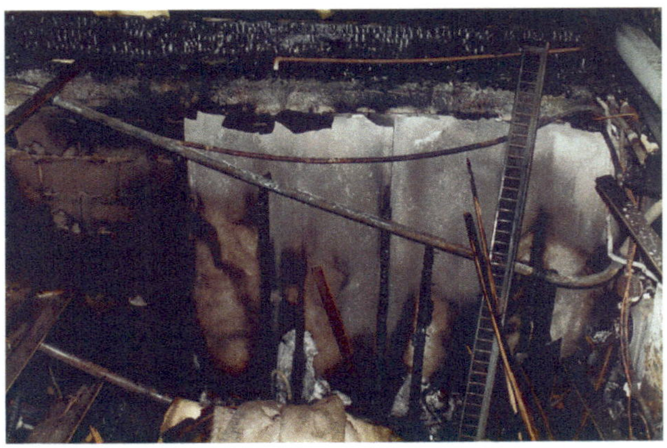

Photograph 47 Burn patterns on the north wall beneath the first pew.

Photograph 48 Electrical copper and aluminum wiring was found passing through the area of fire origin.

Photograph 49 Fused disconnect switches located in the basement of the church supplied electrical power to the wiring which passed through the area of origin.

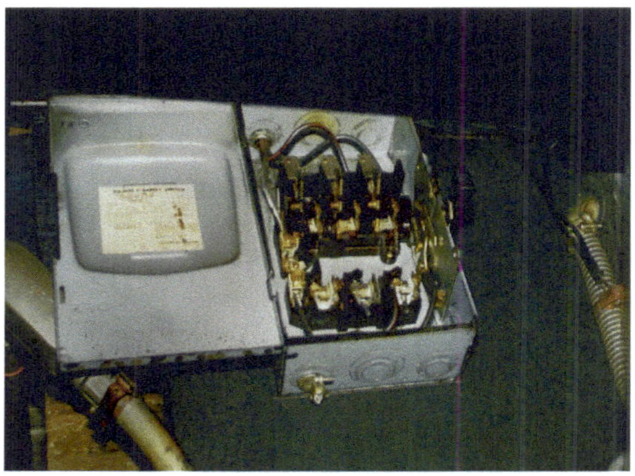

Photograph 50 One fused disconnect switch was missing its fuses and the aluminum wiring previously attached to it had been cut off.

Photograph 51 Heat damage was found on the cut-off wires.

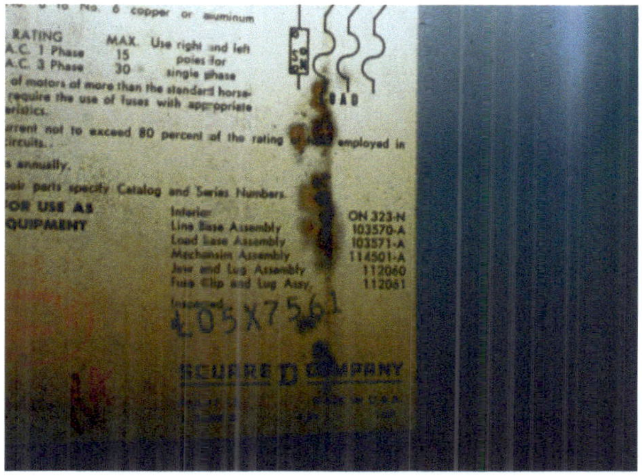

Photograph 52 The manufacturer's label on the inside of the fused disconnect switch was clearly heat-damaged.

When the junction box was opened, a distinct odor of burning electrical insulation was detected and evidence of overheating was found on the spliced wiring (**Photograph 53**). The insulation of the aluminum wiring attached to the receptacle showed evidence of overheating (**Photograph 54**). The receptacle had a room air conditioner plugged into it that did not have the capacity to cool the gymnasium. Therefore, it would draw its maximum rated current when turned on. Testing of the air conditioner found that its compressor motor had shorted out and it would draw more current than rated when turned on. On the day of the fire, there was basketball practice in the gymnasium and the temperature outside the building exceeded 38 °C.

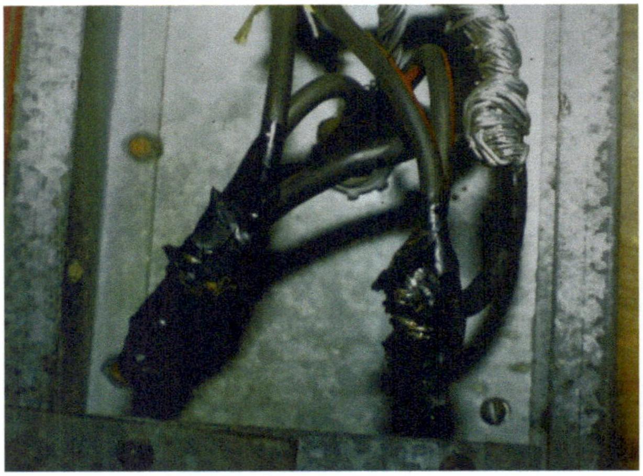

Photograph 53 When the junction box was opened, a distinct odor of burning electrical insulation was detected and evidence of overheating was found on the spliced wiring.

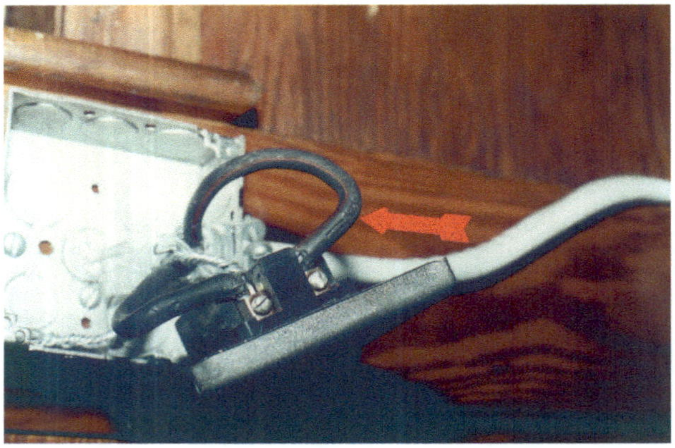

Photograph 54 Insulation of the aluminum wiring attached to the receptacle showed evidence of overheating.

Electricians making repairs in the church were contacted and the fuses previously located in the fused disconnect switch were obtained (**Photograph 55**). The fuses were rated at 100 A and showed evidence of overheating, but they had not opened. The aluminum electrical wiring that they protected was rated at 50 A. The 100 A fuses would not properly protect the 50 A wiring and the resistance of the excessively long circuit limited the fault current. Consequentially, the fire was caused by overheated aluminum wiring that passed through the area of fire origin where heat was entrapped by wooden framing, which eventually ignited.

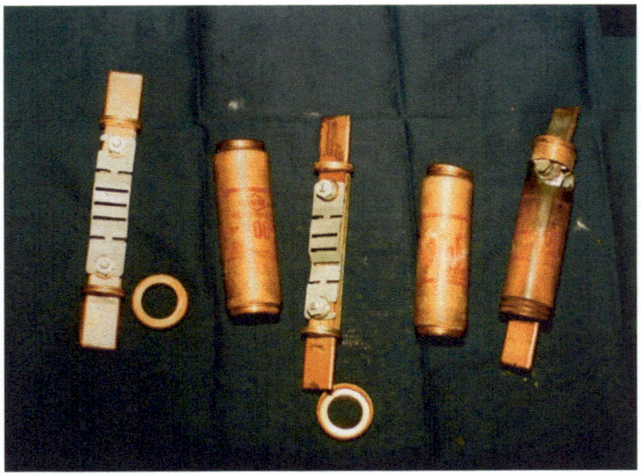

Photograph 55 Electrician smaking repairs in the church were contacted and the fuses previously located in the fused disconnect switch were obtained.

See also: **Engineering:** Human Factors Investigation and Analysis of Accidents and Incidents; **Forensic Medicine/Clinical:** Electrocution and Lightning Strike.

Further Reading

ASTM, 2011. Construction, vols. 04.01–04.13. ISBN:978-0-8031-8503-5.

DeHaan, J.D., 2002. Kirk's Fire Investigation, fifth ed. Prentice Hall, Upper Saddle River, NJ. 0-13-060458-5.

Encyclopedia Britannica, 1978, fifteenth ed., vol. 6, p. 536. ISBN:0-85229-330-5.

Encyclopedia Britannica Micropaedia, 1978, fifteenth ed., vol. 3, p. 1006. ISBN:0-85229-330-5.

Franklin, B., June 1753. Request for Information on Lightning. Gazette, Pennsylvania.

Golde, R.H., 1977. Lightning, vol. 1. Academic Press, pp. 170–172.

Icove, D.J., DeHaan, J.D., 2004. Forensic Fire Scene Reconstruction. Pearson Prentice Hall, Upper Saddle River, NJ, 0-13-094205-7.

IEEE, 1978. Standard Dictionary of Electrical & Electronic Terms, second ed. LCCN, pp. 77–92333.

IEEE, 1995. Circuit Breakers, Switchgear, Substations, and Fuses, C37. ISBN:1-55937-507-8.

IEEE, 1997. National Electrical Safety Code, C2. ISBN:1-55937-715-1.

International Building Code, 2009. 5th Printing, Chapters, pp. 1–35.

Luther, 2008. http://www.Luther.de/en/ and http://chi.gosel.com.net.

Matthews, J., 1998. Fear and Lightning, vol. 1. Chain Reaction Magazine, pp. 22–23. http://chainreaction.asu.edu.

Morse, R.A., 1752. Benjamin Franklin: Papers on Electricity, pp. 276–277, 92004. http://www.franklinpapers.org. and Effect of lightning on Captain Waddel's Compass and on the Dutch Church in New York, from James Bowdoin to Benjamin Franklin, Boston, 2 March 1752.

Nabors, R.E., 1994. Forensic Electrical Engineering & Liability. Lawyers & Judges Publishing Company, Tucson, AZ, 0-913875-07-4.

Nabors, R.E., Fish, R.M., Hill, P.F., 2004. Electrical Injuries Engineering, Medical and Legal Aspects, second ed. Lawyers & Judges Publishing Company, Inc., Tucson, AZ. 1-930056-71-0.

National Electrical Manufacturers Association. ANSI C84.1 Electric Power Systems and Equipment.

NFPA, 1998. National Fire Codes, a Compilation of NFPA Codes, Standards, Recommended Practices, Manuals and Guides. National Fire Protection Association, Quincy, MA, 0-87765-140-X.

NFPA, 2008. NFPA 921 Guide for Fire & Explosion Investigations, 2008 ed National Fire Protection Association, Quincy, MA.

NFPA, 2010. National Electrical Code, International Electrical Code Series, NFPA 70. National Fire Protection Association, Quincy, MA.

NFPA 921, 2008. Guide for Fire and Explosion Investigations, p. 153, 17.4.5.6.

Saferstein, R., 2011. Criminalistics, tenth ed. Prentice Hall, Upper Saddle River, NJ, pp. 28–40. ISBN:13: 978-0-13-504520-6. (Chapter 2).

Underwriters Laboratories, Inc. UL 489, Molded-Case Circuit Breakers and Circuit-Breaker Enclosures.

Wright, R.K., 1997. Investigation of Electrical Injuries and Death. American Academy of Forensic Science Workshop, New York, p. 8.

Relevant Websites

www.aafs.org—American Academy of Forensic Sciences.
www.astm.org—American Society of Testing & Materials
www.cpsc.gov—Consumer Product Safety Commission.
www.craigslist.org—Craigslist.
www.ieee.org/innovate—Institute of Electrical and Electronic Engineers.
www.nafe.org—National Academy of Forensic Engineers.
www.ul.com—Underwriters Laboratory Inc.
www.youtube.com—YouTube.

Investigation and Analysis of Structural Collapses

RT Ratay, Columbia University, New York, NY, USA
DB Peraza, Exponent, Inc., New York, NY, USA

Glossary

Beam A relatively long structural member, usually in a horizontal position spanning between supports, carrying loads transverse to its longitudinal axis.
Bracing Providing support against lateral movement.
Buckling Sudden lateral displacement of a slender structural member under axial compressive loading.
Collapse Total failure, falling down.
Column A structural member, usually vertical and in compression, such as a post or pillar.
Formwork A temporary construction to contain the wet concrete in the desired shape while it is hardening.
Fracture surface The surface along which a material breaks.
Lateral torsional stability Ability to resist buckling which involves lateral deflection and twisting.
Reasonable engineering certainty The term arises from courtroom rules of evidence and, as such, is a legal term. Attained when the investigator satisfies himself/herself of the legitimacy of his/her conclusions and supports his/her opinion with sound technical foundation.
Serviceability The capability of the structure to perform the function for which it was designed.
Space frame A three-dimensional structural framework, such as a roof structure with structural members in more than one plane.
Stress Internal force per unit area, usually given in pounds per square inch, or kilogram per square centimeter.
Understrength Having strength less than required to resist the loads.

Overview

Structural failures may be defined as the unacceptable difference between intended and actual performance. At one extreme, it may be a "catastrophic collapse" with fatalities, and at the other extreme, it may be a serviceability issue, such as excessive deflection/displacement or premature cracking.

Most structural failures are the result of human activities. Errors that can result in failure can occur anywhere along the planning–design–construction–service lives of a constructed facility:

In planning—design concept
In design—detailed design, calculations, drafting
In the design–construction interface—shop drawing detailing, drafting, review, and approval
During construction—erection, inspection, accident
In service—misuse, overload, alteration, adjacent construction, lack of maintenance
During demolition—improper methods

Structural failures, whether they occur during construction or during the service lives of structures, are usually followed by forensic investigations. They often create claims of loss that result in legal proceedings toward the resolution of those claims. The determination of the cause of the failure and the parties responsible for it, as well as the technical presentation of the claim on the one hand, and the presentation of the technical defense on the other hand, require the engagement of structural engineering experts.

The investigator of a failure needs to have an understanding not only of structural behavior but also of codes and standards and of the design–construction process in order to recognize where along the process errors were likely to occur that could have precipitated the failure.

The purpose of a collapse investigation is threefold: (1) to determine the cause(s) of the failure, (2) to identify the party(ies) responsible, and (3) to learn from the mistakes.

The typical process of forensic investigation of a failure, hence the activities of the expert consultant investigating the case, may be coarsely outlined as follows:

- First-response and preliminary assessment
- Development of investigation strategy
- Fact-gathering and document review
- Engineering analyses to determine cause(s) and responsibilities
- Reporting

 http://dx.doi.org/10.1016/B978-0-12-802718-9.00029-6

Figure 1 Marcy Bridge, Utica, NY, USA, collapsed during construction. Courtesy of Robert T. Ratay.

Figure 2 Hartford Civic Center, Hartford, CT, USA, space frame roof collapsed years after construction. From Carper, K.L. (Ed.), 1989. Forensic Engineering. Elsevier Science, New York.

If the failure results in legal action, further investigation may be done as part of overall litigation support, or just in response to the findings of opposing experts.

Except for litigation support, the process is not country dependent. It depends rather on the nature and magnitude of the failure; fatalities, if any; the monetary value of damages and claims; and the time and money available.

Following the October 10, 2002 collapse of a footbridge during construction, shown in **Figure 1**, it was obvious that the omission of internal lateral torsional stability in the design, absence of external bracing during construction, and lack of communication between the designer and contractor contributed to the failure. Very little investigation was required to confirm the initial findings. Following the January 18, 1978 collapse of the large space-frame roof structure, shown in **Figure 2**, under snow load years after its completion, it was clear from preliminary analyses that the primary cause of the failure was an error in the design of the compression members of the space frame, but the engineering investigation of the case continued on for years.

When a collapse occurs, the full scope of work of the investigation can include the following activities:

- First steps
  - Initial field observations
  - Engineering assistance in rescue operations, if any
  - Assurance of public safety
  - Documentation of field conditions
  - Preservation of physical evidence
  - Interviews of eyewitnesses
  - Initial engineering assessment
  - Initial stabilization, repair, and restoration
- Main investigation
  - Familiarization with the project
  - Planning the investigation (document reviews, analyses, tests)
  - Identifying additional expertise required
  - Site inspection and collection of physical evidence
  - Review of design and construction documents
  - Establishing the design, construction, and service history
  - Formulating possible failure scenarios
  - Performing analytical studies and tests
  - Evaluating probability of possible failure scenarios
  - Selecting most probable failure scenario
  - Scrutinizing convergence of evidence
  - Demonstrating validity of opinion
- Reporting
  - Interim communications
  - Oral presentations
  - Written reports
  - Presentation materials and exhibits
- Litigation support—if required
  - Depositions
  - Preparation of demonstrative materials
  - Testimony

First Steps

As with any forensic investigation, the first steps in the investigation of a structural collapse must focus on the preservation of highly perishable physical evidence. There are, however, several aspects that are unique to structural investigations.

First of all, structural collapses usually involve large components. These components often can only be handled using construction equipment. In many cases, not only are the components large but they may also be numerous. A large area will often be needed in order to lay out and securely store these components for further inspection and/or testing. This area, due to its size and the need for the use of heavy equipment to handle the components, often is located outdoors. Items that cannot tolerate exposure to the environment may be placed within a secure trailer or protected in some other manner.

And of course, the nature of the perishable evidence is different from that in other forensic sciences. Following are examples of highly perishable evidence that are associated with structural collapses.

Collapse Configuration

The collapse configuration is a highly perishable evidence. The postcollapse position of the components provides valuable information regarding the sequence of the collapse, which can help identify the area where the collapse originated. It can also provide information regarding the mechanism of collapse.

There are many valid reasons why the physical evidence will need to be moved, modified, or even destroyed. Chief among these is the need to rescue or recover victims who are trapped in the debris. Rescue workers may have to cut structural members and remove debris in order to reach victims. Another reason is the need to prevent further collapse, which could result in additional injuries or property damage. This may require the demolition of certain portions of the structure and/or the installation of shoring or bracing. Finally, there are cost-related reasons, such as minimizing business interruption or the need to complete a construction project by a given date.

It is therefore important to quickly capture the collapse configuration. There are a number of useful techniques for doing this, such as aerial photography, photogrammetry, three-dimensional (3D) laser scanning, and personal observation.

Aerial photographs help provide an overall view of the postcollapse location of components, which can be invaluable. Likewise, overall photographs from various angles and vantage points can provide valuable perspectives.

Software is available to help convert photographs into 3D digital models using the principles of photogrammetry. Digital models created using this technique can be very useful for visualization. Although the dimensional accuracy is variable, it is sufficient for many purposes. The main advantage of this technique is that no special field equipment or personnel is needed, since photographs can be taken with an ordinary camera. In many instances, the technique can be applied "retroactively" to photographs that were not taken with this technique in mind.

A technique for quickly capturing accurate dimensional information of a large site is the use of 3D laser scanning. This is a powerful technology that is capable of capturing the coordinates of thousands of points per second, over a 360° field of view. Scanners typically have a range of nearly 1000 ft and an accuracy of approximately ¼ in. The technique involves the use of a tripod-mounted laser scanner and is often performed by firms with expertise in land surveying. The data are postprocessed with specialized software to create a 3D digital model that can be viewed from any angle.

It is important that the engineer investigating a collapse be present while the structure is dismantled. High-quality field notes, photographs and possibly video are invaluable tools for documenting the collapse configuration and the removal process. In case of large collapses, it will be important to use a nomenclature system for uniquely identifying the various components as they are removed. If drawings for the structure are available, they will provide an obvious starting point for a nomenclature system.

Fracture Surfaces

The fracture surfaces of steel components can contain features that provide clues as to the type of load that caused the failure. These features can be obscured, or altogether destroyed, if the surface is not protected from the elements and is allowed to corrode. A common technique to protect these fracture surfaces from corrosion is to spray paint them with a pigmented coating. The color provides visual confirmation of which surfaces have been protected. The coating can be readily removed once the item has reached the laboratory to allow study of the surface.

Curing Concrete

If a concrete building collapses during construction, one potential cause that must be investigated is whether the formwork was removed prematurely, before the curing concrete had reached the required compressive strength. It is therefore important to determine, as closely as possible, the strength of the curing concrete at the time of failure. The strength of recently placed concrete can change dramatically in the first few days and weeks of its life, so it is important that this be done as quickly as possible. The most reliable technique for doing so is to locate the standard test cylinders that were cast from the batch of concrete of interest when it was placed and to have these tested immediately, as opposed to waiting 7 or 28 days as is customary.

Snow and Other Loads

When roofs collapse under snow load, it is important to try to quantify the weight of the snow that was on the roof at the moment of collapse. The weight and distribution of the snow will change as it melts, it slides, and/or it is blown by the wind. The most direct and reliable method is to take measurements on the roof itself, using containers of known size and scales. Measurements of the snow weight on nearby roofs and on the ground next to the collapsed roof are also valuable.

If there are other loads that may have played a role, such as heavy construction materials that were stockpiled on a floor that was under construction or items stored in a warehouse, it will be important to estimate their weights and location before they are removed.

The Investigation

Once the perishable evidence has been secured, the focus of the forensic evidence turns to the actual investigation.

Investigations of collapses are usually led by structural engineers. However, depending on the nature of the collapse, additional expertise may be needed to perform a comprehensive investigation. The investigative team may need other types of engineers, scientists, and specialists, such as geotechnical engineers, metallurgical engineers, petrographers, chemists, fire engineers, wood scientists, polymer scientists, and crane engineers. It is important to identify the needed expertise early and to engage the appropriate people.

There are a number of fundamental questions that a structural investigation must address.

What was the trigger of the collapse? Every collapse has a trigger. The trigger is the event that caused the collapse to occur at that particular moment. In many cases, but not all, the trigger is obvious. Triggers fall into one of two broad categories: (1) added load applied to the structure or (2) a reduction in the strength of the structure. Examples of added load may be wind load, weight of freshly fallen snow, placement of heavy construction materials on a floor, and impact from a vehicle. Examples of reduction in strength include the removal of a structural member during renovation, corrosion of steel members, and deterioration of wood members.

What was the status of the construction at the moment of collapse? This is an important question if the structure was being constructed or renovated. In some cases, it takes considerable effort to determine this, since documentation of the progress of construction is often incomplete. A related question is what activities were being performed at that moment? These questions often require the input from eyewitnesses, who may be reluctant to share what they know. Whenever possible, eyewitness statements should be collaborated by physical evidence.

What external loads were acting at the time of the collapse? External loads include environmental loads, such as wind load, snow load, and seismic load. Other external loads include live load and loads such as people, furniture, materials, or vehicles. The self-weight, or dead load, of a structure is also usually considered an external load. Some of these loads, such as dead load, are well defined for a given structure and can be determined relatively accurately. Others, such as the load from an impacting vehicle, cannot be determined as accurately.

What were the internal stresses or forces that were acting? Once the external loads have been estimated, they are converted into internal stresses or forces using the principles of mechanics of materials and of structural analysis. For simple structures, hand calculations may be sufficient. For more complex structures, specialized software is often used to create a digital structural model and to perform the analysis. The structural model includes the geometry of the structure, the mechanical properties of each component, the restraint conditions, and the external loads. The structure can be modeled as a wireframe using 1D elements, as surfaces using 2D finite elements, or as a solid with 3D finite elements. The model may be of an entire building, including columns, beams, and floor slabs, or it may be of a single connection. The geometry may be based on design drawings for the structure, if observations of the as-built structure indicate it was built in substantial accordance with the design. Otherwise, the as-built configuration must be used.

What was the resistance of the structure? The resistance of the structure is determined using principles of structural engineering and mechanics of materials. For example, the buckling strength of a compression member depends on the material strength, the member length, the cross-sectional geometric properties of the member, end restraints, and other factors. Laboratory testing may be needed to determine the actual material strength or the strength of an assembly.

Where did the collapse originate? Most collapse initiate at a singular point, and knowing the origin of the collapse is helpful for determining the sequence and progression of the collapse. The collapse configuration, which was discussed under "First Steps," can often provide valuable clues regarding the origin and the sequence. The results of the structural analysis, discussed above, will also help identify the "weakest links." There are some cases where the precise origin is immaterial. For example, if a series of roof girders were all constructed with the same defect, and they all collapsed, it may not be important to determine which of the girders failed first.

What was the "primary cause" *of the collapse?* This is the deficiency that, by itself, is sufficient to cause the collapse. Another term commonly used, primarily for failures in industrial facilities, is "root cause." Examples of primary causes include understrength material, an omission by the design engineer, failure to construct in accordance with the design, and inadequate inspection. The primary cause is typically determined by synthesizing all the available information: eyewitness accounts, physical evidence, laboratory testing, and results of structural analyses. As with other forensic investigations, the "process of elimination" plays an important role in determining the primary cause(s). This involves the following:

1. Identify all possible causes.
2. Eliminate those causes that are not probable.
3. Investigate the remaining causes to determine which one(s) best satisfies the physical and engineering evidence.

4. Determine, if possible, the most probable cause to a level of reasonable engineering certainty. In some cases, there may be multiple causes that are highly probable.

What were the contributing factors to the collapse? A contributing factor is one that, by itself, is not sufficient to cause the collapse, but in combination with other factors can cause the collapse to occur at a particular moment. Or, once the collapse is initiated, a contributing factor can allow the collapse to progress further or more completely than it would have otherwise. Determining these factors also requires evaluating all the available information. It also often requires determining the relative contribution of each of these factors to the collapse, which may require performing multiple hypothetical analyses.

Who was responsible? If there is litigation involved, it may be necessary to go to the next level, which is to assist the attorneys in determining liability. In order to do this, it is necessary to understand the roles and responsibilities of the various parties, to be familiar with the applicable building codes, to be aware of industry practices, and to obtain and meticulously review all the project documents. When opining about responsibility, the investigators have to compare the activities of the various parties to some accepted standards of performance. The performance of engineers is compared to the "standard of care," which is defined as "That level or quality of service ordinarily provided by other normally competent practitioners of good standing in that field, when providing similar services with reasonable diligence and best judgment in the same locality at the same time and under similar circumstances." The performance of contractors is usually defined by the "duty to perform," that is, strict adherence to plans and specifications.

Several reports of seminal collapse investigations are suggested for further reading at the end of this chapter. They illustrate the variety of approaches that are used to answer these crucial questions.

The Investigator/Expert

There are several published definitions of the expert investigator/witness. What they all include are that the expert has the knowledge and skill necessary to evaluate the technical aspects of the case, can use that knowledge and skill to conduct the evaluation, can form an opinion with "reasonable engineering certainty," should be unbiased and free from conflicts of interest, should be truthful, and, if the case winds up in a court of law, should assist the trier of fact (the judge and the jury) in understanding the technical aspects of the case.

Professional opinion cannot always be with 100% certainty. The courts in the United States, for example, require that the opinion be only "with reasonable degree of engineering certainty." The phrase arises from courtroom rules of evidence and, as such, is a legal term. This suggests, among other things, that all mutually exclusive modes of failures need not be investigated if evaluation of the most reasonable causes leads to a definitive conclusion. This may pass the legal test, but might fall short of the definition of a thorough investigation.

Engineers should be thorough in their investigations and careful with the statements of their opinions. A credible, honest, ethical investigator satisfies himself/herself of the legitimacy of his/her conclusions and supports his/her opinion with sound technical foundation.

Often, different investigators arrive at different conclusions. This may be because two engineers working independent of each other may have access to different data, approach the problem from different directions, put unequal emphases on the same evidence, and/or be influenced by their past experiences, preexisting opinions, and biases.

Experts should be aware that their services can be subject to claims of negligence and the liability may extend far beyond the fee for services. A client who used an expert may find the basis to sue that expert, if he/she lost the case or believes that the expert did not perform adequately. Simply being an expert on the losing side of a case is not a sufficient cause for a client to bring suit against an expert; however, if the expert's performance was below the applicable standard of care, he/she may be held liable for the loss of the case.

Postinvestigation

To the credit of the structural engineering profession, failures have been and continue to be used to improve design and construction practices.

A valuable peripheral benefit of a laborious investigation is a clearer understanding of structural behavior and a better appreciation of pitfalls in the current practices. These can provide information and material to bring about changes in design and/or construction practices, codes, standards, oversight, and regulatory procedures that can prevent reoccurrence of similar collapses.

See also: **Engineering:** Materials Analysis and Failure Analysis.

Further Reading

ASCE, Technical Council on Forensic Engineering, 2012. Guidelines for Forensic Engineering Practice, second ed. ASCE, Reston, VA.

Carper, K.L., Feld, J., 1997. Construction Failure, second ed. Wiley, New York, NY.

Cuoco, D.A., Panariello, G.F., 2010. The engineering investigation process. In: Ratay, R.T. (Ed.), Forensic Structural Engineering Handbook, second ed. McGraw-Hill, New York, NY (Chapter 6).

Delatte, N.J., 2008. Beyond Failure: Forensic Case Studies for Civil Engineers. ASCE, Reston, VA.

National Transportation Safety Board, November 2008. Collapse of I-35W Highway Bridge, Minneapolis, Minnesota, August 1, 2007. National Transportation Safety Board, Washington, DC. http://www.dot.state.mn.us/i35wbridge/ntsb/finalreport.pdf.

Peraza, D.B., 2007. The first steps after a failure. In: Proceedings of ASCE Structures Congress, Long Beach, CA, USA, May 2007.

Peraza, D.B., 2010. The first steps after a failure. In: Ratay, R.T. (Ed.), Forensic Structural Engineering Handbook, second ed. McGraw- Hill, New York, NY (Chapter 5).

Ratay, R.T., 2009. Forensic structural engineering in the USA. In: Proceedings of the Institution of Civil Engineers. Special Issue, May 2009, London, UK.

Ratay, R.T. (Ed.), 2010. Forensic Structural Engineering Handbook, second ed. McGraw-Hill, New York, NY.

US Department of Commerce, May 1982. Investigation of the Kansas City Hyatt Regency Walkways Collapse. National Bureau of Standards, US Department of Commerce, Washington, DC. http://www.fire.nist.gov/bfrlpubs/build82/PDF/b82002.pdf.

US Department of Commerce, January 2010. Final Report on the Collapse of the Dallas Cowboys Indoor Practice Facility, May 2, 2009. Building and Fire Research Laboratory, National Institute of Standards and Technology, US Department of Commerce, Washington, DC. http://www.bfrl.nist.gov/investigations/pubs/NISTIR7661_January%202010.pdf.

Biomechanics of Human Gait—Slip and Fall Analysis

TE Lockhart, Virginia Tech, Blacksburg, VA, USA

Glossary

Friction demand (μ_d) Friction demand (μ_d) is a ratio between horizontal and vertical ground reaction forces during the heel contact phase of the gait cycle. It represents the overall requirement for the foot to not slip significantly.

In order to analyze the principles of body stability and the mechanism of slips and falls, it is necessary to understand the dynamic principles of human locomotion. In this chapter, the biomechanics of human gait is presented to elaborate on the "human" aspects of slip and fall accidents. This information may provide a comprehensive tool for forensic scientists to better understand the relationship between the "biomechanics of human gait" and "slip and fall accidents" and may assist in formulating cogent expert opinions.

Biomechanics of Human Gait

The purpose of this section is to assist in the understanding of the gait mechanisms involved in walking, which reflect the dynamic principles of each body segment in locomotion that are relevant for investigating slip and fall accidents. The purpose is also to describe the motor control strategies associated with walking to reveal the complex aspects of initiating events leading to fall accidents and its relationship to posture and balance in human gait. Understanding the motor control strategies associated with walking may provide us with information relevant to decouple the system into identifiable pieces for the scientific reconstruction of accidents. Here, we begin with some basic concepts in human gait.

Locomotion, an inclusive characteristic of all animals, is defined as the process by which an animal moves itself from one location to another. Walking is characterized as a method of locomotion involving the alternate use of the two legs to provide both support and propulsion. Finally, "gait" can be described as the mannerism or style of an individual's walking pattern. Human locomotion falls into a general category known as "striding bipedalism," a locomotion activity in which the center of gravity (COG) is carried alternately over the right and left foot. Normal walking depends on a series of reciprocal movements involving the alternation of the function of each leg between supporting the body and advancing into the next position. These complex tasks are governed by open- and closed-loop motor control systems.

The mechanical definition of human walking and especially the function of gait may help us to better understand human locomotor control. In essence, the main purpose of walking is to transport the body safely and efficiently across the terrain. In order to do so, five major functions of gait must be performed during each step: (1) maintaining upper and lower limb support (such as preventing collapse of the leg during stance phase of the gait cycle), (2) maintaining upright posture and balance of the body, (3) foot trajectory control (such as acquiring safe ground clearance and heel contact), (4) mechanical energy generation to maintain or increase forward body speed, and (5) mechanical energy absorption to decrease forward speed of the body. All of these major functions must be performed within the biomechanical constraints of the body and the physical constraints of the environment. Additionally, the central nervous system (CNS) must also integrate efferent feedback from sensory organs to generate the correct patterns of moments of force at each joint to compare and represent the internal model against the real-world interactions.

Body Segments in Locomotion

Apart from the multiple variations that may occur between different individuals or within the same individual (for instance, as a result of changes in the speed of walking), there are certain observable events that are shared by all. This is because the mastering of the erect bipedal type of locomotion is a learned process and associated with certain personal peculiarities superimposed on the basic pattern of erect bipedal locomotion. In the following subsections, the process of walking in terms of body segments (i.e., lower and upper extremities) in locomotion is further reviewed.

Lower extremities: Gait has been studied using the "walking cycle," which is the time interval between successive floor contacts of each foot. The activity of one leg can be divided into a short swing phase and a longer stance phase.

 http://dx.doi.org/10.1016/B978-0-12-802718-9.00030-2

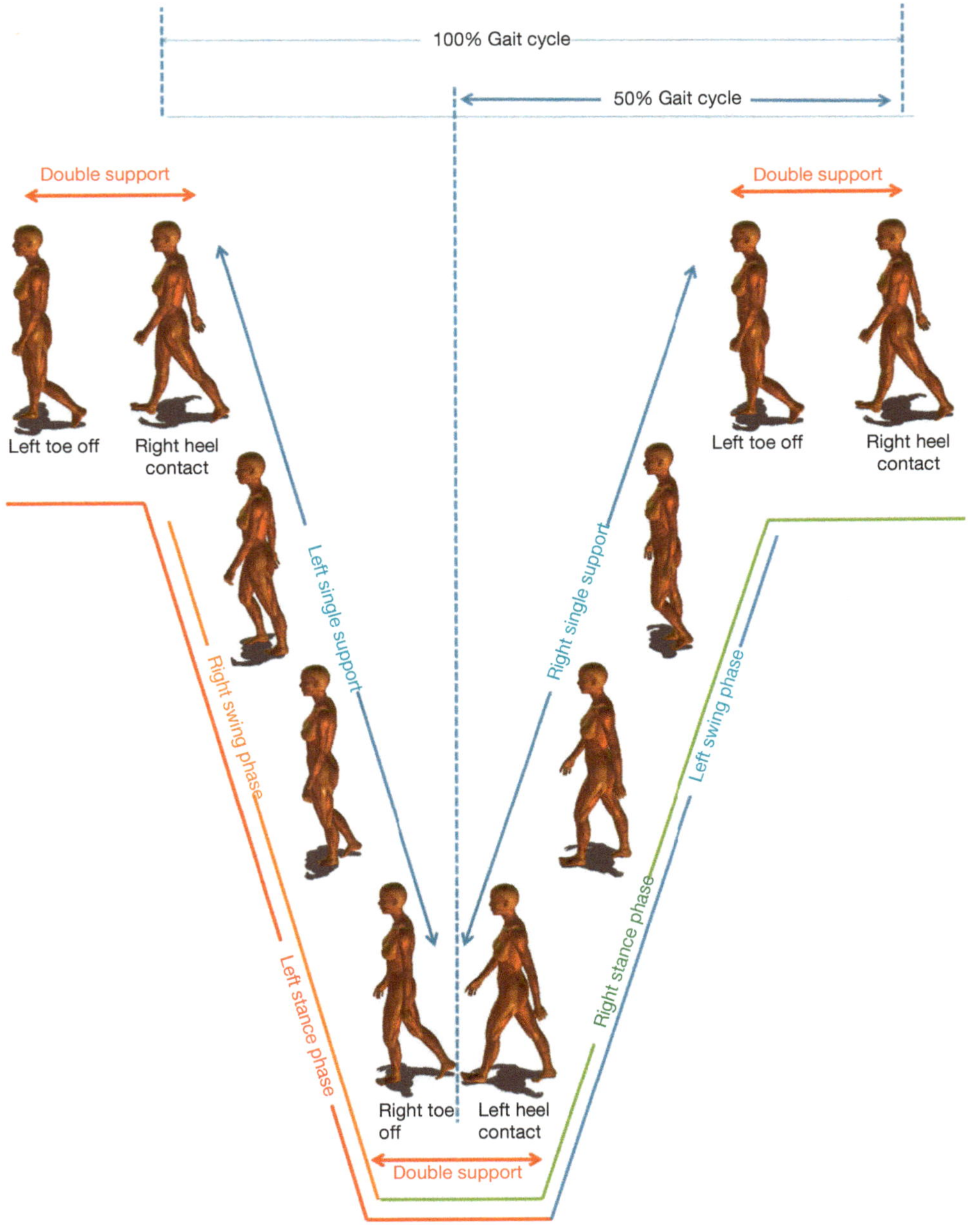

Figure 1 The time dimensions of walking cycle.

The stance phase occurs when the foot is in contact with the floor (starting at heel contact (passive) and ending at toe-off (propulsive)) and the swing phase occurs when the foot is advancing forward to take the next step (**Figure 1**). During each walking cycle, there are two periods of single-limb support and two brief periods of double-limb support (while one limb is about to begin the swing phase and while the other has just ended the swing phase).

Forward walking is achieved by pushing off the stance leg while swinging the other leg forward. At the time of the heel contact phase of the gait cycle, the forward-moving heel contacts the ground and as the limb is kept relatively straight, deceleration of the foot converts to acceleration of the hip. During this phase, the hip and knee extend. Continued forward motion of the body results in the forefoot coming to the ground, and the propulsive part of the support phase begins. At this phase (the muscles plantar flex the foot, flex the knee, and extend the hip), the heel is raised and pushes the foot backward. This is associated with fixation and elevation of the pelvis by the abductors as well as tilting of the body toward the swing leg that allows it to land in a line anterior to the stance leg. The backward (propulsion) force is resisted by friction under the sole.

During the swing phase, the leg is flexed and slightly rotated externally at the hip, flexed at the knee, and dorsiflexed at the foot. Throughout the remainder of the swing phase, the limbs move under the influence of gravity alone and finish in a position which allows direct entry into the next step.

Heel velocity: An important lower extremity variable such as heel velocity during walking provides a linkage between motor control and slip severity as the control of the foot during swing is essentially a ballistic and positional task. Horizontal velocity builds up gradually after heel-off and reaches a maximum velocity late in the swing phase and drops rapidly to near zero just prior to heel contact.

The vertical trajectory during the mid- and late swing phases drops rapidly, but, just before the heel contact event (based upon the stride period), the vertical drop is arrested about 1 cm above the ground level. During the last 10% of swing, the heel is lowered very gently to the ground as horizontal velocity decreases rapidly to near zero. **Figure 2** shows the body kinematics at HC (Planck's constant) for a typical trial. It can be seen that the forward velocity of the body's COG was 1.6 m s^{-1} and the heel velocity was reduced to 0.4 m s^{-1} horizontally and a mere 0.05 m s^{-1} vertically.

The significance of the heel velocity before the HC is that at the end of the swing phase, the heel velocity must be reduced sufficiently so that a dangerous slip will not occur. Walking speed influences heel contact velocity, that is, faster walking speed will increase the contact velocity. In order to reduce the forward velocity of the foot prior to heel contact, the foot is slowed through earlier and/or increased activation of the hamstring muscles. These muscles become active at the termination of swing phase by elongating as they act to decelerate the swinging leg (this is a very effective use of muscle). Decelerating the swinging leg at an appropriate time requires attentional resources via sensory modality.

Upper extremities: Corrective postural movements are made by the upper body, arms, and shoulder. In walking, arm swing is used to offset some of the rhythmical acceleration and deceleration of the trunk by the leg movements, and also to damp out the rotational forces on the trunk. The arm swing varies considerably with variation in the speed of the walk. In the unladen state, the arm swings forward from the shoulder with the arms hanging more or less relaxed in a sagittal plane, but may move slightly toward medial plane. In the laden state, however, these dampening effects are not available. Load carrying (in front) also displaces the body center of mass (COM) anteriorly, placing it closer to the forward edge of the supporting base, thus requiring additional rotational torque at

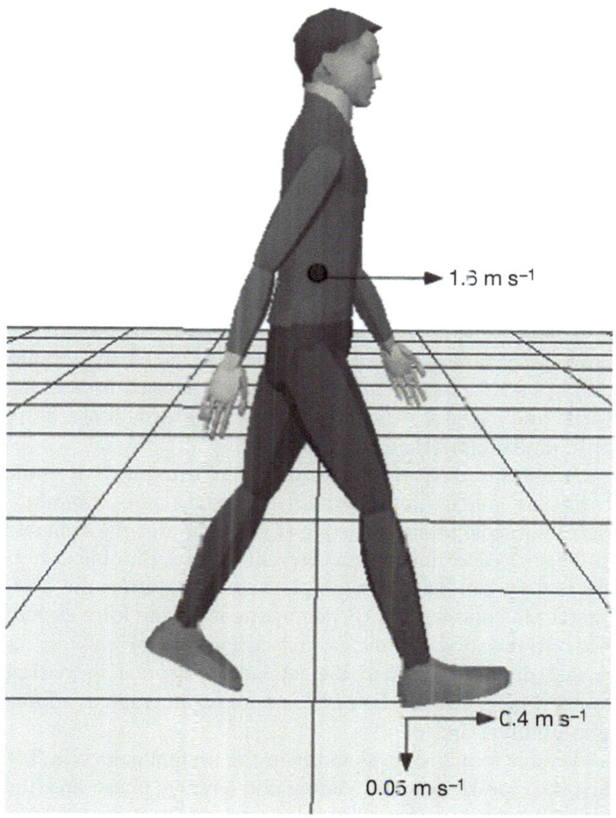

Figure 2 Body kinematics at HC for a typical walking trial. Adapted from Winter, D.A., 1991. Biomechanics and Motor Control of Human Movement. John Wiley & Sons Inc., New York, NY.

the foot–ground contact. Increasing the rotational torque during heel contact phase of the gait may increase the friction demand and increase slip severity. Additionally, blocking the visual path with a load may further influence the awareness of an impending slip perturbation.

Ground Reaction Force

The mechanics and the forces involved in slipping are important in understanding fall accidents. The forces applied by the foot to the floor at the point of foot–floor contact act in three directions: vertical (F_V), horizontal (F_H) in the direction of body motion, and horizontal transverse (F_T) to the direction of body motion. Note that by Newton's third law, the ground reaction forces exerted by the floor on the foot are equal and opposite to the forces exerted by the foot on the floor. During heel contact, there is a resultant forward thrust component of force on the swing foot against the floor. This results in anterior/posterior shearing forces (F_H) acting at the foot–floor interface. Walking speed, which is the product of cadence and step length, affects the magnitude of forward horizontal force (F_H). Forward horizontal force (F_H) increases with increasing step length and cadence; however, the effects of cadence are more pronounced than those of the step length.

Lateral transverse force (F_T) is the result of the lateral momentum during the gait. This lateral momentum exists due to an out-toeing walking pattern. However, the force component F_T can be ignored in normal level walking due to the relatively small transverse forces compared to the other ground reaction force components observed in locomotion experiments.

Vertical force (F_v) is the result of the body weight and the downward momentum of the swing leg against the ground during heel contact. Vertical force (F_v) is affected by walking speed and cadence, which, as previously stated, has a more pronounced effect than step length.

Required coefficient of friction (*RCOF*): The required dynamic coefficient of friction (DCOF) represents the minimum (dynamic) coefficient of friction (COF) that must be available at the shoe–floor interface to prevent forward slipping at heel contact. Perkins used a force platform to measure the horizontal (F_H) and vertical (F_v) components of the force exerted between the shoe and the ground during normal walking. An analog divider calculated the ratio of horizontal to vertical forces (F_H/F_v) and displayed this on an oscillograph as a function of time (**Figure 3**).

Perkins found six peak forces in the normal gait cycle. The first four peaks occurred during the landing phase and the remaining two peaks occurred during the takeoff phase. Peaks 1, 3, and 4 are caused by a forward force, whereas peaks 2, 5, and 6 are caused by a backward force on the force platform.

Peak 1 is caused by the force of impact of the heel tip against the force platform and has a forward direction as a result of the approach angle of the heel to the ground. However, this peak has been found to be inconsistent due to low vertical force during this phase (peak 1).

Peak 2 is caused by a backward force exerted on the heel of the shoe shortly after contact. This force has been noted by several investigators but no reason for its existence has been suggested.

Peaks 3 and 4 are caused by the main forward force which retards the motion of the body and leg. During peaks 3 and 4, the vertical force has risen and a significant proportion of the body weight is being applied through the heel tip (less than 0.1 s after heel contact). Therefore, the error in F_H/F_v ratio is relatively small. As more of the body weight is progressively transferred to the contacting foot, the COG of the body moves over the now stationary foot and the forward force causing peak 4 to decrease. During the takeoff phase, the F_H/F_v (again) increases due to the force (peaks 5 and 6) exerted by the foot propelling the body forward. The significance of the ratio (F_H/F_v) is that it indicates where in the walking step a slip is most likely to occur. If the magnitude of F_H/F_v exceeds the COF between the two surfaces at a particular moment in time, a slip will occur. In this view, there are two critical gait phases in normal level walking from the viewpoint of slipping: (1) shortly after the heel contact, when only the back edge of the heel is in contact with the ground (peaks 3 and 4 in **Figure 3**). Peaks 1 and 2 are not considered hazardous because F_v is quite small at peak 1 and because F_H is directed backward at peak 2. (2) At the moment of toe-off, when only the forepart of the shoe is in contact with the ground (peaks 5 and 6 in **Figure 3**)—here, the required static COF demand is important to thrust the body forward. Theoretically, forward slip at peaks 3 and 4 during landing is more hazardous since the forward momentum of the body will continue to apply the body weight on the slipping foot. Conversely, backward slip at peaks 5 and 6 is less likely to be hazardous, as most of the body weight has been transferred forward from the slipping foot to the opposite leading foot. Backward slip at peak 2 shortly after landing appears to be hazardous, but the likelihood of slip continuing in a backward direction is less as the force rapidly changes direction.

Static versus Dynamic Coefficient of Friction (Tactile)

Tisserand found that rank-order correlation with subjective judgments of slip resistance was negligible for static COF measurements and very high for dynamic COF measurements. Additionally, Harris and Shaw found a high correlation between (walker's) opinions of slipperiness and the kinetic friction of wet floors. Their results are in close agreement with Strandberg and Lanshammar's walking experiments demonstrating that kinetic friction values, measured with sufficient sliding velocity, correlated better with the actual risk of falling than low-speed kinetic friction values or static friction values. Another study by Swensen et al. compared the perceived slipperiness of steel beams by both professional ironworkers and students. They were asked to rate and rank the slipperiness of each beam that was either uncoated or coated with contaminants (water, clay, and oil) after walking across the beams. The results of this study indicated that there is a strong correlation between dynamic COF values and subjective ratings of floor slipperiness. **Figure 4** shows the result of a series of experiments using ten subjects and five different shoes. The mean subjective ranking and the ranking using the dynamic COF are very similar.

Visual versus Tactile Sensation

Cohen and Cohen further explored the perceptual and cognitive factors involved in the perception of floor tile surface slipperiness. The results of this study also demonstrated that tactile cues are most sensitive to physical measurement of dynamic COF. A follow-up field study of the psychophysical assessment of the perceived slipperiness of floor tile surfaces concluded that people tended to make predictions about the slipperiness of walking surfaces and verified these expectations

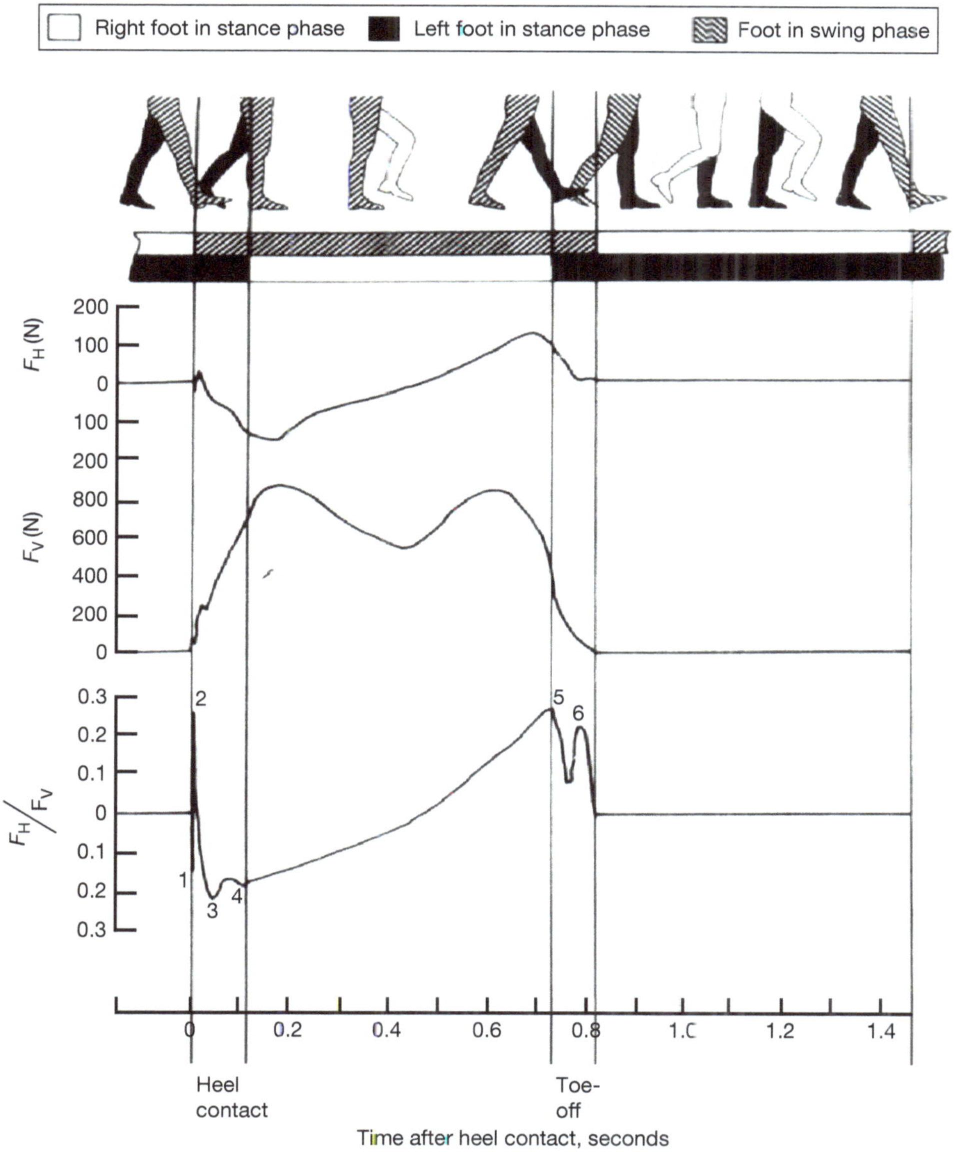

Figure 3 Gait phases in normal level walking with typical horizontal, vertical, and their ratio for one step. Reproduced from Gronqvist, R., Roine, J., Jarvinen, E., Korhonen, E., 1989. An apparatus and a method for determining the slip resistance of shoes and floors by simulation of human foot motions. Ergonomics 32 (8), 979–995.

as they crossed them. The results suggest that visual cues to slipperiness are inferior to tactile sensation. They also suggested that in real-world conditions, the perception of walking surface slipperiness is probably the result of tactile cues, with visual impressions being confirmatory. Thus, in unfamiliar conditions, people may rely on the primary but inferior visual information about a surface's traction until they actually walk on it. The potential for an accident can be created due to misjudgment of slipperiness based on initial visual sensing and the limited time available to make immediate adjustments in gait to accommodate for the hazardous condition.

Slip Resistance Measuring Devices

Many devices have been developed by individuals, organizations, and federal agencies (e.g., OSHA, NIOSH, and NBS) to

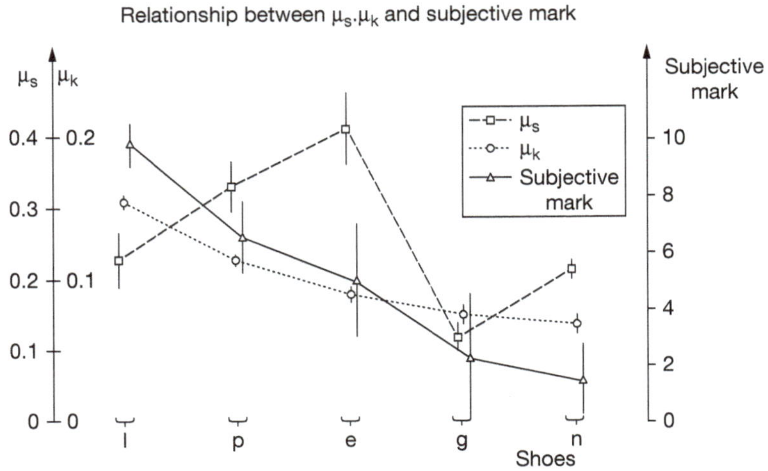

Figure 4 Comparison of mean subjective ranking by 10 subjects and ranking using static and dynamic coefficient of friction for five shoes. Subjective binary choice test. Reproduced from Tisserand, M., 1985. Progress in the prevention of falls caused by slipping. Ergonomics 28 (7), 1027–1042.

quantify the slip resistance of floor surfaces (i.e., COF). At least 70 different meters have been cited in the literature; however, none of these devices are universally accepted. In general, most devices fall into three categories: drag/towed-sled, pendulum, and articulated strut types.

There are two approaches to measuring friction. In the direct approach, the test device measures or indicates the horizontal and vertical forces. These include drag-type meters and articulated strut devices. The indirect approach calculates the frictional force by observing an energy loss in the test device. The most common example of the indirect approach is a pendulum-type device. These devices are limited in their ability to simulate biomechanical factors for an accurate DCOF measurement.

Strandberg and Lanshammar stated that on the basis of tribology and biomechanical analysis of slips and falls, the slip resistance meter should reproduce the following operating variables from crucial gait phases: (1) contact time and normal force time derivative, for the influence of surface patterning and its drainage capability; (2) foot angle, for testing the most critical part of the shoe; (3) contact force application point at the shoe; (4) vertical force, for correct pressure in the contact area; and (5) sliding velocity, for correct dynamic friction forces.

The versatility of different meters has been compared and commented on by many investigators. Due to its heavy weight and bulkiness, the James Machine can be used only in a controlled laboratory situation for testing sample floor material, not the actual floor. The NBS-Brungraber Tester, on the other hand, is easier to handle and can be considered as a portable James Machine. The Tortus provides a permanent trace record when connected to a chart recorder, whereas most devices provide only a visual reading of the peak COF (analogy meter or gauge). However, drag-type devices have overcome most limitations and generally met with the greatest acceptance among both researchers and practitioners. The BOT 3000 digital tribometer can also provide both static and dynamic COF measures.

Trajectory of the Total Body Center of Gravity

The body's COG is a key factor in human gait analysis as it reflects the motion of the whole body. The COG is the theoretical point about which the mass of that body is evenly distributed. Reduction of the partial body masses into a common COG or mass simplifies the movement dynamics to a point where the effect of the moving forces upon the mechanism of gait as a whole can be deduced. In the human body, the location of this theoretical point will depend upon several factors including the distribution of segmental masses and the location of those segments.

In the standing position, the COG is situated centrally in the pelvis, approximately at the level of the second sacral vertebra. However, during forward walking, the equilibrium is lost with the takeoff of the propelling foot when the body's COG momentarily lies beyond the anterior border of the supporting surface and regained as soon as the swinging leg is extended forward and the heel touches the ground. Additionally, the legs move forward and backward relative to the trunk (torso), the arms swing, and the trunk moves up and down and from side to side during locomotion. Consequently, the COG progresses forward and at the same time moves up and down and from side to side.

In normal level walking, the COG describes a smooth sinusoidal curve when projected on the plane of progression

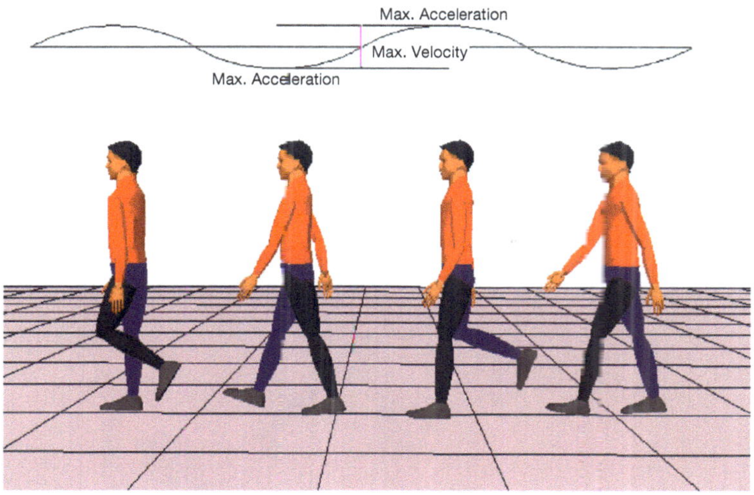

Figure 5 The path curve of the common center of gravity in the sagittal plane.

(**Figure 4**). The path curve of the common center of gravity in the sagittal plane moves up and down. This motion is sinusoidal, with a period of approximately 50 mm in extent for adult males at normal walking speed. The summits occur at the middle of the stance phase of each side and the lowest points occur during double support when both feet are on the ground (**Figure 5**). At the peak of vertical movement, the horizontal velocity reaches a minimum (maximum acceleration). Conversely, the reverse is true at the time of lowest trajectory. The magnitude of the vertical excursion during free-cadence walking correlated significantly with the length of the stride and the peak-to-peak vertical oscillation increased as cadence increased.

In the frontal plane, the common COG moves from left to right as each leg alternately becomes weight bearing. It follows a smooth sinusoidal curve, but with only one oscillation per stride as opposed to the two vertical oscillations in the sagittal plane (**Figure 6**). The COG attains its greatest distance from the midline shortly after the support of the standing leg is shifted to the whole sole. At this point, there is an inversion of motion (the velocity is zero and the acceleration is maximal). Afterward, the COG follows the period of double support and the velocity increases as the COG again approaches the midline.

Shimba and MacKinnon reported that the lateral path of the COG passes forward along the medial border of the foot (sometimes slightly outside that border: ±2 cm; **Figure 7**). The lateral movements represent automatic postural adjustments, shifting the line of gravity alternately toward the eccentrically placed bases of support in keeping with the demands of stability.

In the transverse plane, the COG carries out forward and backward movements (U shape). In this plane, the COG moves forward commensurably with the movements of the pelvis. The maximum forward point is reached as the heel is set to the ground and the maximum backward movement occurs at the takeoff phase of the gait. The amplitude of these backward and forward oscillations is about 12 mm.

Balancing the Center of Gravity of the Human Body

Achievement of even sinusoidal displacements depends on smoothly coordinated angular displacements of the various segments of the lower limb. Various investigators used a series of simple models to illustrate how this smooth sinusoidal displacement pathway is achieved in bipedal locomotion. They show that this type of locomotion requires that the COG be elevated to a height equal to the COG in the standing position. Thus, it will result in a severe jolt at the point of interaction of each two arcs where there is an abrupt change in the direction of movement of the COG. As such, decreasing the total elevation, depressing the COG, and smoothing the series of interrupted arcs require coordinated movements involving all the joints of the lower extremities.

The balance of the human body or its parts requires that all gravitational forces be completely neutralized by counterforces. These counterforces are supplied by the resistance of the supporting surface of the body. However, when gravitational forces fall outside of the supporting surfaces, the translatory force of gravity is not neutralized. In order to neutralize the rotatory forces, the line of COG must also fall in the supporting surface. In general, two factors influence stability: (1) the broader the supporting area (area over which one object is supported by another; in the standing position, the base of support of the human body is the area bounded by the contact points of the

Figure 6 The path of the common center of gravity in the frontal plane.

feet with the floor) the greater is the force necessary to destroy the balance by throwing the line of COG beyond the supporting surface and (2) the lower the COG is located, the greater is the arc that an unbalancing force must describe before it can bring the center to fall outside the supporting area (i.e., more stable). In the human body, the constant mass is supported in standing on a small base. Additionally, the area of the base and the position of the COG are subject to constant and rapid changes and, therefore, require a complex reflex system involving the integration of sensory nerves and the motor nerves controlling the muscles to maintain balance in any given posture of the body. The deviation of the COG is constantly monitored by the following methods:

1. Sensory mechanoreceptors in the capsules and ligaments of joints, which provide information about their position and rate of movement.
2. Stretch receptors in muscles (muscle spindles), which give information on the amount and rate of muscle stretching.
3. Pressure receptors (exteroceptors) in the skin, which provide information about the amount of pressure on the skin of the soles of the feet.
4. The vestibular apparatus, including the semicircular canals found in the inner ear within the temporal bone of the skull, gives information of the motion of the head in all planes.
5. The visual system giving information about the position of the body in relation to objects and surfaces that can be seen.

All this information is processed in the CNS (principally in the cerebellum and the brain stem) and signals are sent to skeletal muscles to contract appropriately to adjust the position of the body to maintain the COG over the base. This process involves unconscious prediction of body motion so that the adjustments are not merely responding to the existing body position, but arranging that the COG to base relationship is appropriate for subsequent movements (i.e., the next step in walking).

Balance task of walking versus *standing*: Human bipedal locomotion (walking) provides a challenging balance task to the CNS and appears to be completely different from the balance task during standing. During standing, the CNS is challenged to keep the body's COG safely within the borders of the two feet (or one foot if balancing on one foot). Studies of balance and posture during quiet or perturbed standing have identified the ankle muscles (plantarflexors/dorsiflexors and invertors/evertors) as dominant. However, during locomotion, ankle muscles are no longer important because the balance task has changed. As explained in the last section (lateral oscillations of the COG), the lateral path of the COG passes forward along the medial border of the foot (even slightly outside that border). Thus, during single support, the body is in a continuous state of falling down because its COG is outside the foot. The only way that recovery is achieved is to position the swing limb so that during double support the CNS can make any rebalancing adjustments.

Gait model: This recovery is a challenging balance task that requires a complex interplay of neural and motor control mechanisms. Motor control is directly linked to the CNS's processing of sensory inputs (vision, vestibular, and

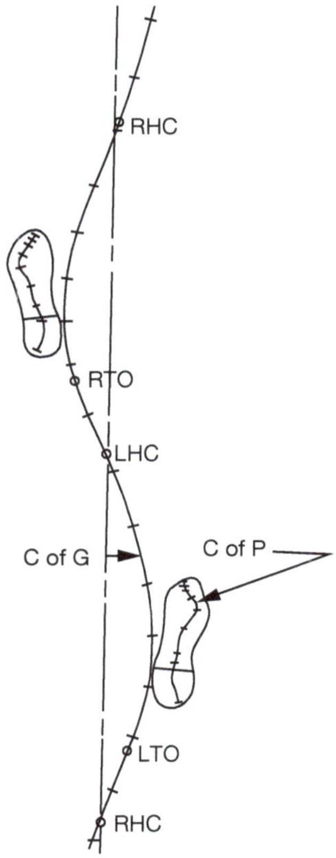

Figure 7 Lateral pathway of center of gravity passing forward along the medial border of the foot. Reproduced from MacKinnon, C.M., 1990. Control of Whole Body Balance and Posture in the Frontal Plane during Human Walking. University of Waterloo.

proprioceptive systems). The brain constructs internal representations of the world by integrating information from the different sensory systems. In other words, the transformation from sensory signals to motor commands is processed within the CNS. The motor system transforms neural information into physical energy by transmitted commands from the brain stem and spinal cord to skeletal muscles. A system model that mimics the behavior of a natural process is known as an internal model, an important theoretical concept in motor control. The internal models include two main components: inverse model and forward model. An inverse model functions as a motor command computation to calculate the desired states. A forward model acts as a predictor to estimate the next state (e.g., future position and velocity). The sensory systems send inputs to an online controller to make an adjustment in real time. In essence, we use the inverse model to modify our walking behavior; that is, walking off the sidewalk to cross the street and stepping back onto the curb requires online controller. Additionally, the internal model is used to predict, and adapt to, the next step (i.e., forward model); thus, we are able to feel momentum changes, for example, when walking onto a broken escalator. The importance of understanding the gait model above is to provide a better understanding of the processes associated with slips and falls. Here, it is clear that "expectancy" is required to walk—that is, during walking we expect the ground to be stable and, as such, we modify our gait to traverse the terrain with appropriate force and speed to safely ambulate; however, if the ground is not stable, there will be a motion perturbation (i.e., expectancy and reality did not match). This perturbation, if not controlled, could lead to slips and falls.

Gait Characteristics Influencing Slip Initiation, Detection, and Recovery

The process of slips and falls can be categorized into four levels as shown in **Figure 8**. The environmental phase considers the effects of contamination. As noted by Chaffin et al., "any fluid contaminant between two sliding surfaces will provide lubrication and thereby lower the DCOF values." Therefore, the presence of contamination (oil, water, etc.) will reduce the available DCOF of the floor surfaces. In terms of slip-induced falls, friction demand characteristics between the shoe sole surface and the floor surface has been implicated as an important predictor variable related to severity of falls. It was stated that most of slip-induced falls occurred when the frictional force (F_{μ}) opposing the direction of foot movement is less than the horizontal shear force (F_h) of the foot immediately after the heel contact on the floor. The RCOF is defined as the ratio of horizontal ground reaction force to vertical ground reaction force. It represents the minimum RCOF between the shoe and floor interface to prevent slipping. Consequently, slip is initiated by the combination of lower DCOF and higher RCOF. A static COF of 0.5 on a level walking surface has been commonly recommended by standard organizations and by individual authors. Dangerous forward slips that lead to falls are most likely to occur 70–120 ms after the heel contacts the ground.

The sensorimotor degradation in older adults often leads to altered gait characteristics affecting the slip-initiation process. Lockhart et al. reported that older adults' heel velocity was faster than their younger counterparts at the heel contact phase of gait cycle. Increases in heel velocity during critical weight transfer may increase the potential of a slip-induced fall if the floor is slippery. In addition, the friction demand of older adults was found to be higher than their younger counterparts. It has also been suggested that whole body COM velocity relative to the base of support may be a factor related to RCOF. Slower whole body COM velocity and COM transitional acceleration (velocity changes from heel contact to shortly after

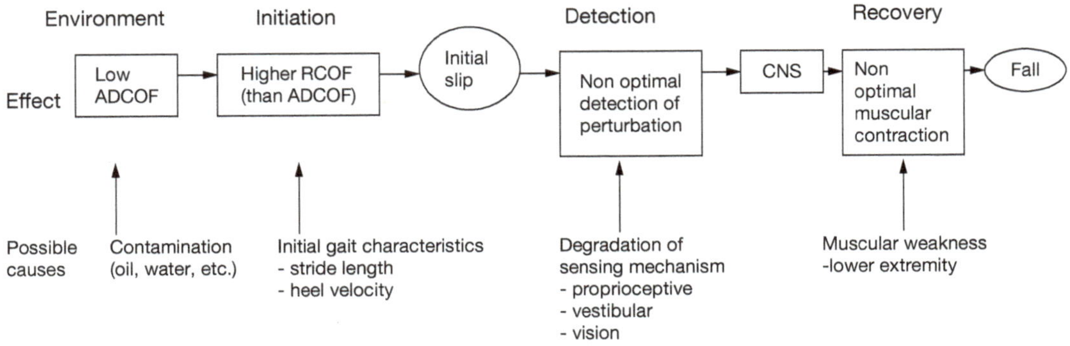

Figure 8 The process of initiation, detection, and recovery of inadvertent slips and falls with possible causes and effects. *RCOF*, required coefficient of friction; *CNS*, central nervous system. Reproduced from Lockhart, T.E., Woldstad, J., Smith, J., 2003. Effects of age-related gait changes on the biomechanics of slips and falls. Ergonomics 46 (12), 1136–1160.

heel contact) were reported in older adults. Alterations in these factors can increase the risk of slip initiation.

During the detection and recovery phases of the slip and fall process, the CNS control plays an important role. The CNS must undertake certain processing stages (detection phase) if a fall is to be avoided or compensated for (recovery phase). During the detection phase, a trigger must be sent through the sensory feedback to the motor control regions of the CNS. This process may be initiated by one or more of the following sensory inputs: somatosensors, vision, and vestibular function. At the input stage, any disruption in the quality of the input signal may increase the likelihood of slips and falls. The somatosensors are responsible for proprioception, the sensing of joint and limb motion. The vestibular organs located at the inner ear (semicircular canals and otoliths) detect angular velocity of the head and act as linear accelerometers. Visual cues provide information about the position and motion of the head with respect to surroundings. In the case of posture control, the relevant signals are processed by motion detection circuitry not only in the retina but also in the visual cortex. These sensory signals are fed back to a series of hierarchical feedback loops to generate motor commands.

The reactive recovery phase involves bringing the body COM within stability limits quickly after a slip is initiated. This is achieved through changes in various kinematic, kinetic, and muscle coactivity mechanisms. One of the important mechanisms during reactive recovery is to reduce the displacement of the slipping foot through increased coactivity of the muscles of the lower extremity. Electromyography recorded from the lower extremity showed an increased coactivation of the rectus femoris and hamstrings, along with the tibialis anterior and gastrocnemius muscles, while recovering from a slip. The postural responses generated by distal perturbed leg muscles are of short latencies (65–110 ms), significant magnitude (two to three times higher than normal walking), and long duration ($\approx$150 ms). Numerous studies have linked slower muscle activation rates in older adults as an indicator of increased risk to slip-induced falls. Kim et al. found a decreased hamstring activation rate in older adults and related it to higher risks of slip-induced falls. The postural activity of bilateral leg and thigh muscles and the coordination between the two lower extremities were found to be key to reactive recovery balance control.

An Example of an Expert Witness Report

In this section, opinions and basis for expert opinions will be further described using an example of a case involving a slip and fall accident. First, a general description of the accident is presented—usually, gathered information at this time is in the form of "depositions" from various individuals involved either directly or indirectly in the accident. In addition to witness statements and/or discussions with the potential client, experts also inspect the area (either via photograph, video, or a visit to the accident site) in question and make measurements (e.g., COF, slope, and illumination).

Afterward, the opinion of the expert is presented using a variety of supporting materials linking human locomotor control and fall accidents. Here, we start with a general understanding of the accident at the time of a group's arrival at a restaurant.

General understanding of the accident

On the afternoon of October 26, 2012, Mrs AC visited the C's restaurant in Parkins, Texas, for a lunch after a salon visit with a church group. Mrs AC and the church group members arrived at the restaurant in a van. It was raining steadily and heavily at the time of their arrival (to the restaurant). As such, the van was parked close to the entrance to accommodate for the rain.

Mr Priggs (Assistant Manager of the C's at that time) held the front door open as they were entering the restaurant. The first group of ladies entered the restaurant. When Cynthia (a member of the group who witnessed Mrs AC's fall from the posterior view) and Mrs AC entered the restaurant, they were among the last of the group to enter the restaurant.

Mrs AC entered the restaurant and walked toward the end of the mat to look for the group. She located the group by identifying the "backs of their heads." She walked from the end of the last mat onto the exposed tile floor, where she slipped and fell to the left side of her body fracturing her pelvis. Mrs AC was assisted by the restaurant employees until the ambulance transported her to the hospital.

Opinion

The exposed tile floor surface between the "mat" and the carpeted surface of the dining area at the C's restaurant at Parkins, TX, created an unreasonable hazard and risk of harm to pedestrians.

In order to maintain dynamic balance during locomotor activities and to help avoid slip-induced falls, gait parameters are adjusted to correct for contaminated or slippery conditions. In other words, when confronted with impending slip and fall situations, our walking style (i.e., the gait parameters) is adjusted accordingly to avoid slipping (note: online controller at work). Persons who have a prior knowledge of a contaminated walkway adjust their gait by reducing friction demand characteristics—for example, reducing heel contact velocity and step length by modifying the support leg's muscle stiffness. For younger individuals, this adaptation is quicker (within one step) than for older adults (at least two steps)—that is, it will take them at least one to two steps to adjust their gait to traverse even readily recognizable slippery floor surfaces. Mrs AC is an older adult.

As indicated, the support leg is modified in muscle stiffness according to the walking conditions to progress the whole body smoothly without abrupt transition of friction demand. The ability to modify the muscle stiffness characteristic of our supporting leg (i.e., make rapid and adaptive adjustment of gait from walking on a nonslippery floor surface to a slippery floor surface) depends upon the quality of the user's perception of the surrounding environment. Hale and Glendon and Tisserand supported that a cognitive process occurs prior to action (locomotor activities) in order to safely traverse over slippery floor surfaces (i.e., online controller). The process starts with expectation—expecting the contaminant (i.e., water, grease, etc.) on the walkway, the area is surveyed (scanned) visually with appropriate thoroughness related to one's expectations. As such, the lack of obvious visual cues to elicit expectation can influence detection of slippery floor surfaces and modification of adaptive locomotor responses.

Furthermore, tactile cues to elicit expectation can also be important. Tisserand suggested that frictional values are estimated and memorized unconsciously from preceding steps (one's own model of slipperiness), and this information is updated whenever the subject feels the floor conditions are different from expected (reality) (note: adaptive controller). In the author's opinion, the cues necessary to correctly detect the prevailing slippery condition were neither obvious nor effective and resulted in a slip and fall accident. In other words, there were no obvious and effective cues to inform Mrs AC that the exposed tile floor surface between the "mat" and the carpeted surface was dangerously slippery and as a result Mrs AC did not adjust her gait to compensate for the slippery floor surface.

Mrs AC's expectation was influenced by several factors. Any visual cue that may distract a pedestrian's attention away from the slippery floor surface could be potentially dangerous. In this case, visual cues to distinguish the slippery floor surface were not available.

Furthermore, the assumption that the floor will be dry in the area where the fall occurred is lacking support. For example, in the case of heavy pedestrian traffic and heavy rain, water will accumulate on the mat and proceeding steps can track moisture onto the exposed tile floor surface making the floor surface very slippery. (Using a slip meter, the available (relevant) COF can be measured.) This statement is further supported by one of the witnesses to the accident—"The way I looked at it, the way that it even looked like this, I could see the skid [moisture trail] on the floor like her heel had hit instead of like the flatness of her foot." This statement further corroborates that not only was the tile floor surface wet, but Mrs AC's gait was also not adjusted accordingly to walk on the slippery floor surface—that is, the usual gait characteristic associated with walking on a slippery floor surface is toe–heel gait to modify friction demand (more vertical ground reaction force than horizontal ground reaction force—as in walking on icy surfaces). In the author's opinion, effective tactile cues necessary for gait adjustment were also lacking.

As indicated above, tactile cues are import in eliciting protective response (i.e., gait adjustments) to walk on slippery floor surfaces. Frictional values are estimated, evaluated unconsciously from preceding steps, and updated readily to traverse the area safely. Walking on a slip-resistant mat first and then stepping onto a slippery surface may create a misperception leading to fall accidents. In other words, the slip-resistant mats provided a secure surface condition for normal walking with no need to modify gait (i.e., heel–toe gait) to traverse safely for Mrs AC; however, when walking on a slippery surface with secure gait, her heel slipped, and she fell.

In summary, visual and tactile cues necessary to correctly detect the prevailing slippery condition (at the exposed tile floor surface between the "mat" and the carpeted dining area) were not obvious and created an unreasonable hazard and risk of harm to Mrs AC. In the author's opinion, the cues necessary to correctly detect the prevailing slippery condition were neither obvious nor effective and resulted in a slip and fall accident.

There are a number of measures that the C's restaurant could have taken to increase pedestrian expectation of slippery floor surfaces and/or reduce the likelihood of falls. One solution would have been to extend the mats covering the tile floor surface all the way to the carpeted area. This would have permitted Mrs AC to get to the dining area without walking on the slippery tile floor surface. Warning signs such as "Watch Your Step" and "Caution—Wet Floor," which are industry standards for hazardous floors, could have been posted (although warnings can be important, proper mats are the preferred safety measure). More probably than not, these measures would have prevented Mrs AC's fall and injuries.

C's restaurant failed to institute reasonable measures to reduce the likelihood of falls and breached its duty to maintain its premises in a reasonably safe condition for patrons such as Mrs AC.

In summary, although biomechanical aspects of gait and posture have been discussed in detail in this chapter, applicable codes and standards should also be referenced. As in all forensic investigations, careful and thorough investigations and analyses must be performed before scientifically credible expert opinions can be rendered.

See also: **Engineering:** Forensic Engineering/Accident Reconstruction/Biomechanics of Injury/Philosophy, Basic Theory, and Fundamentals; Human Factors Investigation and Analysis of Accidents and Incidents.

Further Reading

Brady, R., Pavol, M.J., Owings, T., Grabiner, M., 2000. Foot displacement but not velocity predicts the outcome of a slip induced in young subjects while walking. Journal of Biomechanics 33 (7), 803–808.

Chaffin, D., Woldstad, J., Trujillo, A., 1992. Floor/shoe slip resistance measurement. American Industrial Hygiene Association Journal 53 (5), 283–289.

Cham, R., Redfern, M.S., 2001. Lower extremity corrective reactions to slip events. Journal of Biomechanics 34, 1439–1445.

Cohen, H.H., Cohen, D.M., 1994a. Psychophysical assessment of the perceived slipperiness of floor tile surfaces in a laboratory setting. Journal of Safety Research 25 (1), 19–26.

Cohen, H.H., Cohen, D.M., 1994b. Perceptions of walking slipperiness under realistic conditions utilizing a slipperiness rating scale. Journal of Safety Research 25 (1), 27–31.

Cohen, H.H., LaRue, C.A., 2000. Forensic Human Factors/Ergonomics. International Encyclopedia of Ergonomics and Human Factors. Talyor & Fracisis, London.

Gronqvist, R., Roine, J., Jarvinen, E., Korhonen, E., 1989. An apparatus and a method for determining the slip resistance of shoes and floors by simulation of human foot motions. Ergonomics 32 (8), 979–995.

Hale, A.R., Glendon, A.I., 1987. Individual Behaviour in the Control of Danger. Elsevier Science, Amsterdam.

Hanson, J.P., Redfern, M.S., Mazumdar, M., 1999. Predicting slips and falls considering required and available friction. Ergonomics 42, 1619–1633.

Harris, G.W., Shaw, S.R., 1988. Slip resistance of floors: users' opinions, Tortus instrument readings and roughness measurement. Journal of Occupational Accident 9, 287–298.

Horak, F., Macpherson, J., 1996. Postural Orientation and Equilibrium. In: Handbook of Physiology, vol. 1, pp. 255–292.

Inman, V.T., Ralston, H.J., Todd, F., 1981. Human Walking: Application to Lower Limb Prosthetics. Williams and Wilkins, Baltimore, MD.

Kawato, M., 1999. Internal models for motor control and trajectory planning. Current Opinion in Neurobiology 9 (6), 718–727.

Lin, L., Chiou, F., Cohen, H., 1995. Slip and fall accident prevention: a review of research, practice, and regulations. Journal of Safety Research 26 (4), 203–212.

Lockhart, T.E., Kim, S., 2006. Relationship between hamstring activation rate and heel contact velocity: factors influencing age-related slip-induced falls. Gait & Posture 24 (1), 23–34.

Lockhart, T.E., Woldstad, J., Smith, J., 2003. Effects of age-related gait changes on the biomechanics of slips and falls. Ergonomics 46 (12), 1136–1160.

Lockhart, T.E., Smith, J., Woldstad, J., 2005. Effects of aging on the biomechanics of slips and falls. Human Factors 47 (4), 708.

MacKinnon, C.M., 1990. Control of Whole Body Balance and Posture in the Frontal Plane during Human Walking. University of Waterloo.

Nashner, L., Black, F., Wall, C., 1982. Adaptation to altered support and visual conditions during stance: patients with vestibular deficits. Journal of Neuroscience 2 (5), 536.

Nashner, L., 1980. Balance adjustments of humans perturbed while walking. Journal of Neurophysiology 44 (4), 650.

Noy, I.Y., Karwowski, W., 2005. Perceptual-cognitive and biomechanical factors in pedestrian falls. In: Cohen, H.H., LaRue, C.A. (Eds.), Handbook of Human Factors in Litigation. CRC Press, Boca Raton, FL ch. 19.

Pai, Y., Patton, J., 1997. Center of mass velocity-position predictions for balance control. Journal of Biomechanics 30 (4), 347–354.

Perkins, P., Wilson, M., 1983. Slip resistance testing of shoes – new developments. Ergonomics 26, 73–82.

Perkins, P., 1978. Measurement of slip between the shoe and ground during walking. In: Anderson, C., Senne, J. (Eds.), Walkway Surfaces: Measurement of Slip Resistance, ASTMSTP 649. American Society for Testing and Materials., West Conshohocken, PA, pp. 71–87.

Redfern, M.S., Andres, R.O., 1984. The analysis of dynamic pushing and pulling: required coefficients of friction. In: Proceedings of the 1984 International Conference on Occupational Ergonomics, Ontario. Human Factors Association of Canada, Mississauga, Canada, pp. 569–571.

Shimba, T., 1984. An estimation of center of gravity from force platform data. Journal of Biomechanics 17 (1), 53–57, 59–60.

Stelmach, G., Worringham, C., 1985. Sensorimotor deficits related to postural stability. Implications for falling in the elderly. Clinics in Geriatric Medicine 1 (3), 679.

Strandberg, L., Lanshammar, H., 1981. The dynamics of slipping accidents. Journal of Occupational Accidents 3 (3), 153–162.

Swensen, E., Purswell, J., Schlegel, R., Stanevich, R., 1992. Coefficient of friction and subjective assessment of slippery work surfaces. Human Factors 34, 67–77.

Tang, P., Woollacott, M., Chong, R., 1998. Control of reactive balance adjustments in perturbed human walking: roles of proximal and distal postural muscle activity. Experimental Brain Research 119 (2), 141–152.

Tisserand, M., 1969. Criteres d'adherences des semelles de securite (Disseratation). Institute de Recherche et de Securite, Nancy.

Tisserand, M., 1972. Adherence statique et dynamique des chaussures de securite. Rapport d'etude No. 11/Re. Institute de Recherche et de Securite, Nancy.

Tisserand, M., 1985. Progress in the prevention of falls caused by slipping. Ergonomics 28 (7), 1027–1042.

Winter, D.A., 1991. Biomechanics and Motor Control of Human Movement. John Wiley & Sons Inc., New York, NY.

Wolpert, D., Ghahramani, Z., Jordan, M., 1995. An internal model for sensorimotor integration. Science 269 (5232), 1880.

Forensic Chemical Engineering Investigation and Analysis

DD Perlmutter, University of Pennsylvania, Philadelphia, PA, USA

Introduction

Chemists and chemical engineers are often asked by attorneys to offer scientific and/or technical assistance on matters in controversy when the details involve damage to persons or property from chemicals, fires, or explosions. In addition, if the subjects under examination reach into the court systems, plaintiff or defense arguments may call for expert opinions to support or to refute some position.

Because it is recognized that chemical engineers play an essential role in the design and operation of every chemical and petroleum process, they are often called upon in the aftermath of an industrial accident to evaluate the equipment and practices that were used before and during the event. Their expertise is also sought when fires or chemical events occur in homes or other residential settings. The procedures of evaluation have much in common, but of course, they must be tailored to the details of the equipment being used. In the factory or chemical plant, it will be necessary to examine any mixers, reactor vessels, distillation columns, or separators of various kinds, whereas in a residence, the focus will be on rooms, appliances, and any chemicals in use. In the best modern installations, safety considerations will already have been put in place, but sources of ignition or sources of contamination will still need to be identified.

Huge quantities of natural gas and petroleum liquids are moved every day between refineries and residences through pipelines that crisscross every state and most of the big cities in the United States. Usually the pipelines are buried, emerging only when it is necessary to cross some barrier or for servicing or pumping to the next station en route. Some of the piping is quite old and, occasionally, the welds or seals are inadequate to hold pressure. If leaks are not identified and corrected early, there is a high risk of fire and/or explosion of flammable material. Some very dramatic fires occurred as recently as 2010 in Philadelphia and Texas; the latter event caused one death and eight reported injuries. Chemical engineers are appropriate experts for any investigation of the causes of such events. The 2010 gas release and explosion from the BP Deepwater Horizon rig offshore in the Gulf of Mexico was in the news for months because it was very costly in terms of lives and property, and potentially more damaging to marine life. Eleven men died and many were injured in that event.

There are tens of thousands of chemicals in commerce around the world, and each case is potentially unique in some respect; however, there are some classes of substances and some features of accidents that occur often enough to provide recognizable patterns. With the objective of providing an overview of the wide range of approaches to presenting cases, this chapter identifies some of the important subclasses and their characteristics and offers examples typical of each one.

In each case being considered, it is essential not only to identify a potentially offending substance but also to consider the pathways and combinations that produce the hazard by the nature of their interactions. This means, in effect, that all the pertinent science and technology must be factored into the responses. Depending on the specifics, this may call for basic science such as chemistry, physics, or biology; transport properties for heat, mass, or momentum; or design and operations such as are central to any manufacturing process.

Many times an expert will need to find an acceptable and respected standard of practice by which to judge whether a particular device or decision is appropriate for a given circumstance. For such cases, the American Society for Testing Materials (ASTM) has published a series of indexed volumes that may easily be found in libraries or over the Internet. When fires and/or explosions are under examination, the references of the National Fire Protection Association (NFPA) provide standards and codes of behavior to be followed by experts. The Environmental Protection Agency (EPA) provides information on what is allowed in air and water and the Consumer Product Safety Commission is a source of listed hazards in their area of responsibility. Data on chemicals and their properties as well as on thermal and transport properties, heats of reaction, and chemical equilibria are listed in reference handbooks. The forensic expert will find all these useful to evaluate the crucial events in various circumstances.

More detailed discussions of these different matters are divided in this chapter under the general headings of Fires and Explosions; Pollution and Toxic Substances, and Unexpected Hazards and Unexpected Consequences. Specific examples may also be found in the author's Website (www.chemexpertwitness.com.)

Fires and Explosions

In examining the causal chain that might lead to a fire, one must seek not only the flammable substance but also the source of oxygen (or other oxidizing agent) and the source of ignition,

Table 1 Selected autoignition temperatures (in °C)

Substance	Autoignition temperature (°C)
Acetaldehyde	175
Acetonitrile	524
Acetylene	305
Ammonia	651
Benzene	560
Carbon disulfide	90
Cyclohexanol	300
Decane	210
Diesel fuel	210
Dimethyl sulfoxide	215
Ethanol	365
Ethylene	490
Methane	580
Nitromethane	379
Pentane	260
Propane	480
Propylene	458

Table 2 Selected values of flammability limits

Substances	Lower flammability limit Vol% in air	Upper flammability limit Vol% in air
Acetaldehyde	4	57
Acetyl chloride	7	19
Acetylene	3	82
Ammonia	15	28
Benzene	1	8
Carbon disulfide	1	50
Cyclohexanol	1	9
Decane	1	5
Diesel fuel	1	8
Dioxane	2	22
Ethanol	3	19
Ethylene glycol	3	22
Methyl acetate	3	16
Nitromethane	7	22
Octane	1	7
Propyl acetate	2	8
Propylene	2	11
Toluene	1	7
p-Xylene	1	6

be it a spark, a torch, a pilot light, or a cigarette. The flammable material can be in solid, liquid, or gaseous form. Under extraordinary circumstances, it is also possible to get a flame without an external ignition source, but only at an elevated temperature. This *autoignition* temperature is different for each substance. Some examples are shown in **Table 1**.

If an explosion is under study, it may have been caused by some well-known agents such as nitroglycerine or trinitrotoluene, but it may also have occurred as the result of oxidation of a more common flammable material, if there was some barrier that contained the chemically reacting substances long enough for a pressure wave to violently rupture the barrier. Moreover, not every explosion is the immediate effect of a chemical reaction; it may be caused, for example, by a pressure buildup attributable to gas accumulation in a weak or weakened containment vessel. It is the bursting of the container that qualifies as an *explosion*. If the movement within the combustion zone propagates at a velocity greater than the speed of sound in the vapor mixture, it is referred to as a *detonation*.

Flammability Limits

Typically, combustion occurs after a volatile gas or liquid has evaporated sufficiently to produce a flammable mixture in the air. In evaluating such a process, it is most important to take account of the *lower and upper flammability limits* (abbreviated as LFL and UFL). At concentrations below the LFL, there is not enough fuel in the mix to sustain a flame, and at concentrations in the air that are above the UFL, the supply of oxygen in the air is insufficient.

As the mix is not flammable if either the LFL or the UFL limits are exceeded, fuel concentrations too low or too high as the case may be, it is possible to take advantage of this characteristic by controlled cleaning or by dilution. Before releasing a flammable mixture to the air or to contact with an ignition source, it can be cleaned by passing it through a scrubber, or it can be diluted with nitrogen or some other nonflammable gas.

An alternative description of the hazardous concentration range is sometimes expressed in terms of a *lower explosive limit* (LEL) and an *upper explosive limit* (UEL). This terminology is more precise if the mixture being considered is especially close in composition to the stoichiometric ratio of the fuel to the oxidizer, but as a practical matter, the two definitions are ordinarily treated as identical. It should be noted that organic dust clouds can also have the LEL and UEL characteristics, but the range of sensitivity depends more on particle size than on particular chemical composition.

As shown in **Table 2**, the LFL and UFL values vary over a very wide range, depending on the specific components involved; however, it should be noted that many common hydrocarbons can burn in air only if they are present to an extent between approximately 1 and 7 vol%. The tabulated values are for room temperature and atmospheric pressure, but the limits also depend on temperature and pressure. The flammability range widens with increasing temperature, but both the lower and higher values increase with pressure.

Flash Points

Another measure of the degree of flammability of a substance is its *flash point*, a temperature defined by a standardized empirical test. By following a procedure set down by the ASTM,

Table 3 Selected values of flash points (°C)

Substance	*Flash point (°C)*
Acetaldehyde	−39
Acetonitrile	2
Acetylene	−18
Ammonia	11
Benzene	−11
Carbon disulfide	−30
Cyclohexanol	68
Decane	46.1
Diesel fuel	>62
Dimethyl sulfide	−49
Ethanol	12.8
Ethylene glycol	111
Methyl acetate	−10
Nitromethane	35
Octane	13
Propyl acetate	13
Propylene	−108

a liquid is heated in a specified manner from below in either a closed or an open cup with an ignition source above. As the liquid heats, it volatilizes, and when the vapor mix with the air is sufficient to ignite, the temperature of the liquid is said to be at its flash point. As with other measures of flammability, the flash point is different for each substance tested. Some typical values are given in **Table 3**. The NFPA Standard 704 has assigned flammability ratings to liquids, depending on their flash points, as follows:

Rating 0: Substance will not burn.
Rating 1: Flash point equal to or greater than 200 °F.
Rating 2: Flash point between 100 and 200 °F.
Rating 3: Flash point between 73 and 100 °F.
Rating 4: Flash point less than 73 °F.

Pollution and Toxic Substances

If the subject of investigation is the release of a toxic material, it must be asked how the chemical entity was released. Did a vessel leak? Did some container crack, break, or explode? Was the release accidental or intentional for some reason? Where did the substance go and by what path? Some leaks are to the air, some to the ground, some into water tables, or to other aboveground water courses. Finally, what injury resulted from the events and their consequences?

Air Pollution

Toxic ingredients can enter the air from a variety of possible sources. Some are intentionally released if it is believed that the emission is harmless when it is diluted in the air to a low enough concentration. The Clean Air Act, which empowers the EPA, was last amended in 1990, but standards of acceptability change over time, almost always in the direction of more restrictive expectations. This is easily seen when one reviews the limitations summarized in **Table 4**. The allowable ozone level in city air, for example, has been reduced in the decade between 1997 and 2008 from 0.080 to 0.075 ppm. A new rule proposed in 2011 would require power plants to reduce emissions of mercury by 91% within the next 5 years. More recently, controversy has arisen also on whether the EPA can and should restrict greenhouse gas emissions, especially carbon dioxide.

Some changes in standards can also come about by responses from industry or consents unforced by governmental regulation. By international agreement, for example, the venting of chlorofluorocarbons was phased out to protect the stratospheric ozone layer. This transition required substitution for the chemicals used in refrigerants and foaming agents, among other applications.

Sometimes obnoxious material is released from a chemical operation not because it is thought safe, but rather because the hazard is even greater if the toxic material is not released. This is the case, for example, if the buildup of pressure within a vessel would lead to a catastrophic explosion. A safety blow-off valve could then relieve the internal pressure by venting gas to the atmosphere. Most recently, after the huge earthquake and tsunami that damaged so much of the construction in Japan, the operators of a threatened nuclear power plant attempted to flood the reactor with sea water and vented radioactive steam in order to prevent an even worse meltdown within the reactor.

In any of the aforementioned cases, the wind in the region and the natural diffusion from higher concentrations to lower levels spread the toxins and the associated risks. Given information on the amount dispersed and the local conditions, an expert will be able to estimate the potential damage at various distances from the place of the release, allowing parties in controversy to better assess their positions.

Another class of air standards that applies in the workplace is set forth by the Occupational Safety and Health Administration (OSHA). In order to protect workers from lead poisoning, for example, OSHA has set a *permissible exposure limit* of lead in workplace air at 50 μg m^{-3} averaged over an 8-h workday for workers in general industry. In addition, the standard specifies the frequency and extent of medical monitoring, and other responsibilities of the employer. If controversy arises, a chemical engineering expert could provide information on fumes produced in the process as well as on air monitoring and possible engineering controls.

Ground and/or Water Pollution

When offending chemicals are freed from containment, they can also pass directly into the ground. This is especially noticeable if buried tanks are corroded by long exposure,

Table 4 National ambient air quality standards

Pollutant	*Primary standards* *Level*	*Averaging time*	*Notes on Averaging time*
Carbon monoxide	9 ppm (10 mg m^{-3})	8 h	Not to be exceeded once per year
	35 ppm (40 mg m^{-3})	1 h	
Lead	0.15 μg m^{-3}	Rolling 3-month average	Final rule signed on October 15, 2008
	1.5 μg m^{-3}	Quarterly average	
Nitrogen dioxide	53 ppb	Annual (arithmetic average)	
	100 ppb	1 h	To attain this standard, the 3-year average of the 98th percentile of the daily maximum 1-h average at each monitor within an area must not exceed 100 ppb (effective January 22, 2010)
Particulate matter (PM_{10})	150 μg m^{-3}	24-h	Not to be exceeded more than once per year on average over 3 years
Particulate matter ($PM_{2.5}$)	15.0 μg m^{-3}	Annual (arithmetic average)	To attain this standard, the 3-year average of the weighted annual mean PM2.5 concentrations from single or multiple community-oriented monitors must not exceed 15.0 μg m^{-3}
	35 μg m^{-3}	24 h	To attain this standard, the 3-year average of the 98th percentile of 24-h concentrations at each population-oriented monitor within an area must not exceed 35 μg m^{-3} (effective December 17, 2006)
Ozone	0.075 ppm (2008 std)	8 h	To attain these standards, the 3-year average of the fourth-highest daily maximum 8-h average ozone concentrations measured at each monitor within an area over each year must not exceed the given limit (effective May 27, 2008)
	0.08 ppm (1997 std)	8 h	To attain these standards, the 3-year average of the fourth-highest daily maximum 8-h average ozone concentrations measured at each monitor within an area over each year must not exceed the given limit (effective May 27, 2008)
	0.12 ppm	1 h	The standard is attained when the expected number of days per calendar year with maximum hourly average concentrations above 0.12 ppm is ≤ 1
Sulfur dioxide	0.03 ppm	Annual (arithmetic average)	24 h not to be exceeded once per year
	0.14 ppm	24 h	
	75 ppb	1 h	Final rule signed on June 2, 2010. To attain this standard, the 3-year average of the 99th percentile of the daily maximum 1-h average at each monitor within an area must not exceed 75 ppb

The EPA Office of Air Quality Planning and Standards has set national standards for six "criteria" pollutants. The units of measure are parts per million (ppm) by volume, parts per billion (ppb) by volume, milligrams per cubic meter of air (mg m^{-3}), and micrograms per cubic meter of air (μg m^{-3}).

whatever the setting. Fuel tanks in gas stations are frequently seen being replaced or removed entirely, and the same is sometimes also part of an agreement governing the sale of a home that has buried oil storage. Some industrial operations that need the disposal of unwanted by-products store the waste in an artificial lagoon, that is, in a lake created for the purpose by building confining dams of earth or concrete. With time, these dams can also leak. If they do, the hazardous stored material enters the ground or stream or river in the vicinity. An expert in chemical hydrology can be called upon to estimate effects on the underground water table or on the moving stream above ground.

Storage of Hazardous Materials

Many chemicals are held in storage in anticipation of later use, either as intermediates for further chemical reaction or simply as inventory for later sale. Often, the materials being stored are toxic to humans and animals and, sometimes, they are solvents or diluents that are flammable. When this is the case, special care is needed in providing vessels that will not leak or be corroded by the materials being held in them, but when these precautions prove to be inadequate, the results are catastrophic.

One of the best known serious examples of leaked material occurred in 1984 at a Union Carbide facility in Bhopal, India, when a storage vessel holding methyl isocyanate released its contents. The chemical was being stored as an intermediate in pesticide manufacture. It is a highly toxic gas and its release killed and seriously injured thousands of nearby residents. Accusations of negligence led to both civil and criminal cases in India, ending in fines and sentences of imprisonment. Improved design and practice now call for smaller inventories of toxins, backup safety procedures, and greater removal from population centers.

Unrecognized Hazards and Unexpected Consequences

A trained eye will sometimes recognize a hazard that is not recognized by an untrained user. Consider, for example, the following instances.

Fire in a Closed Room

It is common practice in construction and renovation to finish a new floor or other wooden surface by coating it with a protective finish such as polyurethane. These finishes are typically applied as a solution or suspension in a flammable hydrocarbon solvent. Upon completion of all or part of the task, the work is allowed to "dry," that is, the diluent solvent is allowed to evaporate in order to leave behind the finished protective surface. When the task is completed at the end of a working day, it is commonly convenient to leave the working surface unattended in order that the drying will occur overnight.

If, as is often the case, the room or house being worked on is closed and locked for the night, there is ample time for the evaporating solvent to fill the air in the room to a level that is above the lower flammable limit, creating an accident that is waiting to happen. All that is missing in such a case is a source of ignition, perhaps a pilot light left burning, or a sparking thermostat.

Fire in an Empty Container

When a container holding a flammable liquid is emptied of its contents, it is not truly empty. In fact, it is filled with the air that replaced the withdrawn liquid and with any vapors that evaporated from the liquid. When the liquid is especially volatile, the vapor concentration in the trapped air can exceed the LEL. The mixture will be harmless unless an ignition source is introduced, but this final element would be present if a torch is used to disassemble or clean the vessel.

Spontaneous Combustion

Many chemical substances liberate heat if they are allowed to oxidize in air. Usually this process occurs slowly enough that the heat escapes to the surrounding air, no visible flames are seen, and the process is harmless. Sometimes, however, the heat is trapped before it can escape. This can happen, for example, if the layer of labile substance is deep in comparison to the path of heat escape, or if the oxidizing chemical is held in a closed container.

In such circumstances, when the heat cannot escape quickly enough, the internal temperature rises. At the higher temperature, the oxidation reaction speeds up, liberating more heat even faster than before. The temperature rises still higher and this feedback snow-balling effect can run away and out of control. Eventually, this process leads to a fire, so-called spontaneous combustion, as the autoignition temperature is reached.

Charcoals and supermarket chars are commonly found to be vulnerable to such fires, especially if they are in closed or insulated containers or closets. Such materials are quite porous and can oxidize in air even when left uncovered in garden centers. Oily rags used to clean up paint spills or to wipe down greasy griddles are similarly in danger of what appears as a spontaneous fire. The recent emphasis on low trans fat oils in restaurant and fast food cooking has exacerbated this hazard, because chemically unsaturated oils are especially easy to oxidize. Even commercial quick laundering may not sufficiently remove the offending substances from the flammable rags. In such a case the detergents being used may be supplemented by added enzymatic agents.

Flammable or Toxic Solids

It was noted earlier that volatile materials, those that exhibit large vapor pressures under room conditions, are especially higher risks for fires because their evaporation makes them more likely to create concentrations in the air that exceed the LFL. Most often, these materials are liquids, but it should be noted that some solids, such as naphthalene or mothballs (para-dichlorobenzene), also show elevated vapor pressures at moderate temperatures and are correspondingly hazardous. An expert is able to make a quantitative estimate of the vapor composition.

An Explosive Pharmaceutical

Because chemical agents are used in such a wide variety of applications, the intended use in a particular application may obscure some secondary hazard. In some medical treatments, for example, nitroglycerine in small doses is a pharmaceutical taken by mouth or applied to the skin in the form of a patch. If it is stored in larger quantities, however, it is an unstable hazardous explosive. It was only after Nobel found that nitroglycerine could be stabilized by diluting it in a solid mixture that it could be used commercially as dynamite.

Contamination of Food in Shipment

Much of the world's food is shipped to market over great distances in the holds of freighters. To prevent spoilage, the storage spaces are usually temperature and humidity controlled by the circulation of conditioned air. A unique hazard arises when the ship is fueled, because the vapors of a volatile liquid fuel can enter the circulating air stream and contaminate the food being carried. Careful assessment of both risks and prevention techniques are called for.

Corrosive and Reactive Chemicals

Among the myriad substances that circulate in modern commerce, there are some that are immediately recognized as being corrosive to common metals as well as to people. These include all the concentrated acids (such as sulfuric, nitric, hydrochloric, and acetic), alkalis (such as lye—caustic soda and ammonium hydroxide), and gases (such as chlorine). These are all corrosives that can attack and chemically destroy exposed body tissues. They begin to cause damage as soon as they touch the skin, eyes, respiratory tract, or digestive tract. The more concentrated the corrosive material is and the longer it touches the body, the worse the injuries will be. When these materials are present in a workplace, everyone who works with them or near them must be made aware of their hazards and how to deal safely with them. A fully prepared material safety data sheet is essential information and will be asked for by an expert.

In fact, the list of reactive materials is much longer if one recognizes that significant chemical reactions can occur between many pairs of substances, including some materials that are commonly thought of as inert. Plastics can be changed in their properties by exposure to some solvents or alcohols, causing crazing, cracking, or softening, for example. Even some so-called noble metals (gold, silver, and copper) will react with thiocyanides to form soluble compounds. In each case, the result depends on the pair in question, as well as the conditions of temperature, pressure, and concentration, and an expert may be needed to make the essential distinctions.

Epilog

As a conclusion to this chapter, it is essential to add a caveat because no single individual can be equally expert in all areas of any field. The forensic chemical engineer must assess any given assignment in the light of his personal skills and specialized knowledge and be prepared to recuse himself if the matters at hand are not part of his repertoire.

The chemical engineer's skill is most effective when it is applied in direct conjunction with an attorney's view of legal constraints. An example may serve to illustrate the point. Most forensic matters that come to chemical engineers are related to civil suits, focusing on property damage or personal injury. Such cases are either settled out of court or adjudicated with monetary results. On occasion, however, a question arises that has implications for a criminal case. One such matter was raised when instructions posted on the Internet were followed by a teenager who constructed and set off a homemade "bomb" via the chemical reaction of aluminum foil with an acidic toilet cleaner.

The punishment for this teen's behavior depended on whether or not the device he put together should properly be classified as "an explosive device" and an expert was called on to testify on this issue. The resolution was based on the detailed understanding of the chemistry and physics involved, but it was the collaboration with the attorney in the case that brought out the distinctions on which the chemical engineering expert based his responses. In forensic matters, it should always be remembered that each profession enhances the contribution of the other.

See also: **Chemistry/Trace/Environmental Analysis:** Overview, Analysis, and Interpretation of Environmental Forensic Evidence; **Chemistry/Trace/Explosives:** Commercial; Explosions; Explosives: Analysis; **Chemistry/Trace/Fire Investigation:** Analysis of Fire Debris; Chemistry of Fire; Interpretation of Fire Debris Analysis; Physics/Thermodynamics; Thermal Degradation; **Chemistry/Trace/Forensic Geosciences:** Soils; **Investigations:** Contamination; Explosions; Fire Scene Inspection Methodology; Types of Fires; **Legal:** Expert Witness Qualifications and Testimony; **Management/Quality in Forensic Science:** Standard Methods; **Methods:** Analytical Light Microscopy; Chromatography: Basic Principles; Gas Chromatography; Gas Chromatography–Mass Spectrometry; Mass Spectrometry; Microscopy (Electron); **Pattern Evidence:** Tools; **Toxicology:** Interpretation of Results; Toxicology: History.

Further Reading

Carslaw, H.S., Jaeger, J.C., 2000. Conduction of Heat in Solids, second ed. Oxford University Press, Oxford.

Crank, J., 1975. The Mathematics of Diffusion, second ed. Oxford University Press, Oxford.

Crowl, D.A., Louvar, J.F., 1990. Chemical Process Safety: Fundamentals with Applications. Prentice Hall, Upper Saddle River, NJ.

Perlmutter, D.D., 1986. Self heating and spontaneous combustion. In: Proceedings of the Conference on Gas-Solid Reactions in Pyrometallurgy, W. Lafayette, Indiana.
Perry, R.H., Green, D.W. (Eds.), 1997. Perry's Chemical Engineers' Handbook, seventh ed. McGraw-Hill, New York.
Roberge, P.R., 2000. Handbook of Corrosion Engineering. McGraw-Hill, New York.

Relevant Websites

http://www.chemexpertwitness.com.
http://www.astm.org/Standards/D56.htm; http://www.astm.org/Standards/D93.htm.
http://www.astm.org/Standard/index.shtml.
http://www.atsdr.cdc.gov.
http://www.chemexpertwitness.com.
http://www.cpsc.gov.
http://www.epa.gov.
http://en.wikipedia.org/wiki/Flammability_limit.
http://www.engineeringtoolbox.com/fuels-ignition-temperatures-d_171.html.
http://en.wikipedia.org/wiki/Flammability_limit.
http://www.epa.gov/air/criteria.html.
http://www.nfpa.org.
http://safety.science.tamu.edu.

Materials Analysis and Failure Analysis

AD Micheals, San Jose State University, San Francisco, CA, USA

The Role of Materials Analysis in Determining Causation of Failure

Materials analysis techniques are key tools in the arsenal of the materials failure analyst. Materials analysis seeks to characterize the intrinsic properties of a material in order to understand the extrinsic properties of the material, that is, how the material should behave in a given context. These intrinsic properties include morphology, composition, crystal structure, atomic bonding, elemental distribution, or, indeed, any characteristic that affects the extrinsic properties of the material. Of these, analysis of morphology and composition frequently provides information critical to failure analysis.

Why did a failure occur? This is the fundamental question to be answered by engineering failure analysis. By understanding and quantifying a failed material's properties, and comparing those values with either the specified properties or an exemplar, the materials analyst can gain an understanding of the role the material played in the failure, whether primary, or secondary to another primary failure.

Materials analysis may help the failure analyst determine:

- When a material failed in relation to a sequence of events. Did it fail and cause an event, or did it fail as a result of the event?
- If design or manufacturing was a factor.
- If the choice of material was a factor.
- Whether inadequate maintenance was a factor.

This chapter focuses on a selection of important materials failure analysis techniques used to determine morphology and composition (**Table 1**). Selected applications are presented; a list of useful materials analysis references, including sources for analysis protocols, is provided for further background.

Techniques for Determining Morphology

The characteristics of a material that pertain to its form are collectively termed morphology. Morphology may describe both the external and internal structure of a material. The general external form, including the shape, surface features, and surface texture, as well as the segregation of elements into different regions or phases, and even the internal arrangement of atoms, are important to materials analysis because the structure of a material is intimately related to its properties. The most common morphological characterization technique is microscopy. Various types of microscopies are available. Each technique has different strengths and limitations; therefore, the appropriate method must be selected depending on the information required and the feature size to be observed. The two microscopy techniques covered here are optical and scanning electron microscopy.

Microscopy in Materials Analysis

With the techniques of microscopy, external structure can be magnified and examined using light or electrons. The choice of imaging technique is based on the material being examined the feature size, and the information desired.

Optical Microscopy

In the analysis of materials failures, information about the root cause may be obtained from visual examination. This is especially true when a fracture surface is available. However, magnification is frequently needed to observe the pertinent features of a fracture surface, the surface morphology, or the distribution and morphology of different phases.

In optical microscopy, examination is done using light incident on the specimen and reflected back into the eye of the investigator or to a camera. Transmitted light may also be used with appropriately thin or transparent materials. Traditionally, the light used is in the visible spectrum; however, infrared (IR)

Table 1 Some materials analysis techniques used to determine morphology and composition

Morphology
Optical microscopy
Scanning electron microscopy
Composition
Energy-dispersive spectroscopy
Wavelength-dispersive spectroscopy
X-ray fluoroscopy
Absorption spectroscopy
Auger electron spectroscopy
Mass spectrometry
Differential scanning calorimetry
Differential thermal analysis
Thermogravimetric analysis

 http://dx.doi.org/10.1016/B978-0-12-802718-9.00032-6

or ultraviolet (UV) light frequencies may also be used in conjunction with appropriate detectors and equipment to display images in a viewable format.

Because of the limitations of their resolving power, the typical magnification range of optical microscopy runs from 1 to 400 times the original size; 1000 times is the upper limit of the magnification range for conventional optical microscopes. When viewing fracture surfaces, which are typically rough, the depth of field, which is the ability of the microscope to keep the entire field of view in focus, is frequently a limiting factor. Depth of field is a function of the optics of the microscope and magnification (depth of field decreases with increasing magnification), as well as of the behavior of light in general.

Optical microscopy is typically performed with physical glass lenses. In common usage, a single-lens microscope is called a "magnifying glass," while "microscope" refers to a compound microscope. Compared to the single-lens microscope, the compound microscope uses multiple lenses along the optical path to achieve a range of specific magnifications as well as improved magnification and optical quality. A compound microscope suitable for viewing solid specimens is shown in **Figure 1**.

In materials analysis, there are two specific configurations of optical microscope that are commonly used: the inverted microscope (**Figure 2**) and the stereomicroscope (**Figure 3**).

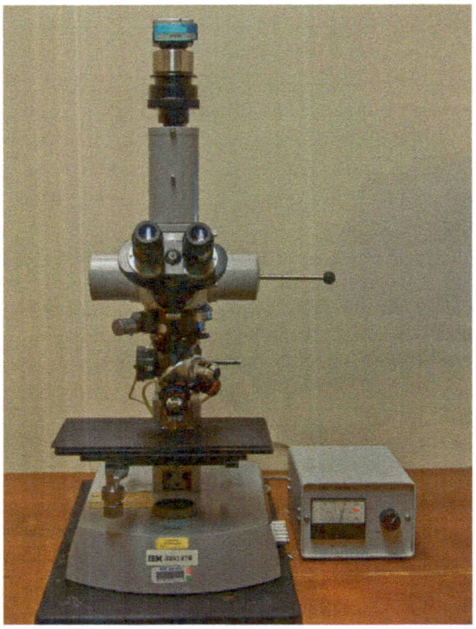

Figure 1 A compound microscope for viewing solid specimens. The optical path goes from the sample through multiple magnifying lenses and is then split, sending an identical image to each of the dual eyepieces.

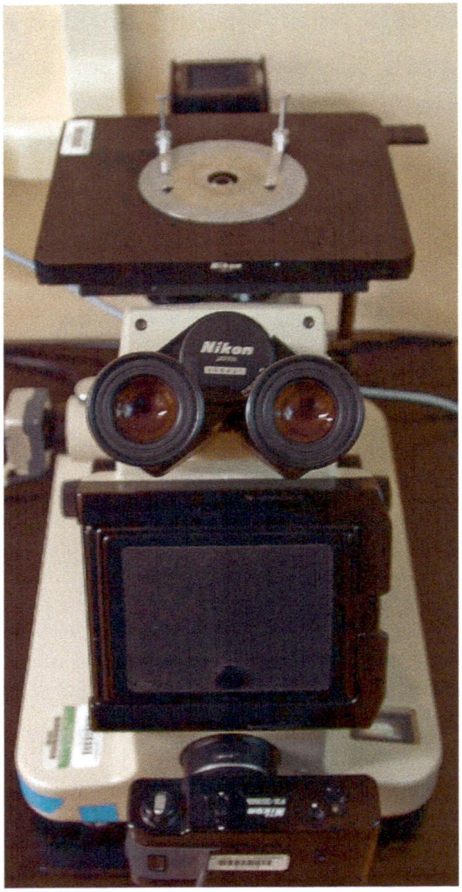

Figure 2 An inverted microscope for examining flat (polished) specimens at high magnifications. Note that the specimen stage is above the optics or inverted from the "normal" orientation.

The inverted microscope is used to image flat, polished specimens. The specimen is placed above the magnifying lenses in an inverted position. Because the specimen is flat, the limited depth of field is not an issue; magnifications of 1000 times can be achieved with clear focus in the plane of the specimen. An appropriately prepared specimen can evidence transitions between different phases, dislocation motion, history of heat treating and mechanical deformation, processing defects, voids, and material composition.

The stereomicroscope consists of two distinct compound microscopes, with offset optical paths terminating at separate eyepieces. The distinct images displayed in the two eyepieces allow for perception of depth. Because the depth of field is inversely related to magnification, magnification is typically limited to 100 times, for the sake of a larger working distance and maintaining an adequate depth of field. Stereomicroscopes

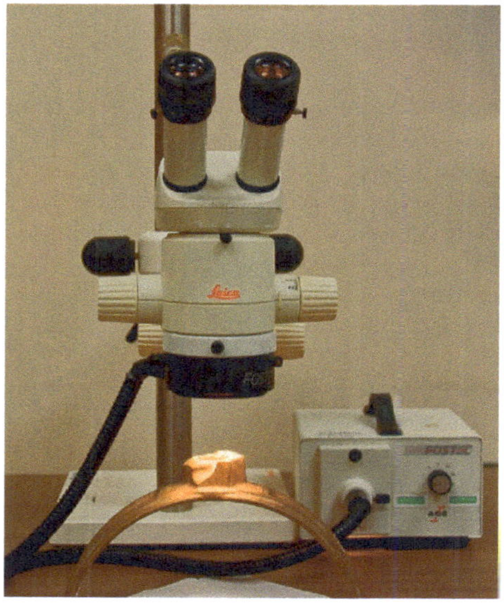

Figure 3 A stereomicroscope used for viewing objects with height differences at low magnification.

are useful for viewing large areas of fracture surface and specimens with large height differences. Stereomicroscopy may often be used to distinguish between various types of fracture, for example, ductile versus brittle fracture, overload fracture, and fatigue fracture.

Specimen Preparation for Microstructural Examination

When analyzing a materials failure, it may be of interest to examine the internal microstructure. Microstructure arises from the segregation of different elements or impurities into different regions; these regions form different structures, called phases, within the material. These phases and their relative, spatial configurations have a significant effect on the properties (both desirable and undesirable) of the material. Many of these structures can be observed and identified with an optical or electron microscope. Specimen preparation procedures have been developed to prepare material specimens for examination of their microstructure. The specimen preparation techniques are collectively grouped under the term metallography. Further information can be found in the References section.

Analysis of failed structural steel using optical microscopy

1040 steel in a structural application failed unexpectedly. The failed steel as well as two 1040 steel samples subjected to different heat treatments were prepared for microscopic examination by being polished and chemically etched. The differences brought about by the heat treatments are visible at the relatively low magnification of 400 times. **Figure 4(a)** shows large grains typical of slow cooling, where time at temperature allows the atoms to move to low-stress positions on the crystal lattice. **Figure 4(b)**, quenched and tempered at a relatively low temperature, is little changed from the fine grain structure typical of martensitic steel; the temperature was too low for the atoms to move and form stress-free grains. The slow cooled sample will have a much lower strength than the quenched sample.

On comparing the failed steel (**Figure 4(c)**) to the samples with known history it was found that the failed steel had not been strengthened by the appropriate heat treatment.

Scanning Electron Microscopy

For greater magnification, and improved depth of field, an electron microscope can be used. The scanning electron microscope (SEM) is a common instrument, which provides images that are intuitive and easy to understand. Magnification ranges from approximately 20 times to over 20,000 times. Depth of field depends on magnification but can be as much as 10 mm at low magnification.

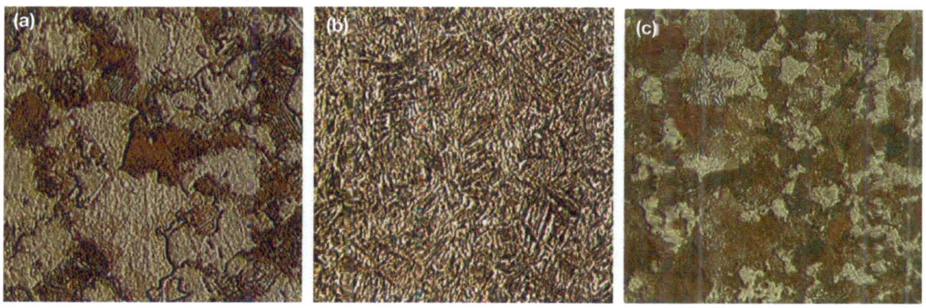

Figure 4 1040 steel, (a) slow cooled, (b) quenched and tempered for 1 h at 205 °C, and (c) failed material of unknown heat treatment. Samples were polished to 0.05 μm alumina abrasive and then etched for 10 s with 2% Nital etchant (2% nitric acid in ethanol). The micrographs were taken with an inverted microscope at 400× magnification.

Whereas an optical microscope illuminates the entire specimen at the same time and creates an image by focusing the reflected or transmitted light, an SEM image is formed by focusing the electrons used to illuminate the specimen onto a very small area and moving (scanning) the focused beam across the specimen's surface. The "brightness" of the energy reflected as each small area of the specimen is illuminated is measured and displayed as the brightness of a pixel on an image.

A schematic of an SEM is shown in **Figure 5**. Electrons are created by heating a filament (tungsten or lanthanum hexaboride) and are then accelerated by an applied voltage of up to 30 kV. The electron beam is focused by electromagnetic lenses to bombard the specimen surface. The beam of these primary electrons (PEs) is scanned across the surface of the specimen. As the beam is scanned, some of those electrons are reflected after scattering, and other additional electrons may be emitted from the specimen. PEs reflected from the specimen are referred to as backscattered electrons (BSEs). BSEs have relatively high energy, up to that of the incident beam. If a PE interacts with a specimen atom and transfers some of its energy to that atom's electrons, one or more of the atom's electrons may be emitted as secondary electrons (SEs). SEs are low-energy electrons and are distinguished from BSEs as having less than 50 eV of kinetic energy.

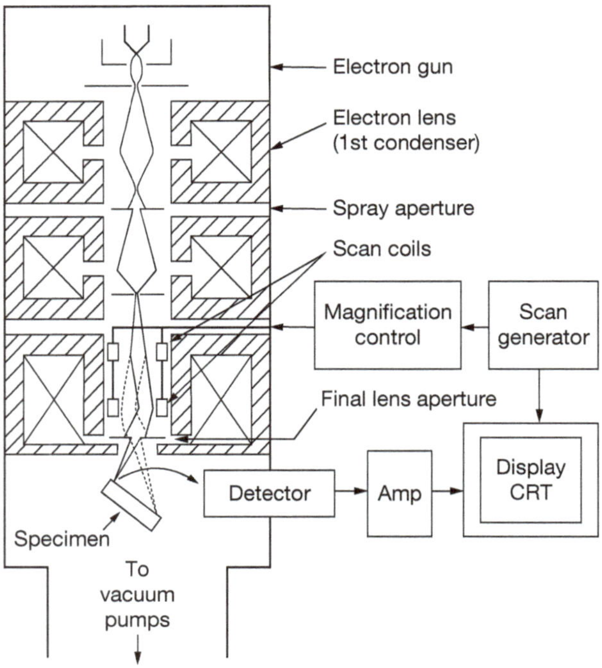

Figure 5 Cross-sectional schematic of the column of a scanning electron microscope. Reprinted from Goldstein, J., Newbury, D., Joy, D., et al., 2003. Scanning Electron Microscopy and X-ray Analysis. Springer, New York, p. 23. Copyright 2003 by Springer Science + Business Media, with kind permission from Springer Science + Business Media B.V.

Specialized detectors are used to sense the electron signals. The detected electron intensity is converted into an electrical signal that is displayed as a gray scale image where brightness is proportional to the detected electron intensity. The image is sent to a computer, to be displayed, printed, or saved in a digital format. Magnification occurs when the PE beam is scanned across a small area of a specimen, and the image of the scanned surface region is displayed on a larger computer screen.

The imaging process usually takes place under high vacuum, around 10^{-6} torr. The vacuum prevents the PE beam from losing its energy in collisions with gaseous atoms and allows the emitted signal to reach the detectors.

The resolution of the SEM, while theoretically limited by the wavelength of the primary electrons, is more practically dependent on two factors: the incident beam diameter, or spot size, and the region of the specimen that the primary electrons interact with, called the interaction volume (**Figure 6**). The beam diameter is determined by how tightly focused the electron beam is. It is controlled by focusing electromagnetic lenses and can be adjusted from microns to nanometers. The dimensions of the interaction volume depend strongly on the incident electron energy. The width and depth may range from hundreds of nanometers up to approximately 5 μm. While a high-energy means that the electrons have a shorter wavelength, the electrons in the beam also have high speed and momentum, which drives the electrons farther into the material. The interaction volume also depends on the atomic number Z of the specimen. As Z increases, the number of outer shell electrons in the atom increases and the incident electrons are more strongly repelled by Columbic interaction with the atomic electric field. So, for small Z, as with carbon, the interaction volume is large; for large Z, as with lead, the interaction volume is small.

As can be seen from the interaction volume, SEs are most representative of the specimen surface, while BSEs result from bulk interactions. Therefore, for best resolution of surface features, imaging is typically derived from SEs. For contrast differences derived from elemental composition, the BSE signal is typically used.

Analysis of a ductile iron pipe failure with scanning electron microscopy

A ductile iron pipe failed unexpectedly during a pressure test. Samples from the failed pipe and from a pipe that had survived the pressure test were examined in the SEM. Both specimens were prepared by grinding and polishing to a flat surface. The specimens were chemically etched to create a slight but observable height difference between the iron and carbon phases. In this image, the primary contrast is from the BSE

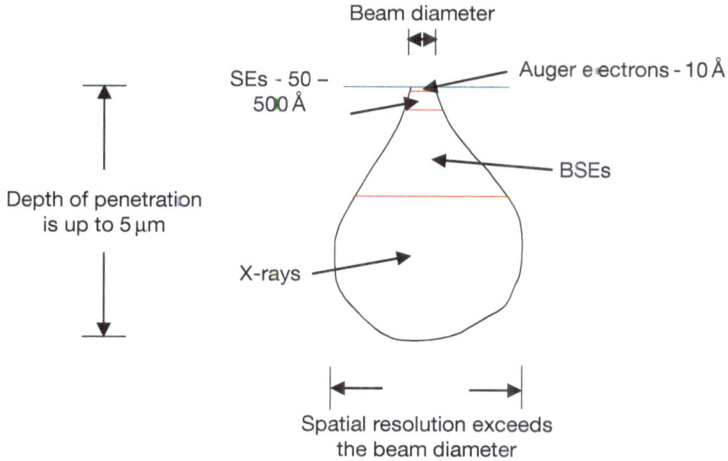

Figure 6 The interaction volume is the three-dimensional portion of the specimen that the primary electrons travel through. Its size varies with the kinetic energy of the primary electrons and with the atomic number of the specimen.

signal. The incident electrons penetrate the low *Z* carbides, with little BSE emission. The higher *Z* iron phase reflects more electrons, emitting a higher intensity signal. Therefore, the iron regions appear bright compared to the carbides.

Ductile iron microstructure typically consists of an iron matrix with distributed spherical carbides. The spherical carbides intercept any cracks in the iron matrix and prevent them from propagating. **Figure 7(a)**, the "good" pipe, shows the expected ductile iron microstructure. **Figure 7(b)**, the failed pipe, shows poorly formed, nonspherical carbides. As a result, the material of **Figure 7(b)** had a lower-than-expected strength compared to that of **Figure 7(a)**. Additional analysis determined the cause of the poor carbide morphology to be errors in the material composition.

Techniques for Determining Composition

Along with morphology, composition plays an inseparable role in the behavior of a material. While the major elements present in a material obviously affect its macroscopic properties, small amounts may also have a strong effect on properties. An arguably important example of this is carbon steel. Changing the steel carbon content from 0.1% carbon to 0.2% carbon, along with appropriate heat treatment, can increase the tensile strength by orders of 10–100; hardness will increase as well.

The compositional techniques discussed here are X-ray spectroscopy, absorption spectroscopy, mass spectrometry (MS), and thermal and thermogravimetric analysis (TGA).

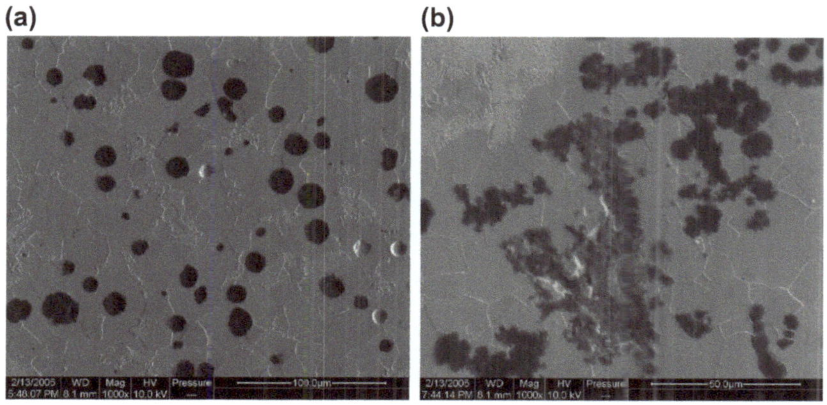

Figure 7 Scanning electron micrographs of two ductile iron specimens. (a) Shows the expected microstructure of the iron matrix with dark spherical carbides; (b) contains poorly formed carbides, which caused lowered material strength.

The Physics of Spectroscopy

Phenomena arising from the electronic structure of an atom, and specifically from the arrangements of the electron levels and interactions between them, are unique for each element and may be used to distinguish the elements present in a specimen. Spectroscopy is the method of detecting the wavelengths of radiation emitted or absorbed by a material and analyzing the results to determine the composition of a specimen. Spectroscopy techniques may determine the types of atoms present, their relative amounts, the type of atomic or molecular bonding, and the ionization state, if any.

Our normal perception of color is a form of spectroscopy that informs us of the composition of what we are viewing, albeit in a limited fashion. While the basis for spectroscopy is rooted in quantum mechanics, it was first described by Newton as he looked at the spectra of the light from the sun separated by a prism. For the purposes of this discussion, however, the shell model of the atom is sufficient to explain the phenomena used in the materials analysis techniques presented.

In the shell model of the atom, a nucleus is surrounded by electrons in specific shells or orbitals. Each orbital corresponds to a particular energy state. When an inner shell electron is removed from the atom, the atom is perturbed from its normal state, where all inner shells of the atom are filled. In an attempt to return to the normal, or ground, state, an electron from an outer shell will move to the inner vacant electron position. This position, closer to the nucleus, is a lower energy position; to make the transition, the electron must give up energy. While excess energy can be removed through increased atomic vibration, the electron can also emit electromagnetic radiation (visible light, UV, IR, or X-ray) or transfer a portion of its energy to another electron (**Figure 8**).

By the quantization of energy theory developed by Planck, the distance between electron energy levels in an atom depends on its atomic number Z, where Z is the number of protons in the nucleus. As Z increases with increasing nuclear charge, the distance between energy levels changes proportionately. For each element, therefore, the energy, or frequency, of the emitted radiation is specific to that element. By detecting and measuring the emitted radiation energy/wavelength, the atomic number can be determined.

The spectroscopic techniques below exploit the relationship between atomic number and emitted radiation frequency to determine composition. To create the emitted radiation, the inner shell electron is removed by a transfer of kinetic energy from an incident particle (a high-energy, high-momentum electron) or from incoming electromagnetic radiation. The method of exciting the emitted signal and the nature of the emitted signal are the primary differences in the techniques.

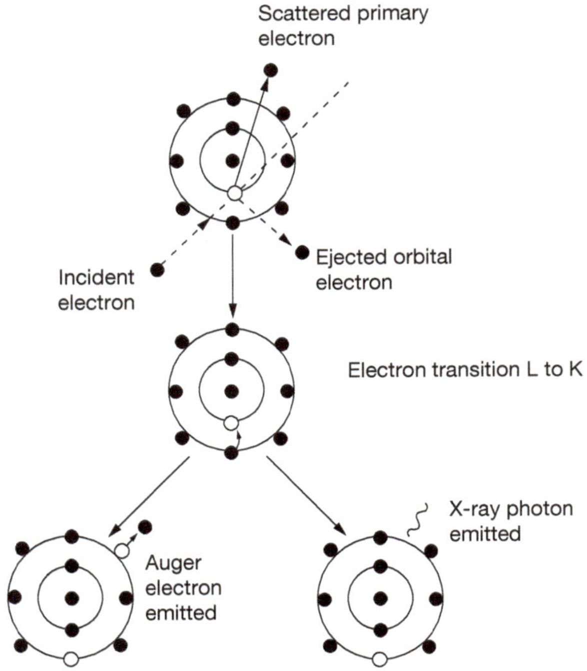

Figure 8 Excitation and relaxation of an atom; excess energy in this case is given off as an X-ray and an Auger electron. Adapted from Goldstein, J., Newbury, D., Joy, D., et al., 2003. Scanning Electron Microscopy and X-ray Analysis. Springer, New York, p. 23. Copyright 2003 by Springer Science + Business Media, with kind permission from Springer Science + Business Media B.V.

X-ray Spectroscopy

When an inner shell electron is stimulated and ejected, and an outer shell electron drops down to fill the hole, an X-ray is emitted. This X-ray is called a characteristic X-ray; it has an energy specific to the element from which it was emitted and to the specific levels transitioned by the electron.

Two techniques are used to detect the emitted X-rays: energy-dispersive X-ray spectroscopy (EDS) and wavelength-dispersive X-ray spectroscopy (WDS).

There are multiple techniques for stimulating the X-ray emission. In X-ray fluoroscopy (XRF), X-rays are used to stimulate X-ray emission. An electron may also be emitted along with an X-ray; this electron is called an Auger electron, and it can also be used to determine element composition. A useful side effect of the electron beam used in electron microscope systems is the stimulation of X-ray emissions. Hence, EDS and WDS detectors are frequently used with electron microscopes.

Each of these techniques has different detection limits and spatial resolution (**Table 2**), which arise from the different incoming and emitted signals exploited by each.

Table 2 Detection limits and spatial resolution for X-ray spectroscopy techniques

Technique	*Elements detected*	*Detection limit (ppm)*	*Minimum spatial resolution*	
			Lateral	*Depth*
Energy-dispersive spectroscopy (EDS)				
Wavelength-dispersive spectroscopy (WDS)	Atomic number >4 (beryllium and higher)	1000 (EDS), 10 (WDS)	1 μm	1 μm
X-ray fluoroscopy			10 μm	10–100 mm
Auger electron spectroscopy	Atomic number >3 (lithium and higher)	1000	0.02 μm	0.003 μm

Energy-dispersive X-ray spectroscopy

In EDS, the X-ray photon from the specimen impacts a detector, which may be a lithium-doped silicon semiconductor or a silicon drift detector. An amount of charge proportional to the energy of the X-ray is created in the detector (3.8 eV creates one electron in a silicon–lithium detector). This current is converted to a voltage proportional to the initial X-ray energy. The output spectrum is a histogram displaying the number of X-rays detected with respect to energy (**Figure 9**).

Wavelength-dispersive X-ray spectroscopy

In WDS, the wave nature of the X-ray is exploited to determine the wavelength. A single crystal, with known atomic plane spacing d, is used as a diffraction grating. An X-ray incident upon the crystal will diffract when it meets conditions set forth by Bragg. When the diffraction crystal is rotated with respect to the specimen, the angle of incidence of the X-ray changes, along with the wavelength needed for diffraction. In this way, a specific range of wavelengths can be scanned and differentiated. Often crystals with different d-spacings are used as needed, depending on the wavelength range to be detected and the sensitivity required. Once X-rays of a given wavelength are being passed by the diffraction crystal, they are counted in a gas proportional counter. The incoming X-ray ionizes the gas in the detector; the resultant free electrons are detected as a current. The current is converted to a voltage; the output is a histogram

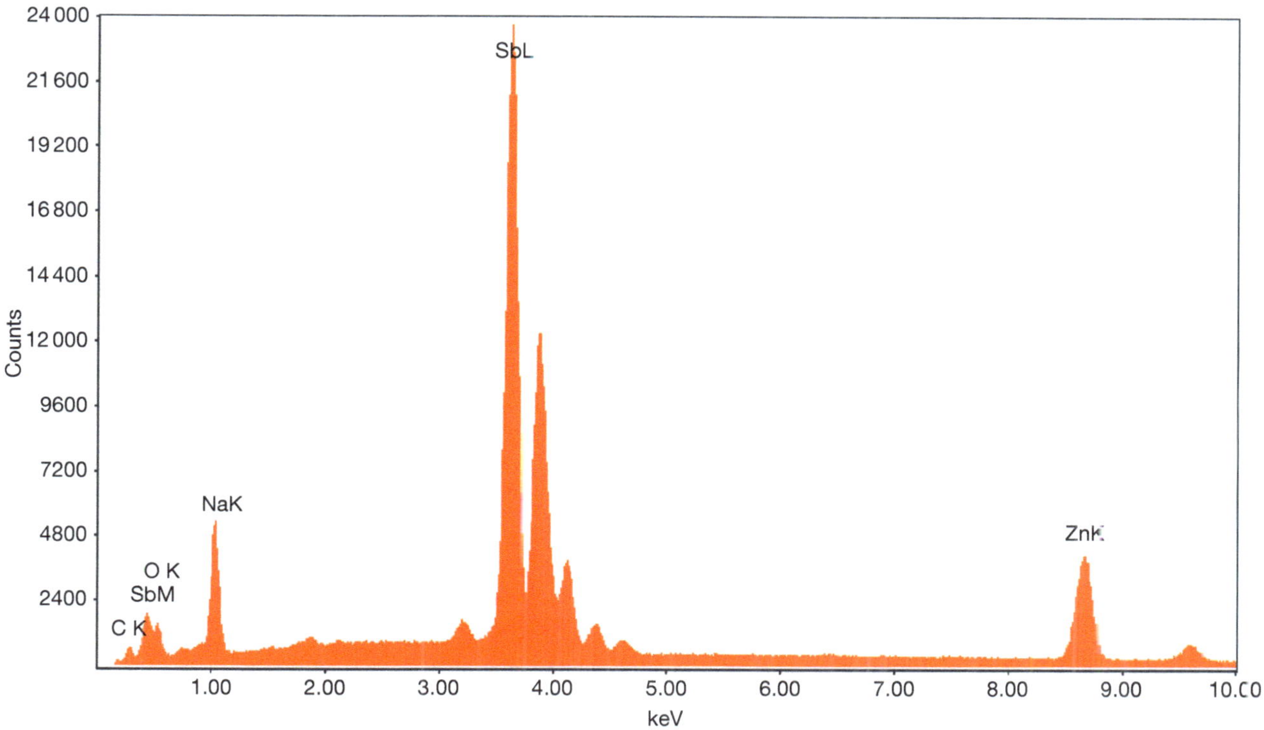

Figure 9 The energy-dispersive X-ray spectroscopy spectrum of a zinc antimonide semiconductor device.

with number of X-rays versus either wavelength or the corresponding energy.

Choosing between EDS and WDS

EDS measures the energy of the X-ray photon; WDS determines the wavelength of the X-ray. Because the energy of an electromagnetic wave is inversely related to its energy, according to Planck's law, this difference is trivial. The more relevant difference is in the way that each looks at a spectrum. EDS looks across a broad range of energies in the spectrum, while WDS looks at specific regions within the spectrum. With an unknown sample, EDS is more efficient; where the concentration of specific, known elements is desired, WDS is more efficient.

Both EDS and WDS can be accessories to electron microscopes. WDS are primarily found on specialized electron microscopes called electron probe microanalyzers, which are designed to collect very accurate compositional data. Electron microprobe configurations are available with limited or full SEM functionality.

While WDS is slower than EDS for large energy ranges, over a short energy range covered by a single crystal, the total scanning time is reduced. Additionally, WDS detectors have a higher throughput than EDS detectors; so WDS can collect more X-rays than EDS in the same amount of time.

Both EDS and WDS have the same spatial resolution; for the same incident electron energy, the volume of material that is sampled is identical. However, WDS has better spectral resolution; the width of the energy peak with WDS may be as low as 5 eV, compared to around 125 eV with EDS (**Figure 10**). The finer resolution of WDS is useful for eliminating problematic peak overlaps that may occur in EDS, for example, between manganese and iron, between oxygen and vanadium, and between sulfur, molybdenum, and lead. A dedicated WDS system is especially useful for routine examination of specimens containing those elements, such as ferrous alloys.

WDS also can detect smaller elemental fractions than EDS. While the ability to separate overlapping peaks certainly helps sensitivity, the benefit is primarily seen in trace element analysis. In WDS, the wavelength is set to that of the desired element; a fast acquisition rate is then used to obtain a high-intensity peak. The higher the peak, the higher the peak-to-background ratio, and the more accurate the quantification will be.

Using EDS to identify an unknown material

A small object, found in commercially purchased processed meat, damaged a consumer's tooth. From its appearance, it was suspected that the object was a bit of hard plastic. To differentiate between animal bone, which may be expected in meat, and a foreign object, the object was examined with SEM and EDS. The chemical composition is shown in **Figure 11(a)**. Calcium, expected in bone, was not present, but a high level of

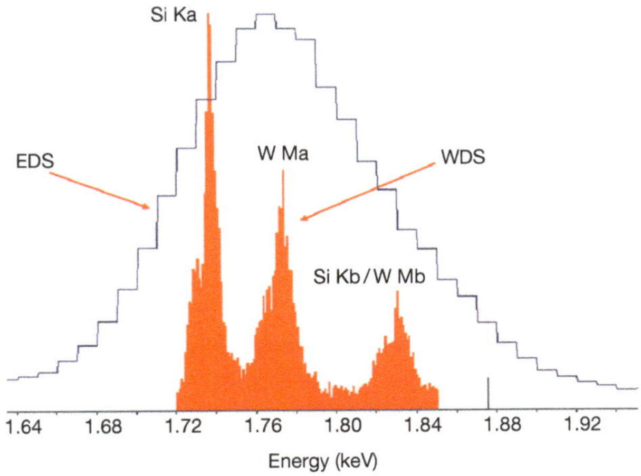

Figure 10 A comparison of the spectral resolutions of energy-dispersive X-ray spectroscopy (EDS) and WDS. The blue EDS spectrum shows a single peak of width 260 eV, which identifies the presence of silicon. The superior resolution of WDS reveals the presence of tungsten as well as silicon. The widths of the wavelength-dispersive X-ray spectroscopy (WDS) peaks are less than 40 eV. Figure courtesy of EDAX Inc.

sulfur was present. For comparison, two known samples, animal bone (**Figure 11(b)**) and two-part epoxy (**Figure 11(c)**), were also analyzed. The unknown object was determined to be a foreign material, a type of glue or epoxy, rather than bone.

X-ray fluoroscopy

XRF determines bulk composition using X-rays to stimulate characteristic X-ray emission. In XRF, the maximum incoming X-ray energy varies with manufacturer and instrument configuration, but typically ranges from 50 to 60 kV. Detection of characteristic X-rays is done by one of the methods described above, either EDS or WDS, and with the same attendant advantages and disadvantages.

In XRF, specimen size, sampling volume, and preparation are additional considerations. XRF instruments typically require a large amount of specimen (1 g or more), with a sampling depth ranging from microns to millimeters, depending on the incident X-ray energy and atomic makeup of the matrix. In the electron microscope, EDS/WDS can be performed on smaller amounts of material, and the X-rays are emitted selectively from the top microns of the specimen surface. However, this selectivity may be disadvantageous in circumstances where the overall composition of an inhomogeneous specimen is needed. Due to the higher sampling depth, XRF is preferred for bulk analysis. In XRF, since morphology is not determined, specimens are commonly ground and pelletized to ensure a homogeneous mixture and a smooth surface.

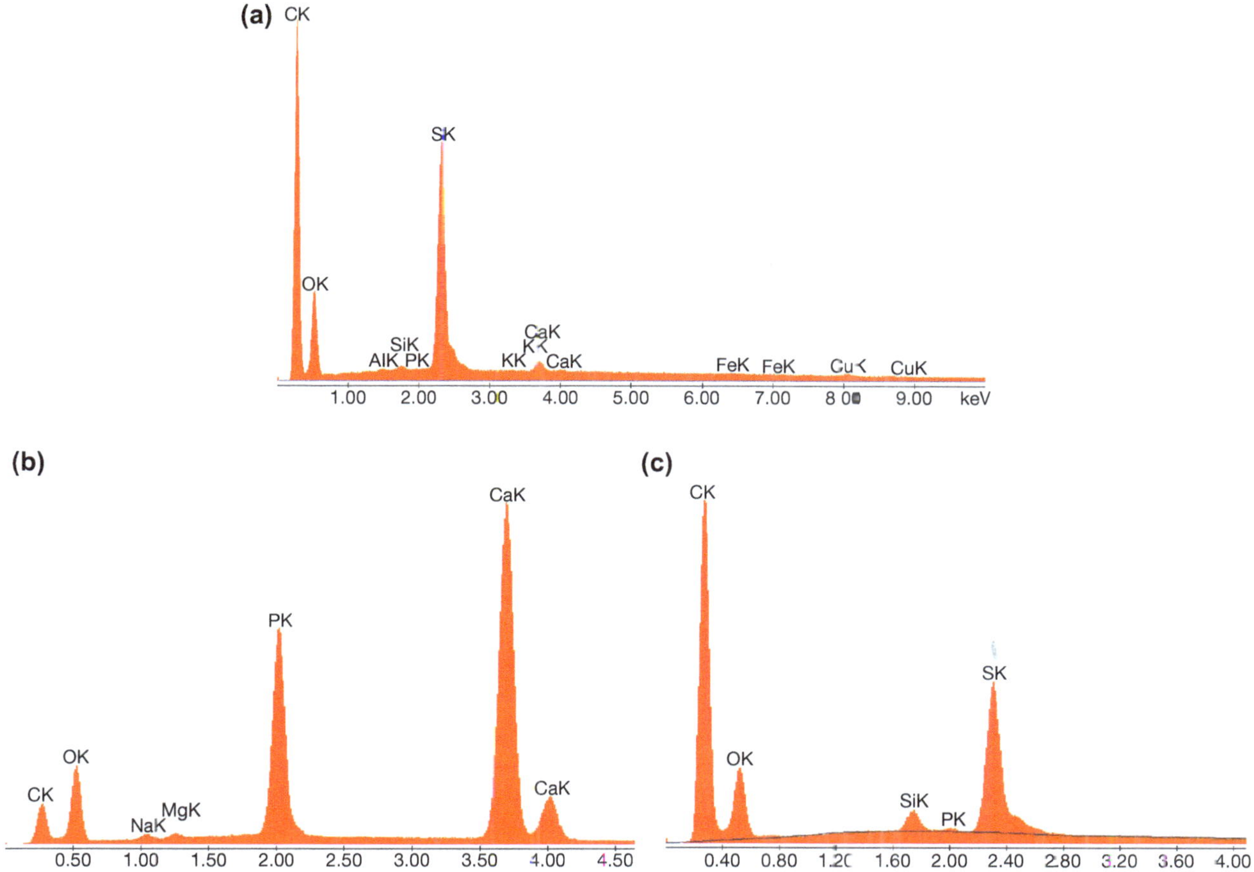

Figure 11 Energy-dispersive spectroscopy spectra for (a) unknown sample, (b) bone, and (c) two-part epoxy.

Auger Electron Spectroscopy

The Auger electron spectroscopy (AES) technique is similar to EDS in signal production, but, instead of X-rays, it detects the low-energy electrons that are emitted during relaxation of an atom after ionization. When an outer shell electron drops down to fill an empty inner shell position, one way it can give up its excess energy is by transferring a portion of that energy to another outer shell electron. This low-energy electron (~20–2500 eV) is then ejected from the atom as an Auger electron.

AES is performed in ultra-high vacuum, around 10^{-9} torr (compared to 10^{-6} torr used in electron microscopy). The ultra-high vacuum keeps the specimen surface clean and allows the low-energy Auger electrons to travel to the detector without colliding with gaseous atoms. An electron beam is directed at the specimen surface, and the surface is oriented so that the emitted Auger electrons are directed into a detector. The most common detector for AES is a cylindrical mirror analyzer (CMA). The CMA consists of two concentric cylinders, each held at a different voltage, creating an adjustable electric field. Fine-tuning of this field directs high-energy electrons away from the detection apparatus, while aiming electrons in the desired low-energy range at an electron multiplier, where they are counted. A typical AES spectrum is shown in **Figure 12**.

AES is a true surface analysis technique; weak Auger electrons will not escape from the specimen except at the top atomic layers of the surface, that is, the top few nanometers. Coupled with an ion gun for removing surface material, AES is an extremely sensitive technique for determining depth concentration profiles.

Absorption Spectroscopy

In absorption spectroscopy, electromagnetic radiation, generally in the UV-visible (UV-Vis)–IR range of the spectrum, is directed at a specimen. Transmitted, scattered, and/or reflected radiation is detected, and changes in intensity with respect to

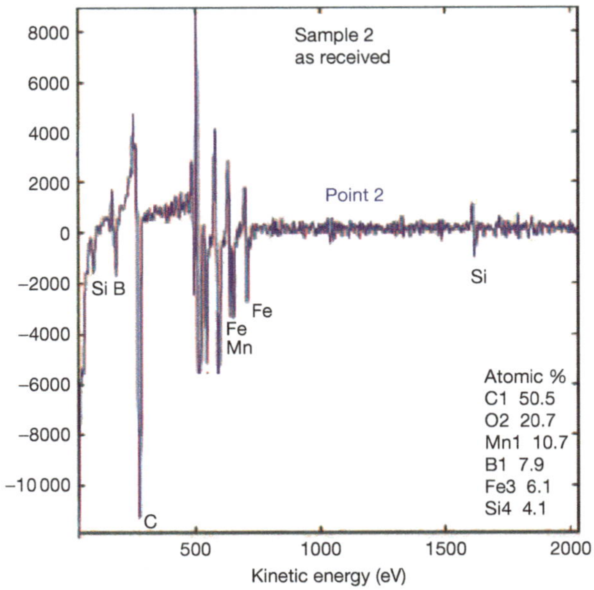

Figure 12 Auger electron spectroscopy spectra of doped silicon nanowires. Figure courtesy of D. Paul, Physical Electronics USA.

wavelength are determined. A diminution in intensity signifies atomic absorption (**Figure 13**). As absorption corresponds to quantum-level changes in the specimen, for example, transitions between specific atomic and molecular energy levels or changes in rotational or vibrational state, the elemental or compound composition may be determined from the specific wavelengths absorbed.

Generally, spectroscopic techniques track changes in light intensity with respect to the incident light intensity (**Figure 14**). The measureable quantities are the incident light intensity I_{Incident} and transmitted light intensity $I_{\text{Transmitted}}$. The fraction of light that is absorbed at a given wavelength is calculated from the measured incident and transmitted light intensity, by

$$I_{\text{Absorbed}}(\lambda) = (1 - I_{\text{Transmitted}}) \times I_{\text{Incident}}$$

Of course, reflected ($I_{\text{Reflected}}$) and scattered ($I_{\text{Scattered}}$) light intensity may also be determined. Various detectors are used depending on the wavelength of light being detected; photodiodes are typically used for visible and UV light.

Many spectroscopic techniques using infrared, visible, and UV light are available; these include IR spectroscopy, inductively coupled plasma atomic emission spectroscopy, UV-Vis spectroscopy, Raman spectroscopy, atomic absorption spectroscopy, and X-ray photoelectron spectroscopy, also called electron spectroscopy for chemical analysis. Further information sources for spectroscopic techniques are provided in the References section.

Mass Spectrometry

In MS, a gaseous sample is ionized then analyzed according to the mass-to-charge ratio (m/Z) of its components. Various techniques may be applied for sample vaporization, ionization, and ion detection, depending on the initial specimen and

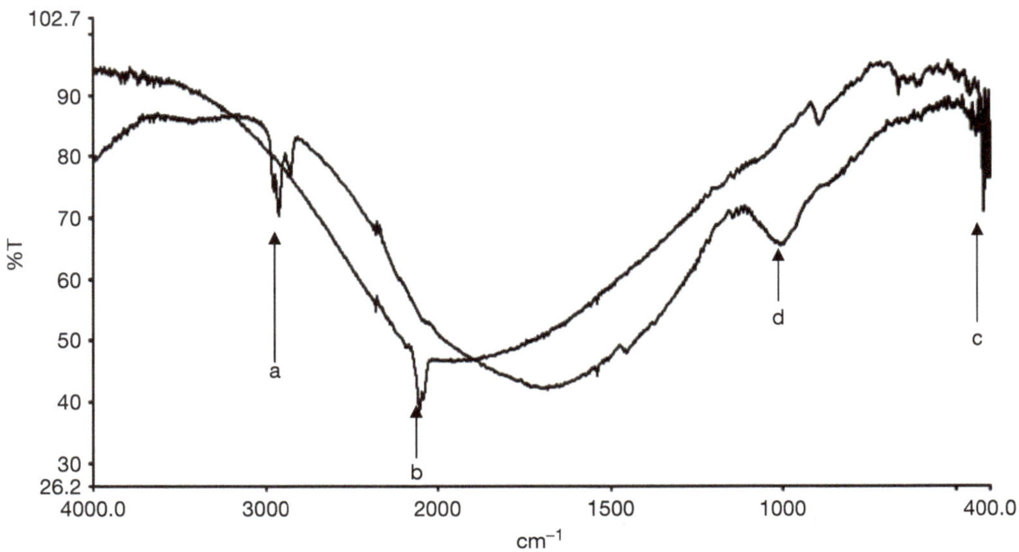

Figure 13 Fourier transform infrared spectra illustrating 1-octene hydrosilation. The sample is silicon with 1-octene. (a) Octane C—H stretch, (b) Si—H$_X$ stretch X = 1–3, (c) possible Si—O—Si stretch, (d) CH bend. Figure courtesy of R. Terrill, Department of Chemistry, San Jose State University.

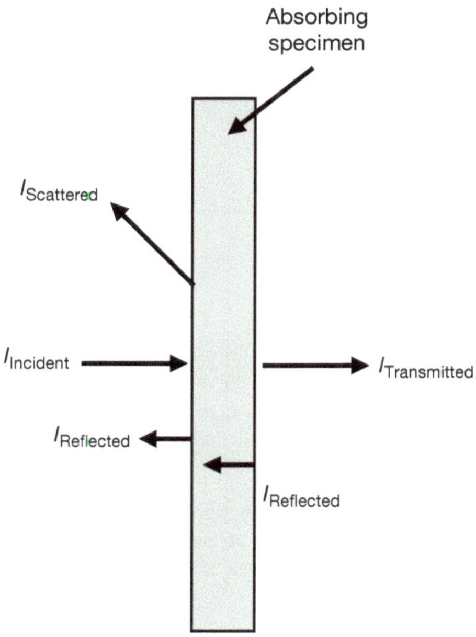

Figure 14 In addition to being absorbed within the specimen, light is reflected from the front and back surface and scattered at surface imperfections. The amount of absorbed light is calculated from the incident and transmitted light intensities.

the desired information. In environmental forensics, for example, typical specimens include sorbent from an air sampler or a material suspected as a source of air contamination. In these cases, the samples are heated in an inert atmosphere, and any resulting effluent gas is ionized by impact with high-energy electrons emitted from a hot tungsten filament. The ions enter the mass spectrometer and are acted on by a magnetic or an electric field; the resulting change in the speed and trajectory of an ion is due to the applied field strength and the mass, charge, and velocity of the particle. Detecting these changes allows m/Z to be calculated.

For further discrimination of chemical species, a common configuration is the gas chromatograph–mass spectrometer (GC–MS). The effluent is passed through a gas chromatograph, where a column packed with a high surface area material absorbs and desorbs chemical species at different rates. The column provides a temporal spread to the species before they enter the mass spectrometer to be enumerated according to their mass-to-charge ratio.

As with spectroscopy, additional spectrometry techniques are available, with different schemes for vaporizing and ionizing the sample and analyzing the mass-to-charge ratio. These include inductively coupled plasma MS, time-of-flight MS, and secondary ion MS. Information sources are provided in the References section.

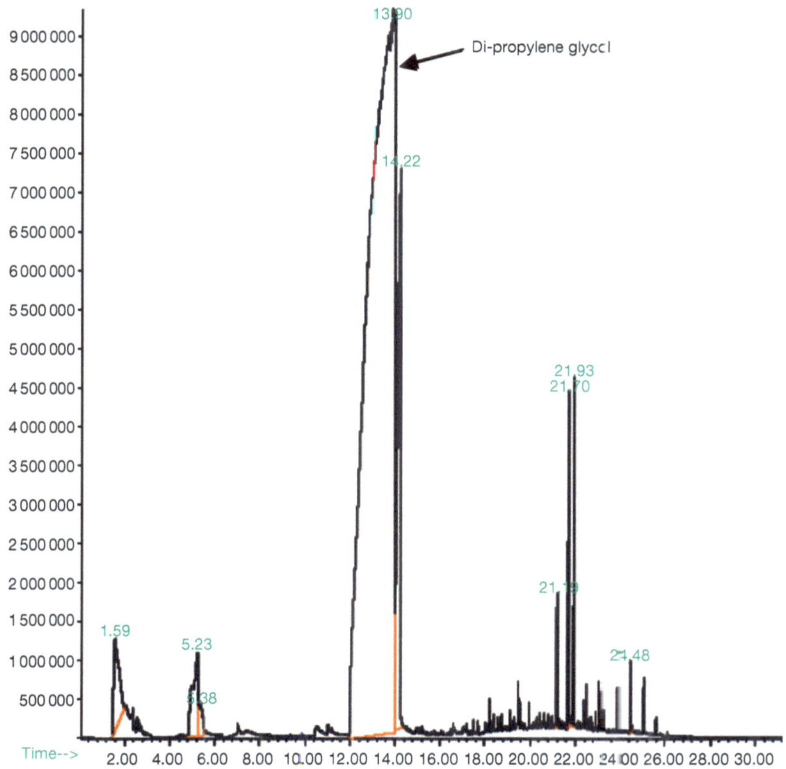

Figure 15 Spectrum obtained via heating and mass spectrometry. The major component of the gaseous effluent was dipropylene glycol.

Analysis of process effluents using MS

To determine whether blueprint paper could be a source of air contamination during a printing process, a sample of the paper was subjected to testing by heating; the resulting effluent gas was analyzed in the mass spectrometer. The resulting spectrum is shown in **Figure 15**. A number of organic compounds were identified; the largest component was dipropylene glycol.

Thermal and TGA

Differential thermal analysis (DTA) and differential scanning calorimetry (DSC) measure the heat input required to increase a specimen's temperature. The term differential is used since changes in the specimen are measured with respect to a standard reference material. In DTA, the differential temperature change between the specimen and reference for a fixed amount of heat input is measured (**Figure 16**). In DSC, the differential heat input required to increase the temperature of both the specimen and the reference is measured (**Figure 17**). In both cases, the specimen and reference are insulated from the environment in an adiabatic calorimeter.

In the related technique of TGA, a small amount of specimen (milligrams) is heated in an inert or reacting atmosphere, and any resulting mass changes are measured with respect to temperature. For example, absorbed water will evaporate by around 100 °C. Bound or adsorbed water will evaporate by around 400 °C, so the degree of hydration may be determined. Decomposition reactions will occur at various temperatures depending on the material composition and atomic bonding.

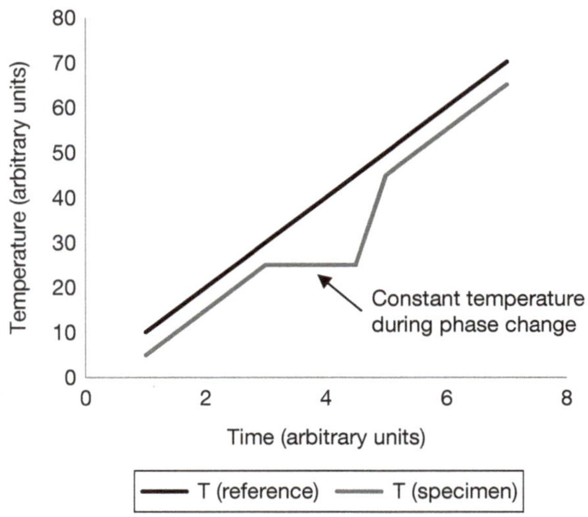

Figure 16 Differential thermal analysis measures the differential temperature change between the specimen and reference for the same amount of heat input. Here the specimen is observed undergoing a phase transition.

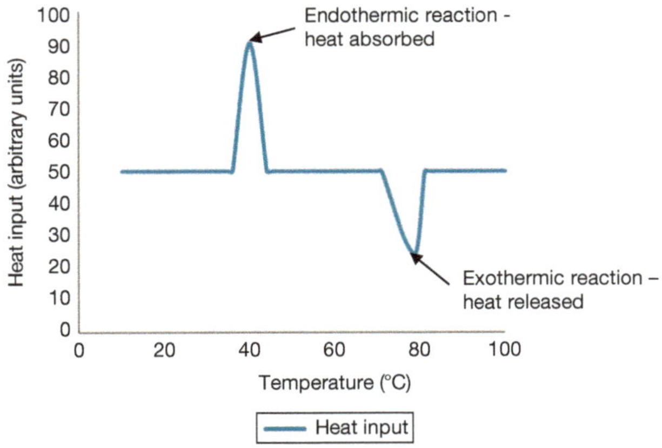

Figure 17 Differential scanning calorimetry measures the differential heat input required to increase the temperature of both the specimen and the reference. Here the specimen undergoes both an endothermic and an exothermic reaction.

TGA-DTA/DSC techniques maybe performed simultaneously to determine the range of thermal characteristics with a single sample/run.

Selecting the Appropriate Technique

Only a small, or limited, selection of useful materials analysis techniques is presented herein; however, many more methods are available, which may be more or less appropriate to a particular investigation. The materials failure analyst must use his or her education, training, and experience to determine which techniques are required. In addition to selecting the appropriate technique, the analyst must give weight to all relevant considerations, including the use of destructive or nondestructive analysis, required specimen size, and specimen composition and characteristics.

See also: **Engineering:** Forensic Engineering/Accident Reconstruction/Biomechanics of Injury/Philosophy, Basic Theory, and Fundamentals; **Methods:** Analytical Light Microscopy; Mass Spectrometry; Microscopy (Electron); Spectroscopic Techniques; Spectroscopy: Basic Principles.

Further Reading

ASM Handbook Committee, 1987. ASM Handbook. In: Fractography, vol. 12. ASM International, Geauga, OH.

ASM Handbook Committee, 1996. ASM Handbook. In: Fatigue and Fracture, vol. 19. ASM International, Geauga, OH.

Callister, W., Rethwisch, D., 2009. Materials Science and Engineering: An Introduction. Wiley, Chichester.

Davis, J. (Ed.), 1998. Metals Handbook. ASM International, Geauga, OH.
Goldstein, J., Newbury, D., Joy, D. (Eds.), 2003. Scanning Electron Microscopy and X-ray Microanalysis. Springer, New York.
Leng, Y., 2008. Materials Characterization: Introduction to Microscopic and Spectroscopic Methods. Wiley, Chichester.
Shipley, R.J., Becker, W.T. (Eds.), 2002. ASM Handbook. Failure Analysis and Prevention, vol. 11. ASM International, Geauga, OH.
Vander Voort, G. (Ed.), 2004. ASM Handbook. Metallography and Microstructures, vol. 09. ASM International, Geauga, OH.
Vickerman, J.C. (Ed.), 1997. Surface Analysis – The Principal Techniques. Wiley, Chichester.
Whan, R. (Ed.), 1986. ASM Handbook. Materials Characterization, vol. 10. ASM International, Geauga, OH.

Relevant Websites

http://www.asminternational.org—American Society of Metals International (ASM).
http://www.astm.org—American Society of Testing and Materials (ASTM).
http://www.iso.org—International Organization for Standardization (ISO).

Key Terms

Air quality standards, Analog tape authentication, Arc mapping, Arcing, Audio enhancement, Audio forensics, Beams, Biomechanics, Bridges, Buckling, Buildings, Chemical Engineering, Chemicals, Circuit breaker, COF, Collapse, Columns, Composition, Construction, Contamination, Corrosion, Crushing, Defects, Deflection, Digital audio, Digital audio authentication, Electrical, Electrocution, Engineering, Eutectic, Expert, Explosions, Failure, Failure analysis, Fall accidents, Falls, Fatigue, Fire, Flammable, Forensic, Forensic audio, Fracture, Fuse, Gait, Gait characteristics, Girders, Hazards, Human gait, Human locomotion, Ignition, Investigation, Locomotion, Materials analysis, Materials characterization, Materials engineering and science, Microscopy, Morphology, Multimedia evidence, Panel board, Pollution, Power, Receptacle, Recorded evidence, Required COF, Rupture, Safety, Service, Short-circuiting, Slip and fall accident, Slips, Slips and falls, Spectrometry, Spectroscopy, Spontaneous combustion, Structural, Toxic substance, Utility, Venting, Walking, Wildland, Yield.

Review Questions

1. What kinds of audio evidence are there? How would you characterize it?
2. What do Grigoras and Smith mean when they say, "most forensic audio recordings are made in non-ideal environments"?
3. What is the difference between analog and digital media? Why does this matter?
4. What does compressing an audio file "losslessly" mean?
5. Who was one of the first electrical incident investigators?
6. Of what use are X-rays in an electrical incident investigation?
7. Where is the most likely initial location for the commencement of an electrical event?
8. What is the definition of a structural failure?
9. Ratay and Peraza state that "Most structural failures are the result of human activities." Compare this statement to the earlier article on HFACS. Are Ratay and Peraza correct? Why or why not?
10. What are some of the factors that could cause a structure to collapse during construction?
11. What is the difference between the trigger of a collapse and the primary cause of the collapse?
12. What are the phases of the "life" of a structure? What could cause failure at each one?
13. What kind of locomotion do humans use? Can you think of another animal that uses it?
14. What is RCOF?
15. What does it mean that a strong correlation between dynamic coefficient of friction values and subjective readings of floor slipperiness?
16. What are the two approaches to measuring friction?
17. What do forensic chemical engineers do that is different from forensic engineers?
18. How are fires and explosions similar? How do they differ?
19. What is flash point? How is it defined?
20. List at least five examples of the types of evidence or investigation that might require a chemical engineer.

Discussion Questions

1. Is audio enhancement a separate forensic technique or is it digital forensics? What is digital forensics? If the audio file is on a computer, what kind of evidence is it? Why is this area of forensic applications so difficult to define?

2. In the sixth example (fire in a church) in Shefchick's article on electrical accidents, how would you begin the investigation to determine if arson was the cause? Write out at least five hypothetical scenarios, ranging from accidental to criminal, and list what evidence would be needed to support or refute each scenario; a table is a good method to use.
3. How are structural failures like arson? Why are these types of scenes potentially more difficult to process and investigate?
4. What role do visual cues play in slip-and-fall incidents? How, as an investigator, could you reconstruct what the person saw and how that related to the incident?
5. Why are forensic engineers not a part of a typical forensic laboratory? How often are their services required? Are they more involved in criminal or civil cases? Why?

Additional Readings

Gentili, F., Petrini, F., 2016. On the role of the numerical analyses in forensic investigations of fire-induced progressive collapses of tall buildings. International Journal of Forensic Engineering 3 (1–2), 45–68.

Grigoras, C., Smith, J.M., Morrison, G.S., Enzinger, E., 2013. Forensic audio analysis–review: 2010–2013. In: Proceedings of the 17th International Forensic Science Mangers' Symposium, Lyon, pp. 612–637.

Palmisano, F., Ratay, R., September 2015. The practice of forensic structural engineering in IABSE member countries: review of a survey. In: IABSE Symposium Report, vol. 105 (33). International Association for Bridge and Structural Engineering, pp. 1–4.

Rumsey, F., 2014. Forensic audio analysis. Journal of the Audio Engineering Society 62 (10), 690–694.

Yang, F., 2016. Slip and fall risk assessment. In: Handbook of Human Motion. Springer International Publishing, pp. 1–22.

> A pessimist says the glass is half empty, an optimist says the glass is half full, and an engineer says the glass is too big.Scott Edward Shjefte.

Dean Gialamas, past President of the American Society of Crime Laboratory Directors, often says that "Ethics mean nothing to me. Ethics is knowing right from wrong and prisons are filled with people who know that; they chose badly. What is needed is integrity, the will to do what's right." Headlines proclaim the end of forensic science because of failures on the part of organizations or people who choose the wrong thing: Stealing drugs, falsifying reports, lying during testimony, fudging science. Listing them would be depressing and, using Gialamas' perspective, pointless: We already know they chose poorly. The real question is, how to get people to choose well, to do what's right? It starts with education and understanding the connection between what a scientist does in the laboratory and the effect that has on individuals and society. In fraud investigations, the acronym "ABC" is used, meaning fraud can originate with a single bad *A*pple, a bad *B*arrel of apples, or a bad *C*rop of them. Again, from Gialamas' view, it could also mean a bad farmer, a manager who sends the wrong messages, does not do what is right for the organization or its employees, and lets bad actions slide. Which of these is worst? Hard to say, because they each can lead to the downfall of a laboratory or a discipline.

As a profession, forensic science needs to see outside its historically limited perspective and look to other professions for insight, advice, and models to improve its standing and reputation. "The way we have always done it" does not work anymore and is an insufficient answer for what ails forensic science. If forensic science does not takes itself seriously, why would anyone else? This section lays out basic issues every forensic scientist faces, as an individual, an employee of an organization, and a member of a professional community.

Crime Scene to Court

K Ramsey and E Burton, Greater Manchester Police Forensic Services Branch, Manchester, UK

Glossary

CBRN Chemical, biological, radiation and nuclear incidents.

CCTV Closed circuit television (cameras or evidence from).

CPD Continuous Professional Development.

CPS Crown Prosecution Service (UK).

HTCU Hi-tech crime unit (examination of hardware/software/data/images from any system or device).

L2 Level 2 investigations, specific skills required for e.g., covert operations, deployment and substitution of items, forensic markers.

LCH Local Clearing House (firearms).

NaBIS National Ballistic Intelligence Service (UK).

NCA National Crime Agency (UK).

NOS National Occupational Standards.

T1/2/3 CSI Skill tiers defined for crime scene investigation officers, with 1 being the most basic level of training (usually volume crime offences only), 2 being the range of volume, serious and major crime investigations, and 3 being trained in crime scene management/the coordination of complex investigations.

VSC/ESDA Video Spectral Comparison—the analysis of inks, primarily in fraudulent documents; Electro Static Detection Analysis—the examination of (writing) indentations on paper.

 http://dx.doi.org/10.1016/B978-0-12-802718-9 00033-8

Introduction

A multitude of disciplines evolved within forensic science during the twentieth century, resulting in highly specialized fields of evidential opportunities to support criminal investigations. Many of the more traditional disciplines, for example, footwear analysis and blood pattern interpretation, now have well-established principles and methodologies that have been proven in a criminal justice context; developments in these areas are largely confined to technical support systems and information sharing through databases. The very rapid rate of development of DNA profiling techniques during the 1980s and 1990s led to the emergence of national and international DNA databases; however, the pace of change has now significantly reduced. Conversely, the end of the twentieth century and the early part of the twenty-first century have seen an explosion of new forensic evidence types that are less established in court—disciplines such as CCTV, mobile phone, computer analysis, and the use of digital images, and social media are collectively referred to as e-forensics.

Owing to the highly specialized nature of each forensic discipline and the varied rate of evolution, forensic science effectively represents a composite of interrelated, and often distinct, opportunities to support criminal investigations.

Most current models of forensic service delivery, especially where part of a wider organization, for example, police forces and enforcement agencies, have arisen over time by bolting on additional elements and clustering together within related fields. If the current capability of forensic science were to be designed from scratch as an effective entity, it is certain that a more integrated, and hence effective, structure would be proposed.

In addition, there has been a professionalization of forensic science in the workplace and increasing requirements for regulation; as recently as the 1980s, crime scene investigation, for example, was widely undertaken by police officers and was largely restricted to recording/recovering visible evidence; this was used in a limited capacity to support that particular investigation without scope for wider intelligence development. Now, crime scene investigation is predominantly undertaken by specialist staff employed to exclusively undertake these duties.

To practice in a forensic discipline, specialized training, qualifications, and competency levels are required. The range of evidence types that have potential to support investigations has widened considerably. Some disciplines lend themselves to cross-skilling.

Public expectations of what forensic science can deliver have been heightened by highly popular mainstream television programs, both documentary and fictional. Often, the expectation of what can be delivered exceeds what is either possible or financially sensible. This leads to a requirement on service providers and users to make informed (evidential and financial) decisions regarding the best use of forensic evidence in support of investigations.

This chapter considers options to optimize the use of forensic evidence types recovered from crime scenes in the context of the different models available to criminal justice systems; the concept of integrated case management is outlined and discussed.

Task

To bring together all potential forensic evidential opportunities, holistically review their significance to the investigation, prioritize the progression of work, deliver the best evidence to the court for testing (complying with all continuity, integrity, and quality requirements), and ensure the best value for money when determining spend on forensic evidence.

Internationally, there are variable constraints and opportunities due both to the different criminal justice models and the commercial market situation at state/regional and country levels.

Models

1. All forensic evidence sourced within a law enforcement agency, for example, a police laboratory.
2. All forensic evidence provided by external specialists contracted to a law enforcement agency.
3. Composite of (a) and (b).

Forensic Strategies

The recovery of evidence from the crime scene is only the start of the forensic process. Once the evidence has been collected, packaged, and preserved, it needs to be analyzed in order to provide meaningful information to the investigation and subsequently the courts.

Forensic examinations are carried out in order to implicate or eliminate individuals and also in order to establish what has occurred during the commission of an offense or incident.

Deciding what analysis is required can be a complex process. Some of the issues for consideration include the following:

- Is it necessary to examine all the evidence that is recovered?
- Should every possible test be carried out?

In an ideal world, it would be preferable to carry out every possible analysis; however, in reality, it is likely that this will be neither practicable nor financially viable. In addition, carrying out every possible analysis would overload forensic laboratories.

When making decisions about what forensic analysis should be carried out, it is vitally important that consideration is given to both the potential prosecution and defense cases. An impartial approach must be taken to assessing examination requirements. It is often not necessary to carry out an examination of every item of evidence recovered, but examinations should be directed to where value could potentially be added to an investigation.

A forensic strategy should be developed around every case where forensic evidence plays a part, and may relate to an overall case or to an individual item of evidence. A forensic strategy should be developed in a holistic manner taking into account all potential evidence types and should direct and coordinate the forensic examinations/analyses that are required.

Forensic strategies can be developed in different ways by one or more of the following:

- Investigating officer
- Crime scene investigator (CSI) or crime scene manager
- Forensic scientist/forensic specialist
- Forensic Submissions Officer: It is a role that can be variably named; this role relates to an informed individual within a police force or law enforcement agency who uses knowledge and expertise to advise on forensic analysis and who has decision-making authority and control of the budgetary spend. May also be known as forensic advisor, scientific support manager, etc.
- Legal Representative
- Pathologist

Forensic strategies are generally initially developed and applied by individuals involved in the prosecution aspects of a crime. Although this is the case, it is vitally important that a balanced and unbiased approach is taken to the development of a strategy and consideration given to information that may support the defense case as well as the prosecution case. Examinations that are likely to add value or provide information to an investigation (irrespective of whether it will support or weaken the prosecution case) should be carried out and all results must be disclosed to the defense team. Defense should also be given the opportunity to carry out any review of strategies, examination processes, and/or results that they require and be provided with access to any items of evidence that they want to examine themselves in order to build the defense case.

In order to develop the forensic strategy and make appropriate decisions about which forensic examinations will be of value to the investigation, the following are necessary:

- To be able to gather as much information as possible about the circumstances of the case
  - circumstances of evidence recovery
  - accounts given by victim(s), witnesses, suspect(s), etc.
- To have an understanding and knowledge of forensic science and its application to investigations

A forensic strategy meeting is a useful way of ensuring that all relevant parties are aware of the full circumstances of the case and enables a "multiagency" discussion about the processing of all exhibits to optimize evidential potential in a comprehensive and coordinated manner.

It can often be the case that police officers do not have a full understanding or knowledge of forensic science, likewise forensic scientists historically have had a relatively poor understanding of police and investigative processes; this can lead to miscommunication and confusion in relation to the application of forensic science to meet investigative needs. A joint approach to the development of forensic strategies helps to improve the communication and understanding on a case-by-case basis.

A formal forensic strategy meeting is often required only in more serious cases; however, the general approach can be applied to any investigation. Even in the most simple of cases, it is often beneficial for discussions to take place between the investigating officer, the CSI, the forensic advisor/budget holder/decision maker, and the prosecutor. Alternatively, generic strategies can be implemented, for example, for a particular crime type or modus operandi.

When making an assessment regarding the potential examination of a particular item and the development of a forensic strategy, the requirements of the investigation are the primary concern and consideration should be given to the following issues:

- The type and nature of the item/exhibit
- The context of the item
  - Exactly where and when it was recovered
  - Condition of the item, that is, wet, damaged, etc.
- The integrity of the item
  - Has it been appropriately recovered, handled, packaged, and preserved?
  - Is the security and continuity of the item intact?
- The potential evidence that may be obtained from the item, for example, DNA, fingerprints, fibers, footwear marks
- The information these evidence types may provide to the investigation
- Whether this potential information is likely to add value to the investigation
  - Is it possible that new information will be provided?
  - Is it possible that an account given by a witness, victim, or suspect will be supported or refuted?
  - Will the information help to establish what has occurred?
- Whether there is a conflict between potential evidence types, and if so, which evidence type will be of most value under the circumstances
  - For example, swabbing/taping for DNA may damage fingerprints, but where the DNA is likely to be at low levels and requires specialized low-template DNA analysis, the presence of DNA may not necessarily prove

contact with an item, whereas fingerprints will always prove that contact has occurred

- The chances of success, that is, obtaining a result/information of value to the investigation (this may be inclusive or exclusive)

Much work has historically been completed in relation to developing and understanding the success rates relating to DNA profiling; however, relatively little work has been undertaken to fully understand the success rates associated with other forensic evidence. This is largely due to the fact that other evidence types, such as fibers, gunshot residue, footwear marks, etc. are generally more complex to interpret than DNA. In relation to DNA profiling, success rates are generally based on the chances of obtaining a DNA profile; however, with the other evidence types, the value of the outcome is very much dependent on the circumstances of the investigation. For example, when searching an item of clothing taken from a suspect for glass, the absence of glass or the presence of glass could both be of value to the investigation depending on the circumstances. The presence of glass on the clothing that matches control sample(s) from the crime scene is only of value if its presence cannot be accounted for in any legitimate way; conversely, the absence of glass on the item of clothing may lead to a conclusion that the suspect was not involved in the offense, depending on the circumstances of the offense and arrest.

In addition to being able to understand and evaluate the chances of being able to obtain a meaningful result, it is also vital that the value of the overall contribution to the entire case is understood. This involves being able to understand the value and contribution of the forensic examination to the detection of the offense as well as the outcome of the court process. This is an even more difficult issue to evaluate and understand than the chances of being able to obtain a forensic test result.

Because the value of forensic evidence is so dependent on the individual case circumstances, decisions about examinations must be made on an individual case basis. There have been recent developments in some agencies/forces to better understand the chances of success of different types of forensic evidence and the value to investigations; this will help to better inform decisions about evidential potential and examination viability as well as assisting to achieve value for money. This approach is best described as *forensic effectiveness*.

The forensic strategy should also take into account the timescales associated with the investigative process and the criminal justice system, and it should be ensured that forensic analysis can meet the requirements of the criminal justice process, including court dates and any requirements to disclose appropriate information to the defense team(s).

Each police force/law enforcement agency will have its own approach to the submission of exhibits for forensic examination/analysis; irrespective of whether the analysis is carried out in an internal police laboratory, external commercial company, or government-owned laboratory, these approaches can be applied to all examinations and all evidence types.

These approaches help to ensure that decisions are made based on scientific knowledge, viability, and evidential value taking into account all aspects of the investigation. They will help to ensure that the best evidence is obtained while considering value for money and that it can be applied to any investigation irrespective of the seriousness of the offense or the scale of the investigation.

Integrated Case Management

The concept and use of forensic strategies in directing investigations is not new, but is often limited by the evolved structure of forensic disciplines within investigative agencies. Classically, DNA and fingerprint evidence from volume crimes will be independently submitted at the same time by different routes and this often results in wasted effort/spends and duplicated results. The development and use of forensic intelligence has been variable. Emerging thinking includes organizational redesign of forensics to better integrate with related functions such as intelligence collection, targeted deployment of resources, and prioritized forensic submissions.

The concept of integrated case management draws together informed operational deployment (e.g., of crime scene investigators) followed by a more holistic approach to submissions for testing. The strategy takes greater account of supporting intelligence and desired outcomes. Regular reviews and trigger points are included for the staged submission of potential evidence, and communication with investigators is enhanced so allowing for a more responsive and directed investigation.

Ultimately, the production of *intelligent identifications* can be better achieved by having an integrated process that links the enforcement priorities, available resources, potential forensic evidence, intelligence, and prosecutor requirements; this model provides flexibility to respond to changing demands and gives an increased likelihood of efficient and effective spend on forensic support to investigations. There is no single way to achieve this, but an illustration of how to rethink some of the traditional silo-based forensic disciplines is provided in **Figure 1**.

Summary

The single biggest challenge to the forensic science community during the twenty-first century is to modernize delivery of integrated services in support of investigations. This must:

- build on the previous development of each discipline;
- accommodate the new and emerging technological disciplines;
- meet the regulatory requirements;

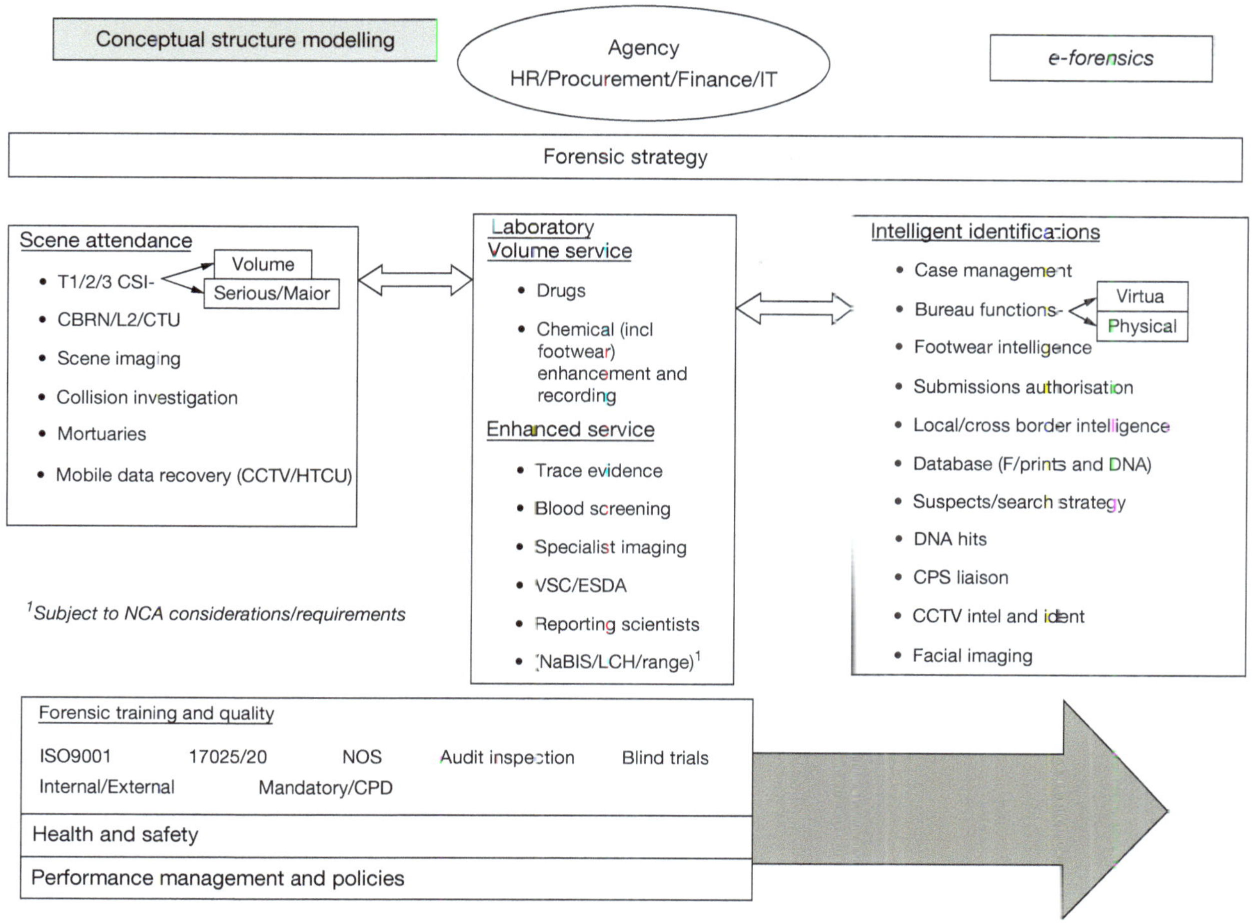

Figure 1 Conceptual structure modeling.

- reflect the changing workforce and skills;
- deliver the best evidence to courts in support of investigations.

See also: **Foundations:** Forensic Intelligence; History of Forensic Sciences; Principles of Forensic Science.

Further Reading

Faigman, D.L., et al., 2006. Modern Scientific Evidence: The Law and Science of Expert Testimony.

Fisher, B.A.J., Fisher, D.R., 2012. Techniques of Crime Scene Investigation, eighth ed. CRC, Boca Raton, FL.

Houck, M., Crispino, F., McAdam, T., 2013. The Science of Crime Scenes. Elsevier.

Innocence Project, 2011 http://www.innocenceproject.org/Content/Facts_on_PostConviction_DNA_Exonerations.php.

Kirk, P.L., 1974. In: Thornton, J.L. (Ed.), Crime Investigation, second ed. Wiley, New York. (1985 reprint ed. Malabar, F.L., Krieger Publishing Company).

NAS, 2009. Strengthening Forensic Science in the United States: A Path Forward. NAS Report: Committee on Identifying the Needs of the Forensic Sciences Community. National Academies Press, Washington, DC.

White, P., 2010. Crime Scene to Court: The Essentials of Forensic Science. Royal Society of Chemistry, Cambridge, ISBN 978-1-84755-882-4.

Forensic Laboratory Reports

J Epstein, Widener University School of Law, Wilmington, DE, USA

There is no precise formula, dictated by law or science, as to what a forensic laboratory report must contain when it reports test results or analysis outcomes. Its content may be determined by the individual examiner's predilections, internal laboratory policy, the law of the jurisdiction, accreditation organization standards, or the reason(s) for its production. What can be said with certainty is that by understanding the current criticism of the practice of producing forensic laboratory reports and trends in standards for reports, and by considering the use to which the report may be put in the court process and the legal and ethical commands regarding reporting and, more generally, the duties of the forensic scientist, a model for forensic laboratory reports can be identified.

Before discussing these factors, it bears mention that the term "report" itself lacks clarity, as it may refer to the complete case file documenting the examination or just to the compilation of results. For this chapter, the term "report" denotes the latter—the document prepared for the consumer (the investigator, counsel, or court official who directed that the examination and testing be conducted). Even this report may vary in degree of detail, as there can be the summary report advising the requesting party of the outcome; a more formal report prepared for disclosure to the court or opposing counsel as part of pretrial discovery; an amplification of the initial discovery-generated report when it is determined that the expert will in fact testify; and a report that will be presented in lieu of actual testimony. Additional documentation may include an administrative or dispositional report detailing the receipt or return of the item(s) sent for analysis.

What must also be acknowledged is that the expert's role in the adjudicative process is in some ways defined by whether the system is adversarial, with the expert being called by the party seeking to establish a point, as in the United States; or inquisitorial/"common law," where the expert is a court witness, presumed to be neutral, and without allegiance to a particular party, as in France, Belgium, and Germany. These demarcations are not always adhered to, as American law permits a trial judge to appoint and take testimony from a "court" expert under Federal Rule of Evidence 706, and in some cases involving offenses of fraud and falsification, France permits competing experts. These differing roles, however, do not alter the necessary components of a forensic laboratory report (and, as is detailed below), both ethical and legal considerations as well as a commitment to the role of science may require the report to be neutral and to acknowledge any limitations and/or weaknesses.

Contents of a Report—A "Science" Standard

At least in the United States, there has been substantial criticism of forensic laboratory reporting. This is found in *Strengthening Forensic Science: A Path Forward*, the 2009 report of the National Research Council of the National Academy of Sciences. After reporting that forensic laboratory reports lack precise terminology, it concluded that most laboratory reports do not meet the standard it proposed: As a general matter, laboratory reports generated as the result of a scientific analysis should be complete and thorough. They should describe, at a minimum, methods and materials, procedures, results, and conclusions, and they should identify, as appropriate, the sources of uncertainty in the procedures and conclusions along with estimates of their scale (to indicate the level of confidence in the results). Although it is not appropriate and practicable to provide as much detail as might be expected in a research paper, sufficient content should be provided to allow the nonscientist reader to understand what has been done and permit informed and unbiased scrutiny of the conclusion.

This criticism does not stand in isolation. A 2011 British court decision also expressed concern over the sufficiency of detail and documentation in a forensic (latent print) prosecution. After noting the failure of the examiner to contemporaneously record "detailed notes of his examination and the reasons for his conclusions[,]" the court added that [t]he quality of the reports provided by the Nottinghamshire Fingerprint Bureau for the trial reflected standards that existed in other areas of forensic science some years ago, and not the vastly improved standards expected in contemporary forensic science.

The NRC standard is more detailed than that of various forensic organizations. ASCLD/LAB, for example, requires that only written reports be generated for "all analytical work" and must contain conclusions and opinions and a clear communication of "the significance of associations made...."

Other standards address the need for full documentation, but do not distinguish between a laboratory's bench notes and the final product. For example, International Organization for Standardization's ISO/IEC Standard 5.10.5 requires that "the laboratory shall document the basis upon which the opinions and interpretations have been made" without specifying where that information is to be recorded. Similar language is used for ballistics reports, as recommended by the Scientific Working Group on firearms (SWGGUN) requiring that "[w]hen opinions and interpretations are included, the laboratory shall

 http://dx.doi.org/10.1016/B978-0-12-802718-9.00034-X

document the basis upon which the opinions and interpretations have been made. Opinions and interpretations shall be clearly marked as such in the test report."

Yet, the more detailed mandate urged by the NRC Report is not unique. Scholars and agencies have articulated similar or at least substantial standards. A publication of *The Royal Society of Chemistry* in 2004 suggested the following information as appropriate for inclusion in an expert report:

- A summary of the event to contextualize the scientific test(s);
- An outline of the scientific work conducted;
- A listing of items examined;
- Description of the work performed;
- A statement interpreting the findings; and

an overall conclusion.

The RSC text also urges that the report identifies the assistants in the testing and the role each played and includes appendices with tables or similar displays of test results.

For DNA analysis, the Federal Bureau of Investigation's standards for DNA laboratories require reports to include a description of the evidence examined and of the technology, results and/or conclusions, and a "quantitative or qualitative interpretative statement."

One final scientific issue regarding the contents of a report is the concern over bias. Research has shown that information received by the analyst might affect his or her judgment, as when the examiner receives domain-irrelevant information such as the fact that the suspect whose fingerprints are being examined "confessed to the crime" or when the verification is not "blind." Documentation of such information in a laboratory report (or the bench notes) is one responsive action, as is an internal laboratory policy to reduce analyst or verifier exposure to potentially biasing information.

Contents of Report: Legal Standards

That which science requires is to some extent mirrored in legal requirements for expert reports. These vary from nation to nation and within nations when states or regions have their own authority to legislate criminal practice.

In the United Kingdom, Rule 33.3, Criminal Procedure Rules 2010 mandates contents of a full report, that is, one for submission in court, as follows:

1. the findings on which they have relied in making the report or statement;
2. details of which of the findings stated in the report or statement are within their own knowledge, which were obtained as a result of examinations, measurements, tests, etc., carried out by another person and whether or not those examinations, measurements, tests, etc., were carried out under the expert's supervision;
3. the identity, qualifications, relevant experience, and any certification of the person who carried out the examination, measurement, test, etc.;
4. details of any statements of fact, literature, or other information on which they have relied, either to identify the examination or test requirements, or which are material to the opinions expressed in the report or statement or on which those opinions are based;
5. a summary of the conclusions and opinions reached and a rationale for these;
6. a statement that if any of the information on which their conclusions or opinions are based changes, then the conclusions or opinions will have to be reviewed;
7. where there is a range of opinion on the matters dealt within the report or statement, a summary of the range of opinion, and reasons for the expert's own opinion;
8. any information that may cast doubt on their interpretation or opinion; and
9. if the expert is not able to give an opinion without qualification, what the qualification is.

Much less specific is the legislated mandate for federal criminal prosecutions in the United States. Under Federal Rule of Criminal Procedure 16, the Government must permit the defense to inspect and to copy or photograph the results or reports of any scientific test or experiment and must produce before trial a written summary of any proposed expert testimony that describes the witness' opinions, the bases and reasons for those opinions, and the witness' qualifications. Defense counsel in criminal cases has a reciprocal disclosure requirement. Despite the seeming generality of these terms, American courts have at times interpreted them to require some greater detail in the reports, such as underlying documentation.

In the United States, an additional requirement derived from the Constitution's guarantee of Due Process of Law may affect what must be included in a laboratory report issued by a police or other government agency. The prosecution must disclose information that is "favorable to the accused" and "material either to guilt or to punishment" as well as "evidence that the defense might have used to impeach the Government's witnesses by showing bias or interest." This extends to "evidence affecting credibility[.]" This information is generally denominated "Brady material."

The applicability of these rules to official (police or state) laboratories is settled. The US Supreme Court has held that the disclosure obligation extends to police agencies working with the prosecution, and this has been extended to forensic examiners. Hence, in a report or some other communication, a forensic examiner in government employ must ensure that "Brady material" is disclosed.

What remains to be defined are the terms "exculpatory" or "impeachment" information. The core of each is easily described. Evidence is "exculpatory" if it tends to reduce the degree of guilt or question proof of culpability; "impeachment" information is proof of a bias or interest, or otherwise information that could be used to contradict or attack the credibility of the analyst or report. This type of disclosure parallels that of forensic laboratory reports imposed by the United Kingdom's evidence code. The code requires inclusion in the report of "a summary of the range of opinion and reasons for the expert's own opinion; [] any information that may cast doubt on their interpretation or opinion; and if the expert is not able to give an opinion without qualification, what the qualification is."

Reports: Stand-Alone Evidence or Support for a Testifying Expert

Whether a laboratory report may stand on its own as evidence in a trial, or instead must be accompanied by testimony of the forensic analyst, is a function of the law of the jurisdiction in which the case is tried. In the United States, a prosecution expert's report may not be admitted on its own, as this is deemed to violate the defendant's right to confront adverse witnesses. The Supreme Court in Melendez-Diaz versus Massachusetts held that a certificate of analysis fell within the core class of testimonial statements because it was a solemn "declaration or affirmation made for the purpose of establishing or proving some fact." In the 2011 follow-up of the Melendez-Diaz decision, the Court further held that another lab analyst may not come in to testify to the report's contents, at least where the other analyst neither supervised nor observed the initial testing. (This applies only to prosecution expert reports, as in the United States only the defendant has a guarantee of the right to confront witnesses. Admissibility of a defense forensic report without examiner testimony would be determined by the state's rules of evidence, but is generally unheard of.)

At the same time, the confrontation right does not mean that the analyst must testify. A state may create a notice and demand statute under which the prosecution notifies the defendant of its intent to use an analyst's report as evidence at trial, after which the defendant has a specified period of time in which to demand the expert's live testimony. A defendant's failure to "demand" waives the need for the analyst's presence and allows use of the report. As well, an accused may always agree to stipulate to the report's content, eliminating the need for any live testimony.

The Melendez-Diaz approach is not followed uniformly on an international basis. Canada permits proof by means of an expert report, without live testimony, where the proponent of the report has provided it to the opposing party and the trial court recognizes the author as a legitimate expert. The court retains discretion to mandate the expert's appearance for cross-examination. Australia's Evidence Act of 1995 similarly authorizes expert proof by certificate, but the opposing party may require the offering side to "call the person who signed the certificate to give evidence." In the United Kingdom, expert reports are themselves admissible as evidence subject to the judge's discretion in requiring the analyst or examiner to appear.

Ethical Considerations and Forensic Reports

The decision of what to include in a forensic laboratory report, beyond that required by law or by science, may be informed by ethical considerations. Forensics organizations often have ethical codes, but they may be silent as to the particulars of report writing. Illustrative is the Code of the American Board of Criminalistics, which only asserts general obligations such as "[e]nsure that a full and complete disclosure of the findings is made to the submitting agency[.]" Other codes may not mention reporting at all but instead address only the delivery of information without distinguishing between the written report and a courtroom presentation of evidence. An exception is that of the Australian and New Zealand Forensic Science Society, Inc., which requires that a report be nonpartisan when results are ambiguous. "Where test results or conclusions are capable of being interpreted to the advantage of either side in a legal proceeding, each result or conclusion should be given weight according to its merit."

Ethical considerations may also be imposed by law. In the United Kingdom, the expert is deemed to hold only one allegiance, that to the court, regardless of the party who retained the individual. Specific ethical obligations are imposed for written reports. First, where there is a range of opinion, the expert must summarize the various positions. Second, if the opinion rendered cannot be given without qualification, the expert must disclose that and state the qualifying aspects or concerns.

Conclusion

Within and across nations, there is no clear standard for forensic reports intended for court use, except where prescribed by law. What should be manifest is that the more detailed the report, and thus the more it is capable of rigorous assessment by an independent expert evaluator, the more credibility will be attributed to both the results and the examiner.

See also: **Legal:** History of the Law's Reception of Forensic Science; Legal Aspects of Forensic Science; Legal Systems: Adversarial and Inquisitorial; **Management/Quality in Forensic Science:** *Sequential Unmasking:* Minimizing Observer Effects in Forensic Science; **Professional:** Ethics.

Further Reading

Codes of Practice and Conduct for Forensic Science Providers and Practitioners in the Criminal Justice System 44–45 (United Kingdom). http://www.homeoffice.gov.uk/publications/police/forensic-science-regulator1/quality-standards-codes-practice?view=Binary.

Dror, I.E., Cole, S., 2010. The vision in 'blind' justice: expert perception, judgment and visual cognition in forensic pattern recognition. Psychonomic Bulletin and Review 17 (2), 161–167.

Dror, I.E., Rosenthal, R., 2008. Meta-analytically quantifying the reliability and biasability of forensic experts. Journal of Forensic Sciences 53 (4), 900–903.

National Research Council, 2009. Strengthening Forensic Science in the United States: A Path Forward. National Academies Press, Washington, DC. http://www.ncjrs.gov/pdffiles1/nij/grants/228091.pdf.

Quality Assurance Standards for Forensic DNA Testing Laboratories, Standard 11.2. http://www.cstl.nist.gov/strbase/QAS/Final-FBI-Director-Forensic-Standards.pdf.

Reviewing Historical Practices of Forensic Science Laboratories, September 29, 2010. http://www.ascld.org/.

Rothwell, T., 2004. Presentation of expert forensic evidence. Chapter 15. In: White, P. (Ed.), Crime Scene to Court: The Essentials of Forensic Science, second ed. RSC, Cambridge, pp. 430–432.

Spencer, J.R., 2002. Evidence. European Criminal Procedures. Chapter 15. Cambridge University Press, New York, pp. 632–635.

Relevant Websites

http://www.criminalistics.com/ethics.cfm—American Board of Criminalistics, Rules of Professional Conduct.

http://www.ascld.org/—The American Society of Crime Laboratory Directors.

http://www.ascld-lab.org/—The American Society of Crime Laboratory Directors Laboratory Accreditation Board.

http://www.afte.org/AssociationInfo/a_codeofethics.htm—Association of Firearms and Toolmarks Examiners, AFTE Code of Ethics.

http://www.anzfss.org.au/code_of_ethics.htm—Australian and New Zealand Forensic Science Society.

http://www.forensicdna.com/Media/Bias_FS.htm—An Extended List of Articles on the Issue of Bias in Forensic Examinations.

http://www.iso.org/iso/home.html—International Organization for Standardization.

http://www.swggun.org/swg/index.php?option=com_content&view=article&id=25:transition-from-ascldlab-legacy-to-isoiec- 17025&catid=10:guidelines-adopted&Itemid=6—SWGGUN, Transition from ASCLD/LAB Legacy to ISO/IEC 17025.

http://webarchive.nationalarchives.gov.uk/+/http://www.justice.gov.uk/criminal/procrules_fin/contents/rules/part_33.htm—United Kingdom, Criminal Procedure Rules 2010.

Health and Safety

N Scudder and B Saw, Australian Federal Police, Canberra, ACT, Australia

Glossary

Clandestine laboratory ("Clan labs") Setting up of equipment or supplies for the manufacture of illegal compounds such as drugs or explosives.

Confined space An enclosed or partially enclosed space that is not intended or designed primarily for human occupancy, within which there is a risk of one or more of the following: (a) an oxygen concentration outside the safe oxygen range; (b) a concentration of airborne contaminant that may cause impairment, loss of consciousness, or asphyxiation; (c) a concentration of flammable airborne contaminant that may cause injury from fire or explosion; and (d) engulfment in a stored free-flowing solid or a rising level of liquid that may cause suffocation or drowning.

Dynamic risk management The continuous assessment of risk in the rapidly changing circumstances of an operational incident in order to implement the control measures necessary to ensure an acceptable level of safety.

Hazard The potential for a substance to cause adverse effects.

Hierarchy of control measures Ranking of measures taken to prevent or reduce hazard exposure according to effectiveness from the most effective measures that eliminate hazards to the least effective that achieve only limited protection.

OHS policy A policy document indicating an organization's commitment to OHS, its intentions, objectives, and priorities and identifying roles and responsibilities.

Risk The likelihood of injury or illness arising from exposure to any hazard(s) and the magnitude of the adverse effect.

Occupational Health and Safety Policy

The legislation in many countries places the onus of responsibility on employers to provide a healthy and safe working environment under occupational health and safety (OHS) legislation and common law. Employers should ensure that all managers, supervisors, and staff are aware of their OHS responsibilities. Management leadership can positively influence OHS outcomes for an organization.

Workplace health and safety is an ongoing process. Subject to the legislative requirements of each jurisdiction, in most instances a documented OHS policy is required. The development of such a policy requires the commitment of both staff and management. Once commitment has been achieved, the OHS policy should be developed with involvement from all stakeholders and promulgated.

The OHS policy should:

- articulate the organization's commitment to OHS;
- indicate that sufficient resources (both financial and personnel) will be provided to promote and maintain OHS standards and meet OHS requirements;
- outline the organization's intentions, objectives, and priorities OHS;
- describe in broad terms the means by which the objectives will be met;
- identify the roles and responsibilities of management, supervisors, and staff in meeting OHS requirements; and
- be signed off by the most senior manager of the organization, reflecting the importance of the policy.

The OHS policy should be reviewed periodically to ensure its currency.

The OHS policy is, however, only one part of an appropriate OHS strategy for a forensic organization. The OHS policy must be underpinned by risk assessments and incident/accident reports that enable the organization to assess its OHS exposure to meet legislative requirements such as reporting obligations and to respond to risks appropriately.

An organization can develop a list of the main hazards that its staff are likely to be exposed to in the course of their duties, utilizing OHS reports, incident/accident reports, and previous risk assessments. Prioritizing the main health and safety issues

 http://dx.doi.org/10.1016/B978-0-12-802718-9.00035-1

allows the organization to develop appropriate action plans to meet the objectives of its OHS policy.

Forensic organizations may consider integration of some OHS requirements with their quality assurance system. Many laboratories effectively use their quality system to embed OHS requirements in their documented procedures to review OHS hazards as part of a periodic audit program or to manage elements of their OHS action plans through their corrective action system. OHS, like quality, can then be viewed as an important yet integrated component of an effective management system.

Risk Assessments

Once potential OHS hazards have been identified, forensic organizations should evaluate the likelihood of injury from the interaction to the hazard and the magnitude of the adverse effect. The process of risk assessment will be very useful for managing potential OHS hazards within the facility and the expected external work environment. The purpose of the risk assessment process is to ensure that all workplace hazards have been identified, recorded, assessed, controlled, and reviewed. The desired outcome of this process is to eliminate, as far as practicable, the risk of injury or illness to personnel, damage to property, and damage to the environment. The process of developing risk assessment is often better suited to the known work environment. An OHS assessment of an office or laboratory can quickly identify specific hazards that may require attention. Obviously, this works well for the office and laboratory environment within one's control; however, each external scene will be different.

It is important that the range of potential hazards in external crime scenes and work environments is considered. While some risks can be grouped and managed collectively, the specific hazard and risk mitigation and control will vary from scene to scene given the circumstances. Given this, forensic practitioners should have an ability to undertake dynamic risk assessments or "risk on the run" as it is known in some jurisdictions.

Dynamic Risk Management

Dynamic risk assessments are conducted by a forensic practitioner as part of the attendance and examination process. In some instances, such as attendance at a clan lab, a person may be designated as the Site Safety Officer and have carriage of this as well as health and safety for all personnel at the site. Practitioners should be trained to assess the risk given the circumstances at the time, considering the actual hazards present at a crime scene.

A designated forensic practitioner or Site Safety Officer should undertake a quick reconnaissance of the crime scene to ensure the safety of forensic practitioners and others working at the scene. A review of the scene should be repeated whenever the situation at the scene changes. This could involve a visual inspection without entering the crime scene and asking a number of questions. For example:

- Does the crime scene involve structures that are now unstable?
- Has confirmation been obtained from the Fire Brigade or other emergency responders that power, gas, and water to the site have been turned off?
- Is there adequate shelter so that practitioners can rest without succumbing to environmental stressors such as heat, cold, wind, or rain?

It is important to close the loop and incorporate any strategic elements of each dynamic risk assessment in OHS policy and planning. After each incident, any relevant information obtained during the dynamic risk assessment should be recorded and collated for strategic analysis.

Hierarchy of Control Measures

Within OHS, there is a "hierarchy of control" designed to mitigate or resolve a risk deemed unacceptably high.

The hierarchy of control is a sequence of options which offer a number of ways to approach the hazard control process. Various control options may be available. It is important to choose the control that most effectively eliminates the hazard or minimizes the risk in the circumstances. This may involve a single control measure or a combination of different controls that together provide the highest level of protection that is reasonably practicable.

1. Eliminate the hazard. If this is not practical, then:
2. Substitute the hazard with a lesser risk. If this is not practical, then:
3. Isolate the hazard. If this is not practical, then:
4. Use engineering controls. If this is not practical, then:
5. Use administrative controls, such as safe work practices, instruction, and training. If this is not practical, then:
6. Use personal protective equipment (PPE), such as gloves, eye protection, boots, and respirators.

It is important that management and staff discuss and consult, where possible, during all phases of the hazard identification, risk assessment, and risk control process.

Examples

1. If an organization is considering purchasing a piece of analytical equipment and two products have the same capabilities but substantially different noise levels during operation, the organization may consider the noise level of the equipment during procurement and opt for the quieter system. This example demonstrates the principle of

eliminating the hazard at source, which is the most effective control measure, when compared to training and provision of PPE such as hearing protection.
2. In the case of a fire scene of a building, applying a hierarchy of control approach, it is first necessary to consider the elimination or substitution of hazards. In a fire scene, this is not possible. It is, however, possible to isolate the scene to prevent danger to the public and to maintain the integrity of the scene. Power, water, and gas to a building should be disconnected prior to entering the site. A structural engineer's opinion may be necessary prior to entry to the building. Safe entry and exit to the site can be established. Other administrative controls, such as briefing practitioners and maintaining records of the entry and exit of personnel, may be applied. Finally, practitioners can be prevented from entering the fire scene unless utilizing the appropriate PPE.

Specific Laboratory Hazards

The likely hazards within a laboratory environment include the following.

Chemicals

Chemical exposure may occur through inhalation, skin absorption, or direct ingestion and, once absorbed, are either stored in a particular organ or tissue, metabolized, or excreted. The effect of a chemical on a person is dependent on a number of factors such as duration and frequency of exposure, concentration of the chemical, and an individual's metabolism. A synergistic effect may occur when the undesirable effects of one substance are intensified if exposure has occurred to another substance.

Some nanomaterials exhibit different chemical properties compared to what they exhibit on a macroscale. As this is a relatively new field, there is insufficient knowledge regarding the hazards posed by nanomaterials. The potential hazards associated with nanomaterials may include increased reactivity because of their increased surface area to volume ratio, the ability to cross some of the body's protective mechanism, and the lack of the body's immunity against such small particles. Because of this lack of knowledge, the suggested control strategy to be used when working with nanomaterials should be "as low as reasonably achievable" approach to reduce exposure.

The effects of chemicals on the body may be categorized as follows:

- poisonous or toxic chemicals are absorbed into the body and exert either an acute or short-term effect, such as headache, nausea, or loss of consciousness, or a long-term effect such as liver or kidney damage, cancer, or chronic lung disease;
- corrosive chemicals burn the skin, eyes, or respiratory tract;
- irritants can inflame the skin or lungs, causing conditions such as dermatitis or bronchitis;
- sensitizers may exert long-term effects, especially to the skin (such as contact dermatitis) and to the respiratory tract (such as occupational asthma) by inducing an allergic reaction; and
- explosive or flammable substances pose immediate danger of fire and explosion, causing damage to the body through direct burning or through inhalation of toxic fumes emitted during combustion.

Safety Data Sheets (SDS), also known as Material Safety Data Sheets, are designed to provide relevant information regarding the identity, physical characteristics, safe storage, use, disposal, first aid treatment, and spill management of substances that are handled in the workplace. The information includes whether the substance is deemed to be a hazardous and/or a dangerous goods item. At a minimum, the SDS should be consulted before the first use of a chemical or other substance within a laboratory or if practitioners are unfamiliar with the product. Copies of SDS should be retained according to legislative requirements. In some jurisdictions, electronic SDS management systems can allow an efficient way of accessing up-to-date SDS information.

Sharps

Sharps are objects that have sharp edges or points that have the potential to cut, scratch, or puncture the skin. Sharps can cause physical injury and have the potential to introduce infectious and toxic agents through the wounds created in the skin. Examples include hypodermic syringes and needles, knives, or broken glassware.

All forensic practitioners have a responsibility to handle and package sharps safely. Particular care should be given to ensuring that sharps are appropriately labeled when packaged. Sharps such as knives could, for example, be packaged in clear plastic tubes, making it easier for a person opening the item to identify the contents and the direction the sharp items is facing. Forensic Labs should be encouraged to develop policies that encourage forensic practitioners and others that submit items to develop safe-packaging procedures.

Biological Material

Examples of "biological material" commonly encountered in forensic examinations include body tissue, blood, and body fluids (urine, saliva, vomit, pus, seminal fluid, vaginal fluid, and feces). Biological material is potentially hazardous as it may contain infectious agents such as viruses, bacteria, fungi, and parasites that cause a variety of communicable diseases.

Hair, fur, and items of clothing that have been in close contact with humans or animals may also harbor parasites such as fleas or nits.

When examining plant material such as cannabis, consideration should be given to the presence of *Aspergillus* sp. mold. If the *Aspergillus* spores are inhaled into the lungs, a serious, chronic respiratory or sinus infection can result. If mold is visible, the cannabis should be treated as a biological and respiratory hazard.

It is impossible to determine the prevalence of infectious or communicable diseases in the environment in which forensic practitioners work. Consequently, practitioners should adhere to recommended procedures for handling biological material and adopt an approach known as the "standard precautions." This approach requires practitioners to assume that all biological material is a potential source of infection, independent of diagnosis or perceived level of underlying risk.

Vaccinations should be offered for practitioners. The types of vaccinations given may depend on whether work is confined to the laboratory or whether work is performed in the field, as well as whether forensic practitioners are likely to be deployed overseas where other diseases may be more prevalent.

Firearms

Forensic practitioners may retrieve firearms from crime scenes. All personnel who may be required to handle firearms, in the field, in the laboratory, or in support roles such as property or exhibit stores should be trained on how to render a firearm safe. As with the "standard precautions," it is important to consider all firearms as potentially loaded and adopt the practice of never pointing a firearm in the direction of another person, even after it has been rendered safe.

Firearms examiners, who undertake firearms investigations including test firing and bullet recovery, will be exposed to hazards such as noise and lead. They should have their hearing and blood lead levels monitored on a regular basis to ensure that hearing protection is being worn and is functioning correctly, and any exposure to lead from the firearms is quickly identified and addressed.

Computer Forensics Laboratory

Computer forensic examiners specialize in obtaining, analyzing, and reporting on electronic evidence stored on computers and other electronic devices. Crimes involving a computer can range across the spectrum of criminal activity from child pornography to theft of personal data to destruction of intellectual property. Potential hazards involve static postures, occupational overuse, and stress from viewing graphic images.

Some suggestions to minimize the stress from viewing graphic images are as follows:

- psychological assessment before and after viewing graphic material, and periodically;
- exposure to only one medium, for example, visual material only, rather than examining both sound and visual material simultaneously;
- specifying limits as to the amount of time spent examining explicit material in a day, and
- ceasing any examination of explicit material at the end of their shift to allow themselves time to refocus attention away from this stressor.

Electrical/Machinery

Forensic laboratories use a wide range of electrical equipment and machinery. Practitioners need to ensure that any inherent risk from electric shock is mitigated. The use of residual current devices (safety switches) is an appropriate strategy, as is visual inspection and periodic testing and tagging of power cords, to detect obvious damage, wear, and other conditions which might render it unsafe by a person qualified to do so under the legislation in effect in the jurisdiction.

Fume Cupboards

Fume cupboards are integral to minimizing the risk of exposure to chemical and biological hazards. Not all fume cupboards are suitable for all hazards. Fume cupboards should be maintained and inspected periodically. During maintenance, attention should be given to the following:

- The fume cupboard itself, including flow rates and replacement of absorbents or filters.
- In the case of externally vented fume cupboards, the ductwork, and location of external vents. This is particularly important during any building maintenance or refurbishment.
- Fume cupboards must be used for all operations that have the potential to release hazardous fumes, mists, or dusts.
- Before commencement of work, ensure that the fume cupboard is clean and free from contamination.
- Ensure the minimum of equipment is stored in the fume cupboard and is placed toward the back of the cupboard to reduce disturbance to the air flowing into the fume cupboard.
- Lower the sash as far as practicable during use to improve fume containment.

Recirculating fume cabinets rely on filtration or absorption to remove airborne contaminants released in the fume cabinet before the exhaust air is discharged back into the laboratory. They are suitable for light to moderate use with a known range of substances. The range of substances for which each cabinet can be used is limited by the need for compatibility with the chemicals in use as well as with the particular type of absorbent or filter fitted to the cabinet.

Robotics

The introduction of automated robotic platforms has significantly enhanced the efficiency of forensic analysis. The use of robotics is becoming more common and is very useful for a range of repetitive laboratory tasks. Besides saving time, robotics overcomes the need for repetitive work involved in pipetting, eliminating musculoskeletal injuries.

Hazards associated with robotics include the risk of exposure to the chemicals used in the work, electrocution, and cutting, stabbing, or shearing from the moveable parts of the robot. The interlocks on the robots should not be bypassed.

X-rays

X-rays are used in analytical and imaging instrumentation. Potential exposure to X-rays is generally localized to specific parts of the body, usually the hands or fingers. Depending on the X-ray energies delivered, effects may range from erythema (redness) at point of exposure, blood changes, cancer through to death. Depending on the legislative requirement in each country, practitioners working with X-ray equipment may be required to use dosimeters to assess radiation dose.

Lasers

Lasers span the visible and nonvisible electromagnetic spectrum and have many applications in forensic science, including Raman spectroscopy. Lasers are generally classified according to the level of risk they represent. Damage from laser beams can be thermal or photochemical. The primary sites of damage are the eyes and skin. Hazards associated with laser work may include:

- fire,
- explosion,
- electrocution, and
- inhalation of contaminants from laser interactions.

Precautions for use of lasers include the following:

- Display the class of laser in use.
- Appropriate protective eye wear with side protection and appropriate attenuation for the wavelength(s) in use must be worn.
- Interlocks on the laser should not be bypassed.
- Keep the laser beam path away from eye level whether one is seated or standing.

High-Intensity Light Sources

High-intensity light sources such as the Polilight provide a range of colored light bands and white light for forensic work.

- Care should be taken that high-intensity white light is not directed onto any object at short distances from the end of the light guide, as this can cause severe heat damage to the object and may result in a fire.
- The light beam should never be directed at eyes as the light can cause permanent damage.

Manual Handling

Manual handling refers to any activity that involves lifting, lowering, carrying, pushing, pulling, holding, restraining, or the application of force. Only a very small number of manual handling injuries are caused by the lifting of heavy weights alone. Actions such as reaching, twisting, bending, or maintaining static postures contribute to injury affecting the muscle or skeletal systems of the body. These musculoskeletal injuries predominantly involve the neck, back, or shoulder or arm muscle, tendon, ligament, or joints.

Injuries may be caused from activities such as maintaining static postures while working at fume cupboards, repetitive keyboard and mouse work, pipetting, and prolonged use of comparison microscopes.

Some preventative strategies include the following:

- Seeking further assistance to have the activities assessed to minimize the manual handling risks inherent in the activity.
- Planning tasks so that rest breaks are scheduled.
- Choosing the best tools for the tasks.
- Alternate hands while using a mouse, if possible.

There is a move to make instruments smaller and more portable for use at crime scenes. While this has significant benefits, including potentially reducing the number of exhibits collected, moving equipment can also raise manual handling concerns.

General Laboratory Management

Housekeeping is important in laboratories. It is important to maintain clear passageways and have proper labeling of chemicals, clean and uncluttered work areas, and appropriate storage. The handling of powders is a potentially hazardous operation and good housekeeping can help minimize airborne contamination from spilled materials. Having a planned preventative maintenance program and regular inspections of the workplace, plant and equipment are essential for the smooth running of the laboratory.

Handling of Exhibits in Court

Each evidential item must be appropriately packaged and sealed, if this is not already the case, before it is exhibited in court. Items such as clothing which are normally stored in paper may need to be repackaged in clear plastic allowing the item to remain sealed and minimizing the risk of cross contamination when handled in court. Caution should be exercised against opening exhibits in court, in case any hazards such as mold or irritant fumes are released.

Hazards in the Field

Forensic practitioners are often required to work or train in the field. Consideration should be given to managing hazards which may affect practitioners, including:

- environmental hazards such as heat, cold, humidity or wet weather, the terrain, and fauna or flora at the scene;
- the type of operation, for example, working in a clandestine laboratory often involves quite specific hazards;
- the possible presence of offenders or other security risks such as booby traps at a scene; and
- the availability of first aid and emergency response domestically and overseas.

The risks from these hazards should be considered within the scope of the exercise or operation. Some possible responses to hazards, which may be considered in a dynamic risk assessment, include the following:

- Designating a location for emergency equipment, such as a crime scene vehicle, and ensuring that disinfectants, antiseptics, and a first aid kit are easily accessible;
- Planning an emergency exit from the scene and ensuring that this is communicated to all personnel present;
- Establishing a decontamination point if there is exposure to chemical or biological material;
- The use of appropriate PPE including sunglasses, sunscreen, and hats when working outdoors;
- Depending on the external temperature, work activity, duration, and PPE worn, practitioners should have access to shade for rest and adequate fluids if required during hot weather to prevent heat stress. The wearing of PPE including chemical suits and respirators requires longer and more frequent periods of rest break for recovery in hot temperatures and humid environment;
- In cold weather, provision should be made to have adequate warm clothing and a sheltered area;
- The risk of animal or dog bites while attending a crime scene should not be discounted. If practitioners are searching in vegetated areas, the risk of snake or tick bites should be considered, along with possible exposure to plants such as poison ivy or stinging nettles.

Confined Spaces

Forensic practitioners may have to enter confined spaces. Due to the high risks associated with entering the confined space, many jurisdictions mandate that entry into a confined space must not be made until a confined-space permit has been issued. Practitioners must receive specific training before work or entry into confined spaces.

Chemical Biological and Radiological and Nuclear Incidents

Forensic practitioners may be required to attend a chemical biological and radiological and nuclear (CBRN) incident. CBRN incidents where forensic practitioners may attend and conduct examinations include:

- chemical (warfare agent, toxic industrial chemical);
- biological (weaponized agent, natural disease);
- radiological (discrete or wide area contamination); and
- nuclear.

Depending on the response agency protocol in place, forensic practitioners may be working closely with the Fire Brigade and other emergency first responders. Entry must not be made into the "warm" or "hot" zone of the scene without consultation with the other emergency first responders.

Clan Labs

Clan labs pose a significant threat to the health and safety of police officers, forensic practitioners, the general public, and the environment. There are many hazards associated with clan labs including:

- flammable materials and/or explosive atmosphere;
- acutely toxic atmospheres;
- leaking or damaged compressed gas cylinders; and
- traps and hazards deliberately set to cause injury or death to police and other responders.

As a result of the frequency at which clan labs are encountered and the severe and variable risks associated with the investigation, many jurisdictions have developed specific policies and procedures concerning clan lab investigations.

For forensic practitioners to deal with clan labs requires a high level of fitness as well as technical expertise. Practitioners have to understand:

- illicit drug chemistry;
- how to neutralize the risks of explosions, fires, chemical burns, and toxic fumes;
- how to handle, store, and dispose of hazardous materials; and
- how to treat medical conditions caused by exposure.

Practitioners must also wear full protective equipment including respirators and may be required to move equipment at the clan lab in the process of collecting evidence. The storage and handling of unknown chemicals from clandestine laboratories or seizures should also be considered. Preliminary identification should take place, before its storage or disposal.

When unknowns such as "white powders," chemicals (in liquid, solid, or gas state), or biological materials are encountered in the field, it is prudent to be cautious and obtain up-to-date intelligence to shed more light on what is at the scene. It may be an explosive material or contain anthrax spores or ricin or something as innocuous as talc.

Some precautions include:

- wearing the appropriate level of protective clothing/equipment for the activity;
- avoiding direct contact with the substance, even if only in small quantities;
- not smelling or tasting anything from the scene;
- noting physical characteristics such as color, form, and consistency;
- where it is safe to do so, looking for hazard symbols on packaging or labels if available; and
- seeking specialist advice if unable to identify the substance.

Potential Hazards During an Overseas Deployment

Forensic practitioners can be required to work overseas to assist with large scale disasters. An example was the Thailand Tsunami Victim Identification process involving forensic practitioners from 30 countries working to recover and identify bodies. Forensic practitioners need to be mindful of hazards likely to be encountered during an overseas deployment depending on the location, magnitude of the operation, and how many practitioners are deployed. Some hazards to be considered include:

- climatic demands;
- remote and sometimes dangerous terrain;
- different cultural sensitivities;
- security requirements;
- different levels of infrastructure support at the locality;
- logistics, including the transport of large quantities of equipment, manual handling, setting up, and packing up;
- different hygiene levels;
- diseases that can be transmitted by insect and or animal vectors;
- the possibility of infectious diseases; and
- asbestos and other hazards in buildings.

Work-Related Stress

Practitioners at work may experience work-related stress. There are some specific stressors unique within forensic work. Forensic practitioners may experience workplace-related stress due to their attendances at morgues, violent crime scenes, Disaster Victim Identification, or from requirements to view explicit or graphic material or images.

Indicators of stress include changes in eating habits, tiredness due to changes in sleep patterns, frequent absences from work, reduced productivity, concentration, motivation, and morale. Physical symptoms may include headaches, abdominal pains, diarrhea, constipation, high blood pressure, insomnia, anxiety state, and depression.

Many organizations offer programs to provide assistance to employees, including counseling to help practitioners to deal with work-related stress or resilience training to manage work-life balance.

See Also: **Management/Quality in Forensic Science:** Principles of Quality Assurance; Risk Management; Principles of Laboratory Organization.

Further Reading

Clancy, D., Billinghurst, A., Cater, H., 2009. Hazard identification and risk assessment – understanding the transition from the documented plan to assessing dynamic risk in bio security emergencies. In: World Conference on Disaster Management, Sydney, Australia. http://www.humansafety.com.au/getattachment/da338cb7-29b0-4d3a-8a06-d7dc0b569a87/C2D.aspx.

Furr, K., 2000. Handbook of Laboratory Safety, fifth ed. CRC Press, Florida.

Green-McKenzie, J., Watkins, M., 2005. Occupational hazards: law enforcement officers are at risk of body fluid exposure. Here's what to expect if it happens to you. Law Enforcement Magazine 29 (9), 52–54, 56, 58.

Hanson, D., 2007. Hazardous duty training officers to tackle hazmat emergencies. Law Enforcement Technology 34 (4), 80–85.

Haski, R., Cardilini, G., Bartolo, W., 2011. Laboratory Safety Manual. CCH Australia Ltd., Sydney.

Horswell, J., 2000. The Practice of Crime Scene Investigation. CRC Press, Florida.

Jackel, G., 2004. The high cost of stress. AUSPOL: The Official Publication of the Australian Federal Police Association and ALAJA 1, 4–37.

Mayhew, C., 2001a. Occupational health and safety risks faced by police officers. Australian Institute of Criminology. Trends and Issues in Crime and Criminal Justice 196, 1–6.

Mayhew, C., 2001b. Protecting the occupational health and safety of police officers. Australian Institute of Criminology. Trends and Issues in Crime and Criminal Justice 197, 1–6.

Rothernbaum, D., 2010. Exposed: an officer's story. Clandestine Laboratory Safety Alert 7 (2), 1–2.

Smith, D., 2005. Psychosocial occupational health issues in contemporary police work: a review of research evidence. Journal of Occupational Health and Safety, Australia and New Zealand 21 (3), 217–228.

Tillman, C., 2007. Principles of Occupational Health and Hygiene: An Introduction. Allen & Unwin, Crows Nest.

Whitman, M., Smith, C., 2005. The culture of safety: no one gets hurt today. Police Chief LXXII (11), 20–24, 26–27.

Winder, C., 2011. Hazard Alert: Managing Workplace Hazardous Substances. CCH Australia Ltd., Sydney.

Witter, R., Martyny, J., Mueller, K., Gottschall, B., Newman, L. 2007. Symptoms experienced by law enforcement personnel during methamphetamine lab investigation. Journal of Occupational and Environmental Hygiene 4, 895–902.

Relevant Websites

http://www.ccohs.ca/oshanswers/occup_workplace/labtech.html—Canadian Centre for Occupational Health and Safety (CCOHS).
http://www.cdc.gov/niosh/—Centers for Disease Control and Prevention (CDC).
http://www.forensic.gov.uk/html/company/foi/publication-scheme/health-and-safety/—Forensic Science Service, Health and Safety.
http://www.hse.gov.uk/services/police/index.htm—Health and Safety Executive (HSE).
http://www.police.qld.gov.au/Resources/Internet/rti/policies/documents/QPSForensicServicesHealth_SafetyManual.pdf—Health and Safety Manual, Police Forensic Services, Queensland Police.
http://www.londonhealthandsafetygroup.org/archive.html—London Health and Safety Group.
http://www.osha.gov/—Occupational Safety & Health Administration.
http://www.ccohs.ca/oshanswers/occup_workplace/police.html—What Do Police Do?

Key Terms

Best evidence, Bias, Brady material, Confrontation, Discovery, Ethical considerations, Forensic, Forensic laboratory report, Hazard, Health, Holistic approach, Integrated case management, Investigations, Occupational health and safety (OHS), Risk, Safety, Strategy, Submissions, Testimony.

Review Questions

1. Are the models offered for forensic service provision suggested by Ramsey and Burton the only ones available? What other models might there be?
2. What is a "forensic strategy"? What questions are part of developing one? Why is it important to a case and to a laboratory?
3. Why is integration of forensic cases "the single biggest challenge to the forensic science community"?
4. What is currently in forensic reports? How do these differ from other scientific reports? How do they differ from published articles in peer-reviewed journals? What *should* be in forensic reports? Make a table listing each component and compare.
5. Who is the "audience" of forensic reports? Are the police using the technical information to investigate the case or other scientists who must review the technical information? Is it both? How does this affect the content of forensic reports? Should it?
6. How do the legal requirements of the contents of a forensic report differ between the UK and the US?
7. Can forensic reports stand on their own (that is, without an expert's testimony)? What determines this?
8. Are there ethical obligations to the format and content of a forensic report? If so, what are they?
9. What is "Brady material"?
10. What is "OHS"? What should this policy articulate?
11. What is a risk in OHS terms? How is it prevented or mitigated? What is the hierarchy of control measures?
12. List some specific laboratory hazards. What are their risks, how are they managed?
13. Why is good housekeeping important in a forensic laboratory?
14. Review the article on clandestine laboratories. In light of OHS policies, what *additional* hazards and risks do they pose?
15. Why is stress a health and safety issue? What types of analysis are particularly prone to mental or stress issues?
16. Are there any specific health and safety dangers for a forensic engineer?
17. Given the complexity of forensic engineering reports (physics, math, materials), how should they be written? How should the testimony be provided? Remember, juries are made up of people who do not necessarily have a background in science.
18. What additional health and safety hazards would there be at the scene of a structural collapse? How might the type of structure affect these?
19. Most, if not nearly all, forensic engineers work privately. What additional burdens are there to supporting health and safety by being self-employed or a small business?
20. How would you incorporate the use of software visualization (say, of a building collapse) into a forensic engineering report? How about testimony?

Discussion Questions

1. How should forensic examinations and analyses be "integrated"? How should forensic engineering work be coordinated with traditional laboratory work? What are the benefits to doing this? What are the problems?
2. Why are forensic reports formatted they way they are? Is there a better way to present this information? Who are they written for and who *should* they be written for?

3. Are some types of forensic science more prone to hazards and health risks than others? If so, list them and explain why they are; if not, why not?
4. What are the health and safety risks for forensic engineers? What kind of cases carry more risks?
5. What should the educational requirements be for a forensic materials analyst? Given such a wide range of industries, topics and materials, what could properly encompass the range of evidence that could be encountered?

Additional Readings

Slosson, J.E., Shuirman, G., 2014. Forensic Engineering: Environmental Case Histories for Civil Engineers and Geologists. Academic Press.

INDEX

'*Note*: Page numbers followed by "f" indicate figures, "t" indicate tables.'

A

Abbreviated injury scale (AIS), 27
Acceleration, 27–28
Accident investigation, 121
Accident reconstruction experts
  accident analysis
    collision model, 38
    postcrash movement, 38
  applications, 20
  contact area, material properties in
    compression phase, 39
    deformations, 38, 39f
    restitution, coefficient of, 39
  crush damage calculations, 25
  driver's reaction
    action time, 43
    definition of, 44f
    lag, 43
    prevention analysis, 44
    reaction time, 43
  eccentric impacts, 39–40
  EDRs, 24
  EES, 41–43
    EBS, concept of, 41
    precrash movement, 43
    stiffness-based and mesh-based impact models, 42
    vehicle's deformation, 41
  full impact, 41
  injuries, biomechanics of, 27–28
  intersection collision, 22–23, 22f
  linear momentum, 22
  Newton's second law, 23
  occupant kinematics and related concepts
    descriptions of, 26
    occupant anthropometry, 25–26
    second collision, 25–26
  POI, 22–23
  results, 45–46
  sample case, 44
  scene data, collection of, 37
  sliding impact, 41
  targets, 37–38
  velocity components, 24
  ΔV vectors, 24–25
Adversary truth, 80
Air France Flight 447
  criminal investigation, 87
  technical investigation, 87
Air Line Pilots Association International (ALPA), 83
Air-bag-related injuries
  automotive industry, 89–90
  forensics of, 95, 96f
  historical context, 89
  human injuries, 90
  sample cases, 92–95
  specific injury patterns
    cervical spine, 91
    cranial and intracranial, 91, 91f
    extremities, 92, 92f
    face and head, 91, 91f
    ocular, 91, 91f
    respiratory, 92
Aircraft flight data recorders
  adversary *vs.* technical truth, 80, 80f
  current recorder state-of-the-art, 85

Aircraft flight data recorders (*continued*)
desired knowledge
approximate *vs.* proximate cause, 81
limitation, 80
seeking causation, 81
single *vs.* multiple causes, 81
digital *vs.* analog evidence, 82
history
black boxes, 82
cockpit voice recorder hostility, 82
flight data recorders, installed location of, 83
government regulation, 83–84
recorder survivability, 83
recording processes, evolution of, 82–83
ULB, 83
voice recordings, public access to, 84–85
recorders, cultural repercussion of
Air France Flight 447, 85–87
EgyptAir Flight 990, 85
recorders, expanded employment of, 87
Airplane crashes, 131
Alcohol, Tobacco and Firearms (ATF), 182
Alignable differences, 16–17
American Society for Testing Materials (ASTM), 261
Analog tachograph chart analysis
digital tachographs, 119–120
falsifications and diagnostic signals, 117
forensic applications, 109
innermost recording, 109
instrument, 111
principles of, 111–113
route tracing, 118–119
speed record, accuracy of
calibration, route-trace method of, 115
instrument, response of, 116
interpretation, problems of, 115
low-speed behavior, 117
tachograph installation, calibration checks on, 114–115
time intervals, 115–116
tire slip effects, 116–117
vehicle, impacts to, 116
work-mode field, 109
Antilock braking systems (ABS), 22–23
Association for the Advancement of Automotive Medicine (AAAM), 27
Atlanto-occipital dislocation, 94
Audio forensics
audio evidence, 212
authentication. *See* Forensic audio authentication
competency and proficiency, 213
enhancement. *See* Forensic audio enhancement
evidence handling, 213
recordings, 212
technology, 212–213
Auger electron spectroscopy (AES), 277

B

Backscattered electrons (BSEs), 272
Blunt force, 122
Bureau d'Enquetes et d'Analyses (BEA), 85
Burn pattern. *See* Fire patterns

C

Car occupants
frontal collisions
airbag injuries, 125
seat belt injuries, 124–125
rear-end collisions, 125
rollover collisions, 125
side collisions, 125
Cause-in-fact, 81
Central nervous system (CNS), 249
Chemical biological and radiological and nuclear (CBRN) incidents, 298
Chronology principles, 5
Coalition of Airline Pilots Associations (CAPA), 84
Cockpit voice recorder (CVR), 82
Coefficient of friction (COF), 252
Collision model, 38
Conduction
convection, 148
electrical sources, 148
friction, 148
heating appliances, 148
radiation, 148
sparks
electrical sparks and arcs, 148–149
material sparks, 148
Crash data retrieval (CDR), 104
Crash hypotheses, 38
Crash investigations
ECU NVRAM data, use of
use, retrieving ECU NVRAM data for
CDR, 104
direct-ECU umbilical method, 104
EDR data, 102–104
EEPROM component, 104–106
hexadecimal notation, 102–103
post-crash-event investigations, 103
vehicle diagnostic communication port, 104
Crime analysis, 63
evidence dynamics, 64–65
phases of, 63–64

physical evidence, role of, 65
reconstruction, 63
experience-based reconstructionist, 66
historical perspective, 65–66
methods of, 66–68
Crime reconstruction, 63
experience-based reconstructionist, 66
historical perspective, 65–66
methods of, 66–68
Crime scene investigator, 59
Crime scene, court
forensic strategies
consideration, issues for, 284
DNA profile, 286
examination processes, 285
forensic effectiveness, 286
forensic examinations, 284
issues, 285–286
multiagency, 285
police force/law enforcement agency, 286
potential evidence types, 285–286
integrated case management, 286
models, 284
public expectations, 284
task, 284
Crime scenes, 32–33
Crime, 10–11
Criminal investigation
case management, 61–62
examination records, 61
initial assessment, 59–60
occupational health and safety, 61
potential evidence
search and recovery of, 61
systematic collection of, 61
scene control and coordination, 59
scene security, 60
stages of, 59
systematic approach, 59
Cuvier's method, 3

D

Decision errors, 50
Delta-v, 19
Deoxyribonucleic acid (DNA), 153
Department of Transportation (DOT), 81
Diagnostic trouble codes (DTCs), 100
Differential scanning calorimetry (DSC), 280
Differential thermal analysis (DTA), 280
Digital and multimedia evidence (DME), 211–212
Disaster victim identification (DVI), 131
autopsy, 132–133
fingerprints, 132
forensic anthropology, 133
forensic odontology, 133
molecular genetics, 133
personal effects, 132
radiology, 133
Distortions, 214–215
Dynamic coefficient of friction (DCOF), 252

E

EgyptAir Flight 990, 85
Egyptian Civil Aviation Authority (ECAA), 85
Electric network frequency (ENF), 219–221
Electrical accidents
aluminum wiring, 226
analysis, 227
casework/examples, 227–240
electrical arcing/short-circuiting generates, 226
electrical conductor overheating, 226
low-voltage electrical shock, 227
medical records, 227
on-scene electrical investigation, 225
photographing electrical wiring, 226
practical electrical knowledge, 225
visiting accident scene, 225
Electrically erasable programmable read-only memory (EEPROM), 104–106
Electron impact (EI), 189
Electronic control units, 100
Electronic data recorders (EDRs), 24
crash investigations. *See* Crash investigations
ECUs, 100
categories, 100–102
DTCs, 102
Electronic instruments, 117
Energy-dispersive X-ray spectroscopy (EDS), 275
Environmental Protection Agency (EPA), 261
Epidural hematoma, 123
Epistemology, 2
Equivalent barrier speed, 41
Event data recorder (EDRs), 87
Evidence, forensic classification of
class-level information, 8–9
methods of
approaches, 8
manufacturing, 8
set theory, 7
taxonomy, 7–8
relationships and context, 10–11
uniqueness and individualization, 9–10

F

Failure analysis
appropriate technique, 280
determining morphology, techniques for
materials analysis, microscopy in, 269
microstructural examination, specimen preparation for, 271
optical microscopy, 269–271
SEM, 271–273
determining composition, techniques for
absorption spectroscopy, 277–278
AES, 277
MS, 278–280
spectroscopy, physics of, 274
thermal and TGA, 280
X-ray spectroscopy, 274–276
materials analysis, role of, 269
Federal Aviation Administration (FAA), 82
Federal Motor Vehicle Safety Standards (FMVSS), 27
Fire debris analysis
evidence collection–sampling containers
bags, types of, 185
laboratory analysis, 186
solvent-free nylon bags, 185
extraction and sampling techniques
direct headspace analysis, 187
extraction/concentration methods, effectiveness of, 188
passive/dynamic headspace analysis, 187–188
gas chromatography, 188–189
GC-MS data, 189–190
instrumental analysis, 188
interpretation of
aromatic products, 193, 197f
ASTM class and subclass, 200, 200t
classification, 192–193
crude oil-refining processes, 193
findings, significance of, 203
gasoline, 193
goal of, 192
ILRs, 201–202
isoparaffinic products, 193, 195f
miscellaneous products, 200
n-alkane products, 193, 198f
naphthenic paraffinic products, 193, 196f
oxygenated solvents, 193–200, 199f
petroleum-based products, 193
systematic approach, 202–203
mass spectrometer, 189
preliminary examination of, 186–187
Fire investigations
chemical reaction
carbon dioxide, 139
carbon monoxide, 139
complete reaction, 139
ethane, combustion of, 139
flaming fire, 140
oxygen-controlled fire, 139–140
smoldering fire, 139–140
conditions for
fire triangle, 138, 138f
fuel, 138
oxidizer, 138
source of heat, 138–139
fire point, 141
flammability limits, 141
flash points, 141
heat source and transfer, 141
ignition, 141–142
phase change and pyrolysis, 140
thermodynamics, role of
heat, 146
physical properties, 146
temperature, 146
Fire patterns
clean burn, 180, 181f
confinement patterns, 175–176, 178f
electrical damage, 179–180
intensity patterns, 181–182
irregular pattern, 176–178, 179f
movement patterns, 176, 179f
overview, 173
plume-generated patterns
columnar patterns, 174, 174f
plume pattern generation, 175
semicircular pattern, 177f
triangular pattern, 174, 174f
U-shaped pattern, 174–175
V pattern, 174–175, 176f
spalling, 178–179, 179f
ventilation-generated fire patterns, 182–184
Fire scene inspection methodology
avoiding spoliation, 169
cause determination, 170–171
documentation, 168
first assumptions, 167
initial evaluation, 168
inventory, 169
origin determination
categories, 169
origin hypothesis, 170
scientific method, 170

overview, 167
planning investigation, 167–168
reconstruction, 168–169
Fire scenes
clothing and shoes, 164–165
comparison, liquids for, 165
comparison samples, 163
evidence collection
electrical malfunctions, 165–166
forensic analyses, 166
spontaneous heating, 165
overview, 161
packaging options, 163–164
sample selection and documentation
accelerant-detecting canine, 162–163
disposable latex/nitrile gloves, 162
doughnut pattern, 161–162, 161f
gasoline-powered tools, 162
plastics and synthetic fibers, 162
soils, 162
vinyl floors, 162, 162f
Fires
fire triangle
classes of, 159–160
fire tetrahedron, 159
phases of, 160
fuel, physical states of
gases, 158–159
liquids, 158
solids, 158
theory of
combustion, 157
conduction, 158
convection, 158
decomposition, 158
heat, 158
pyrolysis, 158
radiation, 158
smoke, 157–158
Flame ionization detectors (FIDs), 187
Flight data recorders (FDRs), 80, 82
aircraft flight data recorders. *See* Aircraft flight data recorders
black boxes, 82
genesis of, 82
Forces, 20
Forensic audio authentication
analog tape authentication, 219
digital audio authentication
butt-splice detection, 222, 222f
compression level analysis, 220f, 221–222
DC-level analysis, 222, 223f
ENF analysis, 221f, 222
hex data, 221
power spectral analysis, 219f, 221
Forensic audio enhancement
audio processing, 215–218
adaptive filters, 217
clipped audio, 216
conversations, 218
convolved noise, 217
digital filters, 216–217
lossy compression artifact, 216
mobile phone interference, 216
spectral subtraction, 217
stereo source separation, 217–218
time domain distortions, 216
critical listening and analysis, 214–215
file preparation, 213–214
output file, preparation of, 218
Forensic chemical engineering investigation
fires and explosions
flammability limits, 262
flash points, 262–263
natural gas and petroleum liquids, huge quantities of, 261
pollution and toxic substances
air pollution, 263
ground/water pollution, 263–264
hazardous materials, storage of, 264–265
unrecognized hazards and unexpected consequences
closed room, fire in, 265
corrosive and reactive chemicals, 266
empty container fire in, 265
epilog, 266
explosive pharmaceutical, 265–266
flammable/toxic solids, 265
shipment, contamination of food in, 266
spontaneous combustion, 265
Forensic engineering
accident reconstruction
crush damage calculations, 25
EDRs, 24
injuries, biomechanics of, 27–28
intersection collision, 22–23, 22f
linear momentum, 22
Newton's second law, 23
occupant kinematics and related concepts, 25–26
POI, 22–23
velocity components, 24
ΔV vectors, 24–25
basic principles
biomechanics, 20
Newton's laws, 20–21
defined, 19
methodology, 21–22

Forensic evidence
  collection techniques
    casting, 33
    PPE, 33
    swabbing, 32
    tape lifts, 32–33
    vacuuming, 33
  control samples, 33
  custody, chain of, 33–34
  scene examination, 32
Forensic investigator, 59
Forensic laboratory reports
  ethical considerations and forensic reports, 291
  legal standards, 290–291
  science standard
    DNA analysis, 290
    ISO/IEC standard, 289–290
    NRC standard, 289
    RSC text, 290
    scientific issue, 290
  stand-alone evidence, 291
Forensic science
  analyses, 3
  basic unit of, trace, 2–3
  comparative method, 16–17
  definition, 2
  evidence, classification of
    approaches, 8
    class-level information, 8–9
    manufacturing, 8
    relationships and context, 10–11
    set theory, 7
    taxonomy, 7–8
    uniqueness and individualization, 9–10
  native principles, 3–4
  nonnative principles, 4–5

G

Gas chromatography–mass spectrometry (GC-MS), 192
Glasgow Coma Scale (GCS), 122

H

Head injuries
  intracranial hemorrhages, 123
  TBIs, 122–123
Head injury criterion (HIC), 27
Human error, 47
  HFACS. *See* Human Factors Analysis and Classification System (HFACS)
  levels, 49–50
  safety management process, 47, 48f
Human Factors Analysis and Classification System (HFACS), 51f, 56–57
  analyst actions
    decision errors, 50
    errors, types of, 50
    perceptual errors, 50
    skill-based errors, 50
    violations, 50–52
  human error, 49
  identify and address threats, 56
  middle management, 54–55
    correct known problems, failure to, 55
    leadership, 54–55
    operational planning, 55
    operational process, 56
    organizational climate, 56
    organizational influences, 55–56, 55t
    resource management, 55–56
    supervisory ethics, 55
  preconditions analyst actions
    adverse mental states, 52–53
    adverse physiological states, 53
    communication and information flow, 53–54
    duty, fitness for, 54
    physical environment, 54
    physical/mental limitations, 53
    technological environment, 54
  productive system, components of, 49, 49f
Human gait, biomechanics of
  environmental phase, 257
  expert witness report
    accident, general understanding of, 258–259
    opinion, 259–260
  factors influence stability, 255–256
  friction, static *vs.* dynamic coefficient of, 252
  functions of, 249
  gait model, 256–257
  gravity, total body center of, 254–255, 255f
  ground reaction force, 251–252
    lateral transverse force (F_T), 252
    RCOF, 252
    vertical force (F_V), 252
  human bipedal locomotion, 256
  human body, balance of, 255–256
  locomotion, body segments in
    heel velocity, 251
    lower extremities, 249
    upper extremities, 251
  mechanical definition of, 249
  reactive recovery phase, 258
  slip resistance measuring devices, 253–254
  slip-initiation process, 257–258
  visual cues, 258
  visual *vs.* tactile sensation, 252–253

Human remains, thermal degradation
blood, 153
bone, 152–153
soft tissue, 152

I

Ignitable liquid residues (ILRs), 161, 201–202
Ignitable liquids (ILs), 192–200
Image recorders, 85
Interpretation/comparative method
analogy, forensic process
cognitive process, 15
features, 16
psychological studies, 15–16
salient traits, 15
three arguments, function of, 15, 16f
Cuvier's Principle of Correlation of Parts, 13–14
forensic science
alignable differences, 16–17
constraints/requirements, 16
cross-mapping, 16
nonalignable differences, 16–17
homologous and analogous traits, 14
Intracerebral hematoma, 123
Inverted cone pattern. *See* Triangular pattern

K

Kirk's individuality principle, 4
Kudlich–Slibar model, 42

L

Lateral continuity principles, 5
Linnaean method, 7–8
Locard's principle, 4, 95
Lower flammability limit (LFL), 141

M

Manufactured evidence, 8
Mass disasters
definition
airplane accident, 129–130
airplane crashes, 131
DVI, 131
forensic pathology, 129, 131
helicopter crash, 131
identification efforts, 129
tasks
bodies, recovery of, 131–132
criminalistic reconstruction, 133
DVI, 132–133
Mass spectrometry, 278–280
Mechanical instruments, 117
Mesh-based impact models, 42
Modular, 111

N

National Electrical Code (NEC), 227
National Electrical Safety Code (NESC), 227
National Fire Prevention Association's (NFPA) 227, 261
National Transportation Safety Board (NTSB), 80
Native principles, 3–4
Newton's laws, 20–21
Nonalignable differences, 16–17
Nonnative principles, 4–5
Nonradar-based event recorders, 72
Nonvolatile random-access memory (NVRAM), 102

O

Occupational health and safety (OHS) policy, 293–295
control measures, hierarchy of, 294–295
dynamic risk management, 294
field, hazards in
CBRN, 298
clan labs, 298–299
confined spaces, 298
overseas deployment, 299
work-related stress, 299
risk assessments, 294
specific laboratory hazards
biological material, 295–296
chemicals, 295
computer forensics laboratory, 296
electrical/machinery, 296
exhibits, handling of, 298
firearms, 296
fume cupboards, 296–297
general laboratory management, 297
high-intensity light sources, 297
lasers, 297
manual handling, 297
robotics, 297
sharps, 295
X-rays, 297
Occupational Safety and Health Administration (OSHA), 263
Officer in charge (OIC), 34

P

Paradigm, 3–4
Perception/reaction time (PRT), 76

Perceptual errors, 50
Personal protection equipment (PPE), 33
Physical thermodynamics
fire investigation
heat, 146
physical properties, 146
temperature, 146
flames, 149–150
ignition and propagation, fire
activation energy, 146–147
combustion speed, 147
thermal power and reaction temperature, 147
ignition sources, thermodynamic classification of
conduction, 147–149
heating, 147
overview, 145
physical systems, 145
smoldering, 149
thermodynamic principles
first law of, 145
second law of, 145
Point of derailment (POD), 75
Point of impact (POI), 22–23
Powertrain control module (PCM), 102
Premise negation, 66–67
Primary electrons (PEs), 272
Probable cause, 81
Proximate/legal cause, 81
Pyrolysis, 140–141
Pyrolytic thermal degradation, 152

Q

Quebec task force (QTF), 125

R

Railroad accident investigation and reconstruction
event recorders
data extraction, 72–73
locomotive event recorders, 72–73
nonradar-based event recorders, 72
human factors considerations–level crossing accidents, 76–77
procedural steps
derailments, 75–76
motor vehicle/pedestrian, collision with, 74
train collisions, 74–75
types of, 72
Random-access memory (RAM), 102
Read-only memory (ROM), 100
Relief First Officer (RFO), 85
Required coefficient of friction (RCOF), 252
Road traffic
car occupants
frontal collisions, 124–125
rear-end collisions, 125
rollover collisions, 125
side collisions, 125
epidemiology, 121
pedestrian collisions, 126–127
relevant injury types
abdominal injuries, 124
head injuries, 122–123
long bones, fractures of, 124
skin injuries, 121–122
thorax injuries, 123–124
two-wheelers, 127
whiplash-associated disorders, 125–126
Rolling road method, 114
Route-trace method, 115

S

Scanning electron microscopy (SEM)
BSEs, 272
ductile iron pipe failure, 272–273
electron beam, 272
interaction volume, 272
PE interacts, 272
resolution of, 272
SEs, 272
Science of individualization, 2
Scuff-type marks, 95
Secondary electrons (SEs), 272
Selected ion monitoring (SIM), 189–190
Set theory, 7
Skill-based errors, 50
Skin injuries
abrasions, 121
decollement, 122
hematoma, 122
laceration, 122
stretch injury, 122, 122f
Solid-phase microextraction (SPME), 188
Speed, 22–23
Static method, 114
Stiffness-based model, 42
Structural failures
catastrophic collapse, 243
collapse configuration, 245
collapse investigation, purpose of, 243
components, 244–245
curing concrete, 245
expert investigator/witness, 247
fracture surfaces, 245

investigation, 244, 246–247
perishable evidence, 245
planning–design–construction–service, 243
postinvestigation, 247
snow and loads, 245–246
Subarachnoid hematoma, 123
Subdural hematoma, 123
Superposition principles, 5
Suspicious deaths
death, cause of, 205
death, manner of, 205
death scene
blood flow patterns, 206
crime scene investigator, 205–206
outdoor scenes, 205
pink-purple discoloration, 206
potential evidence, 206
trace evidence, 206
postmortem examination, 206–207

T

Tamper-evident tape, 34
Taxon, 7–8
Taxonomy, 7–8, 13–14
Technical truth, 80
Thermal degradation
description, 151
effects
human remains, 152–153
polymeric materials, 152
trace evidence, 153–154
wood, 151–152
mechanism of, 151
Thorax injuries
aortic rupture, 124
diaphragm rupture, 124
flail chest, 124
heart contusion, 123
lung contusion, 123
pelvic injuries, 124
pneumothorax, 123
Total ion chromatogram (TIC), 189
Trace evidence
fingermarks, 153–154
ignitable liquids, 154
Trace, 2–3
Traffic injuries and deaths
air traffic, 128
railway, 127–128
road traffic
car occupants, 124–125
epidemiology, 121
pedestrian collisions, 126–127
relevant injury types, 121–124
two-wheelers, 127
whiplash-associated disorders, 125–126
Traumatic brain injuries (TBIs), 122–123
Triangular pattern, 174, 174f
Truncated cone patterns, 173–174

U

Underwater locator beacon (ULB), 83
Upper flammability limit (UFL), 141

W

Wavelength–dispersive X-ray spectroscopy (WDS), 275–276
Whiplash-associated disorders (WADs), 125–126

X

X-ray fluoroscopy (XRF), 274, 276
X-ray spectroscopy
EDS, 275–276
techniques, 274
WDS, 275–276
XRF, 276

FSC
www.fsc.org
MIX
Papier | Fördert gute Waldnutzung
FSC® C083411

Kolibri 360 GmbH
Ferdinand-Jühlke-Straße 7
99095 Erfurt, Deutschland
produktsicherheit@kolibri360.de